W0232201

Spoken Language Generation and Understanding

NATO ADVANCED STUDY INSTITUTES SERIES

Proceedings of the Advanced Study Institute Programme, which aims at the dissemination of advanced knowledge and the formation of contacts among scientists from different countries

The series is published by an international board of publishers in conjunction with NATO Scientific Affairs Division

A	Life Sciences	Plenum Publishing Corporation
B	Physics	London and New York
C	Mathematical and Physical Sciences	D. Reidel Publishing Company Dordrecht, Boston and London
D	Behavioural and Social Sciences	Sijthoff & Noordhoff International Publishers
E	Applied Sciences	Alphen aan den Rijn and Germantown U.S.A.

Series C – Mathematical and Physical Sciences

Volume 59 – Spoken Language Generation and Understanding

Spoken Language Generation and Understanding

Proceedings of the NATO Advanced Study Institute held at Bonas, France, June 26 - July 7, 1979

edited by

J. C. SIMON
Institut de Programmation
Université Pierre et Marie Curie, Paris VI, France

D. Reidel Publishing Company

Dordrecht : Holland / Boston : U.S.A. / London : England

Published in cooperation with NATO Scientific Affairs Division

Library of Congress Cataloging in Publication Data

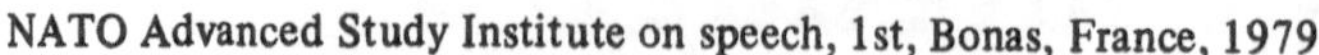
NATO Advanced Study Institute on speech, 1st, Bonas, France, 1979

Spoken language generation and understanding.

(NATO advanced study institute series: Series C, Mathematical and physical sciences; v. 59)
Includes index.
1. Speech processing systems–Congresses. 2. Speech perception–Congresses. I. Simon, Jean Claude, 1923– II. Title. III. Series.
TK7882.S65N37 1979 621.3819'598 80-26841
ISBN-13: 978-94-009-9093-7 e-ISBN-13: 978-94-009-9091-3
DOI: 10.1007/978-94-009-9091-3

Published by D. Reidel Publishing Company
P.O. Box 17, 3300 AA Dordrecht, Holland

Sold and distributed in the U.S.A. and Canada
by Kluwer Boston Inc.,
190 Old Derby Street, Hingham, MA 02043, U.S.A.

In all other countries, sold and distributed
by Kluwer Academic Publishers Group,
P.O. Box 322, 3300 AH Dordrecht, Holland

D. Reidel Publishing Company is a member of the Kluwer Group

Softcover reprint of the hardcover 1st edition 1980

TABLE OF CONTENTS

*Are Review Papers

ACKNOWLEDGEMENTS

In the first place, I wish to thank specially my colleagues J.P. HATON and R. DE MORI, who have helped me to put together this ASI and to make it a success on the spot. Dr L.C.W. POLS has also given valuable advice.

The editing committee should also be thanked for helping with the publication of this book, particularly Drs W. HESS and D.R. HILL.

On the other hand, only through the financial support and the framework provided by the NATO Scientific Affairs Division could such a coherent, high-level meeting be made possible.

The material support of IRIA and Institut de Programmation should also be gratefully acknowledged.

Finally, I wish, as the director of the contributions and also as the animateur of the Centre Culturel de BONAS, to thank all the participants of the ASI for their friendly comprehension during their stay...

Director of the NATO ASI, J.C. SIMON, Institut de Programmation Université Pierre et Marie Curie, 4 place Jussieu, 75230, Paris Cedex 05.

Advisory Committee: J.P. HATON, Informatique, Université Nancy I Case Officielle 140, 54037 Nancy Cedex. France.

R. DE MORI, Università di Torino, Istituto di Scienze dell'Informazione, Corso M. d'Azeglio 42, Torino, Ita.

Editing Committee: J.S. BRIDLE (G.B.), J.P. HATON (Fr.), HESS (R.F.A.) HILL (Can.), DE MORI (It.), J.C. SIMON (Fr.), J.J. WOLF (U.S.A.).

J. C. Simon (ed.), Spoken Language Generation and Understanding, IX.

PREFACE

HOW TO READ THE BOOK: SOME COMMENTS

This book is the lasting result of the first NATO Advanced Study Institute on Speech, held at the Centre Culturel du Château de Bonas, from June 26 to July 7, 1979.

The intent of a NATO ASI is primarily to provide high level tutorial coverage of a field in which research is active; undoubtedly speech generation and understanding is one at the present time.

Thus 12 surveys are offered by some of the best specialists in the field. As a consequence the book may be considered as *a reference book on speech*. The surveys are marked by a * in the Table of Contents.

However, half of the meeting was devoted to discussions and presentations of research. A reviewing Committee decided to ask a number of participants to submit a contribution which would complete the tutorials or would present original work.

A *beginner* in the subject should start by reading the reviews of Hill, Wolf, Bridle, Haton, Woods, Allen, Lienard; preferably in the above order. He would then be familiar with the terms, the problems and the techniques of speech understanding (analysis) and generation (synthesis).

A reader who is already advanced in the field would also profit from the tutorials and will easily find his way among the five sections:

1. An overview, with the emphasis on psychology

An effort was made to bring to the ASI the research results of psycholinguistics. The review of Marslen-Wilson is of prime interest in that respect.

2. Acoustics and Phonemics

Speech recognition is a *multilevel process*. This section describes

J. C. Simon (ed.), Spoken Language Generation and Understanding, xi-xiii.

the very first levels of signal treatment. Many different techniques have been studied and implemented in experimental systems. As in other fields of pattern recognition, preprocessing and the determination of first level features relates more to an art than to a science. A special emphasis has been given to the syntactic approach by De Mori and K.S. Fu.

Though most of the data are extracted from vocoder type systems, some direct measurements on the speech signal itself are also in use, for example, the number of zero-crossings. Hess, Baudry and Dupeyrat give examples of the use of operators on the signal itself to determine the pitch and the formants frequencies.

An agreement seems to exist on what are the essential features to be measured at these first levels:
- for voiced speech, the *pitch peak instants or frequency;*
- *the instant frequency and amplitude of the different formants.*

Further, many efforts have been made to detect the *phonemes*, as they are proposed by the phoneticians. In our own judgement, they seem a concrete reality. But the variability of their representation in the signal and even in the instant spectrum seems very great; cf. Chollet, Allen. This casts some doubts on the possibility of reliable detection.

Still a lot of research work has to be done at these acoustic and phonemic levels to achieve the some reliability as an ordinary human being hearing meaningless words in difficult surroundings (noisy or distorted).

3. Lexicon, Syntax, and Semantic

Being a multilevel recognition process, the upper levels may contribute to the lower level detections. In other words, the meaning of the sentence may allow one to correct faulty phonemic determinations. There is strong evidence that a human being hearing speech under poor conditions does make this sort of restoration. The syntax and the semantics of a sentence may thus help to obtain the correct determination, even if the first level detections were faulty.

A lot of work has been done along these lines; in particular in the U.S.A. Woods' paper gives an excellent review.

Certainly there is a lot more to understanding continuous speech than to understanding isolated words, with or without the word being in a dictionary. But the initial hopes of understanding very faulty phonemic determinations have not been fulfilled; and one of the main results of these important studies is that a better phonemic determination should be obtained in the first place.

4. Speech synthesis

The generation of speech from a written text transformed in a phoneme string is by now quite successful. All the existing problems are not yet solved, for instance the prosodic ones; but some impressive success has been achieved.

Examples of synthetic speech could be heard at the meeting. Allen demonstrated some speech. Rodet gave an excellent example of synthetic singing of ultra high quality. Examples of applications were given by Christinaz and by M-C. Haton.

5. Systems and Applications

A survey lecture by Erman was devoted to a 'Speech Understanding System', relating his work at Carnegie with Raj Reddy. Unfortunately his manuscript was not available for this book.

However, this section gives a number of system descriptions and applications of great interest. Let us point out Nakata's paper, relating some applications in Japan and asking the "good" questions about applications:
- What is a good example of speech recognition application?
- In what direction should we go in speech research with regards to a practical recognition application?
- In this direction, what technical problems should we attack in the future?
- What is the best architecture for real speech recognition machines in the future?

This very last question raises naturally the consequences of the availability of the *microelectronic techniques*. Already the SPEAK & SPELL of Texas Instruments has shown their interest in speech application. It is clear that the next ASI will treat this topic much more extensively. A large number of research labs will switch to microelectronic processors, even themselves designing the "chips".

Another topic should attract attention in the future. Nowadays the speech research community does not seem to be sensitive to *computational complexity*. In other words, as long as a proposed algorithm "works", they do not seem to care if it is *exponential, polynomial or linear*. This attitude seems rather shortsighted from the perspective of possible applications, for which the simplicity and the time delay of recognition are essential.

If, as is probable, more than one algorithm will prove to give satisfactory results, the selection will be made through its complexity; even if computing possibilities are constantly improving and the price is dropping down. But again, this should be an interesting topic for the next ASI on speech!

J.C. Simon

LIST OF PARTICIPANTS

Alinat, P.A.	Thomson-CSF, DASM, Chemin des Travails, 06802 Cagnes sur Mer, France.
Allen, J.	Dept. Elect. Eng. & Computer Science, MIT, Cambridge, Mass 02139, U.S.A.
Baudry, M.	Institut de Programmation, Univ. Paris VI, 4 Place Jussieu, 75005 Paris, France.
Benedetto, M.D.	IBM France, 36 Av Raymond Poincaré, 75016 Paris, France.
Bellissant, C.	Laboratoire IMAG B.P. 53 X, Grenoble 38041, France.
Berthomier, C.	U.E.R. de Physique, Tour 34, Université Paris VII, 2 Place Jussieu, 75007 Paris, France.
Bisiani, R.	Comp. Sc. Dept. Carnegie Mellon Univ. Schenley Park, Pittsburgh, Pa 15213, U.S.A.
Bridle, J.	Joint Speech Res. Unit, Princess Elizabeth Way, Cheltenham, Gloucester, G.B.
Brown, R.S.	Dept. of Linguistics, Univ. of Edinburgh, 40 Georges Sq. Edinburgh, G.B.
Caelen, J.	Cerfia, Université Paul Sabatier, 118 route de Narbonne, 31077, Toulouse, Cédex, France.
Carbonell, N.R.	Dept. Informatique, IUT 2bis Bld Charlemagne, 54000 Nancy, France.
Castellino, P.	Cselt Via Reiss Romoli 274, 10148 Torino Italy.
Chollet, G.F.A.	Inst. de Phonétique, Univ. de Provence, 29 Av. Robert Schuman, 13621 Aix en Provence, France.
Cappelli, A.	Lab. Linguistica Computazionale, CNR, Via S. Maria 36, Pisa, Italy.
Christinaz, D.	Neuropsychiatric Institute, Dept. of Psychiatry and Behavioural Science, University of California, 760 Westwood Plaza, Los Angeles, Cal. 90024, U.S.A.
Darwin, C.J.	Lab. of Experim. Psychol. Univ. Sussex, Brighton BNI 9QC, G.B.

Destombe, F.	IBM France, 36 Av. Raymond Poincaré, 75116 Paris, France.
Endres, W.K.	Forschungsinstitut der Deutsche Bundespost Beim Fernmeldetechnischen Zentralamt, Postfach 50-00, 6100 Darmstadt, R.F.A.
Erman, L.D.	USC-ISI 4676 Admiralty Way, Marina del Rey, Cal. 90291, U.S.A.
de Foras, R.J.	Montjoux, 74200 Thonon, France.
Frediani, S.	Istituto Sc. Info. C.M. d'Azeglio 42, 10125 Torino, Italy.
Fu, K.S.	School El. Eng. Purdue Univ. Lafayette, Ind. 47907, U.S.A.
Gagnoulet	Cnet, Route de Trégastel, 22300 Lannion, France.
Gallenkamp, W.	DBF beim FTZ, Am Kavalleriesand 3, 6100 Darmstadt R.F.A.
Goerz, G.	Univ. Erlangen-Nuernberg RRZE, Martenstrasse 1, D-8520 Erlangen R.F.A.
Gouarderes, J.P.	Cerfia, Univ. Paul Sabatier, 118 route de Narbonne, 31077 Toulouse Cédex, France.
Graham-Stuart, D.	Inst. Phonetic Sc., Univ. Groningen, Grote Rozenstraat 31, Groningen, The Netherlands.
Gregoretti	IET Politec. di Torino, C. Duga Degli Abruzzi 24, 10129 Torino, Italy.
Grossmann	Dep. Bild & Ton, Inst. Heinrich Hertz, Einsteinuferstrasse 37, 1000 Berlin, R.F.A.
Gust, H.	Inst. für Angew. Informatik CIS KU-A, Kurfürstendamm 202, D-1000 Berlin, R.F.A.
Haton, J.P.	Univ. Nancy I, Informatique, Case Officielle 140, 54037 Nancy, France.
Hein, H.W.	Universität Erlangen-Nuernberg, Inst. Math. Martenstrasse 3, 8520 Erlangen, R.F.A.
Hess, W.	Lehrstuhl für Datenverarbeitung, Tech. Univ. München, Arcisstr. 21, Postfach 20-24-20, D-8000 München 2, R.F.A.
Hill	Dept. Comp. Sc. The University, Calgary, Alberta T2N IN4, Canada.
Hodes, P.	Dept. of Education, Mercy College, 8200 W. Outer Drive, Detroit, Mich. 48219 U.S.A.
Hunt, M.J.	Bell Northern Res. Ltd., 3 Place du Commerce, Ile des Soeurs, Verdun, Québec, Canada.
Irwin, M.J.	Joint Speech Unit, Princess Elizabeth Way, Cheltenham, Gloucester, G.B.
Jutier, P.P.	Inserm U88, 91 Bld. de l'Hôpital, 75013 Paris, France.
Kielczewski, G.M.	Uniwersytet Warsawski, Inst. Informatyky, PKiN pok 850-00-901, Warsaw, Poland.
Klaastad, D.	P.O. Box 3696, GB-N, Oslo, Norway (Forsvarets Overkommando).

Laface, P.	Istituto Electrotech. Polytech. Torino, C Duca Degli Abruzzi 24, 10129 Torino, Italy.
Levy, R.	Inst. Programmation, Univ. Paris VI, 4 Place Jussieu, 75005 Paris, France.
Lienard, J.S.	CNRS-LIMSI BP 30, 91406 Orsay, France.
Lingaard	Dept. El. Eng. Queen's Univ., Belfast, N. Ireland, G.B.
Lojacono	Dept. Linguistics, Univ. South. Calif. University Park, Los Angeles, Cal. 90007, U.S.A.
Maitra, S.	Systems Control Inc., 1801 Page Mill Rd., Palo Alto, Cal. 94304, U.S.A.
Mari, J.F.	23 rue du Général Gérard, 54000 Nancy, France.
Mariani, J.M.	CNRS LIMSI, B.P. 30 Orsay 91406, France.
Marslen-Wilson, W.	Max Planck Gesellschaft zur Förderung der Wissenschaften Projektgruppe für Psycholinguistik, Berg en Dalseweg 79, Nijmegen, The Netherlands.
Massimeli, R.	Elettronica S. Giorgio, Elsag SpA Via Hermada 6, 16154 Genova Sestri, Italy
Meli, R.	Univ. Eng. Dept. Trumpington St., Cambridge CB2 IPZ, G.B.
Meier, P,	Digital Control, AG Meiersmattstr. 3-5, 6043 Adligenswil, Switzerland.
Meloni, H.	UER Marseille Luminy, Groupe Int. Artif. Case 901, 70 rte Léon Lachamp 13009 Marseille, France.
Menon, K.	Dept. Comp. Sc. Concordia, Univ. Loyola Campus, 7141 Sherbrooke St. W, Montreal H4B IR6, Canada.
Mercier	CNET Route de Trégastel, 22300 Lannion, France.
Moore, R.	Dept. Phonetics & Linguistics, Univ. College Gower St., London WCIE 6BT, G.B.
de Mori, R.	Istituto Sc. Informazione, C. M.d'Azeglio 42, 10125 Torino, Italy.
Mouradi	Lab. Electronique, Faculté des Sc., B.P. 1014 Rabat, Marocco.
Nakata, K.	Hitachi Central Res. Lab. Kokubunji, Tokyo, Japan.
Nouhen, A.	IUT route de Pennoe, 22300 Lannion, France.
Philipp, J.E.	Inst. für Phonetik Univ. Köln, Greinstrasse 2, D-5000 Köln 41, R.F.A.
Pierrel, J.M.	CRIN IUT Informatique, 2bis Bld. Charlemagne, 54000 Nancy, France.
Potage, J.	ThomsonCSF Div. Télécom., 16 rue du Fossé Blanc, 92231 Gennevilliers, France.

Quinton, P.	Irisa Informatique, Av. du Général Leclerc, 35042 Rennes Cédex, France.
Regel, P.	Univ. Erlangen-Nuernberg Inst. Math. Martenstrasse 3, 8520 Erlangen, R.F.A.
Rodet, X.	Dept. Diagonal, Ircam, 31 rue Saint Merri, 75004 Paris, France.
Saitta, L.	Inst. Sc. Informazione, C. Massimo d'Azeglio 42, 10125 Torino, Italy.
Sakai, M.T.	Dept. Inf. Sc. Kyoto Univ. Sakyo-Ku, Kyoto 606, Japan.
Sanchez, C.	La Nef au Pré Latour 54700 Pont à Mousson, France.
Shevki, H.	Dept. El. Eng. Bogazici Univ., P.K. 2 Bebek, Istanbul, Turkey.
Shirai, K.	Dept. El. Eng. Waseda Univ., 3-4-1 Okubo-Shinjuku-Ku, Tokyo 160, Japan.
Simon, J.C.	Inst. Programmation, Univ. Paris VI, 4 Place Jussieu, 75005 Paris, France.
Slavik, J.	Josef Fraunhofer Inst., Heidenhofstrasse 8, D-7800 Freiburg, R.F.A.
Stenzel, E.	Inst. Nachrichtentechnik, TU Braunschweig, Postfach 3329, 3300 Braunschweig, R.F.A.
Talmi	Heinrich Herz Inst., Einsteinuferstrasse 37, 1000 Berlin 10, R.F.A.
Tran Cao, T.	Inst. Montefiore B 28, Univ. Liège, Sart Tilman, B 4000 Liège, Belgium.
Underwood, M.	ICL Res. & Adv. Dev. Center, Fairview Rd, Stevenage, Herts SG1 2DX, G.B.
Van den Berg, C.G.	Fonetisch Instituut, Oudenoord 6, 3513 ER Utrecht, The Netherlands.
Wolf, J.J.	Bolt Beranek & Newman, 50 Moulton St., Cambridge, Mass. 02138, U.S.A.
Woods, W.	Bolt Beranek & Newman, 50 Moulton St., Cambridge, Mass. 02138, U.S.A.
Yanagida, M.	Inst. Phonetic Sc., Grote Rozenstraat 31, 9712 TG, Groningen, The Netherlands.
Zagoruiko, N.	Institute for Mathematics, 630090 Novosibirsk 90 U.S.S.R.

§ 1

AN OVERVIEW WITH AN EMPHASIS ON PSYCHOLOGY

SPOKEN LANGUAGE GENERATION AND UNDERSTANDING BY MACHINE: A PROBLEMS AND APPLICATIONS ORIENTED OVERVIEW

David R. Hill

Man-Machine Systems Laboratory
Department of Computer Science
The University of Calgary
CALGARY, Alberta, Canada T2N 1N4

ABSTRACT

Speech offers humans a means of spontaneous, convenient and effective communication for which neither preparation nor tools are required, and which may be adapted to suit the developing requirements of a communication situation. Although machines lack the feedback, based on understanding and shared experience that is essential to this form of communication in general, speech is so fundamental to the psychology of the human, and offers such a range of practical advantages, that it is profitable to develop and apply means of speech communication with machines under constraints that allow the provision of adequate dialog and feedback. The paper details the advantages of speech communication and outlines speech production and perception in terms relevant to understanding the rest of the paper. The hierarchy/heterarchy of processes involved in both automatic speech understanding and the machine generation of speech is examined from a variety of points of view and the current capabilities in terms of viable applications are noted. It is concluded that there are many economically viable applications for both the machine generation of speech and the machine recognition of speech. It is also suggested that, although fundamental research problems still exist in both areas, the barriers to increased use of speech for communication with machines are psychological and social, rather than technological or economic.

1. INTRODUCTION: HISTORY AND BACKGROUND

Historically, speech has dominated the human scene as a basis for person to person communication, writing and printing being used

J. C. Simon (ed.), Spoken Language Generation and Understanding, 3-38.

chiefly for record-keeping (especially for legal and government purposes) and long-distance communication. It is only comparitively recently that written language skills have been widely acquired and practised, even in the developed western world, and it is still true that large numbers of people in the world are unable to read or write. With the advent of television and radio speech has started to gain ground again even in the west, and it is a perennial complaint amongst educators that the rising generation is not at ease with the written forms of language. Even universities are finding it necessary to set tests of students' ability to comprehend and write their native language. It is in this sense that speech may be considered a more fundamental means of human communication than other forms of linguistic expression.

Face to face speech allows a freer and more forceful use of language, because it is aided by contextual, situational and all manner of linguistic and paralinguistic features (voice tone, rhythm, the rise and fall of pitch, gestures, pauses and emphasis) that are simply absent from written forms. Grammatical rules of all kinds may be broken with impunity in speech to give a stream of language that would be incomprehensible in written form. Even for careful speech (as in an important meeting) it is sobering to read a verbatim transcript of the proceedings afterwards. However, everyone understands at the time and (perhaps more important) if they don't, they can seek immediate clarification. The planning load for speech communication is less, even over a telephone, because ideas can be added to and expanded as required. A written document must anticipate and deal with all questions and obscurities that could arise for arbitrary readers, whenever they read it, and under whatever circumstances, without all the additional resources inherent in speech. This is why written documents follow such a mass of conventions and rules so precisely. That kind of redundancy and familiarity is necessary to overcome the many disadvantages with which written documents are burdened. But, in this sense, written documents can remain precise despite the test of time and circumstances.

To summarize then: human beings can usually communicate far more readily, conveniently, and effectively using speech than they can by writing *providing there is adequate feedback*. Under these conditions, the human feels more comfortable and is less encumbered. This is what people really mean when they say that speech is more natural and convenient as a means of communication. With reduced feedback, speech must put on some of the formal trappings of written language, which explains the difference in character between a lecture, a seminar and a conversation. With increasing interaction and feedback, the constraints of language may be increasingly relaxed, short cuts taken, and the communication experience greatly improved. Books record, but teachers and lecturers are likely to communicate what is recorded in a more effective (if different) form, depending on their familiarity with the subject(s). In passing, it should also be noted that it is easier to close a book (and do something else) than to walk out of a well delivered lecture or seminar. In this sense, speech also has immediacy and impact.

Thus, in the context of humans communicating within some system, the first reason for requiring a *speech* interface, as opposed to any other, is based in the psychology of the human. It is natural, convenient, and indeed almost expected of any system worth communicating with.

There are, however, a number of very practical advantages:

(a) it leaves the human body free for other activities providing an extra channel for multi-modal communication in the process;
(b) it is well suited to providing 'alert' or break-in messages even for quite large groups and may communicate urgency in a way only rivalled by real flames;
(c) no tool is required;
(d) it is omnidirectional, so does not require a fixed operator position;
(e) it is equally effective in the light or the dark, or, in fact, under many conditions when visual or tactile communication would be impeded such as in smoke, or under conditions of high acceleration or vibration (though the latter may cause some interference with *human* speech *production*);
(f) it avoids the use of keyboards and displays thereby taking up less panel space and avoiding the need for special operator training to cope with the key layouts and display interpretation;
(g) it allows simple monitoring of command/response sequences by remote personnel or recording devices;
(h) it can allow a single channel to serve more than one purpose (e.g. communication with aircraft at the same time as data entry to an Air Traffic Control computer);
(i) it allows blind people, or those with manual disabilities to integrate into man-machine systems with ease;
(j) it allows security checking on the basis of voice characteristics;

and perhaps most important,

(k) it is compatible with the widely available and inexpensive public telephone system.

Against these we must set a number of disadvantages:

(a) it may be subject to interference by noise;
(b) it is not easily scanned for purposes of selecting occasional

items from some large body of information;

(c) special precautions must be taken for verification and feedback, particularly for messages initiating critical responses, for reasons previously discussed;

(d) unrestricted speech input to machines is likely to be unavailable for a long time to come (in fairness, it should be pointed out that there are severe restrictions on other forms of input, the removal of which would involve the solution of similar problems);

(e) speech input to machines seems likely always to require additional expense, compared to written input, simply because of the considerable extra processing power and memory space needed to decode the inherently ill-defined and ambiguous acoustic evidence -- a step more akin to using handwriting for input in place of well defined machine generated character codes. However, these costs may be tolerable, given the falling price and increasing power of the necessary hardware, for the same reasons that very expensive Optical Character Recognition (OCR) equipment is viable because of the high labour costs of the specialised personnel that are otherwise needed and the advantages gained.

It is concluded that we *should* use speech to communicate with machines. People need to interact with machine systems in the most effective way under a variety of circumstances in which speech offers solutions to problems, or communication enhancing advantages. Especially, the use of speech would allow remote access to machine systems of all kinds over the public telephone network. In data collection tasks, this would allow the responsibility for the entry and verification of data to be placed on the generator of the information, cutting costs and ensuring more accurate data. In general, a speech interface would make the many specialised information processing systems more accessible, cutting out middle people with special training and releasing them for more productive tasks, eliminating the need for expensive terminal and modem equipment in many applications of information systems, and opening up further avenues for personal information services of all kinds. With the consequent falling costs for even fairly sophisticated speech interface equipment, and the expanding market, there should be both the resources and incentive to solve the remaining problems in providing really excellent speech communication with machines.

The remainder of this paper will examine the nature of some of the problems, both for synthesis of speech and the recognition and understanding of speech, and attempt an overview of progress to date, including some current hardware and applications.

2. THE SURPRISING NATURE OF PERCEPTION: IS SEEING REALLY BELIEVING?

It may seem strange to be concerned with *seeing* in a paper on speech. The reason for the concern is simple: both involve perception, a tricky subject for the uninitiated for whom some of the pitfalls can more easily be illustrated in the visual domain than in the auditory domain. The essential basic fact about perception in *any* domain is that it is *an active organising process of interpreting incoming sensory data on the basis of experience.* Although different people have large amounts of experience of the real world in common, the details, even quite important details, may vary from individual to individual. And the immediate context is relevant too. This is why different people may honestly disagree in their descriptions of events or situations witnessed together. Anyone familiar with the celebrated Ames distorted room will know the devastating effect on perception of the false assumption that the room is a normal undistorted room. Real people apparently change in size as they walk about the room, *even though the observer is intellectually aware that the room has a strange shape,* because false assumptions are built right into the neural processing that is carried out to interpret the incoming visual data, assumptions learned on the basis of the experience of acting in normally shaped rooms. This occurs because the distorted room is carefully arranged to present the same visual characteristics as a normal room from the viewpoint of the observer.

At a more detailed level, it can be shown that, although the eye acts, in a sense, like the imaging tube of a television system, transmitting data about picture points to the visual cortex of the brain, there are some fundamental differences. The retina is not just a passive relay station for picture points, converting light to electrical impulses in the same pattern arrangement as the pattern of colours cast onto it by the lens, but it is a piece of brain tissue that effects a great deal of processing (e.g. sending signals to the visual cortex about the *differences* in level of illumination within small *areas* of retina, called 'receptive fields'). Thus, although the number of light sensitive cells in the area of most acute vision (the fovea) is about equal to the number of nerve fibres leaving that area for the brain, the connectivity is *not* 1:1. There is good evidence that, once in the brain, this information is hierarchically processed in terms of specific features of the visual space such as lines, edges, movement, and orientation, with colour being coded as the pattern of relative activity between cells differentially sensitive to different wavelengths of light. There is no little man, sitting and interpreting television pictures in the visual cortex -- that would imply an infinite regress anyway. Instead, the appearance of reality is inferred on the basis of cues, little bits of evidence, distributed over the entire visual field, and for which the corresponding spatial relationships have been *learned* in terms of their relationship to other sensory modalities and to motor performance. We see an edge as continuous because our experience with this kind of visual image has been consistent in all respects with this interpretation -- perhaps an oversimplification, but perception is of this nature. Von Senden, in his book (1960), describes the experience of people given sight by corneal transplant, having been blind from birth. Such people do not see very much at all when the bandages come off -- they must painfully (and usually not entirely adequately) learn to interpret the sensory data, not at an intellectual linguistically oriented level ("What is that pink round object with hair on top etc.") but at the level of basic notions of squareness and spatial relationship. Gregory (1969) reports similar cases, happily becoming rare, and notes that one subject did eventually learn to draw pictures but relied heavily on his tactual experiences obtained whilst blind so was never able to draw the front of a bus, or deal as easily with lower case letters as upper case (he had handled toy bricks with raised upper

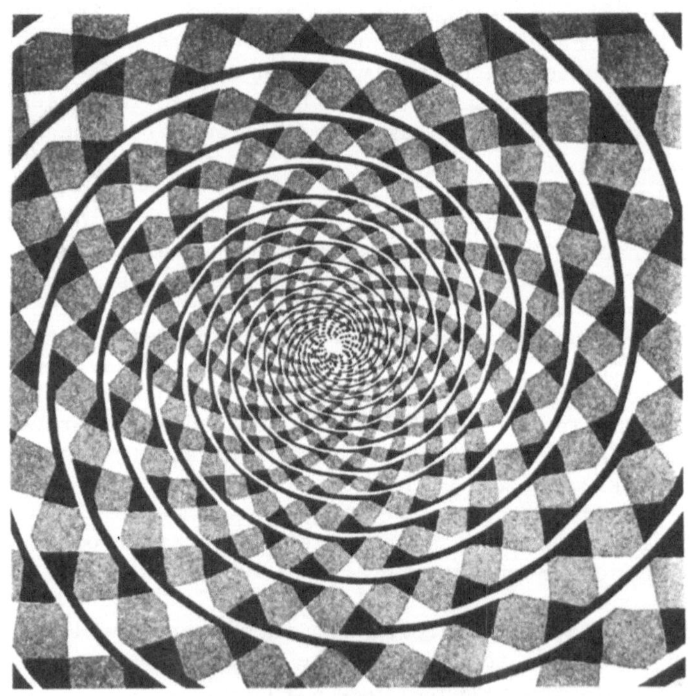

Figure 1: The Fraser spiral

case letters whilst blind, but had not the same experience for lower case letters). The Fraser Spiral (figure 1) gives an excellent example of how an overall field of visual cues can give a completely erroneous idea of the main forms depicted. Because of the slope relationships between neighbouring parts of the figure, the main lines appear to form a continuous spiral, yet if the line is carefully followed at any point, it will be found that the lines actually form a series of concentric circles. It is this piecemeal basis of perception that is exploited by the Dutch artist, M.C. Escher, in studies such as *Waterfall*. The local areas of this figure are valid, it is their overall relationships that are wrong, and are contrived to give an appearance of water falling down a considerable drop and then (magically!) returning to its starting point, again under gravity. It is quite difficult to explain what is wrong with this kind of figure since it is the global relationships that are wrong, any local collection of visual cues can be reconciled with the reality we know. The devil's pitch-fork shown in figure 2 has the same kind of local consistency, coupled with global inconsistency, but on a smaller and more obvious scale. Either end looks alright, but the two ends require incompatible interpretations of some of the edges depicted. Escher's drawings are more subtly incompatible from one part to another, yet we perceive the structure as 'real' and can only intellectualise the global problems by pointing out the inconsistencies, much as a newly sighted patient in Gregory's account intellectualises 'squareness' by counting corners.

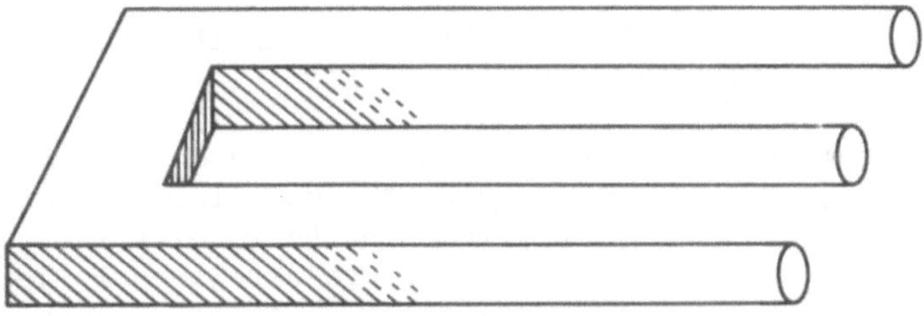

Figure 2: The "Devil's Pitchfork"

It is important to realise that these kinds of effects are possible in all domains of perception. We perceive reality insofar as our perception is useful in dealing with reality. We may produce illusions either by invalidating the normally correct assumptions as in the Ames room, or by exploiting its basic deficiencies as Escher does. But above all, seeing is *not* believing -- there is not an objective reality in our visual input. We see (hear, feel, etc.) what we have found it useful to see (hear, feel, etc.) in connection with our attempts to deal with the environment, based on our experience of what *works*. It is not a matter of what we know intellectually. Moreover the reality that we build, from fragmentary evidence, depends equally upon these often unknown assumptions that embody our previous experience, and which can only be changed by active involvement in new experience where the assumptions no longer work.

At this point, the reader is reminded that we are mainly concerned with speech perception, yet have illustrated some of the fundamental nature of perception with reference to vision. The reason is simple. We have a much more tenuous grasp of what the features of auditory perception might be, especially for speech perception. The only really convincing auditory illusion of which this author is aware is that of continuously rising pitch in a complex sound which actually consists of a series of pure tones each rising over a fixed frequency range, but whose relative amplitudes are dynamically varied to produce the illusion. The extent to which we have unravelled some of the cues to speech perception, and learned to manipulate them is the subject of the next section, but the reader is cautioned that our ability to intellectualise and explain the reality that we experience in the auditory domain is at an even more primitive level than for vision. Auditory constructs seem much less accessible since we are still unsure about the true nature of the relevant auditory cues and the range of relevant relationships between them. To the extent that our knowledge is deficient, we can neither construct speech nor recognise it.

3. SPEECH PRODUCTION AND PERCEPTION: PLAUSIBLE CUES TO THE AUDITORY CONSTRUCTS OF SPEECH

Speech is generated and received as a continuously time-varying pressure waveform. Figure 3 shows the main features of the vocal apparatus. Forcing air through the vocal folds and causing them to vibrate, or forcing air through a relatively narrow constriction anywhere in the vocal tract, acts as a source of energy that may excite resonances in the tube-like vocal tract that stretches from the folds to the lips. A further resonant cavity, leading to the nostrils, may be connected in parallel by opening the velum. The energy due to air flowing through the vibrating vocal folds is termed voiced energy, and the state is termed voicing, leading to voiced sounds. The other kind is termed fricative energy or, if there is no readily identifiable point of constriction in the vocal tract, aspiration (as for the sound representing the letter 'h' at the beginning of 'house'). The two kinds of energy may be mixed, leading to what are termed 'voiced fricatives' or 'voiced h'. The energy supply constitutes the *source* in the so-called 'source-filter model of speech pro-

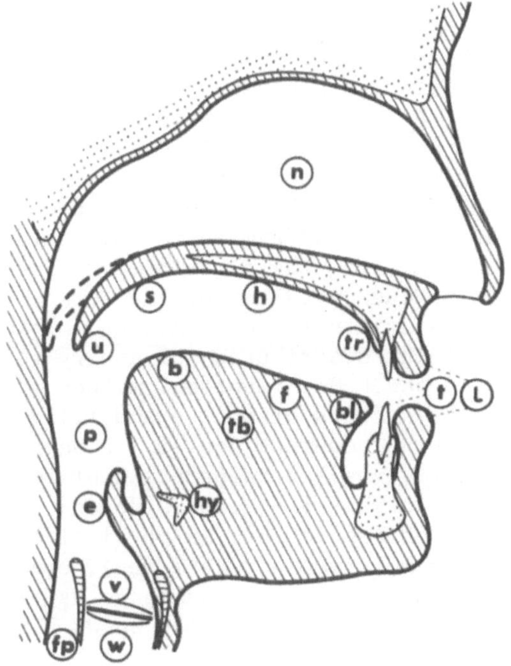

Figure 3: The human vocal apparatus

Key

b	- back of tongue	**l**	- lips	**tr**	- teeth ridge (alveolar ridge)
bl	- blade of tongue	**n**	- nasal cavity	**u**	- uvula
e	- epiglottis	**p**	- pharynx	**v**	- vocal folds (bounding glottis)
f	- front of tongue	**s**	- soft palate (velum)	**w**	- windpipe
fp	- food passage	**t**	- teeth		
h	- hard palate	**tb**	- tongue body		

duction'. The source energy is distributed through the frequency spectrum, falling off with increasing frequency at a fairly uniform rate. In passing through the resonances of the vocal tract, the distribution of energy from the source is modified, peaks being produced at or near the resonant frequencies of the tract. These peaks are termed *formants*. The resonant frequencies, and hence the frequencies of the formants, vary according to the shape of the vocal tract which is determined primarily by the movements of the tongue. Although there is a series of formants, decreasing in amplitude with increasing frequencies, research has shown that only the three formants lowest in frequency are necessary to good speech intelligibility. If artificial speech is generated, the higher formants, together with the effect of high frequency emphasis due to radiation from a rather small orifice (the mouth), may be approximated by fixed filters. The connection of the nasal cavity further modifies the frequency distribution of energy. Thus the overall effect of the vocal tract, mouth and nasal cavity (if connected) is to act as a filter on the source energy -- hence the source-filter model of speech production. There are other ways of looking at speech production, but for our purposes, this is adequate. Irregularities in the source, minor resonances due to side cavities (sinuses), and other effects form subsidiary peaks and troughs in the distribution of energy against frequency. Futhermore, certain configurations of the vocal tract may bring two formant peaks so close together that they appear as one peak. For these and other reasons, there is no simple relationship between peaks in the speech spectrum (i.e. the distribution of energy against frequency) and the frequency values of the formants, although a variety of algorithms exists for determining formant frequencies, or other measures of the distribution of energy relevant to characterising speech sounds. One important method (Cepstral analysis) deconvolutes the effects of source and filter, making the task of formant tracking somewhat easier. Linear Predictive analysis also leads to excellent separation of formant peaks. However, because the formant frequencies depend on the size of the vocal cavities, and people vary in physical characteristics (especially between men, women and children), even if the formant frequencies are known, their relationship to particular speech sounds is still problematical. Figure 4 shows a typical speech spectrum.

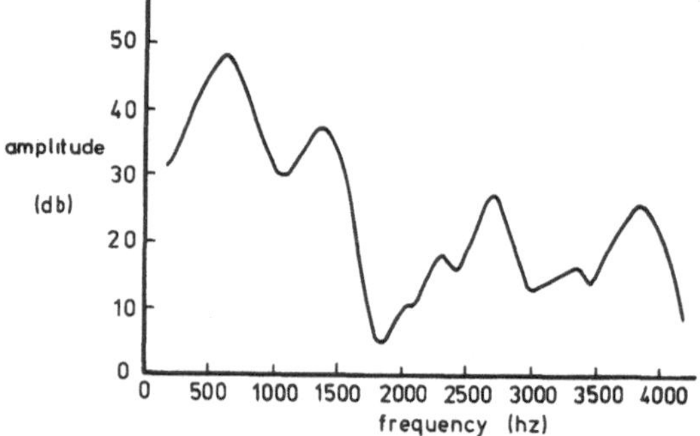

Figure 4: A typical speech spectrum

The invention of the sound spectrograph apparatus in the 1940's represented an important breakthrough in speech research because it allowed a convenient representation of the way the speech spectrum varied with time. Figure 5 shows an output display produced by such an apparatus -- a so-called *spectrogram*. Energy, represented as darkness of marking, is plotted against frequency and time, so that formants show up as wandering dark bars, periods of voicing as vertically striated regions, and periods of fricative or aspirated energy as randomly patterned marking, with, perhaps, overlaid formant patterning. Soon after the invention of this device, another device to perform the inverse transformation of spectrogram-to-speech was invented at the Haskins Laboratory, then in New York, and using these pieces of

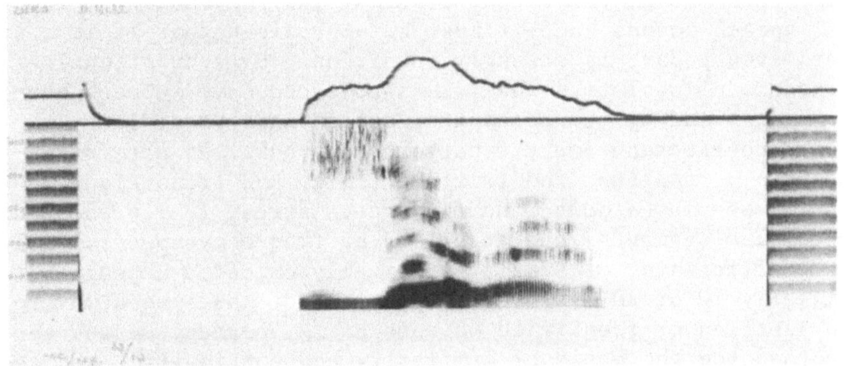

Figure 5: Speech spectrogram of the word "zero"

apparatus, and other ingenious devices, the basis of our present knowledge of speech cues was laid down. Modern technology has produced additional means of characterising the pressure waveform of speech, including methods of Linear Predictive Coding (LPC), and the Fast Fourier Transform (FFT), that are specially well suited to computer processing, but none seems to have had the seminal effect that the spectrograph and pattern playback had. At least, not yet. Those two early pieces of apparatus transformed speech into a perceptual domain that allowed conceptualisation of the relevant speech features, and their manipulation to test theories of their relevance. Modern methods are different only in degree, not kind.

As a result of many years of experiment, especially at the Haskins Laboratories, but in many other places as well, a view of speech cues has been formed in which the sounds of speech representing vowels are characterised by quasi-steady state spectra, whilst other sounds, especially the stop consonants and nasals (b,d,g,p,t,k,m,n, and the sound at the end of running) are characterised by changes to the spectra -- in formant terms, by formant transitions. Some consonants are also associated with various noise bursts due to release of closure in the vocal tract and others with more or less continuous quasi-steady state spectra associated with constriction of the tract in forming the consonant, concurrent voicing, or a mixture of both. However, the transitional spectral changes (in formant terms -- the formant transitions), as the articulators move from one notional posture to the next to string the sounds of speech together, are very important and act as cues to the consonant articulations. The various noise bursts are also important cues, as are the presence and relative timing of voiced energy, the durations of sounds, small shifts in voice pitch, and the spectra produced during consonantal constriction (including, for voiceless stops, relative absence of energy).

These cues are at the lowest level -- the perception of individual speech sounds, and it must be emphasised that it is a view that, even today, is strongly conditioned by conventional methods of linguistics analysis into successive 'speech sounds' or *phones*, and by the conceptualisation made possible by the sound spectrograph and the pattern playback. In attempting to automate perception, the relative timing and transitional cues tend to be played down, and the speech stream is represented by successive segments, characterised by frequency spectral values, LPC coefficients, or similar statically oriented measures. Typically 10 or 20 millisecond portions of the speech waveform are labelled or identified on some basis, perhaps merged and smoothed (on the basis of similarity and continuity), and then used to hypothesise strings of elements nearer to what a phonetician might consider to be the speech sounds (or allophones*) comprising a given utterance. The view of artificial speech production is not too different from this, and the subject is taken up again below.

It is possible (perhaps desirable) that a radically new framework in which to formulate hypotheses of speech structure at the acoustic level of analysis may be invented. In fact, in phonology, a non-segmental approach to analysis is not new while the present author has suggested an alternative 'event-based' approach that might avoid some of the inconsistencies of the segmental approach for purposes of speech analysis at the acoustic level, and hence for describing the perceptually relevant structure of speech (Hill, 1969; 1972a).

However, for the present, both speech perception and speech production are commonly investigated in the context of what we may call 'the segmental assumption'. Attempts to automate the processes involved are greatly affected by this. As a result, in both recognition and synthesis of speech by machines, the effects on sounds of the neighbouring sounds, together with the perceptually relevant transitional cues, are approximated by increasing the segment inventory and using sequences of segments.

4. HIERARCHIES IN SPOKEN LANGUAGE: LEVELS OF KNOWLEDGE

Thus far, we have considered, in outline, some of the facts about speech at the acoustic level in the context of attempting to classify speech sounds. This level is analogous to the decoding of symbols used in a communication system. There are, of course, increasingly higher levels of speech structure that are related the production and perception of speech, and which are therefore

---

* *Phones* are grouped into classes on the basis of functional similarity. The classes are called *phonemes* and allophones are the contextually determined acoustic variants of a given class.

also relevant to the machine generation and understanding of speech. Furthermore, the analogy should not be taken too far because, in a communication system (such as one based on Morse code), the symbols are discrete and the coding both discrete and bearing a one to one relationship to symbols. In speech the symbols interact and overlap and the coding does not bear a simple relationship to the symbols encoded.

The levels of knowledge that are relevant to determining the required response to a spoken message, assuming some kind of approximate phonetic description of the input has been made, are syntactic, semantic and pragmatic. Syntax is concerned with the arrangements of symbols to form words and of words to form sentences. The semantic level is concerned with meanings, in a dictionary sense -- broadly the same for all speakers and listeners; while the pragmatic level takes account of particular experiences and situations of speakers and listeners and may vary widely from individual to individual. In deciding what response is appropriate, given some input, two extreme approaches are possible. The most appropriate responses, given the situation and other contextual constraints, may be listed in order of likelihood. Various alternative ways of eliciting the response may be generated, taking into account semantic and syntactic constraints. These may be instanced at increasingly lower and more detailed levels, until eventually a number of alternative formulations at the acoustic level could be matched against the actual input, the closest match then indicating which of the likely responses was actually required. This would be a top-down or synthetic approach. Alternatively, the given input could be mapped, on the basis of syntactic knowledge and likely errors, onto alternative symbols strings, then lexical items, and subsequently, using semantic knowledge, alternative meanings could be hypothesised with final interpretation to an actual response being based on the particular circumstances under which the input was received. This would be a bottom-up or analytic approach, involving parsing and semantic information structures. Either approach, by itself, would tend to generate very large numbers of alternative possibilities. It seems likely that humans use both approaches in understanding speech, and probably use both in generating speech in the sense that the speech must be generated (synthetic or top-down approach) but is simultaneously monitored and corrected by the speaker hearing his own speech. In any case the numbers problem may be reduced by reconciliation of the alternatives at some intermediate level. The intermediate levels are not independent: they interact. Also probabilistic decisions are involved at all levels and the whole procedure undoubtedly has a flavour of the relaxation techniques used in engineering calculations where solution attempts must be iterated until a self-consistent set of sub-problem solutions is found. Our knowledge of how to organise such procedures in the complicated domains of syntax, meaning and situation, is rudimentary in the extreme and it is in this area that much research has been carried out, in relation to artificial intelligence and computer science as well as speech understanding, over the last eight years. The improvement in our knowledge of how to organise such systems for speech understanding was listed as one of the good things to come out of the $15M five year ARPA speech understanding project (Klatt, 1977).

There is another (perhaps parallel) hierarchy in speech in terms of different levels of acoustic structure. The major division here is between the segmental and the suprasegmental. Even when segmentation is not attempted (and all the major ARPA financed schemes segmented at least into sample segments) any acoustic cues that are extracted from the input waveform may be thought of in terms of at least these two levels. They bear some relationship to the levels of knowledge discussed above, though the relationships in many cases may be tenuous and contentious. Segmental cues are relevant to the level of symbols and, while each must be considered in the context of neighbouring cues because of interactions between successive postures of the vocal apparatus and because of the need to handle transitional information, their relevance and manifestation is comparatively localised. Suprasegmental acoustic cues are concerned with acoustic structure at a higher level, effects such as rhythm and intonation that manifest themselves as patterns of relative variation over larger stretches of speech, although the same or closely related acoustic parameters may be involved as at the segmental level (e.g. a voiced sound necessarily has pitch, but the pattern of pitch variation is intonation). Suprasegmental cues give information about higher level constructs in the syntactical domain, such as syllable, word and phrase structure, and also give important clues as to meaning. Thus, for example, suprasegmental cues are important in helping to distinguish *im*port from im*port*, and the sentence:

It was a pretty little girl's school,

can have a variety of meanings, depending upon the suprasegmental patterning. One highly debated matter at this level (it affects, among other things, rhythm) is *stress*. Because it is often conceptualised in terms of force of utterance, it is associated with acoustic intensity in writings that consider the topic. Now it is true that acoustic intensity is one possible acoustic correlate of stress, but there are others. It is also *relative* acoustic intensity that is involved -- one cannot judge stress for an isolated monosyllable. However, it is also a subjective judgment from the point of view of the producer of an utterance -- certain syllables are stressed. For purposes of understanding speech, of catching the intention of the speaker, either by machine or human, and even for purposes of the machine synthesis of speech, one is interested in the corresponding subjective effect to the listener of *prominence*. A syllable is prominent (or in other terminologies *salient*) to the extent that it stands out from its neighbours as far as the *listener* is concerned. The acoustic cues to prominence are relative acoustic intensity, relative duration, and relative pitch movement (in increasing order of importance). Prominence (the perceptual consequence of stress) is clearly seen as suprasegmental. The word 'import' changes its meaning depending on which of the two syllables is perceived as most prominent, and prominence is one (only one) of the factors that may be manipulated to change the meaning of the sentence about the school.

The acoustic structure in its suprasegmental aspects carries information at yet another level, for it is possible to infer the speaker's state of mind, health, attitude to the situation and so on from acoustic cues of various kinds, some of which are properly called suprasegmental and some of which (for spoken English at least) are properly called paralinguistic. This level is clearly on a par with the pragmatics of the situation.

Finally, it must be noted that the acoustics of speech carry information relevant to identifying the speaker -- whether the speaker is male or female, adult or child, and even specifically who the speaker is. This fact is exploited by, for example, the US Air Force who use the matching of spoken utterances for security checking people requiring access to restricted areas. This is technically *speaker verification* -- it is not inconceivable that a good mimic, who knew the phrases and had a vocal apparatus of similar dimensions as an authorised person, could mimic an authorised user and illegally gain entrance. The more difficult problem of *speaker identification* -- that is unambiguous identification of an arbitrary speaker based only on a recorded sample of speech against the background of an unknown population -- is still a subject for research, to the chagrin of law enforcement agencies. However, some kind of classification of a speaker can undoubtedly be carried out on the basis of his or her speech, and this information equally undoubtedly helps listeners to understand what is said. Machines designed to understand speech can gain advantages in this way. For spectral template matching schemes, such as the commercially available speech recogniser from Threshold Technology in the US, or HARPY, the most successful ARPA project recogniser, the associated 'tuning' to the speaker would seem essential to optimum performance.

5. MACHINE SYNTHESIS OF SPEECH

There has been considerable progress in the machine synthesis of speech over the last six or seven years. A variety of very inexpensive devices for synthesising speech from low bit-rate representations have become available, and several practical schemes for converting ordinarily spelled English text to the parameters required to drive such synthesisers have been published (MacIlroy, 1974; Elovitz, et al., 1976; and Sherwood 1978, for example). It is possible to buy a synthesiser capable of producing almost unrestricted English speech for a few hundred dollars so that even hobby computers and small business systems may use speech output. Speech output is now so compact and inexpensive that a number of calculator firms have produced hand-held calculators that supplement the displayed output with spoken output, with facilities for repeating (thus Telesensory Systems Inc. of Palo Alto in California offer a talking calculator for the blind at $395, based on a single integrated circuit chip for the synthesis). Texas Instruments have also announced a speech synthesiser chip, apparently not for sale to others, which is incorporated into their hand-held *Speak & Spell*, a $50 device designed to help children learn to spell (Electronics Review, 1978). It is possible to buy a PDP-11 computer with full text-to-speech conversion as an option, based on the Federal Screw Works VOTRAX, and using MacIlroy's program running under the UNIX operating system, which was also developed at Bell Laboratories.

There are a variety of bases for making computer speech output devices, but they may be divided into three types. There are those that store, in analog or digital form, the actual time-waveform of speech, and simply reproduce it when required. We may call these *direct recording* devices, though the speech quality and factors like access time, [illegible]y with the actual design. Then there are those that store, in some form, the parameters or other data, that comprise low bit-rate representations of speech, and are used to control synthesisers capable of reconstituting whole speech. Instead of bit rates (or equivalent) of the order of 30K to 100K bits per second, these devices require only 1K to 4K bits per second of speech. We may call these *compressed recording* devices. Then there are those devices that are capable of being driven from quasi-textual data, requiring only 50 to 100 bits per second of speech. It seems fair to call these last types *reading* devices. In cases where the compressed data for driving the appropriate synthesiser may readily be derived from quasi-textual representations of speech, the only difference between a compressed recording device and a reading device may be the provision of software (a computer program) to convert text strings to synthesiser parameters.

The big advantage of direct recording devices lies in the naturalness and intelligibility that is easily achieved. There are disadvantages too. If PCM or some other form of digital recording is used, large amounts of data have to be handled. Otherwise special means of accessing numbers of analog recordings, often on a random basis, have to be provided. In the former case, expense and address space may cause problems, while in the latter case both expense and lack of electromechanical reliability may prove difficult. In either case, it may be necessary to provide some way of changing the working set of speech data. A further problem arises; it is necessary to know in advance, and record (with difficulty and expense from a real speaker) any speech material that is needed. This may lead to special difficulties in keeping quality uniform, in keeping the speech data base up to date, and in being able to meet new situations. Finally, such recordings make it very difficult to string together different utterances, or vary individual utterances, because of the lack of control over the suprasegmental aspects of the speech. The output is liable to sound lifeless and strange. However, for small vocabularies, and fixed message formats, the speech quality and intelligibility can be excellent. Such devices are used for speaking clocks, stock market quotations, creditworthiness checking, and the like. For applications where the constraint of sequential access does not cause access-time problems (sequential teaching strategies, weather reports, tourist information and so on), the devices may be very close to ordinary audio taperecorders, and may function extremely well on the basis of unrestricted vocabulary. Of course, the necessity for recording still exists and any changes are liable to require the whole tape to be re-recorded. Also, even when scripted, reading errors can occur, and are often difficult to spot and correct. Lastly, after repeated access, a contact recording medium will wear.

Apart from random access tape-recorders, which suffer particularly severe problems of electromechanical reliability (or rather lack of it) and may suffer from limited addressing structures and large access times, the first successful commercially available direct recording computer speech output device was the IBM 7770 drum system. This allowed 128 audio tracks on a magnetically coated drum. Each word (or possibly part word) was recorded three times around a given track, and the output could be multiplexed. More compact devices have appeared on the market since, often involving optical recording akin to film soundtracks on glass discs. Although such devices offer faster access, access at random, and less wear than tape-recorder-like devices, the vocabulary is more restricted.

Another IBM device, the 7772 Audio Response Unit, employed compressed recording, the synthesiser being a channel vocoder -- essentially a form of spectral section playback device. About 4,000 bits per second of speech are needed to produce reasonably high quality speech from such a device. Another form of synthesiser, the resonance vocoder synthesiser (of which the first example was Lawrence's PAT, or Parametric Artificial Talker), is based on the source-filter model of speech production. Four parameters control the source (voicing frequency, voicing amplitude, fricative amplitude, and aspiration amplitude) whilst four more control the filter (three formant frequencies, and the main frequency peak for fricative energy). Other versions exist in which the controls are somewhat different, but the overall effect is the same. Some models offer additional controls for things like the nasal spectrum, but only expensive research models allow sophisticated facilities such as control of the source spectrum. Such devices, like the channel vocoder synthesiser, may produce speech on the basis of stored parameters tracks that require one or two kilobits per second of speech. More recently, with the advent of LPC representations of speech, synthesisers have been produced that are essentially special purpose digital signal processors, capable of performing, in real time, the arithmetic needed to reconstitute speech from the stored sequences of LPC coefficients. This is the way that Texas Instrument's *Speak & Spell* generates its speech output. About 1200 bits per second of speech are used for the TI device, though this is probably only half the minimum required for the really excellent speech of which Linear Predictive methods are capable. Other schemes exist for special digital signal processors to produce synthetic speech.

The main advantage of all these devices lies in the fact that they do not require the enormous amounts of data that direct digital recording requires whilst still offering many of its advantages, including: (a) the ability to store speech in ordinary computer memory; (b) avoidance of the need for tricky devices to handle analog recordings; (c) improvement of access characteristics; (d) facilitation of vocabulary change; and (e) ready allowance for duplication. However, they usually suffer the same important disadvantages of inability to handle suprasegmental aspects and of needing to know and record utterances in advance. Although this latter problem may be ameliorated by the fact that large vocabularies may be kept in mass storage and brought into memory when required, it is still a serious problem. Adding to an existing vocabulary can still be especially difficult if the original speaker has gone to work for someone else.

Undoubtedly, the reading machine *par excellence* at the present time is the Kurzweil Reading Machine, or KRM. This device combines an omni-font Optical Character Recognition (OCR) system with text-to-speech speech conversion means. The resulting system is able to accept books of different sizes, using different type fonts, and read them aloud. This impressive performance is not a laboratory demonstration, but is achieved by a machine that is in production, and has been installed in public libraries in the U.S. It is claimed that the current hardware has been very carefully engineered to minimise, almost to the point of extinction, the internal adjustments needed to keep the device operating correctly. In view of the high cost of talking books services, and the dwindling (or non-existent) funds available for these to libraries, such devices are an incredible boon to the blind, both for purposes of study, and doing normal jobs. They are, however, expensive compared to the few hundred dollars required for more restricted machines. The prospective purchaser will need, perhaps, ten thousand dollars or so. Undoubtedly, much of the cost lies in the document reading, and development costs of the integrated, engineered system. One may look for falling costs as sales rise, and the investment is amortised. Apart from the difficult engineering problems that were solved, a variety of speech related problems had to be solved (or ignored). Understandably, the speech quality falls short of more sophisticated systems such as that due to Allen and his coworkers at MIT, but it is an entirely practical product.

The problems solved, at least to an adequate first approximation, include speech synthesis at the segmental and suprasegmental levels by using rules embodying what we know about speech structure based on some phonetic (sound oriented) representation; and the generation of an adequate phonetic representation on the basis of printed text, including some provision for rhythm and intonation. These are the two main levels in the hierarchy from text-to-speech, but clearly there are sublevels. It is these that form the subject of the next section.

6. MULTI-LEVEL RULES FOR DRIVING SPEECH SYNTHESISERS

For English, there is no uniform simple relationship between the normal written form (orthography) of the language and the sounds that represent those written forms. Quite apart from idiosynchratic proper names in British English such as Cholmondeley (pronounced Chumley), and downright misunderstandings (as in the case of the foreign student who gave up when, outside a theatre he saw the advertisement "Cavalcade -- pronounced success"), English spelling is notoriously irregular and its pronunciation problematical. George Bernard Shaw (an advocate of spelling reform) illustrated the problems dramatically by his droll spelling of 'fish'; he suggested that, according to spelling conventions, it should be spelt 'ghoti'. That is 'gh' as in 'enough', 'o' as in 'women', and 'ti' as in 'nation'.

Despite the obvious irregularities the basic component of rules for text-to-speech-sound conversion is a set of letter to sound rules. If these rules are carefully contrived, and context sensitive, quite a small number can cope with a surprising range of apparent irregularities and generate speech that is at least intelligible. Elovitz *et al.* report a set of rules developed for the US Navy comprising 329 rules, each specifying a pronunciation for one or more letters in some context, which could accurately transcribe English text to International Phonetic Association sound symbols for 90% of the words, or 97% of the individual sounds, in an average sample. Their scheme, for American English, was an adaptation of an earlier system due to Ainsworth (1974), but involved no preprocessing of the input.

The next step beyond this simple scheme is to add a dictionary of exceptions, as did MacIlroy (1974), since adding context rules to do the same job amounts to the same thing, but is less efficient. Using a dictionary of exceptions means that an input word (perhaps preprocessed) is first checked against the dictionary. If it is there, the pronunciation is retrieved, and the sound symbols sent for synthesis. Otherwise, it is passed for conversion to the letter-to-sound rules. This addition has the attraction that it is actually the commoner words, in general, that have the most irregular spelling, and one might well wish to store their pronunciation, even if it were regular, to save computer time. MacIlroy's letter-to-sound system, somewhat surprisingly, involved rather more computing than that due to Elovitz *et al.* since it was iterative -- many of the stored rules were rewrite rules, rather then direct sound conversion rules.

There are serious problems in attempting to extend this approach either in terms of improving its performance on a given set of words, or in extending it to cover new words. All manner of normal combinations turn up which trick the most carefully formulated rules and (especially taking into account the variety of inflections of basic forms possible, compound words, and exceptions of all kinds) the lists and rules soon grow to take up uneconomic amounts of memory space for storage, and of computer time for searching the lists and applying the rules. Part of the problem is that the pronunciation of even regularly inflected forms is determined by the underlying form which, though regular when uninflected, is ambiguous when inflected due to the deletion of -- say -- final mute 'e'. Thus, in adding '-ed' we have, for example, 'site'-'sited' versus 'edit'-'edited'. This kind of problem confuses people the reverse way, and they spell 'edited' as 'editted' (c.f. 'pit'-'pitted'). Mute 'e' causes problems in other ways. For instance, it may be difficult to determine whether a non-final 'e' should be pronounced or not as in 'bumblebee' versus 'problematical'. Syllabification may be of help in pronouncing a word as for 'culminate' versus 'calming'. At the higher syntactic level -- that of word arrangements in the language, knowing what part of speech is involved may disambiguate pronunciation, although the effect may be segmental as in 'wind' (noun) versus 'wind' (verb), or suprasegmental as in '*im*port' (noun) versus 'im*port*' (verb). Finally, there are effects on pronunciation that may only be resolved at a semantic or pragmatic level, as for 'lead' (noun) in 'dog lead' versus 'red lead'. Even human beings need help sometimes, and a bomb-disposal expert might wish to 're-fuse' a bomb in order to blow it up, the hyphen indicating the syllabification, and hence the unusual third pronunciation for this letter string*. With the modern tendency to exclude all hyphens from English text, regardless, a reading machine may require to understand what is being discussed to get that one right. Elovitz *et al.* give an excellent example -- 'unionized' which might refer to unions or to ions.

Many of these pronunciation problems may be resolved on the basis of either the grammatical structure of what is being read, or on the basis of the underlying structures of the words in terms of the basic units of meaning. In written form, these units are called *morphs* whilst in spoken form they are termed *morphemes* (the individual written marks used in the orthographic form of language are termed *graphemes* whilst the sound classes used by phoneticians for describing speech at the equivalent level are called *phonemes*, though given phonemes have many different acoustic manifestations).

It was on the basis of this insight that Allen (1968; 1976) and Lee (1968) developed a system for the synthesis of speech on the basis of unrestricted text. The papers referenced give a detailed view of this procedure which is also discussed in the present volume. Four main components are required for the basic determination of pronunciations: (a) a parser, to determine the grammatical structure of the input and generate the parts-of-speech labels for the input words; (b) a decomposition procedure to break down the input words into their morphs; (c) a dictionary of morphs, giving morpheme (target translation) equivalences, morph class, and morpheme parts-of-speech information (entries leading to multiple morphemes, such as 'refuse' having one entry per morpheme); and (d) a procedure for handling any odd-ball cases that cannot be matched. In his 1968 paper, Lee says:

> While the relationship between a written word and a spoken word is very complex, the relationship between a morph and a morpheme is almost one-to-one, barring homographs and homophones.

There are two further important points in the decomposition procedure. First, the morphs are stored in right to left spelling, and the decomposition proceeds from right to left -- this is because when two morphs combine, the changes in spelling occur only to the left morph, so right-to-left decomposition allows such changes to be anticipated. Secondly, in decomposition a longest string strategy is followed, with back-up and retrial in case a match cannot be found for any remainder.

---

* As opposed to '*re*fuse' (noun — garbage) or 're*fuse*' (verb — decline)

The problems of correct pronunciation on the basis of input text do not, however, end there. There are also the problems of rhythm and pitch variation - problems of phrasing and stressing the spoken speech in a natural way. There are still unsolved research problems in the whole area of text-to-speech conversion, but the intonation and rhythm of spoken English is one area where fundamental knowledge is still missing, or is the subject of intense debate, and one where competing theories abound. Lee and Allen worked on the assumption that listeners would use their internalised rules to hear the structure of sentences on the basis of fairly minimal clues, thus betting firmly on the active nature of perception. Some stress rules are incorporated in the morph dictionary, so that the lexical stress for each spoken word is determined. The relative levels of stress are then assigned on the basis of Chomsky and Halle's stress rules (Chomsky & Halle, 1967). The overall intonation contour algorithm is essentially that described by Mattingley (1966), which applies simplified pitch changes on the basis of clause structure and clause type, with pitch levels following Pike (1945). Elovitz *et al.* (1976) found that comprehension of spoken English was improved by using more natural (preferred) stress patterns (prominence patterns). They tested monotone speech; random stress; alternating stress (a simpler realisation than Lee and Allen's); algorithmically placed stress somewhat reminiscent of Lee and Allen's scheme, but adapted to the simpler letter-to-sound conversion; correctly hand-placed stress; and correctly hand-placed stress, with some equalisation of the time intervals between stressed syllables. The algorithmic and hand-stressed speech were not significantly different in subjective preference, but were both preferred to the first three methods, whilst the last method was the most preferred. Theories of English rhythm that propose a tendency towards equal intervals between stressed syllables are termed 'theories of isochrony' and, though such theories are generally out of favour, especially in the US, the present author in collaboration with others, has recently found evidence of such a tendency, although it is not the major rhythmic factor and the actual inter-stress intervals vary in absolute duration over a 5 or 6:1 ratio (Hill *et al.*, 1977). Witten (1978) has described a simple scheme for assigning timing and voice frequency information to spoken British English, based on Halliday's (1970) simplified description of intonation contours for non-native English speakers and a rather literal interpretation of isochrony. Olive (1975) has also considered voicing frequency variation.

With the exception of 'lexical stress' (i.e. determining which syllable, if any, in a given word has the potential for stress, and assigning it), schemes for handling the suprasegmental aspects of speech synthesis depend heavily on hand placed stress placement information, despite being loosely described as 'by rules'. Consider the sentence:

'Bill killed John.'.

Even taken purely as a statement, which would imply a generally falling voicing frequency with the major fall on the 'tonic stress', to know which of the three words was to receive the tonic stress, one would need to know which of three possible questions the sentence was intended to answer:

'Who killed John?',
'What did Bill do to John?',
or 'Whom did Bill kill?'.

If none of thos questions was answered, the stress and intonation contour, as well as the timing, might be different again. Given a parser, enough information about clauses can be determined to make reasonable if not totally natural choices for voicing frequency variation, but in the absence of understanding on the part of the system it is difficult to see how a machine might determine the best placement of tonic stress. Understanding English is an unsolved research problem, and approximate methods must suffice for the time being, methods such as assigning tonic stress to the last 'content' word in the phrase. ('Form' words consist of words association with the grammatical structure, such as conjunctions, prepositions, or articles, while 'content' words carry the information). Without a parser, various regular and extra marks would have to be included in the text.

This then is an outline of the complexity involved in converting text to naturally spoken English. Although by no means all the problems are solved, it should be noted that the Kurzweil machine incorporates enough of the advanced facilities mentioned to provide an engineered, available, working system. In the next section we look briefly at some of the cheaper alternatives, and some current applications.

7. APPLICATIONS AND HARDWARE FOR SPEECH SYNTHESIS

At present, in the absence of really effective and inexpensive speech input devices, the two most significant advantages of machine generated speech are the ability to generate computer responses intelligible to people over unadorned telephones, and the alert or break-in character. This latter aspect is exploited by employing speech in aircraft emergency warning systems. In exploiting the former advantage it is still necessary to use keys (either Touch-Tone® keys on the telephone or equivalent keys on some portable device) to enter the data which invalidates many advantages of speech communication. However, key entry, and audio response, do allow cheap remote access, and do allow blind people the chance to work on such systems.

Because of the high technology base in the US, and also the early public availability of telephones with Touch-Tone® keys, applications of audio response have generally been in advance of those in Europe, where additional portable tone generators must be used in conjunction with standard dial telephones. In either case, after the required connection has been made, audio tones may be used to communicate alphanumeric codes into a computer which is equipped with suitable Touch-Tone® signal decoding circuits, and one of the previously noted methods of audio response generation may be used to carry the computer's share of the dialog. The use of user codes, service codes and passwords allows for secure access by many people for different purposes within one computer system. The users are able to gain access to whatever service they require anywhere there is a telephone and, even if they always access it from the same place, that is *all* they need if the phone has Touch-Tone® dialling. Even with the extra tone generator keyboard, the cost is an order of magnitude cheaper than for a conventional terminal. That, together with the easy user acceptance of a small keyboard and voice response, means that many workers may be put in direct touch with a computer system who would otherwise feel unable to use it. Many of the current applications spring from this fact.

Passive applications for machine speech have been with us as long as recording techniques have been available, and everyone must be familiar with speaking clocks, telephone answering machines, and the like which deliver pre-recorded messages when stimulated, day after day. The speaking clocks normally assemble their messages from a fairly small number of pre-recorded words which can give acceptable suprasegmental control in this restricted case. More recently, the increased flexibility and reliability of computer synthesised speech has allowed the improvement and extension of telephone intercept messages, and at the same time has saved money for the telephone companies. By investigating intelligibility and dialog structure problems for such machine messages, new insight into the nature of human speech communication has been gained; such matters as the need to help listeners anticipate the structure of forthcoming messages by using everyday forms, and by keeping to a uniform structure; the need to allow a pause of several seconds after giving the first three digits of a telephone number to allow rehearsal and recognition before proceeding to the last four digits; and the need to give clues that the message is machine generated. There is still much to be learned, at the very simplest level, about the human factors of speech communication.

However, the advent of good quality computer controlled speech messages, coupled with telephone input means, has allowed interactive applications of speech communication with machines, and it is here that the current boom is occurring. Managers have long realised the economic and data-reliability advantages of putting the information generator on-line to the data collection system, but have simply not had cost effective and acceptable means of doing so until now. Furthermore, increasing numbers of staff are finding a need to have quick and easy access to the various collections of data that are building up. Thus we are now seeing voice response applications in data capture and data-base access: sales order entry by salespeople in the field; credit authorisation services for small businesses; data collection from the factory floor; inventory control in warehouses; ordering and stocking information for customers on a 24 hour basis; debit and credit management, together with customer billing for small businesses and independent professionals, again on a 24 hour basis; stock market quotation services; and a host of similar dialog-based applications. In addition, there are: talking calculators (not only useful to the blind, but useful to sighted people who may, as a result, concentrate their attention on the columns of figures they may be entering); security warning systems in factories and other secure areas that obtain the guard's attention with specific violation information (wherever he is -- including over the radio, if necessary); process-state information systems that assist control engineers in running all manner of automated or semi-automated processes (with similar benefits to those obtained by the security guards); devices to assist the handicapped; and teaching aids of various kinds (it is difficult to imagine machine help with spelling problems in the absence of speech). And, of course, there is the Kurzweil machine, that makes books available to blind readers on more equal terms with sighted readers.

Many of these applications may be satisfied by quite simple hardware of the compressed recording type. However, reading machines offer significant advantages for systems requiring widely varying message formats, vocabularies in excess of a few hundred words, or for systems that users may hear repeating similar messages many times or to which users may listen for long periods. At the current state of the art, adequate solutions exist for all the basic problem areas.

The actual physical devices available, as well as the general price structure of speech synthesis equipment, are changing quite rapidly at present, partly because of the current hardware technology revolution and partly because of the newness of the area. One important factor in device selection is the number of speech output lines that may be handled simultaneously. Many of the digitally based systems can be multiplexed very efficiently, and drum-based direct recording systems can generally handle more than one line. However, analog synthesisers used for compressed recording systems, or as reading machines, cannot be multiplexed. On the other hand, most are cheap enough that each line required may be handled by its own microprocessor-driven synthesiser, although, with the limited computation that would then be possible, reading versions may be restricted to rather unnatural diction.

8. SOME FUNDAMENTAL PROBLEMS IN ASSOCIATING RELEVANT MACHINE RESPONSE WITH HUMAN VOICE INPUT

The difficulty in automating the human ability to listen to a speaker and respond appropriately, even ignoring the nature of the response medium, is two-fold in that it requires the automation of human speech perception and human speech understanding. Neither of these processes is well understood, let alone formalised to the point where they may be incorporated as machine algorithms. A great deal of fruitful and relevant research in both these areas has been carried out over the past 30 years, though the results are scattered over a wide variety of specialist journals in acoustics, neurophysiology, psychology, linguistics, electronics and computing (Hill, 1971). There are still many gaps in our knowledge, and still a lack of the kind of insight required to allow definition of machine procedures to duplicate the amazing human ability to extract meaningful references to real and abstract worlds from acoustic signals of the highly idiosynchratic, variable, and incomplete character that typify spoken language. The problem is usually broken down into a number of sub-problems, representing phases in the progression from acoustic code to response, but recent work under the ARPA speech understanding project (Newell *et al.*, 1973; Klatt, 1977) confirms the educated view that these phases are not sequential steps, but rather parallel activities that combine many sources of knowledge in order to determine what is relevant as a response. Figure 6 attempts to show a diagram of some of the major activities and interactions.

Major unsolved problems still exist at all levels of the speech-input/response-generation task solution. However, over the last eight years considerable progress has been made on certain problems, principally in the computer science areas of: (a) system organisation (combining knowledge sources and controlling interactions); (b) grammar design (to handle large vocabularies with present limited techniques while meeting task objectives and keeping the language 'usable'); (c) continuous-speech decomposition control strategies (avoiding computational cost while allowing for interactions at word boundaries, reasonable numbers of alternatives in the early stages of decomposition, and error recovery); and (d) the incorporation of work from the late 60's on the automatic parsing of English sentences, including Woods' Augmented Transition Network (1976) approach that allows some semantic information to be associated with the syntactic analysis. At the acoustic-phonetic level, methods of verifying proposed words by predicting the expected succession of acoustic-

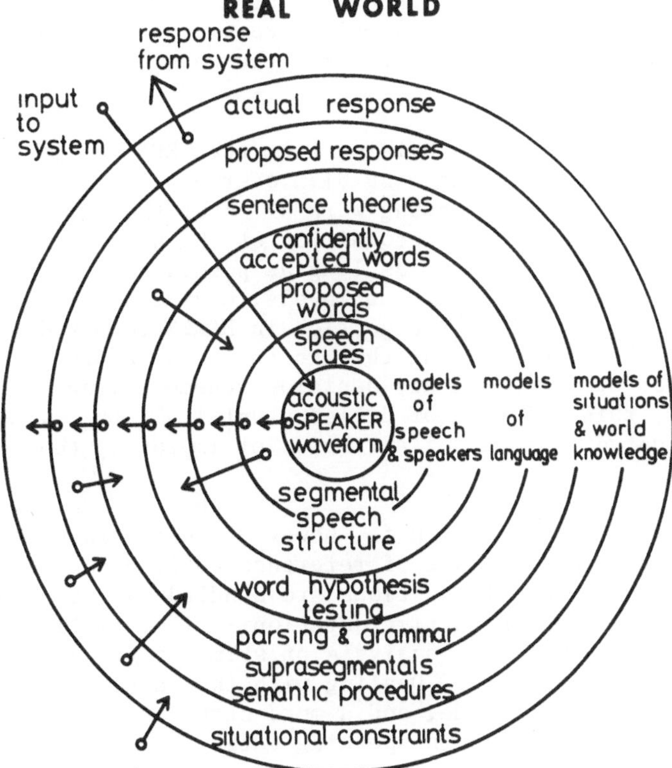

REAL WORLD

Figure 6: Overview of speech understanding systems

phonetic states, allowing for time-warping to cope with varying speaking rates, and matching against the measured input have been developed. There has been a welcome reduction in emphasis on the need to identify phonemes *per se* as an intermediate step in processing. Furthermore, the HARPY system (Lowerre, 1976), not only avoided phonemic labelling (using a network of acoustic-phonetic states that incorporated word boundary interaction effects, grammatical constraints, and served as the basis for word verification in another system), but also provided for the set of acoustic-phonetic spectral templates (98 in total, based on LP analysis) to be gathered automatically by the system during a training phase. To train the system a new speaker would speak 20 sentences. Given the basis of operation of the system, such training would provide an important component of speaker normalisation, although only the templates were updated, not the entire network. At the acoustic-phonetic level, LPC based analysis for the production of spectral samples has emerged as the dominant means of obtaining speech data from the waveform, at least for the ARPA work.

Another noteworthy point arising from the ARPA speech understanding projects (which, consuming $15M over 5 years, have tended to dominate speech understanding work all over the world) is that there has been a greater recognition of the need to maintain probabilistic alternatives as long as possible in recognition strategies used at the various levels, making final decisions as late as possible. Thus recognition in HARPY occurred as the last step, by back-tracking through the network along the path of 'best prior states'. In any decision scheme having a sequential component, this deferral of decision until all the evidence has been evaluated is an important means of improving the performance.

The observant reader will note, however, that this program is still all within the frame of reference built up on the basis of segmental analysis. Transitions are handled as sequences of spectral sections, as is time duration -- i.e. implicitly. It is a preoccupation with the analysis of stimulus frequency rather than the time structure. Dealing with the latter would require explicit handling of time, and a more structured approach to dealing with variations in speaking rate. Little progress has been made in using suprasegmental aspects of the speech signal as an aid to understanding speech either in terms of rhythm (perhaps for the same reason) or in terms of variations in voicing frequency, and perhaps other clues. Furthermore, none of these systems was designed to be a cost-effective solution to any specific recognition problem, the aim being to show feasibility rather than economic viability. HARPY ran on a PDP-KA10 -- a big machine capable of 400,000 instructions per second, using 256K words of memory, each 36 bits. To process one sentence cost $5. This is representative of the costs of these kinds of system.

It might be possible to quibble as to whether any 'understanding' was involved in any of the systems. In fact 'understanding' was defined in terms of producing the correct response, whatever words were hypothesised. Within the very restricted syntax and task domain, this is a fairly weak form of understanding, but it represents a worthwhile innovation in the philosophy of recognition at the semantic and pragmatic levels equivalent to the abandonment of the necessity for phoneme recognition at the lower levels. It is probably analogous to what the human being does in difficult recognition/understanding situations. It also brings up another important aspect of the systems developed under ARPA's umbrella -- that as well as building recognition possibilities from the acoustic domain bottom upwards, they predicted recognition possibilities from the top down, thus bringing semantic and syntactic constraints to bear in a useful way in reducing computation and improving accuracy, but without generating huge numbers of hypotheses at the acoustic-phonetic level (except for the SDC system) by reconciling the analytic (bottom up) and synthetic (top down) approaches at an intermediate level in such a way as to avoid testing all *a priori* possibilities.

In an earlier 'State-of-the-art' review (Hill, 1972[b]), written just as the ARPA speech understanding work was getting under way, the present author suggested that the research would meet its goals. It did -- just ! However, it is interesting to compare the performance of Vicen's 'continuous speech recogniser' that was current at that time, with the performance of HARPY. Vicen's (1969) scheme recognised by keying the decomposition of an input sentence to easily segmented words ('block' -- surrounded by stop gaps for example) and building right and left from these. This was one strategy used in the ARPA work. On the basis of sentences directed at a Robot, such as 'Pick up the large block on the right', Vicen achieved 95% correct word recognition for the three non-redundant 'keywords' in typical sentences. Sentence 'understanding' was, therefore, $(.95)^3$, or roughly 85%. The vocabulary was small and the language syntax very tight, so that only 192 sentences were possible. That system also ran on a PDP-10 computer, giving a cost (then) of 35¢ per second of speech -- or say less than $1 per sentence. HARPY, using a word vocabulary of 1011 words, with an average branching factor of 33 (i.e., on average, a word following at a given point in the sentence might be any one of 33), was able to recognise speech with only 5% semantic error -- i.e. with 95% semantic level recognition, at a cost (today) of only $5. The price ($15M over five years, and $5 per sentence) seems high, but there has clearly been progress. The question is whether such sophisticated systems can now be made cost effective in some way. For the time being the client requiring voice input in preference to a keyboard must be content with very modest performance compared to this, but will find such cut-down systems have arrived in the market-place during the last few years at quite reasonable prices. It is this topic that we now turn to.

9. APPLICATIONS AND HARDWARE FOR SPEECH INPUT TO MACHINES

Under suitable constraints, restricting speech input devices to the identification and use of words spoken in isolation, it is possible to build cost effective devices for the voice control of machines. One may reasonably call such devices *voice buttons* and applications generally amount to the replacement of a simple keyboard. Sophisticated use of such devices depends on: (a) careful dialog design (with provision for feedback and error correction); (b) excellent recognition rates (achieved by the presence of unambiguous word boundaries and training to individual speakers); and (c) reliable rejection of unwanted noises such as coughs, sneezes, asides to human co-workers, and machine noise. The effective implementation of these constraints, which comprise the system aspects of voice input, together with any applications packages that the manufacturer supplies in order to open up those application areas that he sees as viable, are largely what determines the cost of voice-button input. The hobby computer nut, who merely seeks to amaze his friends with a computer that does something (almost anything will do) when spoken to, can obtain a 'voice input terminal' for only a few hundred dollars. With enthusiastic dedication, and some clever programming, such devices can leave the toy class and become quite useful alternate channels for tasks like control of a text editor, eliminating the annoying need to change from command mode to text mode and back at frequent intervals. The devices usually supply simple measures related to the spectral energy content of the input speech and leave timing up to the software.

Two companies that entered the field early with computer based systems, and more sophisticated front end processing hardware, were Scope Electronics Inc., of Reston, Virginia, and Threshold Technology Inc., of Delran, New Jersey. Both these companies offer voice-button systems that meet the constraints for effective applications by commercial end-users. Like the hobby voice inputs, the basic acoustic data consists of a time-segmented spectral description. However, the frequency content is more accurately represented and segmentation is driven by spectral changes so that any word that is input may be characterised by a bit pattern of one to five hundred or so bits that captures enough of the relevant changes in spectral energy with time, with sufficient repeatability, that a user may teach the machine patterns that may be expected for his speech for a fairly arbitrary small vocabulary. A typical vocabulary size would be 50-100 words, although up to 900 words are claimed. With fairly crude data, relatively simple matching algorithms, and little in the way of syntactic and semantic back-up, reliable discrimination is compromised by going for too large a vocabulary. Either the user will end up having to tolerate more errors, or the parameters for decision will have to be set so tightly that a great deal of input will simply be rejected, which would make the system unusable from a human factors point of view. It is also possible that, with too large a vocabulary, there simply will not be the discriminatory power to separate all words, unless care is taken in selecting the words -- an awkward constraint for a supposedly straightforward input means. Although there are a number of voice button input devices currently on the market (including the Threshold Technology device, marketed in the UK by EMI Ltd.) there are not continuous speech recognisers commercially available.

Because of the uncertainties and difficulties associated with the machine recognition of speech, careful dialog design assumes special importance. Apart from being tailored to specific users and applications, it must, as noted, provide a good human-factors oriented interface, and include easy verification, correction and execution facilities. Hebditch (1978) has considered problems of dialog design while another paper in the same volume reports a system using natural language dialog (Harris, 1978). In connection with spoken input, dialog design assumes special importance. It should be pointed out that similar special software techniques had to be developed to allow reliable input of graphical data using light-pens and data tablets which, uncorrected, produce a very uncertain kind of graphical information.

There are a surprising variety of applications for which voice input is currently used. Some applications arise (as for voice output -- in fact, often in conjunction with it) because it facilitates remote use of a computer system, and/or allows data capture at source. However, since there is now no keyboard to use (dialogs should be designed to achieve this) some of the other previously mentioned advantages of the speech interface begin to become important allowing other applications. Thus the TTI and SCOPE devices may be used to enter numeric data on latitude and longitude as well as labelling topographic features when digitising maps. This allows operators to manipulate the digitising cursor and still maintain visual continuity with ease. There is a whole class of inspection problems that are easily solved with the use of voice input, including quality control and production monitoring, in which the inspector can work with hands free, with less fatigue, and actually achieve greater throughput together with few inspection-data entry errors. Many warehousing problems are similar to the parcel-sorting problem. In such tasks, incoming objects are oriented and placed on a conveyor. They are then automatically removed from the conveyor at some later point corresponding to a destination, be it mail bag or outgoing truck. Without the use of voice data-entry, two people are required, one to do the handling, the other to key in appropriate destination data based on marks on the objects sorted. With voice data-entry, one person can manage both tasks. Another class of applications is that associated with entering data from handwritten sheets. These cannot be handled by OCR techniques, especially if torn or crumpled, but sorting through by hand and keying in the data is frustrating because of the difficulty of holding small scraps of paper at the same time as keying data. Voice entry frees both hands, again reducing both fatigue and errors, and raising throughput. In some cases, due to dirt, vibration, the need to wear gloves, and so on, a keyboard or pencil and paper may be out of the question. Furthermore, for some of the applications (e.g. inspection) cutting out pencil and paper stages, even if they are practical, not only gives the advantages mentioned above, but, by putting the inspection process (or whatever) on-line, allows early trends to be spotted and planned for.

One speciality of TTI's, for which they have produced a tailor-made system (the VNC-200), is voice data entry for the production of numerical control data tapes. This not only represents data capture at source, in the sense that the operator with the drawings and knowledge of machine characteristics can, with a suitably designed prompt-based dialog, generate the tapes himself, but the system frees his hands and eyes to concentrate on and manipulate the working blueprints. Again, errors and fatigue decrease, and throughput rises. The VNC-200 vocabulary comprises 60-70 words used normally by NC operatives and the system is self-contained, including the usual post-processor, allowing the voice commands to be converted to the tape required to drive the NC machine. The time from receiving the blueprint to cutting metal is dramatically reduced. The system also includes automatic error detection, pattern storage, nested pattern definition, calculator facilities, and roughing routines, amongst others.

The Scope VDETS (Voice Data Entry Terminal System) is billed as a direct Teletype replacement, complete with RS 232 interface, and one of the options offered with the system is speaker identification. Vocabularies are quoted at 200+ words with templates for up to four speakers held concurrently. A general operating system (VOICE) is offered that allows customers to tailor-make their own applications packages, and synthesised speech feedback is offered, as well as a 'relax/attention' mode switch. One very interesting application noted by Scope as having been tried in mock-up is the checking and control of weapon and electronic systems by the pilot of a military aircraft. Given the density of controls, cramped space, and time constraints with which the pilot of such an aircraft is faced, this kind of off-load, when he is already equipped with a noise-cancelling/excluding microphone, is a very logical step and represents taking up yet another of the advantages of a speech interface mentioned at the beginning of this paper. The 54 word vocabulary used for this application allowed 13,000 distinct control sequences, which would not only save a few switches and lights (with voice response), but would reduce the associated mental load on the pilot who might have to try and find the right switch in an emergency.

Undoubtedly, voice input is here to stay, and more than one writer has commented that the problems of increased use are psychological and social rather than technical. One class of users, not yet specifically mentioned in connection with voice input to machine (those who are severely disabled, but still are able to use their voice), will not find those problems any hindrance. Quite minimal voice buttons could give them a degree of control over their environment, and a useful place in society, that would not have seemed possible 10 years ago. It is to be hoped that there will be equivalent developments in voice input for the disabled as there have been in voice output for the blind.

10. THE FUTURE

The future for speech communication with machines is bright. There is no doubt that it is so convenient to be able to exercise some control over a machine without having to look at it, touch it, or even go near it, and to know in return what the machine is doing, that the idea will catch on, just as convenience foods have caught on, despite the expense and somewhat *ersatz* quality. People in the developed countries seem willing to pay heavily for all manner of conveniences. The techniques will also catch on, however, because of the very practical advantages such as money saving cuts in staff and errors, and money making increases in throughput and customer satisfaction, when the devices are properly applied to tasks for which they are suited.

If there are to be any spectacular technical advances, they are likely to be in the areas of speech structure representation and knowledge structures, though it would be unwise to predict their form. Without such seminal innovations, the next ten years of research, development and applications progress are likely to proceed in much the same way as the last -- namely a steady encroachment of research on the stubborn problems that remain and a continued blossoming of new devices incorporating improved engineering solutions in the market place. The applications seem unlimited, governed only by people's willingness to change. If there is to be any really noteworthy development in the coming ten year period, it will very likely be the effect of voice communication means on the average householder's use of computers. As telephone companies replace obsolescent rotary dial telephones with less costly Touch-Tone® units, the situation will arise in which, without further expense, there is an inexpensive computer terminal in every home, every office, and at every street corner -- at least in North America but to an increasing extent elsewhere. Given the variety of services that might be offered over such a network at the low costs that automation and mass production would offer, there could be a real revolution in the consumer information market. Services could include the determination of the best prices in town for any items coupled with ordering; help with income tax forms; electronic secretarial services by way of phoning to remind people of appointments, birthdays, innoculation boosters and so on; access to a variety of government public service data-bases; the provision of all kinds of services on a 24 hour a day basis; and no doubt many other possibilities that even the most imaginative would overlook. For home use, coupling such a system to the sophisticated cable television services planned could offer home computing and information facilities undreamt of ten years ago.

One thing is certain. The opportunities opened up by the current availability of speech input and output devices represent only the merest trickle before the flood. Whatever else happens over the next ten years, barring only a cataclysm, speech communication with machines will cease being a cause of wonderment and disbelief and become, instead, another widely accepted tool for real work.

REFERENCES

AINSWORTH, W.A. (1974) A system for converting English text into speech. *IEEE Transactions of Audio & Electroacoustics* AU-21, pp 288-290, June.

ALLEN, J. (1968) Machine to man communication by speech Part II: Synthesis of prosodic features by rule. *Spring Joint Computer Conference*, pp 339-344.

ALLEN, J. (1976) Synthesis of speech from unrestricted text. *Proc. IEEE 64(4)*, pp 433-442, April.

CHOMSKY, N. & HALLE, M. (1968) *The Sound Pattern of English.* Harper & Row.

ELECTRONICS REVIEW (1978) Single silicon chip synthesises speech in $50 learning aid. *Electronics Review* 51 (13) June 22, pp 39-40.

ELOVITZ, H.S., JOHNSON, R., McHUGH, A. & SHORE, J.E. (1976) Letter-to-sound rules for automatic translation of English text to phonetics. *IEEE Trans. on Acoustics, Speech and Signal Processing* ASSP-24 (6), December, pp 446-459.

GREGORY, R.L. (1969) *Eye and Brain.* McGraw-Hill: New York.

HALLIDAY, M.A.K. (1970) *A course in spoken English: Intonation.* Oxford University Press: London.

HARRIS, L.R. (1978) Natural language communication. *Infotech State-of-the-Art Report on Man-Computer Communication,* Infotech Information Ltd.: Maidenhead, UK.

HEBDITCH, D. (1978) Programming dialogue. *Infotech State-of-the-Art Report on Man-Computer Communication,* Infotech Information Ltd.: Maidenhead, UK.

HILL, D.R. (1969) An approach to practical voice control. *Machine Intelligence 5* (D. Michie & B. Meltzer, eds.), pp 463-493, Edinburgh University Press: Edinburgh, UK.

HILL, D.R. (1971) Man-machine interaction using speech. *Advances in Computers 11,* (M. Yovitts, ed.), pp 165-230, Academic Press: New York.

HILL, D.R. (1972a) A basis for model building and learning in automatic speech pattern discrimination. *Proc. NPL/Phys. Soc./IEE Conf. on the Machine Perception of Patterns and Pictures,* National Physical Laboratory, Teddington, UK. April 12-14 1972, pp 151-160, Physical Society: London.

HILL, D.R. (1972b) An abbreviated guide to planning for speech interaction with machines. *Infotech State-of-the-Art Report Series 2 "Interactive Computing"* Infotech Information Ltd.: Maidenhead, UK.

HILL, D.R., WITTEN, I.H. & JASSEM, W. (1977) Some results from a preliminary study of British English speech rhythm. *94th Meeting of the Acoustical Society of America,* Miami, Florida, 12-16 December.

KLATT, D.H. (1977) Review of the ARPA Speech Understanding Project. *Journal of the Acoustical Society of America* 62 (6), pp 1345-1366, December.

LEE, F.F. (1968) Machine to man communication by speech Part I: Generation of segmental phonemes from text. *Spring Joint Computer Conference,* pp 333-338.

LOWERRE, B.T. (1976) The HARPY Speech Recognition System. *Carnegie-Mellon University Dept. of Computer Science Technical Report,* April.

McILROY, M.D. (1974) Synthetic English speech by rule. *Bell Telephone Labs. Computing Science Technical Report # 14,* Bell Telephone Labs: Murray Hill, NJ.

MATTINGLEY, I.G. (1966) Synthesis by rule of prosodic features. *Language and Speech* 9 (1) 1-13, Jan-Mar.

NEWELL, A. *et al.* (1973) *Speech Understanding Systems: Final Report of a Study Group.* North Holland: Amsterdam.

OLIVE, J.P. (1975) Fundamental frequency rules for the synthesis of simple English declarative sentences. *Journal of the Acoustical Society of America* 57 (2) pp 476-482.

PIKE, K.L. (1945) *The Intonation of American English.* U. Michigan Press: Ann Arbor.

SHERWOOD, B.A. (1978) Fast text-to-speech algorithms for Esperanto, Spanish, Italian, Russian, and English. *International Journal of Man-Machine Studies* 10 (6) pp 669-692.

VICENS, P. (1969) Aspects of speech recognition by computer. *Tech. Report CS127*, Computer Science Dept., Stanford University: Stanford, California.

Von SENDEN, M. (1960) *Space and sight: the perception of space and shape in the congenitally blind before and after operation.* Free Press: Glencoe, Illinois.

WHITFIELD, I.C. & EVANS, E.F. (1965) Responses of cortical neurones to stimuli of changing frequency. *Journal of Neurophysiology* 28, pp 665-672.

WITTEN, I.H. (1978) A flexible scheme for assigning timing and pitch to synthetic speech. *Language and Speech* 20, pp 240-260.

WOODS, W.A. (1975) Syntax, Semantics, and Speech. *Speech Recognition* (D.R. Reddy, ed.), pp 345-400, Academic Press: New York.

SPEECH UNDERSTANDING AS A PSYCHOLOGICAL PROCESS

William D. Marslen-Wilson

Max-Planck-Gesellschaft
Berg en Dalseweg 79
Nijmegen
The Netherlands

Abstract: An overview of recent psychological research into speech understanding is presented, with an emphasis on spoken word-recognition processes. Human speech understanding is shown to be interactive in character, with speech being understood as it is heard by optimally efficient processing procedures. The flow of analyses through the system is assumed to be controlled by the processing principles of bottom-up priority and of obligatory processing.

1. INTRODUCTION

The purpose of this chapter is to try to convey some of the flavour of the speech recognition and understanding problem as it is viewed from a psychological perspective. To do this I will describe in detail some of the experimental research in this domain, illustrating the kinds of theoretical problems it involves, and the kinds of empirical data that one can hope to obtain. To be able to evaluate psychological research in this area, one has to understand that the problems involved are generally very different in character from those faced by the computer scientist or electronics engineer who is trying to build a speech recognition (or even speech understanding) machine.

The psychologist is, first of all, confronted with a device (the human mind) that has successfully solved the speech understanding problem. His (or her) task is to try to find out what this solution might be. This task is made considerably more complicated by the fact that it is not known *a priori* exactly what the problem is that the human speech understanding system is actually solving.

J. C Simon (ed), Spoken Language Generation and Understanding, 39-67

It is, for example, not certain what is the perceptual target for human speech analysis. Thus a basic part of the psychologist's task is to try to define the nature of the problem to which he is trying to find a solution.

A further set of problems are methodological and technical. The phenomena under investigation -- the processes of human speech understanding -- are mediated by complex and unobservable sequences of mental events. We do not have direct experimental access to these processes, and must therefore use indirect measures which we can hope to relate to the properties of these mental events as they take place in time.

Within the general area of human speech understanding, much of the discussion here will center around the processes involved in spoken word recognition. This is, first, the aspect of speech understanding that presents the least severe difficulties from the point of view of the problems raised above. We can be reasonably confident that "words" (or, rather, their internal mental representations) are psychologically real targets for speech analysis, and the available experimental tasks can be plausibly related to word-recognition processes. In the case of other aspects of speech understanding (such as the proposed phonemic, syntactic, and semantic forms of analysis), it is very unclear how one should characterize the perceptual targets involved, and, therefore, difficult to find tasks that unequivocally tap the assumed underlying process.

A second reason for this focus on word-recognition processes is that word-recognition occupies a central position in the language understanding process. Words form the essential target for the early analyses of the sensory input, and, at the same time, are the essential source for a further mapping onto the structural and interpretative analysis of the message being transmitted. Investigation of word-recognition, therefore, gives us access both to the sensory analyses which feed into this process, and to the higher-level analyses which operate on the products of this process.

A third reason for concentrating on word-recognition is that this is the area of psycholinguistic research into speech understanding that seems most directly relatable to research on speech recognition and understanding machines. In fact, the properties of the human solution to the word-recognition problem may have considerable consequences for the likely properties of sucessful engineering and computational solutions -- especially where these involve the analysis of connected speech.

2. THE INTERACTIVE STRUCTURE OF HUMAN SPEECH UNDERSTANDING

The problem of human speech understanding is just one example of the general problem of characterising the role of knowledge in the perceptual processing of a sensory input. The listener comes to the speech perceptual situation with a variety of different types of stored knowledge (or knowledge sources). This includes not only the listener's general linguistic and non-linguistic knowledge, but also his representation of the current utterance and discourse in terms of these knowledge types. These knowledge sources constitute one side of the speech understanding situation. On the other side stands the acoustic input itself, and the products of the primary sensory analysis of this input. The problem that we face is to characterise the general properties of the relationships between these different aspects of the understanding process.

In particular, what is the nature of the processing environment within which the analysis of the sensory input develops? Is speech analysis a fundamentally bottom-up process, in which information available at higher levels of the system does not interact with analyses at lower levels? Or does the system permit top-down interactions -- for example, can constraints derived from the semantic properties of the message directly affect subsequent decisions about the word-candidates that best fit the sensory input available at some later point in the utterance?

The answer to these questions has profound consequences for the kinds of information we can expect to be carried by the speech signal. If the signal is produced so as to meet the needs of a strictly bottom-up analysis process, then we can expect it to carry sufficient information to unambiguously specify the correct analysis of the input at each of the successive stages of perceptual processing implied by a bottom-up analysis system (e.g., phonetic, lexical, syntactic, semantic, etc.). But if the signal is produced so as to meet the needs of an interactive analysis process, then we should not expect it to carry the information adequate for successful bottom-up analyses, even at the more peripheral stages of the system.

For example, if word-recognition in utterance contexts were a strictly bottom-up process, then we should expect to be able to find invariant properties of the signal that uniquely determined which word corresponded to which portion of the speech input. But if word-recognition were an interactive process, in which sensory and contextual inputs were combined to reach the word-recognition decision, then we could not expect to find the same invariants in the signal. For the signal need only carry sufficient information to specify the correct word-choice *relative* to the contextual environment in which the word in question was ocurring.

There is good evidence that human word-recognition -- and human speech understanding in general -- is in fact fundamentally interactive in nature. This evidence will be described in the next sections of the chapter.

2.1. The immediate interpretation of spoken language

If human speech understanding is interactive, then it follows that there has to be contextual information available with which the analysis of the sensory input can interact. In some earlier theories of spoken language processing (1, 2), it was claimed that there were considerable delays in the development of a syntactic and a semantic interpretation of the speech input. On the strongest versions of this view -- known as the "clausal hypothesis" (for further discussion see (3, 4)) -- it was assumed that syntactic analysis of the input was delayed until a syntactic clause-boundary was reached, and that the semantic interpretation was further delayed until this syntactic analysis was completed. If this theory were correct, then speech processing could hardly be fully interactive, since there would be no syntactic and semantic context available within a clause with which the analysis of the sensory input could interact.

Several experiments show that the clausal hypothesis is not correct, and that the structural and interpretative analysis of the input begins immediately. I will use the term "structural" here to refer to the specifically linguistic (syntactic) knowledge that is necessary for language understanding, while the term "interpretative" applies to those aspects of mental knowledge that are necessary for the construction of an interpretation of the utterance in some possible world or discourse context.

One source of evidence is the *speech shadowing* task. This is an experimental situation in which the subject is required to repeat back speech heard over headphones, with as short a repetition delay as possible. The repetition delay is established by measuring the interval between the onset of a word in the material the subjects are hearing and the onset of the same word in their repetition. Of particular interest here are the so-called *close* shadowers, who can reliably and accurately repeat back normal connected prose at mean delays of around 250 msec, which is a lag of little more than a syllable (3, 5, 6, 7).

This very short delay means that we know exactly how much of the input the subject could have heard at the point that he made his response. There is no question, in interpreting shadowing performance, of "backwards" effects on processing decisions of information that only becomes available later in the sentence. Thus, by determining what variables affect close shadowing performance, we can find out what kind of analysis of the input is available to the listener as he hears it (or, at least, within 250 msec of hearing it).

Several experiment show that close shadowers are performing a structural and interpretative analysis of the input word-by-word as they hear and repeat it. For example, in one experiment I examined the properties of the spontaneous errors that the close shadowers made (3, 5, 6). Many of the errors involved substituting new words for the words that were actually present in the input. If the shadowers' performance was not controlled by the structural and interpretative properties of the text, then the substitution errors they made should not be controlled by these variables. There would be no reason why they should not insert words that were inappropriate to the available sentential and discourse context. It turned out that over 98% of the relevant spontaneous errors were entirely congruent with their context, up to and including the word immediately preceding the error.

Taking a different approach, one can demonstrate the close shadower's use of structural and interpretative information by depriving him of such information (6). The results show that close shadowers perform more poorly (are slower and less accurate) on semantically uninterpretable material than on normal prose materials, and worse still on completely scrambled material (that has no syntactic structure either). In another experiment, single anomalous words were inserted into an otherwise normal sentence that the subject was shadowing. These words were either structurally or interpretatively anomalous with respect to their context. Shadowing performance would only be disrupted if the shadower was processing the material to a level at which such an anomaly could be detected (an interpretative anomaly is only anomalous with respect to a given interpretative framework). The results clearly showed the close shadowers to be sensitive to these different types of anomaly (6, 7).

The close shadowing results show, then, that speech appears to be interpreted as it is heard. This conclusion is also supported by the results of some other experiments using the *word-monitoring* task. This is an experimental situation in which the subject listens to a sentence for a target-word specified in advance, and presses a response-key as soon as he detects the target. The main determinant of response-time in this task is the time it takes to recognise the word in question. Since word-recognition is facilitated by contextual factors (see sections 2.2. and 2.3), we can use the monitoring task to track the availability of different forms of sentential analysis during the processing of a sentence. If an analysis of a particular type is available to the subject, then it should affect his response-time (*via* its effects on word-recognition), relative to a situation in which such an analysis is not available to him.

In the major experiment of this type (8, 9), we examined the time-course of structural and interpretative processing across entire sentences. The target-words appeared at different serial positions in the test-sentences (from the second to the tenth word-positions). The test-sentences were either normal sentences, or were either semantically uninterpretable, or both semantically and syntactically uninterpretable. Examples of each type are given below, with the target-word italicised.

i. The church was broken into last night. Some thieves stole most of the *lead* off the roof.

ii. The power was located into great water. No buns puzzle some in the *lead* off the text.

iii. Into was power water the great located. Some the no puzzle buns in *lead* text the off.

These contrasts between prose contexts enabled us to measure the availability of different types of sentential analysis at different points in a sentence. The most important result was the finding that the normal prose contexts showed a significant advantage over the other two contexts right from the beginning of the sentence. A further experiment (9) showed that this early advantage in normal prose was dependent on the presence of a context sentence (in (i) above the context sentence is "The church was broken into last night"). When the subjects heard the test-sentences (e.g., "Some thieves stole most of the lead off the roof") in isolation, then there was no difference between the normal sentences and the others at the beginning of the sentence.

Taken together, these monitoring experiments show that the listener begins his interpretation of the utterance from its first word onwards, and that these early analyses crucially involve the mapping of the new sentence onto the interpretative context provided by the prior discourse. Although early syntactic analyses must, of course, also be carried out, these are only one component of the dominant frame of reference for speech understanding -- that is, its general interpretative representation. The listener's perceptual goal is the full interpretation of what he hears, and his on-line processing is always conducted with respect to this goal.

Given, then, that the listener attempts to interpret speech word-by-word as he hears it, we now have to show that he uses this knowledge to interact with his processing of the subsequent sensory input. This can be most clearly shown in the domain of spoken word recognition.

2.2. The speed and earliness of spoken word recognition

The argument for the interactive character of spoken word-recognition is primarily based on the evidence that spoken words, in utterance contexts, are recognised so early relative to the sensory information available at the point of recognition, that it is necessary to postulate an additional, contextual input to the recognition process.

This evidence derives, first, from the speech shadowing and word-monitoring experiments described in section 2.1. The mean repetition latency for close shadowers was about 250 msec. Assuming that about 50-75 msec of this response delay is taken up by the time it takes to actually execute the response, then this level of performance indicates that the shadowers were recognising words in normal prose contexts within about 200 msec of their onset. A similar estimate can be derived from the word-monitoring experiments. The response latencies here averaged 275 msec (8, 9). Again subtracting about 50-75 msec for the time it takes to execute the response (pushing the response-button), this also indicates a recognition time of about 200 msec after word-onset.

In the monitoring experiment it was possible to measure the temporal durations of the 81 different target-words. For the words in normal contexts, the average duration was 370 msec. This means that the subjects were successfully discriminating the target words after little more than half of the acoustic input corresponding to the words could have been heard. This claim is supported by two other types of experiment.

One of these used the *phoneme-monitoring* task. This is a task in which the subject is required to press a response-key when he detects a phoneme target specified in advance. When the phoneme is located in a real word, it has been established that the subject's strategy is first to identify the word, then to scan his phonological knowledge of that word to determine whether it contains the phoneme in question, and then to make his response (see section 3.2.). Thus, by subtracting from the overall reaction-time the amount of time required for response execution (50-75 msec), and the time required for a phonological scan (on average about 150 msec -- see (9)), we can arrive at an estimate of the time taken up by the first stage of the task -- that is, identifying the word in the first place.

An experiment by Morton & Long (10) obtained phoneme-monitoring reaction-times of 420 msec to phoneme targets placed at the beginnings of words. Subtracting from the overall monitoring latencies the response execution and phonological scan components, we again arrive at a recognition time of around 200 msec. Since the words bearing the targets averaged 400 msec in duration, less that half of the acoustic input corresponding to the word could have been heard when the word was identified.

Finally, there is a quite different kind of task recently developed by Grosjean (11). This is the so-called *gating* paradigm, in which subjects are presented with successively longer fragments of a word, and are asked at each point to try to determine what the word is. This is not a reaction-time task, but it nonetheless allows a precise control over the amount of sensory information the subject could have heard when he made his response.

Grosjean gated the acoustic signal for each word at successive 30 msec intervals, starting at the beginning of the word. When the word was presented in normal sentence contexts, the average amount of signal the subjects needed to correctly identify the word was 199 msec. This was for words with an average total length of around 410 msec.

Thus evidence from four quite different kinds of experiment converge on a recognition-time for words in normal contexts of about 200 msec from word-onset, at a point when only half of the total acoustic signal corresponding to the word might have been heard. Word-recognition appears to be possible when the subject could only have heard the initial consonant-vowel (CV) or vowel-consonant (VC) sequence of a word. Some recent measurements of the duration of a selection of word-initial CV and CCV sequences in sentence contexts averaged 202 msec (12, 13).

It is possible to roughly quantify how insufficient the available acoustic-phonetic information would be, given recognition within 200 msec. The word-monitoring studies mentioned above involved 81 different word-targets. For each word we determined how many words in the language (American English) would be compatible with its initial CV, CCV, or VC sequence. We followed the pronunciations given in Kenyon & Knott (14), and excluded very uncommon words. The median number of possibilities estimated in this way was 29. There are, for example, about 35 words in American English beginning with the sequence /ve/.

The figure of 29 may be a slight over-estimate, since the phonemic notation used by Kenyon & Knott does not capture various phonetic distinctions that might otherwise have served to reduce the number of words that appeared to begin with the same initial sequence. Nonetheless, this estimate clearly indicates that there were, on average, several possible word-candidates compatible with the sensory information that would be available when the correct word was successfully recognised.

Since this means that the sensory information available at the point of recognition is insufficient to allow the selection of the single correct candidate, then it must be contextual information that provides the additional necessary constraints on the identity of the word intended by the speaker. There is clear experimental evidence to support this inference.

2.3. Contextual interactions with spoken word-recognition

The first type of evidence for early contextual effects comes from the consistent finding that the fast recognition times reported in section 2.2. depend on the availability of contextual (structural and interpretative) constraints. In the shadowing task, the mean shadowing latencies for the close shadowers increased by 57 msec when they were required to shadow semantically uninterpretable material rather than normal prose passages (6). In the word-monitoring experiments (8, 9), monitoring latency increased by 58 msec when the target-words were embedded in semantically uninterpretable test-sentences. Monitoring latencies were still further slowed down when the targets occurred in scrambled strings, which had no syntactic interpretation either.

In the phoneme-monitoring task, a number of experiments (10, 15) have shown increases of about 80 msec to targets in highly probable words as opposed to targets in less probable words. For example, time to detect the phoneme /d/ would be slower in the context "Slowly he opened the *d*ance..." than it would be in the context "Slowly he opened the *d*oor...". Finally, in Grosjean's (11) experiment using the gating paradigm, the average amount of signal necessary to recognise the words increased from 199 msec in the normal context conditions to 333 msec when the gated fragments were presented in isolation, with no constraining context.

A number of other experiments, all using the *mispronunciation detection* task (16), also provide evidence for interpretative interactions with word-recognition processes. In this task, subjects listen to continuous texts for mispronounced words -- that is, words changed by a single consonantal phoneme into a nonsense word. Mispronunciation detection-time is found to interact with the predictability of the original word in its sentence and discourse context. To perform the task, subjects first determine what the mispronounced word should have been (what the original word was), and this process is clearly facilitated by contextual constraints.

These various findings demonstrate, then, that there is a general facilitatory effect of semantic and syntatic context in experimental situations involving word-recognition processes. The fast reaction-time tasks show that these effects are located early in the recognition process. Contextual information can evidently interact with word-recognition decisions at a point in the word at which several word-candidates may still be compatible with the available sensory input.

This conclusion is supported by the results of a further experiment, which directly examined the role of on-line interactions between sensory and contextual inputs during word-recognition (6, 7). A group of close and of more distant shadowers were asked to shadow 120 pairs of normal prose sentences, each of which contained a critical trisyllabic word somewhere in the second sentence of the pair. These words were either normal with respect to their sentential context, or were made either semantically or both semantically and syntactically anomalous. Secondly, within each context condition, either the first, second, or third syllable of the critical word was changed, so as to make it into a non-word. Thus, "president" might be changed to "howident", "company" to "comsiny", and "tomorrow" to "tomorrane". What I was interested in here was the relative frequency, as a function of contextual appropriateness and of location of mispronunciation, with which the subjects restored the mispronounced words to their original form -- that is, repeated back an input like "instrutage" in its original form of "instrument".

The results were clearcut, and are summarised in Table One.

Table One: Word restoration totals

	Contextual Appropriateness		
Mispronounced Syllable	Normal	Semantically Anomalous	Semantically and Syntactically Anomalous
First	2	5	1
Second	21	7	0
Third	32	4	3

Restorations were very infrequent in all conditions except two. Namely, those conditions where the disrupted original word had been normal with respect to its context and where the first syllable of the original word was intact. Equally importantly, this distribution of restoration frequencies held just as strongly for the closest shadowers (with latencies of 233, 258, and 285 msec) as it did for the more distant shadowers (with latencies ranging up to 944 msec). I also ran a control experiment where the same mispronounced words were given to the subjects in isolation, without any sentential context. Under these conditions there were very few restorations, and their distribution did not correlate with the pattern in Table One.

The results of this experiment demonstrate a reciprocal interaction between contextual and sensory inputs. Sensory information plays a determining role, since words are very rarely restored when their first syllable is disrupted. But contextual inputs are also critical, since words are also very rarely restored when the word indicated by an undisrupted first one or two syllables is inconsistent with the structural and/or interpretative context in which it is occurring. This shows rather directly that contextual information is able to control the lexical interpretation assigned to a given acoustic-phonetic sensory analysis.

The data from the close shadowers is crucial here, since it shows that this contextual effect on word-recognition can be traced back to within the first 200 msec after word-onset. These subjects are initiating their word-recognition responses when they could only have heard part of the first syllable of the word, and yet, even at this early point in the word, contextual factors are affecting the way in which they interpret the speech input. The performance of these short latency shadowers also serves to confirm the claim that contextual effects are operating under conditions of genuine bottom-up uncertainty.

Take, for example, the mispronounced word "comsiny", which was frequently restored in context (but never in isolation) to its original form of "company". The initial syllable, /kʌm/, is consistent with a number of words, and there is no way that the sensory input which the close shadower has heard at the moment of response could uniquely select "company" as the word to repeat. In particular, since the second syllable was mispronounced, there would have been no coarticulatory cues in the first syllable that might have pointed to "company" rather than, say, "comfort". Thus, as far as the sensory analysis was concerned, the contextual environment could equally well have led to the selection of "comfort" as the word to be restored. This clearly implies that "comfort", just as much as "company", would have to have been available as a word-candidate until the point at which the "company" option was actually selected. This, in turn, supports an analysis of word-recognition processes in which potential word-candidates start to be generated very early on in the processing of the speech signal corresponding to a given word, and where the on-line use of contextual information allows the correct word to be selected from among these various possibilities.

The experiments described so far in this section provide the most direct evidence for an interactive word-recognition process. There is also considerable indirect evidence for the presence of interactions, which is discussed in detail in some earlier papers (9, 17). Much of this evidence concerns the flexibility of the balance, during human speech processing, between contextual and sensory inputs to the word-recognition process.

An important requirement of an interactive system is that it should be able to adjust its performance to the constantly fluctuating informational conditions under which it must operate.

As an utterance is processed, there will clearly be variations in the strength of the contextual constraints available at different points in the utterance. When contextual constraints are weaker, then the system will be more dependent on the sensory input to achieve a correct analysis.

Evidence for this process of adjustment can be extracted from the word-monitoring experiments already mentioned (see section 2.1.) These experiments were designed so that the same 81 target words were tested, across subjects, in all three context conditions (normal prose, semantically uninterpretable prose, and scrambled prose). This meant that we could examine the ways in which the same acoustic-phonetic inputs, mapping onto the same locations in the mental lexicon, are treated by the word-recognition system under different conditions of contextual constraint. What we looked at here was the relationship between monitoring reaction-time and word-duration, making the assumption that the greater the dependence of recognition performance on the sensory input, then the more of the word that would need to be heard, and, therefore, the stronger the effects of duration on monitoring latency. This relationship between word-duration and reaction-time, as a function of context, was evaluated in a series of linear regression analyses.

These analyses show that the word-duration effects are strongest in the scrambled sentences, with the linear effects of word-duration accounting for 81% of the variance in the monitoring latencies. The effects are somewhat weaker in the semantically uninterpretable sentences, with 55% of the variance accounted for, and weaker still in the normal prose contexts, where less than 25% of the variance is accounted for.

These measurements of the importance of the sensory contribution to recognition performance can be contrasted with the results of a different set of regression analyses -- carried out on the same monitoring latencies to the same words -- which measure the contribution of contextual factors to monitoring latency in each context. This second set of analyses examined the relationship between reaction-time and position of the target-word in the test-sentences (positions varied from the second to the tenth word). These regressions on word-position show the extent to which monitoring latencies are determined by interpretative and structural constraints as they accumulate during the processing of a sentence. Neither type of constraint is available in the scrambled sentences, and only structural constraints are available in the semantically uninterpretable material.

The results show exactly the opposite pattern to the word-duration analyses, and clearly illustrate the flexible balance between the available contextual and sensory inputs to the word-recognition process. The contextual effects are strongest in normal prose, with the linear effects of word-position accounting for nearly 80% of the variance. They are considerable weaker in the semantically uninterpretable contexts, accounting for 52% of the variance, while in the scrambled sentences there is no effect of word-position detectable at all.

In summary, the evidence so far presents a picture of human word-recognition as an interactive process, maintaining a dynamic balance between different sources of processing information. In the next section I will describe an experiment which suggests that similar interactions determine the interplay of structural and interpretative processes during speech understanding.

2.4. Structural and interpretative interactions

In order to understand an utterance, the listener will normally analyse its structural (syntactic) properties as part of the process of determining the proper mapping of the utterance onto its interpretation in context. The question here concerns the nature of the relationship between the structural aspects of processing and the interpretative analysis which is the goal of the system. Some theories of speech processing have made strong claims for the "autonomy" of syntactic processes (e.g., 1, 2, 18). That is, syntactic analyses are assumed to be strictly bottom-up in character, and to be conducted without any possibility of reference to the further interpretatibility of the string.

This hypothesis has been tested in an experiment which examined the effects of different interpretative contexts on the analysis of a structurally ambiguous string (19). The type of ambiguity we looked at is exemplified in the following sentence: "Hunting eagles can be dangerous". The phrase *hunting eagles* has two possible readings; either it is the eagles who are doing the hunting, or, alternatively, someone (unspecified) is hunting the eagles.

It is possible to construct contexts which bias the interpretation of these ambiguities in the direction of just one of these readings. Here are some examples:

1a) If you walk too near the runway, *landing planes* (are)...

1b) If you've been trained as a pilot, *landing planes* (is)...

2a) If you watch them as they swoop down for the kill, *hunting eagles* (are)...

2b) Since it's forbidden by law, *hunting eagles* (is)...

In contexts (1a) and (2a) the preferred reading of the ambiguous fragments implies that it is the planes that are doing the landing, and the eagles that are doing the hunting. This means that a plural verb form (like *are*) would be an appropriate continuation of the fragment. The opposite is true for the (1b) and (2b) contexts, in which the appropriate continuation is a singular verb, such as *is*.

The experiment was designed to find out whether these preferences were computed on-line; that is, whether by the time the ambiguous fragment had been heard, the listener had already established a preference for one reading or another. To do this we presented the context-sentences and the ambiguous fragments auditorily, over headphones. Immediately at the offset of the last word in the fragment (e.g., planes) we presented *visually* a continuation word, which was compatible with one of the readings of the fragment. Thus, following the context clause and the ambiguous fragment in (1a) or (1b), the subject would immediately see either the word "IS" or the word "ARE". The subject's task was simply to say the word as rapidly as possible. Our assumption was that naming latency would be slower when the continuation word was inconsistent with the preferred analysis of the fragment -- but only if the subject had already computed this preferred analysis by the time he saw the contination word.

This assumption was confirmed by the results. Naming latency to inappropriate continuations was significantly slower than to appropriate continuations. Thus, the latency to name "IS" was slower in contexts (1a) and (2a) than in contexts (1b) and (2b), while latency to name "ARE" was slower in the (1b) and (2b) contexts than in the (1a) and (2a) contexts. The speed of the subjects' latencies (500-600 msec) meant that they had little time available for post-stimulus analyses. The effects of sentential context on the subjects' performance could only have derived from analyses that were already available to them when the fragment ended.

These results mean, then, that on-line sentence processing involves the rapid and sophisticated integration of at least three different sources of information -- lexical, structural, and interpretative. No single one of these sources of information could have been sufficient by itself to produce a preference. Consider, for example, the context clause in (1a) -- "If you walk too near the runway,". This context leads to a certain preferred structural analysis of the fragment "landing planes". However, it does not of itself require that a structure of this type must follow. If the fragment here had been "*avoiding* planes", then the alternative structural reading would naturally have been adopted, even though the context stayed the same. The preferences obtained in this experiment must involve the on-line analysis of the meanings of the words and their structural properties, relative to the most plausible interpretation of this information in the light of the scenario set up by the context clause.

Note that this experiment does not completely rule out the possibility that syntactic processing is autonomous, since it is possible to imagine a system in which the intermediate products of autonomous syntactic analysis are made immediately available to subsequent interpretative processes, and that it is the effects at this level that the experiment is tapping. In fact, this alternative proposal can probably never be excluded on the basis of experimental data. We will exclude it here, however, because the proposal for an interactive system accounts for the data in a more straightforward way, and because there is no clear psychological processing evidence that independently requires autonomous syntactic processing.

A general problem here is that it is very unclear at the moment how one should characterise the role of syntactic knowledge in human speech understanding processes. From the point of view of most linguistic theories, an autonomous syntactic level of analysis lies at the core of the whole theoretical enterprise. But this type of theory should on no account be confused with a processing theory. The fact that certain analytic levels are distinguished in a linguistic theory does not require that the same levels are functionally distinct in the speech processing sequence. An adequate processing model will of course have to provide some account of the role of structural syntactic knowledge in human language processing. But the available psychological processing data tend, if anything, to be inconsistent with the notion that structural knowledge participates in processing as a distinct level of analysis of the input, with a distinct computational vocabulary that is stated in strictly syntactic terms (for further discussion, see 4, 9, 20).

3. OPTIMAL EFFICIENCY IN HUMAN SPEECH UNDERSTANDING

Given the claim that the human speech understanding system is not organised on a strictly bottom-up basis, and that it allows for interactions between knowledge sources, we now have to place constraints on the possible organisation of such a system. I will begin in this section by developing a claim about the general design strategy for human speech processing -- namely, that it is constrained to be optimally efficient. This claim can be most clearly defined, and experimentally tested, in the domain of spoken word recognition.

3.1. Optimal word-recognition processes

The general claim I want to make is that the human speech understanding system is organised in such a way that it can assign an analysis to the speech input at the *theoretically earliest point* at which the type of analysis in question can be *securely* assigned. What is meant by the term "securely" here is that the system does not, within limits, make guesses about the correct analysis of the input.

This general claim can be translated as follows into the word-recognition domain: A word is recognised at that point, starting from the beginning of the word, at which the word in question becomes *uniquely distinguishable* from all of the other words in the language *beginning with the same sound sequence.* What is meant here can be illustrated by looking at an example -- that is, at the relationship of a given word to the set of words in the language that begin with the same sound sequence.

Table Two lists the words in British English (21) that begin with the sound sequence /tre/. I will refer to this list as the *word-initial cohort* for /tre/. Assume that the word to be recognised is "trespass" (/trespəs/). Given the information in Table

Table Two: Word-initial cohort for /tre/

treachery	tremolo
treachery	tremor
tread	tremulous
treadle	trench
treadmill	trenchant
treasure	trend
treasury	trepidation
treble	trespass
trek	tress
trellis	trestle
tremble	

Two, we can then determine the point at which /trespəs/ becomes uniquely distinguishable from all of the other words beginning with /tre/. The word is not discriminable at the /tre/, nor at the /tres/, since there are two other words in the language which share this initial sequence (tress, trestle). But immediately following the /s/ these two can be excluded, so that the /p/ in /trespəs/ is the theoretically earliest point at which the word could be recognised.

It is at this point, therefore, that an optimal recognition system should be able to recognise the word. It cannot do so earlier, since there are other words that cannot be excluded. But if it does so any later -- if the word is not discriminated until the /ə/ or the final /s/ -- then the system would not be optimally efficient. It would be using more sensory information to make the recognition decision than was strictly necessary.

We can model an optimally efficient recognition system by

making the following assumptions about the psychological processes involved. The basic assumption is that human word-recognition is based on a distributed processing system, made up of a large array of computationally active recognition elements, with each recognition element corresponding to one word (or perhaps morpheme) in the language. These elements are assigned the ability to become active whenever the appropriate patterns are detected in the sensory input. This means that at the beginning of a word, a subset of the array of recognition elements would become active. This subset is assumed to correspond to all the words in the language, known to the listener, which begin with the same initial sequence. Thus, given the initial sequence /tre/, all the words in Table Two (or, rather, their corresponding recognition elements) would become activated. These elements would continue to actively monitor the incoming sensory input.

The recognition elements are assigned the further ability to de-activate themselves when a subsequent mismatch occurs between the sensory input and their internal specifications of the sensory properties of the words they represent. Thus, as more of the signal for a given word is heard, this signal will diverge from the specifications of an increasing proportion of the members of the word-initial cohort. As the degree of mismatch reaches some criterial level for individual members of the word-initial cohort, these elements will de-activate themselves, and drop out of consideration as word-recognition candidates.

This process of sequential reduction will continue until only one candidate is left. At this point the recognition decision can in principle be made, for words heard in isolation. The claim I am making is that it is indeed at this point that the human word-recognition system does make the recognition decision.

Such a recognition process would clearly be optimal in the sense I have defined. Note that what enables the model to behave optimally is that it allows the incoming sensory input to be constantly assessed with respect to the changing sets of lexical possibilities that are consistent with this input at different points in time. A distributed processing system seems the only way of supporting this type of performance in real time -- where many different possible analyses have to be considered simultaneously.

The next section of the paper summarises two experiments which support this kind of analysis of human word-recognition processes.

3.2. Evidence for optimal processing efficiency

The first experiment here is designed to show that human speech analysis involves the continuous assessment of the input with respect to the words in the language that are compatible with this input. The experiment used an *auditory lexical decision* paradigm, in which the subjects heard isolated sound sequences, and made non-word decisions (pressed a response key) whenever they thought they heard a sound sequence which did not form a word in English. The important variable here was the point in each non-word sequence at which it became a non-word, and the way in which these "non-word points" were determined. The non-word points varied in position from the second to the fifth phoneme in a sequence, in sequences that were one, two, or three syllables in length.

The calculation of the non-word point for each stimulus was based on the cohort structure of the language (General British English). For example, one of the non-words was "stadite" (/stædeɪt/). This becomes a non-word at the /d/, since there are no words in the word-initial cohort for /stæ/ which have the continuation /d/. Another example would be the non-word /trenkər/. This becomes a non-word at the /k/, since there are no words in English beginning with /tren/ that continue with /k/ (see Table Two). A total of 218 non-word sequences were constructed, and were presented to the subjects in isolation, mixed in with an equal number of real English words.

If processing is indeed optimal, then the subjects should have been able to begin to make the non-word decision, in each non-word sequence, at precisely that point where the sequence diverged from the existing possibilities for words in English. The results strikingly confirmed this hypothesis. Decision-times were constant, averaging around 450 msec, when measured from the onset of the non-word point in a sequence. This held true independently of where in the sequence the non-word point occurred, and independently of the total length of the sequence. For example, in the non-word /sθɔɪdɪk/, which becomes a non-word at the /θ/, the mean decision-time was 449 msec from the onset of this phoneme. In the non-word /feðɔ:n/, which becomes a non-word at the /ɔ:/, decision-time was 450 msec measured from this point.

These results show, then, that human listeners are capable of conducting an on-line assessment of the speech input with respect to its possible lexical interpretations. A second experiment shows that they use this capacity in the recognition of real words.

This second experiment used the phoneme-monitoring task. The subjects heard isolated real words, one third of which contained the phoneme /t/, and pressed a response button whenever they detected this phoneme target. The rationale for using this task (see section 2.2.) was the belief that phoneme-monitoring involves first recognising the word bearing the target, and then checking the phonological representation of the word to establish whether it contains the target phoneme. This implies that phoneme-monitoring latency can be used as an indirect measure of word-recognition time.

The first part of the experiment was primarily methodological in intent. I wanted to make sure that the task did behave as if it were sensitive to word-recognition variables. To do this, I varied the position of the phoneme target in the words (which were either one, two, or three syllables in length). The targets could occur either (1) at the beginning of the word, (2) at the end of the first syllable, (3) at the end of the second syllable (of two and three syllable words), or (4) at the end of the third syllable (for three syllable words only).

The assumption here was that if phoneme-monitoring in real word contexts did involve first recognising the word containing the phoneme, then detection latencies should decrease to targets occurring later in the word. For the more of a word one has heard, then the less the delay before one can have access to the word-based knowledge that is assumed to be being used in the phoneme-monitoring decision. As an additional control, I constructed a set of phonologically legal nonsense words, which also contained targets in the same four locations across words of different syllable lengths. Any facilitation of monitoring performance that derived from word-identification factors should not show up in the pattern of responses to non-words. Different groups of subjects were run on the two types of stimuli; subjects either heard only real words, or only nonsense words. All subjects were native speakers of Dutch, and the stimuli were real or nonsense words in Dutch.

The results are clear, and are summarised in Table Three,

Table Three: Phoneme-monitoring latencies for real and nonsense words (measured in msec from /t/ onset)

	Target Positions			
Word Type	1	2	3	4
Real Words	570	474	376	261
Nonsense Words	451	450	460	444

which gives the mean detection latencies for the two types of words at each target-position, collapsing across word-lengths. The /t/ onset was defined as the beginning of the word for the word-initial targets, and as the onset of the silence preceding the /t/ burst when the target occurred either in the middle of the word or at the end.

The response-times for targets in non-words are constant across target-positions, varying around 450 msec. But response-times for targets in real words decrease steadily across the word, by about 100 msec per target-position. This strongly suggests that phoneme-monitoring interacts with word-recognition processes, since the only difference between the two types of stimuli was whether they were words or not. The finding that phoneme monitoring is *slower* in word-initial position for real words than for nonsense words indicates that the subjects in the real word conditions are tending to delay their responses until they find out more about the word in which the target is embedded.

The next step was to investigate in detail the relationship between phoneme-monitoring latency and the point in the word at which an optimal processing theory could predict that the word could be recognised. The 80 three-syllable words used in the experiment were analysed to determine their recognition points -- that is, the points at which they separated from their word-initial cohorts. The criterion used for estimating recognition-point was a morpheme-based one[1], with syllabic stress patterns also being taken into account.

The results of these analyses were used to compute the distances in msec between the phonemes chosen as recognition points and the phoneme targets in the words. Some examples of this are given in Table Four. The recognition points could occur

Table Four: Examples of distances (msec) between recognition points and phoneme-target locations (both points in italics)

	Distance	Monitoring Reaction-time
T OE N *E* M E N	+388	660
A R *T* I *K* E L	+145	518
O P V R *E* *T* E N	-167	365
M O N *U* M E N *T*	-443	175

at up to several hundred msec before or after the location of the phoneme-target. A linear regression analysis was then used to test the regularity of the relationship between monitoring latency and the distance of the target from the recognition point. The results of this analysis showed a very close relationship between the two variables. The correlation coefficient was .96, indicating that the target distance from the recognition point accounted for over 90% of the variance in the 80 mean latencies for the targets. The later the target occurred in the word, relative to the recognition point, the faster the subjects were able to respond. This result strongly supports the cohort theory, since its estimates of recognition points are remarkably accurate predictors of variations in phoneme-monitoring latency.

These two experiments demonstrate, then, that the human word-recognition system is capable of behaving in an optimal manner. That is, it can assess the sensory input, on-line, against the set of possible words with which the input is compatible, and can act immediately when a unique word-choice can be securely identified. We can reasonably assume that this optimal processing property applies to all aspects of the speech understanding process, especially given the speed with which an interpretation of the input can be constructed (see section 2.1.). It remains a problem, however, to define and test this larger claim for domains of analysis other than word-recognition. Not enough is known about these other domains to be able to say precisely what *any* kind of solution would look like, let alone an optimally efficient one.

I would like to add two further comments here. First, that the use of a phonemic notation to specify the stimuli in the two experiments here, and to determine recognition points, etc., should not be taken as a theoretical claim. That is, I am not claiming that the input to the human word-recognition system takes the form of a string of phonemic labels. The reason that a phonemic analysis has been adequate here may mean only that phonemic labels tend to coincide reasonably well with the sequential patterning of informationally important variations in the speech signal. But as far as the cohort model is concerned, this sequential patterning could just as well be delivered to the word-recognition system in the form of spectral parameters or auditory features as in the form of segmental labels.[2]

The second point here is that if speech understanding is going to be truly optimal, then it follows that it must also be interactive. The word-recognition process, for example, will be more efficient in its use of the sensory input if contextual constraints are able to interact directly with the process of selecting a candidate from among the word-initial cohort. If recognition elements can drop out of the word-initial cohort because of mismatches between the properties of the words they represent and the structural and contextual environment, then this will tend to speed up the recognition process. The cohort will be immediately reduced in size, interpretatively plausible candidates can be given a higher weighting, and, in general, less sensory input will be needed to reach a secure decision about the correct word-candidate. This proposal is, of course, consistent with the evidence for early contextual interactions with word-recognition processes that I described in section 2.3.

4. PROCESSING PRINCIPLES AND CONTROL STRUCTURE

In the previous sections I have argued for a picture of human speech processing as an optimally efficient and highly interactive process. This raises the further question of the control structure of such a system; how are the interactions between knowledge sources regulated in such a way that optimal efficiency is obtained, and that the final product of the analysis process remains in secure contact with the original input. If higher-level knowledge sources are allowed too much influence, then speech understanding runs the risk of becoming a process of hallucination rather than of synthesis.[3]

One way of approaching the control problem is to postulate an additional executive process -- as in many computational speech understanding systems -- which acts to schedule and control the flow of information between knowledge sources, with respect to the ultimate analytic goals of the system. One could, therefore, postulate an equivalent separate executive process in human speech understanding. But this seems both unexplanatory and unmotivated. Furthermore, such a postulate may not be necessary, since the basic processing principles governing human speech understanding may be sufficient to provide the system with an intrinsic control structure.

4.1. Psychological processing principles

The operations of human speech understanding can be constrained by two complementary processing principles. First, that perceptual processing operates on the principle of *bottom-up priority*. This means that the initial set of possible analyses, within the domain of any one knowledge source, is always determined by the bottom-up input relevant to that knowledge source. Thus, for example, the universe of possible word-recognition candidates is determined by the properties of the sensory input to the array of recognition elements. This word-initial cohort defines the decision-space within which further analyses, incorporating contextual inputs, can take place.

The second principle is that speech processing operations are *obligatory* in character (9, 18, 22). Given the relevant bottom-up input, a knowledge source must run through its characteristic operations on this input. Again, there are clear instances of this in spoken word recognition. If a speech input can be lexically interpreted, then apparently it must be. In several studies where subjects are asked to focus their attention on the acoustic-phonetic properties of the input, they do not seem to be able to avoid identifying the words involved. One instance of this is the phoneme-monitoring studies described in sections 2.2. and 3.2.

I will assume that these principles of obligatory bottom-up priority apply to all aspects of the speech understanding system, and that they determine the flow of analysis through the system. But I should emphasize that these principles apply only to what we can call *normal first-pass processing*. This is the sequence of highly skilled and automatised routines that the listener runs through in the normal successful analysis of a natural utterance in a natural context. In normal communicative situations, these operations are usually all that are necessary to support successful speech understanding.

Once the obligatory routines have been performed, then the listener may have access to the analysis of the input in many different ways for many different purposes. He can, for example, try to examine the acoustic-phonetic properties of the input on their own, he can focus on the structural relations between words, and so on. But these secondary and "off-line" processes have nothing to do with the normal obligatory processes I have been talking about throughout this paper. I suspect that a great deal of confusion in thinking about human language understanding has derived from a failure to maintain this distinction between on-line and off-line processes.

4.2. Bottom-up obligatory processing

The first benefit of adopting the principle of bottom-up obligatory processing is that it forces the on-line analysis process to stay in contact with the sensory input. The central link between input and interpretation is the word-recognition process, and this process is always circumscribed by its bottom-up input. The scope of the operations of the structural and interpretative sources is determined by the words that are recognised, and the principle of bottom-up priority protects this recognition process from dominance by these contextual sources.

This means that the processing structure of the system does not allow, in normal first-pass processing, for direct top-down effects. That is, it does not allow a higher-level source to predetermine the likely or possible analyses at some lower level. Thus the contextual environment for word-recognition cannot act to pre-select some likely set of word-candidates, in advance of any of the sensory input corresponding to the words in question. These contextual selection criteria can only be applied to the set of possible candidates permitted by the sensory input at any given moment. Given the rapid simultaneous contextual assessment processes permitted by the distributed cohort model, this constraint on the system does not prevent it from being optimally efficient in its operations.

The same constraint can be assumed to apply to the structural and interpretative aspects of processing. A structural knowledge source can only assign an analysis to the string which is compatible with its bottom-up input (the successive words in the utterance). Top-down influences are assumed not to pre-determine possible structural assignments, any more than they are allowed to pre-determine possible word-candidates. Similarly, the operations of the interpretative source are driven from the bottom-up, on the basis of information about the words being recognised and about the structural properties assigned to these words by the structural source. Both sources interact with each other, and with the original selection of the appropriate words (or word-senses) from the recurrent bursts of word-candidates at the onset of each word, but they do so strictly under the control of the environment defined from the bottom-up by the possible lexical interpretations of the sensory input.

Note that the obligatory bottom-up constraints on structural processes do not mean that they cannot be influenced by interactions with interpretative sources -- anymore than the obligatory and bottom-up constraints on word-recognition processes mean that they are not also interactive in nature. In the case of structural analysis, the effects of these interactions are less noticeable, because of the restricted set of structural alternatives that a given sequence of words will usually permit. But when structural ambiguities are not directly resolved by further bottom-up information, then the effects of interpretative interactions can be detected -- as in the experiment described in section 2.4.

The second main consequence of obligatory bottom-up processing is that it ensures that the analysis of the input will always be developed as far as it can be as rapidly as possible. For as soon as the appropriate bottom-up input becomes available to a given knowledge source, then this source must begin to perform its characteristic operations on this input. In the case of spoken word recognition, a subset of recognition elements will become active as soon as their distinctive input patterns become available. This will make information about the syntactic and semantic properties of these word-candidates available. This information in turn will trigger attempted analyses by the structural and interpretative sources. Thus, to the extent that a particular input leads to acceptable further analyses, then these analyses must propagate through the system, and establish consequences at the interpretative level.

This kind of patterning of information flow seems required, in any case, to explain the early contextual interactions with word-recognition processes (see sections 2.2. and 2.3). A word-candidate can only be assessed for its contextual suitability if information is made available about its syntactic and semantic properties. For it is only in terms of these properties of word-candidates that contextual constraints can apply. And since contextual influences are detectable before a single word-candidate could have been selected on bottom-up grounds alone, then syntactic and semantic information must be made available about all of the early word-candidates. This information, in turn, has to be assessed against the various structural and interpretative paths that the utterance and the discourse up to that point will permit. Otherwise, the contextual environment cannot provide a basis on which to determine the relative on-line acceptability of different word-candidates.

Thus the requirements of an interactive system with the temporal parameters I have described fit in very well with the consequences of adopting the bottom-up obligatory processing principle. The result is a model of human speech understanding which permits the speech input to be interpreted as early as the properties of this signal permit. Given that the speech signal is necessarily extended in time, it seem desirable that the speech recognition system should be designed to minimise the delay with which the significance of this signal can begin to be made available to the perceiver.

5. CONCLUSIONS AND IMPLICATIONS

This chapter has presented a view of human speech understanding as an interactive and apparently highly efficient process, that attempts to analyse its input, with respect to an interpretation in a discourse and a world, word-by-word as the signal is heard. The main strategy the system adopts to do this is to bring the analysis of the signal as early as possible into contact with the word-recognition system. For it is at this level that structural and interpretative factors can be brought to bear on the analysis of the signal, and that the signal can start to have direct consequences for the on-going interpretation of the utterance. The primary sensory processes, delivering an input to the word-recognition system, can be assumed to be entirely autonomous. But all available knowledge sources can interact once the word-recognition system is activated, within the constraints imposed by the principles of obligatory processing and bottom-up priority.

This view is clearly only a preliminary sketch for a proper psychological model, and all it can claim to do is to say something about the general processing structure of the system. It has said very little about the actual content of different knowledge sources, and, indeed, it remains uncertain just how many distinct knowledge sources are functionally distinct in processing, and how the stored knoweldge they contain should be mentally represented. The present model requires only that this knowledge should be stored in such a way that it can be made available for an on-line interactive process, which may consider many analysis paths simultaneously.

A final question is whether the present model of human speech understanding has any implications for the enterprise of building speech recognition or understanding machines. My feeling is that if there are any implications, they derive from the requirements that the proposed psychological system would place upon the properties of the signal. As I mentioned in section 2., an interactive system places different demands on the signal than a strictly bottom-up system.

I should add two qualifications here. The first is that if there are any implications, they will apply only to relatively unrestricted systems designed to understand connected speech. They are unlikely to refer to limited application systems, designed to recognise small sets of words uttered in isolation. The problems here can be (and have been) solved by economical signal processing and pattern-matching techniques, without any need to refer to the properties of the human recognition process. The second qualification is that my remarks will depend on the assumption that the human speaker produces speech so as to meet the precise needs of the recognition system for which it is intended. Thus the properties of the recognition process have consequences for the kinds of information the signal will normally have to contain, and for the way in which this information is distributed in the signal over time.

The first point is that human speech recognition is only partially dependent on a bottom-up analysis of the signal. The primary target of speech analysis appears to be the word-recognition system, and, when words occur in contexts, the sensory input is only one (although the most critical) contribution to the recognition process. The bottom-up signal has to be sufficient to trigger the appropriate set of word-candidates, and, within that set, it has to be sufficient to uniquely mark the correct candidate relative to the contextual environment. But the signal does not have to be sufficient to distinguish the correct candidate relative to all of the other words in the language.

The further implication, then, is that it is only at the beginnings of words that the signal can be relied upon to be adequate from the point of view of bottom-up analysis. This raises the problem for the recognition system of determining where words begin (so that it knows where to look for the critical sensory information). For the human listener this may not be a problem, since he can apparently recognise a word, on average, before all of it has been heard. This would clearly help to determine where the next word starts. It may well be, in fact, that speech is uttered by humans on the assumption that successful rapid word-recognition and interpretation is initiated immediately the utterance begins. Thus the analysability of the signal is predicated on the assumption of a continuous successful interpretation of this signal. Crucial bottom-up cues in the signal may, then, only be functional with respect to a certain degree of successful analysis of the preceding input.

We know from our own experience that if we fail to catch onto the beginning of an utterance it is often difficult to segment or identify the subsequent speech stream. Speech understanding systems that do not work strictly left-to-right may be trying to extract information from the signal that is only present from the perspective of a successful sequential analysis process.

A further point concerns the nature of the contextual information that the human listener appears to bring to the speech analysis process. The psychological data (sections 2.1. to 2.3.) show that the interpretative domain onto which the utterance is being mapped functions throughout the speech processing system, and has powerful effects even on word-recognition decisions. Again, if the signal is being produced with regard to contextual constraints that refer to interpretative variables, then it is going to be difficult to successfully recognise speech without incorporating just these constraints into the system.

The implication here, it should be stressed, is *not* that the speech signal is intrinsically noisy and sloppy, and that top-down inputs somehow rescue an interpretation from a sort of bottom-up madhouse (which is how some speech understanding systems appear to operate). It is more plausible to assume that the signal is very well adjusted to the knowledge environment in which it is being produced; that the speaker in general provides just that amount of bottom-up information that is necessary for successful recognition, relative to the on-line contextual environment for analysis. It may be only because the relationships between sensory and contextual variables during processing are not properly understood that the signal appears noisy and inadequate. But it is clear that if we are going to be able to understand these relationships, then we need to know a great deal more both about these contextual variables and about the properties of the speech signal. This would seem to be a suitable area for cooperation between psychologists and speech and computer scientists.

NOTES

1. Recognition-points were assumed to be the points at which morphemic groups separated from each other, rather than individual words. Thus "industry" becomes unique when it separates from "inductive" -- at the /s/ -- rather than when it separates from a word from the same morpheme group, such as "industries". It was also assumed that differences in syllabic stress counted as differences in segmental properties.

2. Unfortunately, research into the acoustic-phonetic analysis of speech has not yet provided a convincing account of what the products of this process might be. As, for example, Nooteboom (23) has argued, at least one reason for this may have been the lack of interest of speech researchers in the relationship between acoustic-phonetic issues and the word-recognition problem.

3. It is worth mentioning, in this connection, the perhaps apocryphal stories about the speech understanding system HEARSAY I. This system's interpretative framework was a chess game and the rules of chess, and it apparently tended to replace what was actually said by what it felt would be a better move in the game.

REFERENCES

(1) Fodor, J.A., Bever, T.G., & Garrett, M.F. *The psychology of language*. New York: McGraw-Hill, 1974.

(2) Forster, K. *The role of semantic hypotheses in sentence processing*. In Colloques Internationaux du C.N.R.S., No. 206. Paris, 1974.

(3) Marslen-Wilson, W.D. *Linguistic descriptions and psychological assumptions in the study of sentence perception*. In R.J.Wales & E.C.T.Walker (Eds), *New approaches to the study of language*. Amsterdam: North-Holland, 1976.

(4) Marslen-Wilson, W.D., Tyler, L.K., & Seidenberg, M. *Sentence processing and the clause-boundary*. In W.J.M.Levelt & G.B. Flores D'Arcais (Eds), *Studies in sentence perception*. New York: Wiley, 1978.

(5) Marslen-Wilson, W.D. *Linguistic structure and speech shadowing at very short latencies*. Nature, 1973, 244, pp.522-523.

(6) Marslen-Wilson, W.D. *Speech shadowing and speech perception*. Unpublished Ph.D. Thesis, Department of Psychology, Massachusetts Institute of Technology, 1973.

(7) Marslen-Wilson, W.D. *Sentence perception as an interactive parallel process*. Science, 1975, 189, pp.226-228.

(8) Marslen-Wilson, W.D., & Tyler, L.K. *Processing structure of sentence perception*. Nature, 1975, 257, pp.784-786.

(9) Marslen-Wilson, W.D., & Tyler, L.K. *The temporal structure of spoken language understanding*. Cognition, in press, 1980.

(10) Morton, J., & Long, J. *Effect of word transitional probability*

on phoneme identification. Journal of Verbal Learning and Verbal Behavior, 1976, 15, pp.43-51.

(11) Grosjean, F. *Spoken word recognition and the gating paradigm.* Unpublished manuscript, Northeastern University, 1979.

(12) Sorenson, J.M., Cooper, W.E., & Paccia, J.E. *Speech timing of grammatical categories.* Cognition, 1978, 6, pp.135-153.

(13) Cooper, W.E., Sorenson, J.M., & Paccia, J.M. *Correlations of duration for nonadjacent segments in speech: Aspects of grammatical coding.* Journal of the Acoustical Society of America, 1977, 61, pp.1046-1050.

(14) Kenyon, J.S., & Knott, T.A. *A pronouncing dictionary of American English.* Springfield, Mass: G. & C.Merriam, 1953.

(15) Dell, G.S., & Newman, J.E. *Detecting phonemes in fluent speech.* Psychonomic Association meetings, Austin, Texas, 1978.

(16) Cole, R.A., & Jakimik, J. *A model of speech perception.* In R.A.Cole (Ed), *Perception and production of fluent speech.* Hillsdale, New Jersey: LEA, 1979.

(17) Marslen-Wilson, W.D., & Welsh, A. *Processing interactions and lexical access during word-recognition in continuous speech.* Cognitive Psychology, 1978, 10, 29-63.

(18) Forster, K. *Levels of processing and the structure of the language processor.* In W.E.Cooper & E.C.T.Walker (Eds), *Sentence processing: Studies presented to Merrill Garrett.* Hillsdale, New Jersey: LEA, 1979.

(19) Tyler, L.K., & Marslen-Wilson, W.D. *The on-line effects of semantic context on syntactic processing.* Journal of Verbal Learning and Verbal Behavior, 1977, 16, pp.683-692.

(20) Tyler, L.K. *Serial and interactive theories of sentence processing.* In J.Morton & J.Marshall (Eds), *Psycholinguistics: Series 3.* To appear.

(21) Gimson, A.C., & Jones, D. *Everyman's English pronouncing dictionary.* London: J.M.Dent, 1977.

(22) Shiffrin, R.M., & Schneider, W. *Controlled and automatic human information processing: II. Perceptual learning, automatic attending, and a general theory.* Psychological Review, 1977, 84, pp.127-190.

(23) Nooteboom, S.G. *The time-course of speech perception.* Ms. 352, IPO, Eindhoven, 1979.

WHAT THE ENGINEERS WOULD LIKE TO KNOW FROM THE PSYCHOLOGISTS

M.J. UNDERWOOD B.Sc, Ph.D, MBCS

ICL RESEARCH AND ADVANCED DEVELOPMENT CENTRE,
FAIRVIEW ROAD, STEVENAGE, HERTS, UK.

1. INTRODUCTION

The emphasis in this NATO Advanced Study Institute has been largely on the techniques of automatic speech recognition and synthesis. There is increasing evidence that the years of research effort are now leading to some practical real-world applications and these are uncovering another set of problems which have received very little attention to date. At ICL's Research and Advanced Development Centre, we are investigating how speech may be used for direct two-way communication between computer systems and their users. It has become clear to us that the technical problems of synthesis and recognition, important and difficult though they are, are not the critical problems that have to be solved in developing practical systems using speech. I would like to suggest that the critical problems lie in the human factors area. The main purpose of this paper is to indicate the nature of these problems and why they are so important, in the hope that this will cause more people to study them.

The main source of those problems can be summarised as follows. Speech has evolved as the most natural mode of interactive communication between people. As part of this, people have developed all kinds of conscious and sub-conscious patterns of behaviour which are suitable for person to person communication, but which may not be appropriate when one of the partners in a conversation is a machine. Although significant progress has been made in designing machines to recognise and generate speech, it will still be many years before we can program them to have really human-like

J. C. Simon (ed.), Spoken Language Generation and Understanding, 69-75.

capabilities for handling speech and language. The crucial question is whether the advantages of being able to use speech to communicate with machines are outweighed by the unnatural constraints that are placed on a man-machine dialogue by the limitations of the machine. This question can only be answered by building machines with speech input and output channels, and evaluating them doing a real job of work.

2. SPEECH SYNTHESIS

As the speech signal conveys so much more information than just the message it carries, e.g. mood, character, nationality, sex of the speaker, it is not surprising that there are many interesting reactions on the part of listeners to synthetic speech. Once intelligibility has been demonstrated, attention and comment often shifts to such factors as naturalness and voice quality. What importance should be given to these aspects of computer-generated speech?

2.1. Naturalness

It is difficult to define what is meant by the naturalness of a voice, but there are several aspects to it which need to be distinguished. One obvious aspect is whether the voice sounds as though it were made by a human being. There is much to be said for not attempting to make a computer-generated voice sound completely natural. As well as being difficult to achieve, it is probably undesirable, particularly in interactive situations, otherwise the listener may respond to it as though it were a real person. Until our machines become much more sophisticated, it will be advisable to remind the listener that it is a machine talking and not a person, so that he conforms to the restricted rules of conversation. On the other hand, some applications of computer-generated speech may need to instill a high degree of confidence in the human listener. In these situations, the voice quality could be extremely important.

Another aspect of naturalness concerns the way in which spoken material is presented. People normally present information verbally in a manner that it makes it possible for the listener to understand and remember it. For example, the quantity 247 is normally spoken as "two hundred and forty seven" rather than "two four seven". Furthermore, experiments at Bell Labs. (1), have shown that spoken sets of digits (e.g. telephone numbers) presented with natural prosodic features (particularly rhythm and intonation) are more intelligible, and more importantly, more easily remembered by listeners than sequences that do not have these features. In this instance, the most important aspect of naturalness is the manner of presentation rather than the subjective naturalness of the voice.

2.2. Type of Voice

The most obvious aspect of this is whether the voice should sound male or female. I am not aware of sound scientific arguments that show that one is preferable to the other. As far as intelligibility is concerned, the male voice with its low pitch yields more detailed information about the formant structure. The female voice with its more frequent excitation of the formants contains more information about the temporal aspects of speech, particularly the all-important formant transitions. Beyond this any preference is likely to be personal or sociological. For example, a fairly normal expectation is that it will be a woman's voice that answers a call to a telephone switchboard. As for personal preference, if a machine voice is to have a mechanical quality, then I find the concept of mechanical man more appealing than that of a mechanical woman!

3. SPEECH RECOGNITION

The use of speech to communicate with a machine is much more complex than the currently used methods of data input. For example, the detection of which key has been depressed on a keyboard is a much simpler and more reliable process than that of recognising which word has just been spoken. Moreover, the keyboard is sensitive only to which key has been depressed; there is no indication of the force or rapidity with which the keys have been struck. With speech however, variations in overall amplitude or rate of speaking do not change the message content, but it is difficult to design speech recognition machines to accommodate such variations easily.

In interactive use, the situation becomes more complicated because failures in machine recognition may lead to frustration on the part of the user, with the result that he changes his speech in a subconscious and therefore uncontrollable manner. Such behaviour is likely to make his speech less recognisable, and lead to a further increase in his frustration, making communication totally impossible. A system designed for practical use will need to be carefully designed to avoid this pit-fall. Clearly an important factor is the error rate of the recogniser; a machine with a low error rate leading to less frustration than a machine with a high error rate, but is it as simple as this?

3.1. Acceptable Error Rate

In order to establish an acceptable error rate, it is necessary to define a means by which it can be measured. It is normal practice in comparing speech recognition machines to quote the percentage of words correctly recognised. This is not adequate for comparing two machines doing real jobs. Consider two machines A and B, each with a 20 word vocabulary, each quoted as giving recognition scores of 95% correct. Machine A consistently misrecognises only one word in its vocabulary, machine B has the errors evenly distributed across all the words in the vocabulary. The decision as to which machine is better will depend upon the nature of the task for which it is being used, and the extent to which the errors occur on critical words, particularly the all-important control words "Erase" or "Cancel". This is an extreme example, but nevertheless indicates that performance depends on more than just a simple measure of error rate. The better machine is the one that enables the user to complete his task with the least effort.

Any properly designed speech recognition system has to be designed so that there is an easy method by which the user can correct errors. Such is the nature of speech that errors are unavoidable. At the very least, the user needs a means by which he can correct his own speaking errors, e.g. speaking one word when he meant another. For those tasks where the data throughput rate is important, any errors will slow the data input rate, and it is desirable for the error correction mechanism to be used as infrequently as possible. If a person is going to be using a speech recogniser for a significant part of his working day, he is unlikely to forget how to use the error correction procedure. If however, a speech recognition machine is being used primarily by people who make very infrequent use of the machine, a case could be made for not allowing the recognition performance to be so high that the errors are so infrequent that the user has forgotten how to correct an error when one occurs. In overall system terms, it might be better to keep the error rate away from zero, in order to keep the user alert. For applications of an interactive rather than a data input nature, it is probably not too important to have a very low error rate. In human conversations, particularly over the telephone, we frequently have to correct what has been said, especially when transmitting numeric information which has very little contextual content. Indeed, we probably cannot remember that we have had to correct an error. The important conclusion from this is that the error correction system should be simple and effective to use.

3.2. Feedback

Feedback of the machine's recognition to the user is an important aspect of building the user's confidence in the machine. The most important questions concern when and how the feedback should be given.

For communication via the telephone system, the most appropriate feedback is by means of speech. If the feedback is given on a word by word basis to an isolated word recogniser, this has the effect of slowing the input rate to around one word every two seconds, This time is made up in the following manner.

a) Man - time to speak word
b) Machine - time to detect word has finished
c) Machine - time to recognise word
d) Machine - time to utter reply
e) Man - time to decide whether machine was correct
f) Man - time to decide whether he should say "cancel" or continue with the next word.

Visual feedback on a character display may be more appropriate in other circumstances and is usually faster. In other situations, e.g. control of machinery, it may be sufficient for the feedback to occur by direct visual observation of the machinery being controlled, rather than verifying each word before it moves the machinery. Clearly the safety of this depends upon the machines being controlled'.

The use of computer-generated speech, particularly synthesised speech, raises interesting questions about the effect of the machine voice on the user's speech. Even with adults, a certain amount of sub-conscious mimicry probably takes place. By making the synthesiser pronounce words in a way that is acceptable to the recogniser might help a human user to speak in a more machine-recognisable manner. Moreover, if the machine is capable of detecting why words are not being recognised, it could provide guidance to the user as to how he might speak in order to be more easily understood. This guidance might be given in a direct form, e.g. "speak more loudly please" or it might take a more subtle form by reducing the output level of the synthesised speech. This would give the user the impression that the communication channel was deteriorating and that he would have to speak up in order to be understood. This type of phenomenon is common experience when using the telephone, yet to my knowledge the extent to which people modify their voices is not characterised at all in quantitative terms. It will be impossible to consider using such techniques until we have an understanding of the conditions under which the effects are likely to take place and their likely magnitude and direction.

3.3. The Effect of Errors

In designing a system that will interact with a man by means of speech, one cannot regard the user as having characteristics that do not change with time. For example, fatigue may effect a user's voice patterns. Another aspect of practical concern is what does a speaker do when he is misrecognised. One's intuitive reaction is that he will repeat the word in a manner that makes it more distinguishable from the word with which it was confused. If the confusion is a common human one, e.g. "one" for "nine", the speaker may feel he knows how to change his voice. Whether this is relevant or helpful to the machine or not is another matter. If the confusion is an unlikely one (because the machine operates in a different way to people), the speaker may not know what to do. Our preliminary findings (2) indicate that the only stratagem adopted by speakers is to speak more loudly. If the design of the machine is such that changes in amplitude do not affect its performance, then such a stratagem will not improve the recognition performance for that user with that machine, assuming that amplitude is all that the user can and does change. Thus a good speech recogniser needs to be responsive to the kind of changes a person is likely to make to his voice when speaking.

The effects of errors on the user's way of speaking are of paramount importance in designing a machine that will tune in to the characteristics of a speaker or telephone line. Adaptive systems are often quoted as the solution to the multi-speaker problem. However, very little work seems to have been published on the control of such systems. Inevitably a motivated user will be trying to speak in a manner that the machine will understand and if the machine is adapting to the characteristics of the man, it is important that the adaptation occurs on samples that are representative of his normal behaviour and not those that occur just after there has been an error. This is a very complex problem, and like most of the others that I have mentioned, can only be tackled by doing experiments with real people in front of real speech recognisers doing real jobs of work.

4. CONCLUSION

The main aim of this paper has been to indicate why I believe that the major problems at the man-machine speech interface are likely to be psychological in nature until the speech processing capabilities of our machines approach that of man himself. Until that happens, it will be necessary to pay careful attention to the way man and machine talk to each other. We need to develop better models of how people communicate so that the potential mis-match between man and machine can be kept to a useable minimum. The only way to develop such models is to set up experiments in real situations and measure how people react. It is encouraging that technological improvements are reducing the cost of such equipment to the point where it will be possible to establish the large body of human factors data required. At present, we take speech so much for granted in our everyday lives, but much of our knowledge about how we use it is intuitive and qualitative. We need quantitative data in order to design useable systems that will allow man and machine to communicate easily with each other by means of speech.

REFERENCES

1. L.R. Rabiner, R.W. Schafer, J.L. Flanagan. Computer Synthesis of Speech by Concatenation of Formant-Coded Words, Bell System Technical Journal Vol. 50 pp. 1541 - 1558 (1971).

2. T.R. Addis Human Factors in Automatic Speech Recognition, ICL Technical Note 78/1, available from the author at the above address.

PERCEPTUAL STUDIES OF SPEECH RHYTHM: ISOCHRONY AND INTONATION

C.J. DARWIN and Andrew Donovan

Laboratory of Experimental Psychology, University of Sussex, Brighton, BN1 9QG, U.K.

ABSTRACT

Three experiments are described which show that, within a tone group, the stressed-vowel onsets in English utterances are perceived as being rhythmically more regular than they in fact are. Between tone groups this does not hold. The implications for speech synthesis are discussed.

1. INTRODUCTION

For the psychologist, programs that synthesize or recognise speech are potential models of how the human brain speaks and perceives. They are particularly attractive models since they are explicit (which psychological models sometimes are not) and are capable of producing an output that can be compared directly with human performance.

The experiments described below and in a previous paper (Donovan & Darwin, 1979) are an attempt to determine which of two current approaches to the synthesis of speech-segment durations is the more psychologically plausible. The two approaches that we contrast both produce speech segment durations that are possible naturally, yet they do so by very different philosophies.

Klatt (1979) initially allocates to each segment (such as (i), say) a duration. Durations may then be shortened (towards a minimum possible duration) or lengthened by the application of rules (modelled on the speech of a particular person) that take into account factors which include how many syllables a word contains, stress, whether a syllable segment is word- or clause-final or not, and how many segments are in a consonant cluster.

J. C. Simon (ed.), Spoken Language Generation and Understanding, 77-85.

Speech synthesised by these rules is about as natural-sounding as speech synthesised with durations copied directly from the natural model, and has about 90% of the word intelligibility (Carlson, Granstrom & Klatt, 1979). The naturalness and the intelligibility of this rule-generated speech improves if the durations are linearly adjusted so that the stressed-vowel onsets occur at the same time intervals as in the original.

The interval between stressed-vowel onsets (or foot) provides the starting point for a different approach to segment-duration synthesis. Following Abercrombie and others, Witten (1977; Witten & Smith, 1977) starts with a sequence of equal temporal intervals between stressed-vowel onsets. Each interval is then subdivided between the syllables and then the segments of a foot according to rules involving vowel quality, word boundaries and syllable structure. Upper bounds are set on the durations of segments and lower bounds on the durations of syllables so that any particular interval may have to be lengthened or shortened to prevent these bounds being transgressed. Feet containing many segments will generally end up being longer than those with few. There has been no formal evaluation of Witten's scheme, though he comments that "it sounds quite good - occasionally very good - and always tolerable."

It is reasonable to assume that, with some effort, Witten's scheme could provide speech of a naturalness and intelligibility comparable to that of Klatt's. So which approach is psychologically preferable must be decided by testing the means by which the durations are arrived at.

The major difference between the two schemes is in their starting point; Klatt has ideal segment durations but no device analogous to the foot, whereas Witten has ideal foot durations but no ideal segment durations. Dispute on the status of the isochronous foot in stress-times languages is not confined to speech synthesis programs. It is clear that the strictly isochronous foot is not a phonetic fact. But it is also clear that despite this the idea of a tendency towards isochrony persists both scientifically and pedagogically. It manifests itself phonetically in the shortening of syllables as more are added to the same foot (Huggins, 1975; Fowler, 1977), and it also appears as a factor in statistical analyses of segment durations (Hill, Witten & Jassem, 1978).

Lehiste (1973, 1977) has argued for isochrony to be partly a perceptual phenomenon. She found that subjects were less accurate in telling which of two feet in an utterance was the longer than they were in making the same judgement about two intervals of lowamplitude noise. A perceptual illusion must have an explanation and it is our contention that, as has been suggested for other illusions, the isochrony illusion reflects our awareness not of the veridical physical stimulus but rather of the underlying object, in this case an underlying regular beat. Our experiments first demonstrate the illusion itself, using a technique more sensitive than Lehiste's; we then use the illusion to study the scope of rhythmic constraints in English.

2. EXPERIMENT 1

The main task for each of the 10 subjects was to adjust the two intervals between three brief click-like bursts of noise until they matched the intervals between the /k/'s in each of three utterances:

"The crew claim cargo."
"The crewmen will claim cargo."
"The crew claiming the cargo."

Subjects could hear the speech or the train of noise-bursts separately as often as they wished, but they could not hear the two together. As well as matching the noise bursts to the rhythm of the /k/'s, subjects also matched them to each of three sequences of brief tones whose onsets were identical in rhythm to the /k/'s in the original speech. To make the speech-matching task easier, subjects were encouraged to adopt the strategy of saying the utterance to themselves while listening to the noise-bursts. We had found that this strategy had been spontaneously adopted by the more reliable of the subjects in previous experiments. To check that subjects were not speaking more isochronously than the originals, their own productions were recorded during the experiment.

Figure 1 shows the foot lengths for the original utterances, the utterances as spoken by subjects and the intervals (and their standard errors) between the clicks matched to the original utterance and to the tone-burst control.

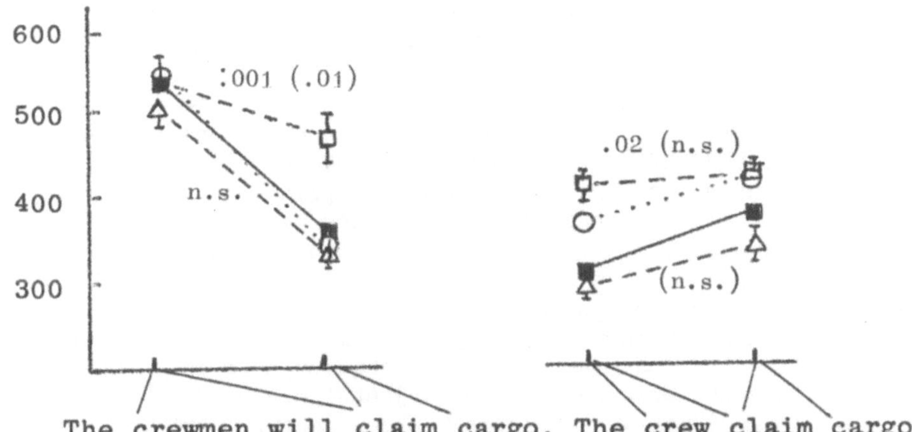

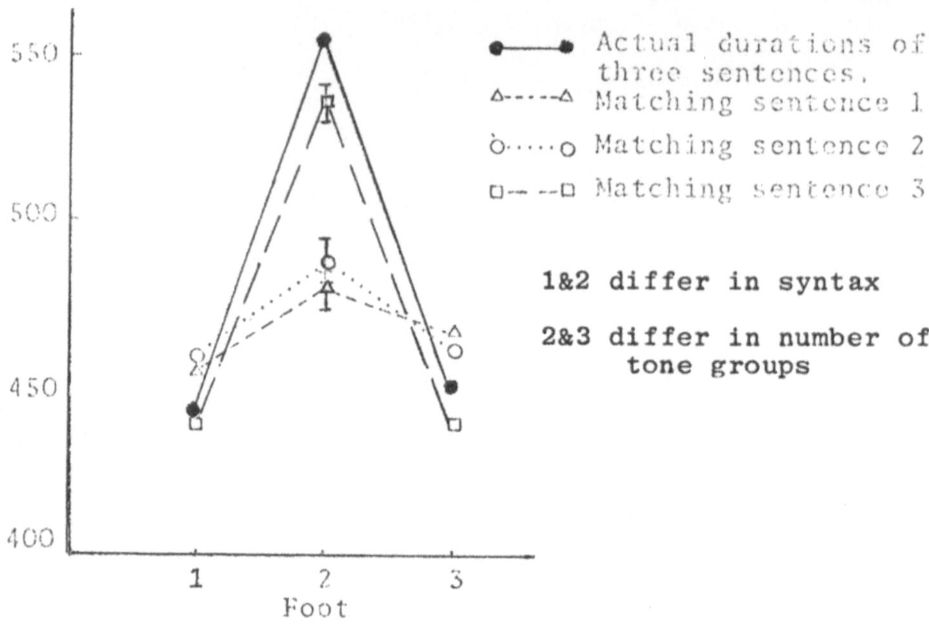

Figure 1. Mean durations (and standard errors) for subjects' matches to speech and non-speech control sounds.

To test for any tendency for the matched durations to be more isochronous than either the original utterance or subjects' own productions, the following expression was evaluated for each of the three utterances:

$$(1-a1/a2) - (1-p1/p2)$$

where:

ai is the actual duration of the 1th foot
p1 is the perceived duration of the 1th foot.

Whether this expression was greater than zero was statistically assessed for the durations matched to speech using either the original or the subject-spoken durations as the 'a' terms, and these two significance levels (with the latter in parentheses) are shown against the appropriate lines in the figure. The significance of the durations matched to the tone control were similarly assessed against the original durations and that level is also shown.

Except for one of the comparisons involving the initially most isochronous utterance, all the utterances are matched by durations that are significantly more isochronous than either the original or the subjects' own productions.

In so far as subjects' perceptions reflect the underlying rather than the surface rhythm of speech, this experiment has provided evidence that the intervals between stressed-vowel onsets are more rhythmically regular in the underlying than in the surface rhythm. To this extent our results support an approach such as Witten's which has the surface rhythm less regular than the underlying rhythm, rather than one such as Klatt's, which has rules that tend to do the opposite.

A number of Klatt's rules are concerned with clause and phrase boundaries. Pauses are inserted before main clauses and phrase-final syllables are lengthened. A change in the distribution of syntactic boundaries within the feet of our three utterances cannot explain the differences in perceived rhythm since all three utterances have their major syntactic boundary in the first complete foot. The location of phrase boundaries cannot then explain why the first foot is sometimes perceived as longer than it is and sometimes not. But we cannot entirely rule out some influence of phrase boundaries on the illusion of isochrony, since in a previous experiment (Donovan & Darwin, 1979; expt. 2) we had found that the illusion broke down for some sentences at a major syntactic boundary. A possible confounding influence in that experiment was tone group structure. Recent work from a number of different approaches (e.g. Rees, 1975; Selkirk, 1978) claims that the tone group or intonational phrase may be not only an intonational but also a rhythmic unit. Rees, for example, takes the domain of isochrony to be the tone group (in order to allow breaks of rhythm to be used as indicators of tone-group boundaries), while Selkirk identifies pre-pausal lengthening with the ends of intonational phrases. Selkirk in particular is at pains to point out that the intonational phrase is not syntactically defined, although its end may often coincide with a phrase boundary.

If indeed the tone-group or intonational phrase is a rhythmic unit, we might expect both production and perceptual constraints to exist only within such a unit. The three sentences used in experiment 1 had all been read as a single tone group (with the tonic on "claim"). The next two experiments examine the effect of the change of the number of tone groups on the perceived rhythm of an utterance. In the first one the intonational change is correlated with a syntactic change, while in the second it is made independently. Both experiments show a significant reduction in the perceptual isochrony illusion between, as opposed to within, tone groups.

3. EXPERIMENT 2

This experiment was similar to the first except that seven subjects listened to two sentences, the first containing both a tone group boundary and a syntactic boundary in the middle foot:

//1 Tell the/<u>town</u>folk a//1 tale of/<u>ter</u>ror

the second containing neither:

//1 Tell the/terrified/town the/<u>tale</u>

In addition the second experiment differed from the first in having a series of four identical synthetic syllables

(/bə/) instead of clicks in the control condition. The bursts of the /b/'s of these syllables matched the foot durations of the second sentence. The durations of the first sentence were so similar to those of the second that it was not worth running a separate control.

The results and the statistical reliability of the tendency towards perceptual isochrony for the two pairs of adjacent feet are shown in Figure 2.

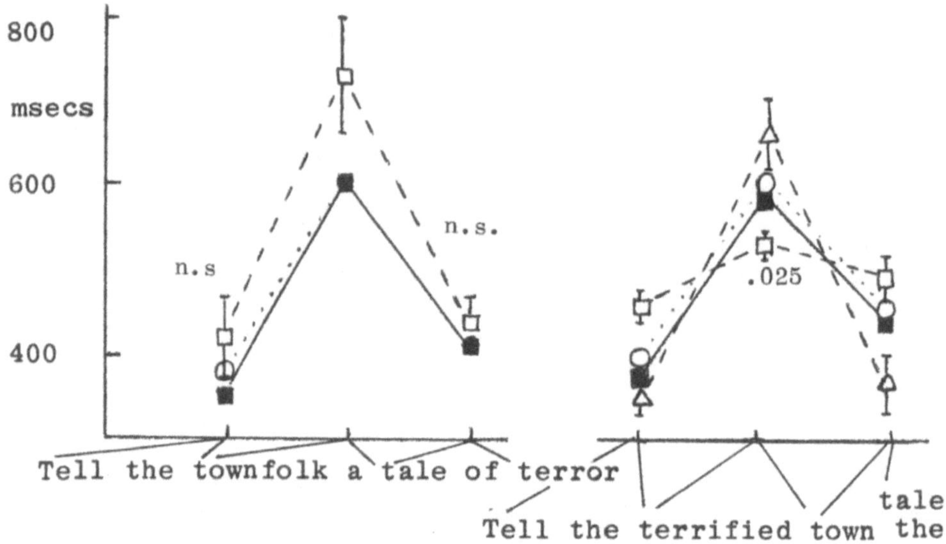

Fig. 2. Results of experiment 2. Symbols as in exp. 1.

There is a clear tendency for the second sentence to be heard as more isochronous than it actually is, but there is no such tendency in the first sentence, which contains a tone-group boundary in the middle foot. Since the presence of a phrase boundary did not prevent the isochrony illusion from being found in experiment 1, the results of experiment 2 support the view that the domain of perceptual isochrony is, like that suggested for production, the tone group.

4. EXPERIMENT 3

The third experiment (which as already been described in Donovan & Darwin, 1979) uses a different technique to tackle the same problem as in the previous experiment. We had observed in earlier experiments that although subjects' own productions of an utterance were sufficiently similar in rhythm to the originals not to vitiate the results, they did differ substantially in intonation, perhaps reflecting their own increasingly frustrated state of mind during the experiment. In order to provide better control over intonation we used a new technique in which the subject was asked to tap out the rhythm of a sentence as he heard it. Three utterances were used. These had very similar phonemic structures and were chosen so as to have identical foot durations, but they differed from each other in their syntactic structure or in their number of tone groups. The three sentences were

(1) //1 Tim's in/Tuscany's/training/troops
(2) //1 Tim's in/Tuscany/training/troops
(3) //1 Tim's in/Tuscany//1 training/troops.

Both (2) and (3), but not (1), contain a major phrase boundary in the second foot, but in (2) this is not marked by the end of a tone group.

For each sentence 15 subjects (who heard the sentences in different orders) were given a plausible context for the sentence such as "Where's Tim and what's he doing?" for (3) or "What's Tim doing with his troops in Tuscany?" for (2) and then they heard the sentence three times. They then heard the same sentence repeated 15 times (with a warning tone 750 msecs before each utterance). Their taps to the next two utterances were ignored leaving 10 trials for analysis for each sentence.

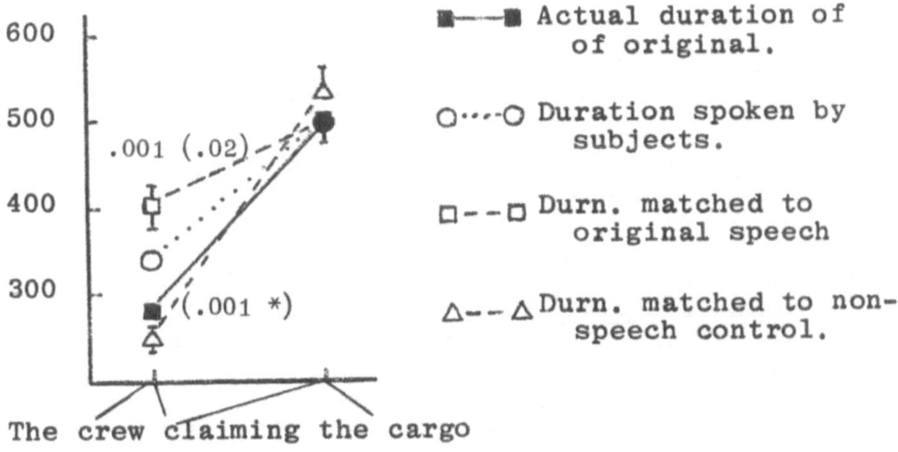

Fig. 3. Mean tapped intervals to the three sentences in experiment 3.

The results of the experiment in Figure 3 showed that the number of tone groups had a distinct effect on the perceived rhythm but the syntactic structure did not. Thus sentences 1 and 2 did not differ from each other significantly and both showed a tendency to perceptual isochrony, but they both differed from sentence 3 which did not show perceptual isochrony.

5. DISCUSSION

The main conclusion from our experiments is that within a tone group, the perceived rhythm of speech is more rhythmically regular than is the speech itself. Between tone groups, this tendency towards perceived isochrony is not found. We interpret our conclusion as supporting models of speech production which recognise an underlying rhythm to speech in which the interval between stressed-vowel onsets is more regular than in the surface form. We also see our results as supporting those linguistic theories that distinguish between prosodic and syntactic structure and that recognise a relation between intonational and rhythmic units.

REFERENCES

Carlson, R., Granstrom, B., & Klatt, D.H.: 1979, Some notes on the perception of temporal patterns in speech. Proceedings of Ninth International Congress of Phonetic Sciences, Copenhagen, vol. 2, 260-267.

Donovan, A. & Darwin, C.J.: 1979, The perceived rhythm of speech. Proceedings of Ninth International Congress of Phonetic Sciences. Copenhagen, vol. 2, 268-274.

Fowler, C.A.: 1977, Timing Control in Speech Production. Ph.D. Thesis, Univ. Connecticut. Bloomington, Ind: Indiana Univ. Linguistics Club.

Hill, D.R., I.H. Witten and W. Jassem: 1978, "Some results from a preliminary study of British English speech rhythm", Research Report Number 78/26/5, Department of Computer Science, University of Calgary.

Huggins, A.W.F.: 1975, On isochrony and syntax. In Auditory Analysis and Perception of Speech, G. Fant & M.A.A. Tatham (eds.). London: Academic Press, 455-464.

Klatt, D.H.: 1979, Synthesis-by-rule of segmental durations in English sentences. Proceedings of Ninth International Congress of Phonetic Sciences, Copenhagen. vol. 2, 290-297.

Lehiste, I.: 1973, Rhythmic units and syntactic units in production and perception. J. Acoust. Soc. Amer. 54, 1228-1234.

Lehiste, I.: 1977, Isochrony re-considered. J. Phonetics 5, 253-263.

Rees, M.: 1975, The Domain of Isochrony. Edinburgh Univ. Dept. of Linguistics, Work in Progress 8, 14-28.

Selkirk, E.O.: 1978, On prosodic structure and its relation to syntactic structure. Paper presented at Conference on the Mental Representation of Phonology. Nov. 1978, Amherst, Mass.

Witten, I.H. & Smith, A.: 1977, Synthesizing English rhythm - a structured approach. Fifth Man-Computer Communications Conference. Univ. Calgary, Alberta, Canada. May, 1977.

Witten, I.H.: 1977, A flexible scheme for assigning timing and pitch to synthetic speech. Language and Speech 20, 240-260.

ESTIMATION OF ARTICULATORY MOTION FROM SPEECH WAVES AND ITS APPLICATION FOR AUTOMATIC RECOGNITION

K. Shirai and M. Honda*

Department of Electrical Engineering,
Waseda University
Toyko, 160 Japan.

ABSTRACT. It is well known that there are several difficulties in the calculation of the vocal tract shape from speech waves, but if possible, the estimated vocal tract shape may be a good feature for speech recognition.

In this paper, it is shown that by introducing an effective model which can represent even delicate articulation with a small number of parameters, the estimation of the articulatory state from speech waves becomes possible and further the difference between speakers can be normalized by a rather simple procedure.

The results obtained by this method agree well with the data observed by X-ray photograghs. An experiment to discriminate Japanese five vowels shows that this scheme has great possibilities not only for speech recognition but also for other fields of speech study.

1. INTRODUCTION

It has been recognized that feature extraction is one of the central problems of speech recognition, especially for a spoken word recognition machine. Even in a speech understanding system which employs information from higher levels such as syntactic and semantic levels, a more efficient feature extraction method at the acoustic levels may help considerably in the improvement of system performance.

---

* M. Honda is now with Musashino E.C.L., N.T.T.
This work was partly supported by the Ministry of Education Grant-in-Aid for Scientific Reseach No.239005.

J. C. Simon (ed.), Spoken Language Generation and Understanding, 87-99.

The main difficulty which exists in extracting feature from speech patterns is the great variability of each utterance and the difference between talkers. Many feature extraction methods have been studied but most of them use acoustic parameters calculated rather directly from speech waves. However, in order to adapt to the wide variability of speech waves, it becomes necessary to apply fundamental knowledge from physiological studies of articulatory mechanisms.

Recently, several authors have studied the estimation of vocal tract shape from the speech wave.[2-6] The results are rather successful for voiced sounds. In this paper, a new technique is presented for estimating the vocal tract shape, which depends on precise analysis of real data for articulatory mechanisms.

First, the articulatory model which constitutes the base of this method is introduced. This model relies on statistical analysis of real data and makes it possible to represent the position and the shape of each articulatory organ very precisely using a small number of parameters; the physiological and phonological constraints are automatically included in this model.

Secondly, this articulatory model is used to estimate articulatory motion from speech waves. Articulatory parameters are estimated by minimizing the spectral error with some constraints. This is a nonlinear optimization problem and is solved by an iterative procedure. The power spectrum of the speech wave is composed of the vocal tract transfer function and the glottal wave characteristics. The spectral shape of the vocal tract transfer function is separated in such a way that the formant structure is preserved. Combined with a procedure for adjusting the mean vocal tract length for each speaker, this method is very successful in estimating the articulatory parameters of vowel sounds.

Thirdly, examples of the results obtained by this method are shown to prove the effectiveness of the estimated set of the articulatory parameters for speech recognition. A preliminary experiment to discriminate five Japanese vowels shows the validity of this feature extraction method.

It is further expected that the detection of nasal sounds and the discrimination between /m/ and /n/ will be possible by using the estimated movement of the velum.

2. CONSTRUCTION OF AN ARTICULATORY MODEL

A set of articulatory parameters is equivalent to a coordinate system for measuring the position of the articulatory mechanism. The selection of the articulatory parameters is related to the purpose behind the construction of the model.[7-10] In this paper, a phenomenological approach is adopted. This method is convenient for representing articulatory motion as precisely as possible by a small number of parameters.

The total configuration of the model is shown in Fig. 1. The jaw is assumed to rotate with the fixed radius FJ about the fixed point F, and the location J is given by the jaw angle X_J. The lip shape is specified by the lip opening height L_h, the width L_w and the protrusion L_p relating to the jaw position on the midsagittal plane. The tongue contour is described in terms of a semi-polar coordinate system fixed to the lower jaw and represented as a 13 dimensional vector $r = (r_1, \ldots, r_{13})$. The variability of the tongue shape seems infinitely large but, in the usual articulatory process, there may be some limitations on account of physiological and phonological constraints. These constraints can be expressed by the strong correlation of the position of each segment along the tongue contour. Therefore, it becomes possible to represent the tongue shape effectively using only a few parameters. These parameters can be extracted from the statistical analysis of X-ray data.

A principal component analysis is applied and the tongue contour vector for vowels r_V may be expressed in a linear form as,

$$r_V = \sum_{i=1}^{p} X_{T_i} v_i + \bar{r}_V \quad , \tag{1}$$

where v_i's $(i=1,2,\ldots,p)$ are eigenvectors, and $\bar{r}_V$ is a mean vector for vowels which corresponds roughly to the neutral tongue contour.

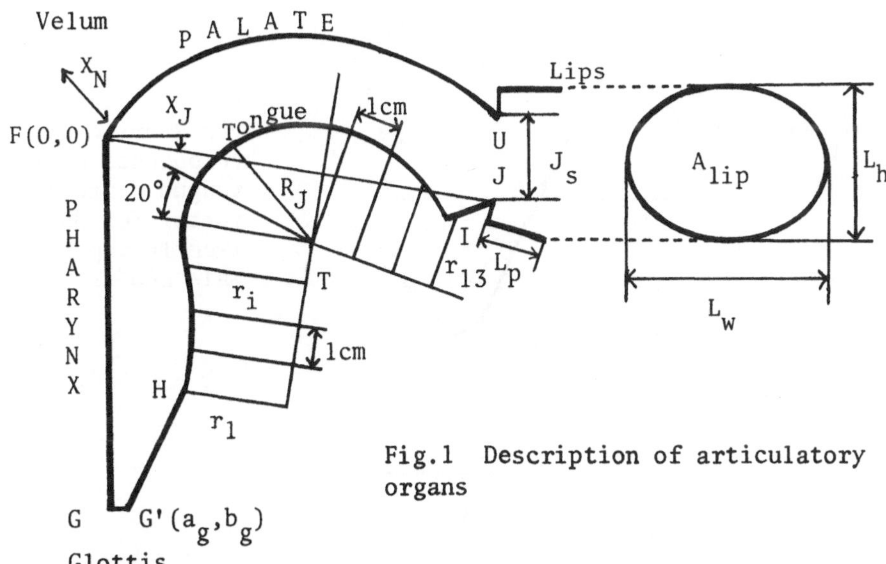

Fig.1 Description of articulatory organs

$$\bar{r}_v = \frac{1}{N}\sum_{k=1}^{N} r_{v_k} \quad , \quad N\text{: number of data} \tag{2}$$

The eigenvectors are calculated from the following equation

$$C_v v_i = \lambda_i v_i \tag{3}$$

where C_v is the covariance matrix which is defined by

$$C_v = \frac{1}{N}\sum_{k=1}^{N} \{(r_{v_k} - \bar{r}_v)(r_{v_k} - \bar{r}_v)^T\} \ , \tag{4}$$

and λ_i is the corresponding eigenvalue to satisfy the characteristic equation

$$|C_v - \lambda I| = 0 \ . \tag{5}$$

The extracted principal component vectors, the mean vector and their cumulative contributions are shown in Fig. 2.

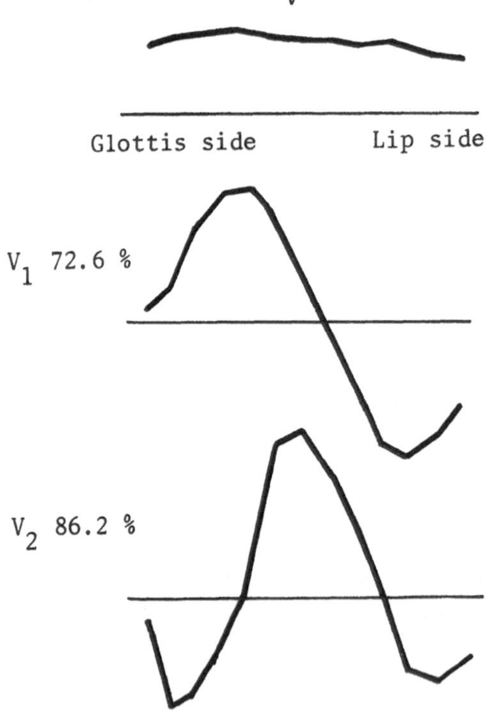

Fig. 2 Principal components of the tongue contour vector for data from 10 Japanese male adults

The same statistical technique as described above may be used for the expression of lip shape.

$$\begin{bmatrix} L_h \\ L_w \\ L_p \end{bmatrix} = \begin{bmatrix} k_1 J_s + \bar{L}_h \\ k_2 J_s^{k_3} + \bar{L}_w \\ \bar{L}_p \end{bmatrix} + \begin{bmatrix} \ell_h \\ \ell_w \\ \ell_p \end{bmatrix} X_L , \qquad (6)$$

where J_s signifies the distance between the upper and the lower incisors. The coefficients k_1, k_2 and k_3 show the effects of the jaw angle. $\bar{L}_h$, $\bar{L}_w$ and $\bar{L}_p$ correspond to the values for the time when the jaw is closed and the lips are in the neutral position. Active movement of the lips is reflected in the second term, and X_L is the lip parameter.

It is assumed that the position of the glottis moves with the lip parameter X_L.

$$g = k_4 X_L \qquad (7)$$

The vector (ℓ_h, ℓ_w, ℓ_p) is obtained by the principal component analysis. The constants in Eq. (6) are $(\bar{L}_h, \bar{L}_w, \bar{L}_p) = (0.5, 1.1, 1.0)$ [cm]; $k_1=0.4$; $k_2=1.61$; $k_3=0.2$; $k_4=0.2$; and $(\ell_h, \ell_w, \ell_p) = (-0.38, -0.91, 0.18)$.

The complete model with the nasal cavity is shown in Fig. 3 and the characteristics of the articulatory parameters are summarized in Table 1. More details about the model are found in the references.[11,12]

3. ESTIMATION OF ARTICULATORY PARAMETERS FROM SPEECH WAVE

In this section, the estimation of the articulatory parameters from the speech wave is considerd in the framework of a speech production model. This problem becomes the nonlinear optimization of parameters under a certain criterion, and it must be solved by an iterative procedure. Therefore, the uniqueness of the solution and the stability of the convergence are significant problems. In our method, these problems are solved by introducing constraints to confine the articulatory parameters within a restricted region and by the selection of an appropriate initial estimate using a continuity property. Tailoring the model to the speaker is, in addition, very beneficial to the estimation.

Let an m dimensional vector y be the acoustic feature to represent the vocal tract transfer function of the model. In this study the cepstrum coefficients are adopted as the acoustic parameters because the distance between two sets of cepstrum coefficients is equal to the distance of log-spectra and the main information of the vocal tract shape is reflected in the cepstrum co-

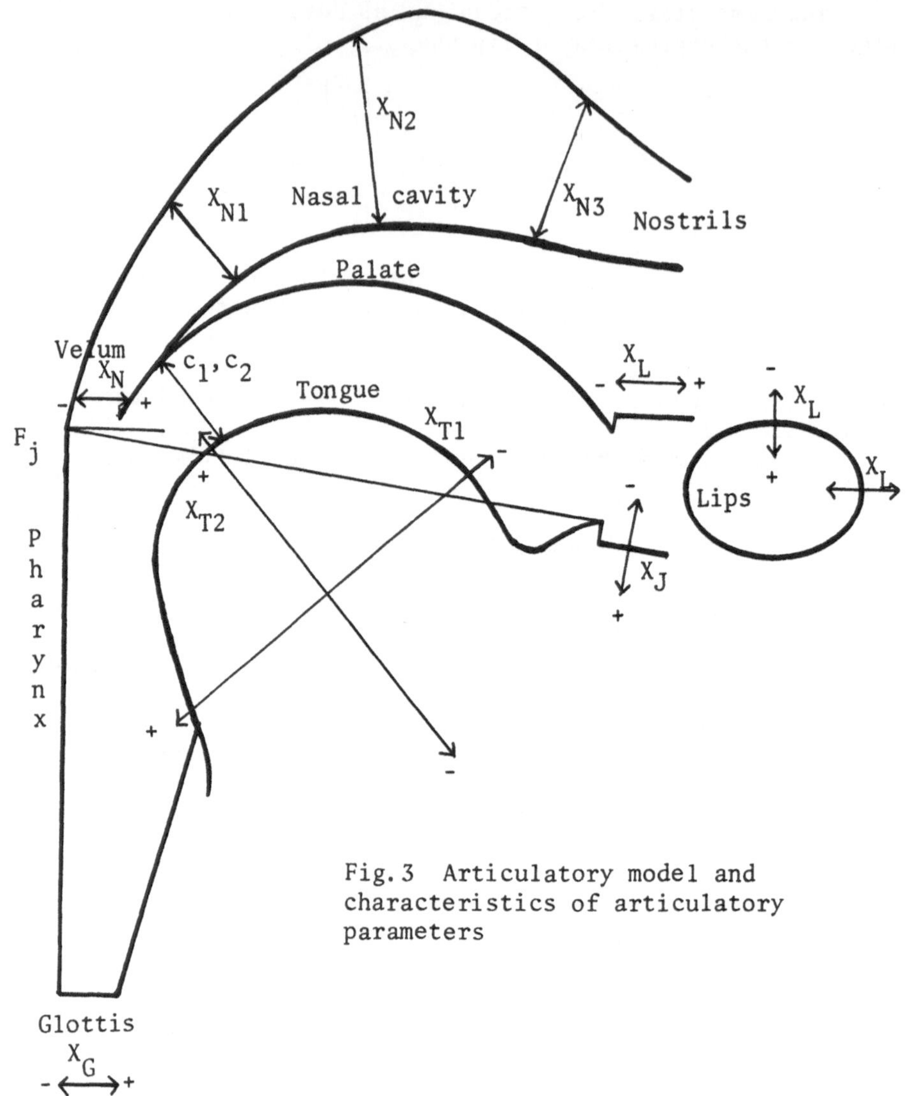

Fig.3 Articulatory model and characteristics of articulatory parameters

Table.1 Articulatory parameters

Articulatory parameters	Tongue X_{T1}	Tongue X_{T2}	Jaw X_J	Lip X_L	Glottis X_G	Velum X_N
+	back	high	open	round	open	open
-	front	low	close	spread	close	close

efficients of the lower order. The acoustic parameters are expressed as a nonlinear function h(x) of the articulatory parameters which involves the transformation from x to y through the speech production model. On the other hand, let ${}_sy$ be the acoustic parameters measured from the speech wave after the glottal and radiation characteristics are removed. Then, the estimate $\hat{x}$ of the articulatory parameters is obtained in order to minimize the next cost function.

$$J(x_k)=\| {}_sy_k - h(x_k)\|_R^2 + \| x_k\|_Q^2 + \| x_k - \hat{x}_{k-1}\|_\Gamma^2 , \qquad (8)$$

where R,Q and Γ are the weighting matrices, k is the frame number and $\hat{x}_{k-1}$ is the estimate at the previous frame. In the cost function, the 1st term is the weighted square erorrs between the acoustic parameters of the model and the measured values. The 2nd and the 3rd terms are used to restrict the deviation from the neutral condition (x=0) and from the estimate of the previous frame, respectively. These terms are also efficient in reducing the compensation of the articulatory parameters.

The solution minimizing the function J is obtained by the following iterative form,

$$\begin{aligned}\hat{x}_k^{i+1} &= \hat{x}_k^i + \lambda_i \Delta\hat{x}_k^i \\ &= \hat{x}_k^i + \lambda_i\{(\frac{\partial h(\hat{x}_k^i)}{\partial x})^T R(\frac{\partial h(\hat{x}_k^i)}{\partial x}) + Q_i + \Gamma\}^{-1} \\ &\quad \cdot\{(\frac{\partial h(\hat{x}_k^i)}{\partial x})^T R({}_sy_k - h(\hat{x}_k^i)) - Q_i\hat{x}_k^i + \Gamma(\hat{x}_{k-1} - \hat{x}_k^i)\}, \end{aligned} \qquad (9)$$

where the coefficient λ_i is adjusted to satisfy the condition $J(\hat{x}_k^{i+1}) \leqq J(\hat{x}_k^i)$, and the weight matrix Q is variable as $Q_i = Q_a c_q^i + Q_b$ $(c_q \leqq 1)$ to improve the convergency of the solution. The iterative procedure starts with a given initial estimate x_k^0. The stability and the speed of the convergence are dependent on the initial estimate. For continuous speech, the estimate at the previous frame can be used. In cases where it is necessary to excute the estimation process at each frame independently, the initial value is obtained by a piecewise linear estimation method.

The speech wave characteristics contain the effects of the glottal wave and radiation, as well as the vocal tract impulse response. With this method we tried to extract information about the vocal tract shape in as pure a form as possible. The procedure has three stages as shown in Fig. 4. At first, the spectral envelope of the speech waveform is flattened by inverse filtering. Next, a piecewise linear estimate of the articulatory parameters is obtained. This estimate can be replaced by the result from the previous frame. The signal s_t'' resulting from the inverse filtering of the estimated vocal tract transfer function G(z), then contains the gross spectral gradient which is not compensated for in the first stage. At the third stage, the vocal tract impulse response v_t is calculated by applying the inverse filter which is obtained from s''_t to the pre-emphasized signal s'_t.

This method is effective for improving the accuracy of the estimation, especially with regard to the lips.

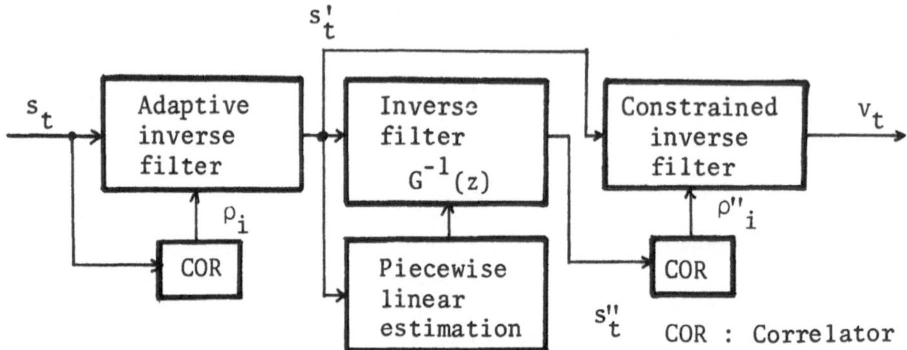

Fig.4 Block diagram of the model-matching inverse filter to separate the vocal tract impulse response from speech waves

4. EXPERIMENTAL RESULTS

The above method was tested through simulation experiments before the application for the real speech. It was found that for a variety of vocal tract shapes, the estimation was quite successfull. And further the method has been proved to work well for real speech data. In this section, only the result of the application for real speech waves is presented.

First of all, the mean length of the vocal tract should be matched for each speaker. Therefore, the length factor $\overline{S}_{FK}$ was introduced to give a uniform change in the vocal tract length. This length factor and the articulatory parameters can be estimated at the same time by the iterative procedure described above. In this case, the velum parameter X_N is excluded and the weighting matrices in Eq.(8) are set as $R=I_{12}$, $Q_a=Q_b$, $C_q=0.2$, $q_{ii}=0.2$ $(i=1,3,4,5)$, $q_{22}=0.4$, $q_{66}=0$, $\gamma_{ii}=0$ $(i=1,\ldots,5)$, $\gamma_{66}=1$. In Fig.5, the results for the convergence of $\overline{S}_{FK}$ are shown.

After the adjustment of the vocal tract length, the articulatory parameters are estimated. In this case, the weighting matrices are put as $R=I_{12}$, $Q_a=Q_b$, $c_q=0.2$, $q_{ii}=0.1$ $(i=1,3,4,5)$, $q_{22}=0.4$, $q_{66}=0$, $\gamma_{ii}=0.2 (i=1,\ldots,5)$, $\gamma_{66}=100$. In Fig. 6, the estimated result for continuous utterance /a i u e o/ is shown.

The movement of each articulatory parameter is quite reasonable and the matching between the original spectral pattern and the reconstructed one from the estimated parameters is good with the exception of the third formant at /i/.

For nasal sounds, the shape of the nasal cavity is adapted

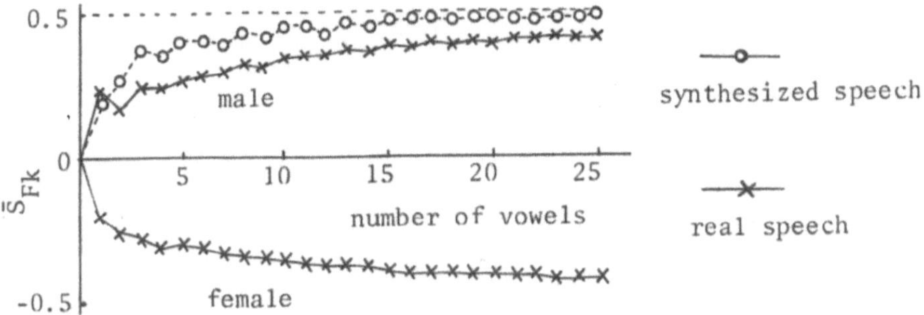

Fig.5 Convergence of the scale factor of vocal tract length

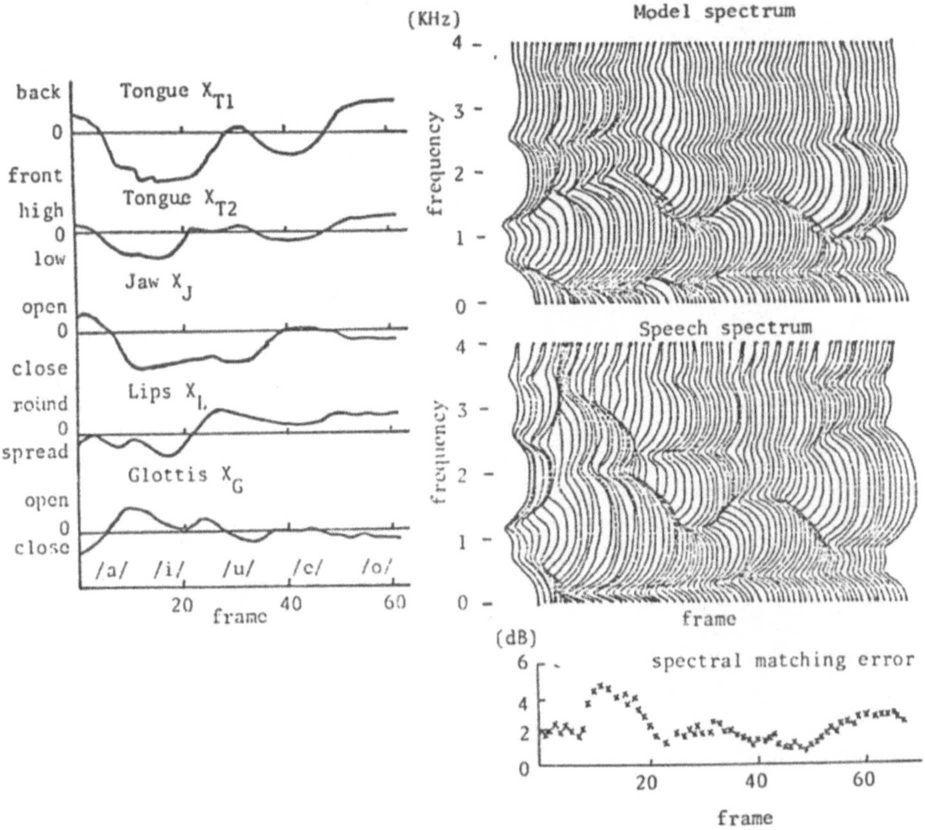

Fig.6 Estimated articulatory motion and spectra for /a i u e o/

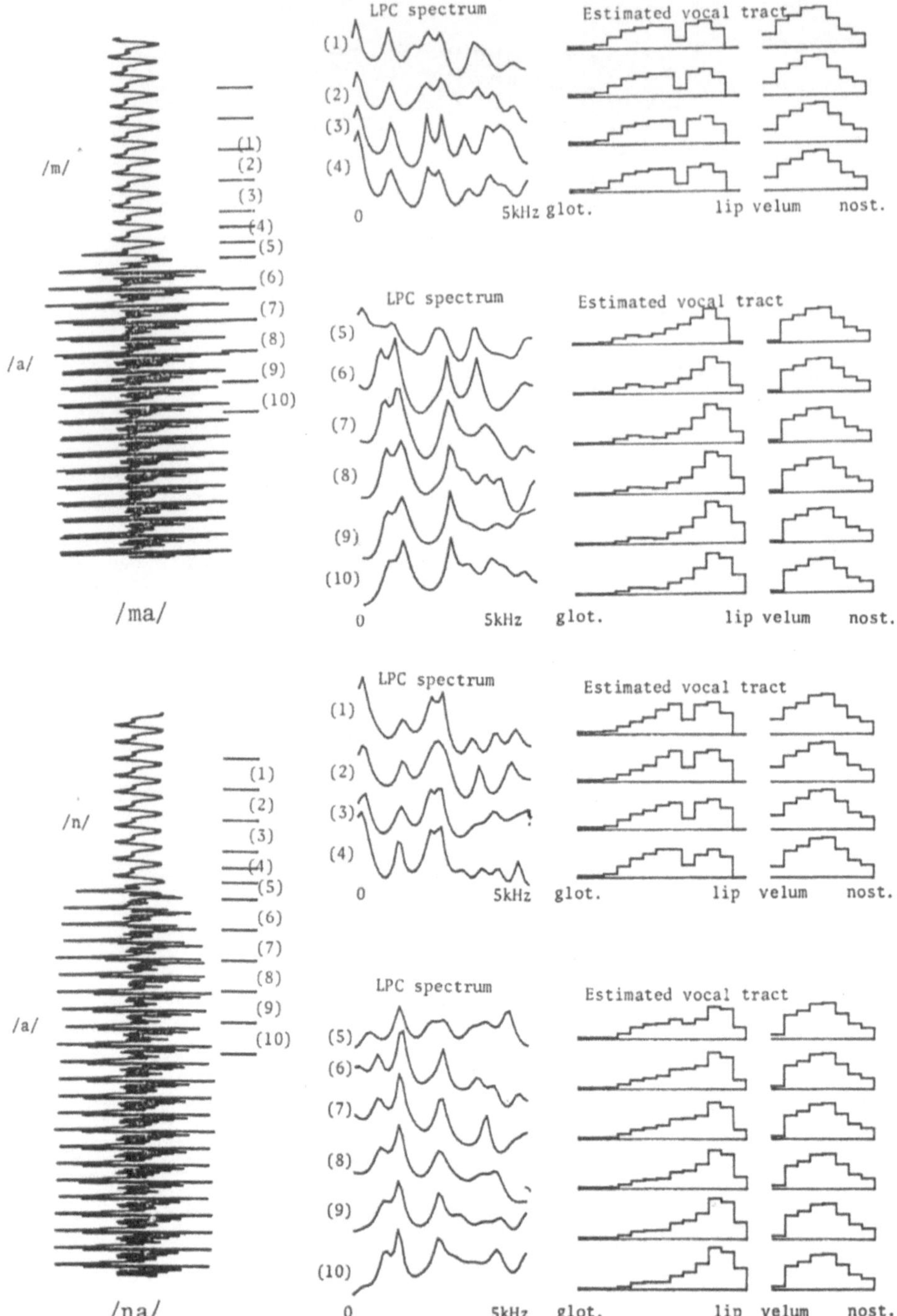

Fig.7 Estimated results for nasal sounds

for the speaker beforehand, and is made by adjusting X_{N1}, X_{N2}, X_{N3} using several nasalized vowel sounds. The results for /ma/ and /na/ are shown in Fig. 7. The detection of nasals can be made from the movement of the velum parameter X_N. And further, the discrimination of /m/ and /n/ becomes possible through the difference of lip opening at the beginning of the transition and the succeeding vowel.

Finally, we report the results concerning the vowel discrimination. The speakers in this experiment were 10 male and 5 female adults. Each speaker spoke Japanese five vowels 5 times separately and 20 kinds of symmetric continuous vowels $V_1V_2V_1$ (V_1 V_2: a i u e o). In Table 2 the results for various combination of learning and test data sets are presented. It is found that in this method the adjustment of the vocal tract length for each speaker compensates well for the difference between males and females, though it is not accurate in detail. Such adaptation may be possible by learning procedure for each speaker.

Continuous three vowels were uttered twice and the first data were used for the learning and the second were for the test. The results are shown in Table 3. As for the learning, two methods were tested, i.e. one was supervised learning for 5 speakers and the other was unsupervised learning for each speaker. In the latter case, the coefficients learned for the five speakers were used in the initial value of the discrimination function. When the unsupervised learning process was used, the average score of correct recognition was 99.3%. In this experiment, the effect of co-articulation was not taken into consideration and there were several cases which essentially could not be recognized correctly.

Table 2 Scores of isolated vowel recognition for the several training sample sets

Training sample \ Test sample	M	F	M & F
M	100.0 %	96.0 %	98.7 %
F	98.0 %	100.0 %	98.7 %
M & F	99.2 %	97.6 %	98.7 %

M : sample set of male speakers
F : sample set of female speakers

Table 3 (a) Confusion matrix for the connected vowels recognition results. : Supervised learning for 5 speakers

Input \ Output	/a/	/i/	/u/	/e/	/o/	Omis-sion	addi-tion
/a/	59	-	-	-	-	-	1
/i/	-	56	1	-	-	-	3
/u/	-	-	59	-	-	1	-
/e/	-	-	-	59	-	1	-
/o/	-	-	1	-	58	1	-

Table 3 (b) Confusion matrix for the connected vowels recognition results.: Unsupervised learning for each speaker

Input \ Output	/a/	/i/	/u/	/e/	/o/	omis-sion	addi-tion
/a/	60	-	-	-	-	-	-
/i/	-	60	-	-	-	-	-
/u/	-	-	58	-	-	2	-
/e/	-	-	-	60	-	-	-
/o/	-	-	-	-	60	-	-

5. CONCLUSION

This report has discussed a method of calculating the vocal tract shape from the speech wave. The method is based on articulatory model which is constructed by a statistical method. Various physiological constraints are effectively reflected in the articulatory model and the articulatory motion can be expressed by a small number of parameters. Following this, the calculation of the vocal tract shape is reduced to an estimation problem for the articulatory parameters.

The estimation was proved to be successful for vowels and nasals. Several preliminary results for vowel discrimination suggest that the articulatory parameters may be good features for recognition, since the difference between speakers can be adjusted easily and those parameters have a more speaker independent nature than others which have been used till now.

REFERENCES

1. K. Shirai, Feature extraction and sentence recognition algorithm in speech input system, 4th Int. J. Conf. on Artificial Intelligence, 506-513, 1975.
2. M. R. Schroeder, Determination of the geometry of the human vocal tract by acoustic measurement, JASA, 41, 1002-1010, 1967.
3. P. Mermelstein, Determination of the vocal tract shape from measured formant frequencies, JASA, 41, 1283-1294, 1967.
4. H. Wakita, Direct estimation of the vocal tract shape by inverse filtering of acoustic speech waveforms, IEEE Trans., ASSP-23, 574-580, 1975.
5. T. Nakajima, T. Omura, H. Tanaka and S. Ishizaki, Estimation of vocal tract area functions by adaptive inverse filtering methods, Bull. Electrotech. Lab., 37, 467-481, 1973.
6. H. W. Strube, Can the area function of the human vocal tract be estimated from the speech wave?, U.S.-Japan Joint Seminar on Dynamic Aspects of Speech Production, Tokyo University Press, 231-250, 1977.
7. K. N. Stevens and A. S. House, Development of a quantitative description of vowel articulation, JASA, 27, 484-493, 1955.
8. C. H. Coker and O. Fujimura, Model for specification of the vocal-tract area function, JASA, 40, 1271, 1966.
9. B. Lindblom and J. Sunberg, Acoustical consequences of lip, tongue, jaw and larynx movement, JASA, 50, 1166-1179, 1971.
10. K. Shirai and M. Honda, A representation of articulatory model, Anual Meeting, Inst. Electronics and Commu. Engineers Japan, No.1585, Aug. 1974.
11. K. Shirai and M. Honda, An articulatory model and the estimation of articulatory parameters by nonlinear regression method, Trans. IECE Japan, 59-A, 8, 35-42, 1976.
12. K. Shirai and M. Honda, Estimation of articulatory motion, U.S.-Japan Joint Seminar on Dynamic Aspects of Speech Production, Tokyo University Press, 279-302, 1977.
13. H. Wakita, Normalization of vowels by vocal-tract length and its application to vowel identification,IEEE Trans., ASSP-25, 183-192, 1977.
14. K. Shirai and M. Honda, Feature extraction for speech recognition based on articulatory model, Proc. of 4th Int. Joint. Conf. on Pattern Recognition, 1064-1068, 1978.

§ 2

ACOUSTIC AND PHONEMIC

SPEECH SIGNAL PROCESSING AND FEATURE EXTRACTION

Jared J. Wolf

Bolt Beranek and Newman Inc.
50 Moulton Street
Cambridge, Mass. 02138
U.S.A.

ABSTRACT

Speech signal processing and feature extraction is the initial stage of any speech recognition system; it is through this component that the system views the speech signal itself. This chapter introduces general approaches to signal processing and feature extraction and surveys the techniques currently available in these areas.

1. INTRODUCTION

The task of an automatic speech recognition (ASR) system is to convert a speech signal into a linguistic representation of its content -- to "recognize" the linguistic message contained in the signal. A common aspect of all ASR systems is that they *process* the speech signal in some way in order to produce a representation of the signal that is better suited to the recognition process. This operation (or set of operations) is the function of the component that can be called "speech signal processing and feature extraction" (SSPFE).

What do we mean by "speech signal processing" and by "feature extraction"? "Speech signal processing" refers to the operations we perform on the speech signal (e.g., filtering, digitization, spectral analysis, fundamental frequency estimation, etc.). "Feature extraction" is a pattern recognition term that refers to the characterizing measurements that are performed on a pattern (or signal); these measurements ("features") form the input to the classifier that "recognizes" the pattern. In many cases in speech

J. C. Simon (ed.), Spoken Language Generation and Understanding, 103-128.

recognition, I find it impossible to separate these functions; they are one and the same.

This chapter will attempt to describe the range of methods used for speech signal processing and feature extraction in automatic speech recognition. The treatment will be more descriptive and evaluative than mathematical. For theoretical and computational details, the reader will be referred to specific references in the literature. Beyond these specific references, several more general ones should be mentioned. Flanagan's landmark book (1) is probably the most complete description of models of speech production, perception, and analysis. Schafer and Markel (2) have assembled into one volume reprints of fundamental articles on speech analysis. The single most useful reference, though, is the excellent new text by Rabiner and Schafer (3), which is devoted entirely to digital speech processing. Digital signal processing in general is the subject of many texts, for example (4-6). The _IEEE Transactions on Acoustics, Speech, and Signal Processing_ (which was named _Transactions on Audio and Electroacoustics_ prior to 1974) and (to a lesser but still important extent) the _Journal of the Acoustical Society of America_ are the primary journals in which new developments in SSPFE have been chronicled. Also to be noted is the institution, beginning in 1976, of the annual IEEE International Conference on Acoustics, Speech, and Signal Processing; the printed proceedings of these conferences (available from IEEE) are valuable sources of the very latest results in all fields of speech processing.

Before proceeding further with the discussion of speech signal processing and feature extraction, we shall review briefly the nature of the speech signal from the viewpoint of the speech production process. Then we shall proceed to a general discussion of SSPFE, followed by discussions of time-domain and frequency-domain analysis methods.

2. THE SPEECH COMMUNICATION PROCESS

Speech is a complex sound produced by the human vocal apparatus, which consists of organs primarily used for breathing and eating: the lungs, trachea, larynx, throat, mouth, and nose. The source of energy for the production of these sounds is the reservoir of air in the lungs. Sound is generated in two ways. If air is forced through the larynx with the vocal folds appropriately positioned and tensioned, it sets them into oscillation, so that they release puffs of air in a quasi-periodic fashion at rates of about 80 to 200 Hz for male speakers, and at faster rates for women and children. This glottal source is rich in harmonics, and it excites the acoustic resonances of the vocal tract above the larynx, which filter the sound. These sharp resonances, called formants, are determined by the shape of the throat, mouth, and, if the velum (soft palate) is open, the nasal cavity. It is through manipulation of the vocal tract shape by the articulators (tongue, jaw, lips, and velum) that we control the formant frequencies that differentiate the various voiced speech sounds. These are the vowels, nasal consonants, liquids (/r/ and /l/), and glides (/w/ and /y/). Only the lowest three or four formants, up to about 4 kHz, are perceptually significant.

The other sound source used in speech is turbulent noise, produced by forcing air through some constriction (such as between the tongue and teeth in th-sounds) or by an abrupt release of pressure built up at some point of closure in the vocal tract (such as behind the lips in a /p/). The spectral peaks associated with these fricated sounds generally lie between 2 and 8 kHz and are primarily determined by the position and the shape of the constriction. Some sounds, the voiced fricatives, such as /z/ and /v/, have both voiced and turbulent excitation.

Distinguishable differences in the voice signal can be produced by quite small changes in the way the vocal tract is manipulated, so a very large number of sounds can be produced. For the communication of language, however, only a restricted number of sounds, or more accurately, sound classes, are used. Words in English are made up of approximately 40 of these phonemes, which correspond roughly to the pronunciation symbols in a dictionary. Since there are relatively few of these elemental sound units, many speech recognition systems choose phonemes or phoneme-like units as the units of recognition.

This concept of elemental units does not reduce speech recognition to the sequential recognition of 40 or so fixed patterns. The acoustic realizations of phonemes in real speech are drastically different from their characteristics in isolated environments. When phonemes are strung together into words, the acoustic characteristics of successive phonemes become overlapped, due to the dynamics of the articulators and the tendency for an articulator to anticipate the position it will next assume. (For example, when you say the /t/ of the word "tin", the lips are spread apart, but in "twin", they are pursed together. Since the lips are not used for the articulation of the /t/, they are free to assume the position for the following phoneme.) These context effects, along with effects due to linguistic stress, speed of talking, the size of a particular speaker's vocal tract, and variation in the way a speaker says a given word from repetition to repetition make the decoding of the speech signal a nontrivial problem. To be sure, the speech wave does contain much information directly related to the phonemes intended by the speaker, but the details of the coding are complex (1,7,8).

The phonemes themselves can be grouped into classes according to the ways they can modify and affect one another when they are in proximity in normal speech. It is convenient to describe these classes in terms of about fifteen binary distinctive features (9-12), most of which have straightforward articulatory and acoustic attributes, such as voiced/unvoiced, nasal/nonnasal, fricated/unfricated, tongue position: back/front, high/low, etc. Distinctive features have proved to be quite powerful in describing the phonological changes that underlie the contextual effects referred to above. For that reason, some speech recognition systems use them as recognition units or as ways of organizing their recognition of phoneme-sized units.

3. SSPFE APPROACHES

Speech signal processing and feature extraction is an area that requires a combination of knowledge and skills dealing with (a) theory of speech, (b) theory of signals, (c) working within constraints of cost and processing time, and (d) cleverness (heuristics for extracting desired properties from the signal). This combination of theory and real-world constraints places SSPFE firmly in the realm of engineering.

Before considering specific SSPFE techniques, this section dwells briefly on several more general aspects of SSPFE.

3.1 Analog and Digital Techniques

There is a major division of SSPFE techniques between analog and digital. Analog signal processing techniques are those performed on the (electrical analog of the) speech signal by means of electronic circuitry. This requires investment in special-purpose electronic equipment, which, although frequently modular, is somewhat inflexible, and furthermore, requires periodic calibration and adjustment. Digital signal analysis, on the other hand, can be performed on a general purpose digital computer, a microprocessor, or special purpose digital hardware. When implemented on a computer, such digital instrumentation is very flexible. Programming turns the computer into a custom-designed instrument; the characteristics of the instrument may be changed or even redesigned without recourse to a soldering iron or supply cabinet, and there is nothing that can go out of calibration. Furthermore, many of the digital techniques now available are simply not realizable in the analog domain. Perhaps the most notable drawback is that while analog instrumentation is inherently real-time, it is not always possible for digital techniques to be performed in real time. This is counterbalanced by the potentially far greater functionality of digital techniques. The same technological advances that have made computers faster and cheaper have done the same for digital processing techniques, and we may expect this tendency to accelerate.

The flexibility and capability of digital SSPFE methods, together with the increasing speed and decreasing costs of computers and microprocessors, have established a trend of increasing importance of digital techniques in all aspects of speech processing. This chapter will treat both analog and digital techniques, but the dominance of the digital domain is unavoidable.

For a signal to be analyzed digitally, it must first be converted from the form of a continuous signal (an acoustic pressure wave or its electrical analog) to a digital signal. A digital signal is discrete in two ways: it consists of measurements of the speech signal amplitude of discrete, regularly spaced instants of time, and each of these _samples_ has been _quantized_ to a certain number of bits of precision. This transformation from a continuous waveform to a list of digital numbers is performed by an _analog-to-digital converter_. Nyquist's sampling theorem (see, for example, (3-6)) tells us that in order to capture the information contained in a bandwidth of B Hz, we must sample at a rate of a least 2B samples per second; furthermore, if the signal contains energy at frequencies higher than B Hz, the result will be distorted unless we first lowpass filter the signal at or below B Hz. For example, an application in which the 0-5 kHz band is of interest may call for lowpass filtering at or below 5 kHz and sampling at 10,000 samples per second, with each sample represented by a 12 bit number.

3.2 Time and Frequency Domain Approaches

Another fundamental division of SSPFE approaches is between _time domain_ and _frequency domain_ approaches. Time domain approaches deal directly with the waveform of the signal, and these representations are often attractive because of their simplicity of implementation. In working with time domain representations, it must be noted that the shape of the waveform for voiced speech depends on (among other things) the pitch period, so we generally shouldn't pay attention to _details_ of the waveform. Also, the perception of speech is only minimally dependent on the phase of the signal, so _phase-independent_ measures are desirable.

Frequency-domain approaches involve (explicitly or implicitly) some form of spectral representation. A spectral representation often displays characteristics not evident in the time domain, and we know that a kind of spectral analysis is performed in the ear. The above observations about pitch and phase have counterparts in the frequency domain. For voiced speech, the spectrum has both a _fine_ and a _coarse_ structure, due to the pitch harmonics and the vocal tract transfer function respectively. We use the magnitude or power spectrum, which does not contain phase information.

A time/frequency dichotomy does not account for all SSPFE approaches. Some are hybrids of the two, and for others the concept of "time or frequency" seems ill-defined.

3.3 Production/Perception Orientation

The first things one usually learns about speech are related to its production - either from an anatomical point of view or via the characteristics of the signal. Most SSPFE techniques speak directly to this production/signal point of view, and for good reason. But this orientation does not speak directly to ASR (as it does, for example, to speech transmission). In ASR, the problem is, of course, recovery of the linguistic message. The speech signal is the carrier of the message, not the message itself.

There is an existence proof for ASR: the human. Therefore, some researchers have reasoned, when it comes to selecting the characteristics of a SSPFE component, one should start with what is known about the processing done in the human auditory system. Such processing should be suitable and adequate. Furthermore, the human speech perceiver finds easy many tasks that machines stumble over, such as coping with speaker differences, perception of speech in the presence of noise, and most of all, the extraction of cues for the decoding of the linguistic message. It is hoped that adopting such a processing mechanism may aid ASR in these tasks.

This is a tempting hypothesis, and one to which more researchers have lately turned. But it is not guaranteed to be an easy one to follow. We must ask if our knowledge of human auditory processing is sufficiently complete to support processing for ASR, and whether in an ASR device, we can simulate human processing well enough. (For example, is a set of 18 1/3-octave filters and envelope detectors a sufficiently complete representation of thousands of auditory nerve channels?) Presently constructable machines cannot begin to simulate the rich interconnections and parallel processing of the human nervous system, so we must content ourselves with imitating only certain aspects of the peripheral processing. When our SSPFE diverges from the human's, as it must, will the difference between the two be significant? Put differently, the "human" way is clearly appropriate for humans, but is it also good for machines? Finally, the human perceptual mechanism is not as open to experimentation and observation as are the production mechanism and the characteristics of the speech signal.

Having said that, I hasten to add that I do not wish to disparage the perceptual style of SSPFE for ASR. It is an interesting and clearly relevant point of view, and it is a relatively recent approach that may become important as we learn more about both human perception and machine recognition techniques.

3.4 The Short-Time Principle

SSPFE is processing of the speech signal for the purpose of making a new representation that is suited to automatic speech recognition. The speech signal waveform itself is not a good representation upon which to base the recognition because it does not display directly the features of the linguistic message. For one thing, the waveform changes <u>much more rapidly</u> than do the speaker's articulatory gestures. For example, it requires on the order of 10,000 samples per second to represent speech with fair fidelity, yet the nominal rate of speaking is on the order of 10 phones per second. This mismatch of rates indicates an underlying assumption of most speech processing techniques: the properties of the speech signal change relatively slowly with time. This assumption leads to the important principle of "short-time" analysis, in which short segments of the speech signal are isolated and processed as if they had come from a steady sound with fixed, not changing, properties. The analysis is repeated as often as desired, usually at regular intervals, so that the <u>sequence</u> of results follows the time-course of the changing speech signal.

The width of each analysis window and the time interval between successive analyses involve tradeoffs between several factors. The analysis window width must not be so long as to smear together successive yet distinct speech events, but it must be long enough to yield sufficient frequency resolution and to produce a reasonably smoothly varying function of time. For spectral analysis (including linear prediction), more than one pitch period is needed, implying an analysis window width of 10-25 ms; for pitch period estimation, more than two pitch periods are required, so 30-50 ms is customary. The interval between analyses must be small enough not to miss brief events such as stop bursts, but not so small as to be a negligible fraction of the window width. Interframe intervals of 5-12 ms are commonly used; successive analysis windows may overlap.

4. TIME DOMAIN METHODS

4.1 Energy

The amplitude of the speech signal varies over time due to the type of speech sound and to suprasegmental factors. The short-time energy is a convenient way to represent these amplitude variations. For a digital signal s(n):

$$E = \sum_{n} [s(n)w(n)]^2 \qquad (1)$$

where w(n) is a window function, which is nonzero over only the analysis time interval. That is, w(n) selects the interval to be analyzed. (For energy determination, the shape of w(n) is not critical; it is often a rectangular window, i.e., 1 over its nonzero extent. For spectral analysis (Sec. 5), we shall note that the shape of w(n) is important.)

Because of the squaring operation, the energy can be sensitive to large signal levels, and the computation can produce overflow on short word length computers. For these reasons, a magnitude (or amplitude) can be calculated instead:

$$M = \sum_{n} |s(n)| w(n) \tag{2}$$

For analog signals, it is more convenient to measure magnitude, by means of a rectifier and lowpass filter. Note that the impulse response of the lowpass filter corresponds to the window function w(n).

Energy or magnitude provides a primary basis for separating speech from silence and, because voiced sounds are generally higher in amplitude than unvoiced, for distinguishing between voiced and unvoiced intervals. They are also useful for delineating syllables and for locating the syllable nucleus.

4.2 Zero Crossings

It is well known that infinitely clipped speech is intelligible (although distorted), so there must be useful information in the zero-crossings of the signal. The zero-crossing rate (ZCR) is obtained by counting the number of times the signal changes sign during a fixed-length analysis interval (13,14). This rate gives a very rough measure of the frequency content of the signal, as may be appreciated by noting that a sinusoid of frequency F gives an average ZCR of 2F sec^{-1}. ZCR is sometimes used after broad bandpass filters (15) for a rough formant estimation, but the fact that the frequency ranges of the formants overlap limits the accuracy of these estimates.

The ZCR also helps distinguish between speech and nonspeech (16) and between voiced and unvoiced sounds (the mean ZCR for voiced sounds is about 1500 sec^{-1} and for unvoiced sounds 5000 sec^{-1}, but the distributions overlap (3)).

ZAPDASH is a relatively coarse characterization of the signal used in the Hearsay-II speech understanding system (17). The name is an acronym for "Zerocrossings And Peaks of Differenced And SmootH waveforms". The processing consists of first forming two processed versions of the speech signal: lowpass filtered at 1 kHz and differenced. As discussed below in Sec. 5, differencing a digital signal is roughly equivalent to differentiating a continuous signal, in that it emphasizes the high frequency region.

Each of these signals is characterized over a 10 ms interval by measurements of the peak-to-peak amplitude and the zero-crossing rate.

In application, the ZAPDASH parameters were used only for segmentation into regions of small acoustic change and subsequent broad labeling. This was followed by linear prediction analysis at the middle of each segment for more precise labeling. Note the SSPFE philosophy of applying cheap processing overall, followed by more expensive processing used sparingly, only once per segment. Although ZAPDASH was performed by a program, it could easily have been implemented in analog or digital hardware.

A related processing technique of "cycle-synchronous" "up-crossing" measurements has been developed by Baker (18). An up-crossing is a zero-crossing with positive slope; the important measurement is the log reciprocal of the time interval between successive up-crossings, which has the form of a log "instantaneous frequency." Unlike the ZCR, up-crossings are not short-time averaged; it is claimed that this technique yields measurements with much greater time resolution than is the case with the usual short-time procedures.

4.3 Pitch Period Estimation

Measurement of the pitch period (or equivalently the fundamental frequency) of voiced speech and separation of voiced from unvoiced intervals are problems for which many techniques have been devised. Some of these techniques are oriented toward the time domain, and these will be described here; others are frequency-domain oriented, and these will be described in Section 5.9. Rabiner, et al. (19) have directly compared several pitch period estimation techniques.

4.3.1 Autocorrelation. The short-time autocorrelation of the signal,

$$\emptyset(k) = \sum_n s(n)s(n+k), \quad k=1,2,3,... \tag{3}$$

peaks when K is equal to the pitch period, even if the signal is only approximately periodic, and this is the basis for autocorrelation methods. But the autocorrelation function also contains other information, enough to represent the entire spectrum. Autocorrelation peaks due to formant structure can be as large as peaks due to approximate periodicity. Therefore, before autocorrelating, it is useful to lowpass filter the speech (to remove the higher formants) and/or to flatten the spectrum (to remove formant information). Spectral flattening can be done by center-clipping (20) or linear prediction inverse filtering (21).

The autocorrelation computation is fairly expensive, so shortcuts have been devised. If the speech is lowpass filtered, it may be sampled at a lower rate than for other purposes (or if already a digital signal, it can be downsampled). Downsampling by a factor of N reduces the computation by N^2 (22). If the center-clipping is followed by infinite peak-clipping, the resulting signal takes on only the values {-1,0,+1}. The autocorrelation of such a signal involves only additions of ± 1, which is easily accomplished in digital hardware by an up/down counter (23).

A related operation is the Average Magnitude Difference Function (AMDF),

$$AMDF(k) = \sum_n |s(n)-s(n+k)|, \quad k=1,2,3,... \tag{4}$$

which shows a deep dip when k is a multiple of the pitch period (24). Because subtractions are usually less expensive than multiplies, the AMDF is substantially less expensive to implement than autocorrelation.

4.3.2 Data Reduction. Miller (25) has described a pitch period marking technique that characterizes the (lowpass filtered) signal in terms of the time and height of the excursions (local peaks) above and below the zero line. Heuristics are used to detect periodic patterns in the excursions and to reduce them to a single marker per pitch period. The method is quite fast, but it is obviously sensitive to the effects of phase distortion and noise.

4.3.3 Parallel Processing Approach. Although the excursions of the speech waveform do show periodicity during voiced sounds, one particular type (e.g., positive peaks or negative peaks) will not be a reliable indicator of periodicity in all speech sounds. The parallel processing approach described by Gold and Rabiner (26) uses six simple local peak detectors, combining their individual outputs in a "coincidence voting" decision scheme to produce a reliable judgement of the pitch period. Because this technique operates directly on the waveform without any short-time function computation (such as autocorrelation) it is a relatively fast method.

5. FREQUENCY DOMAIN METHODS

The representation of signals as the sum of sinusoids or complex exponentials often leads to convenient solutions and insight into physical phenomena. For linear systems, (such as our model of speech production), it is convenient to determine the system response to such a representation. A spectral representation often displays characteristics not evident in the waveform, and we know that a kind of spectral analysis is performed in the ear. For these reasons, frequency domain methods are important in speech analysis.

It is with frequency-domain methods, in which the shape of the spectrum is relevant, that the shape of the short-time window function is important. Windowing is multiplication of the waveform and window function, which appears in the frequency-domain as convolution of their Fourier transforms. Therefore, a narrow main lobe of the window's transform (to minimize "smearing" of the spectrum) and low side lobes (to minimize "leakage" from other parts of the spectrum) are desirable. In these respects, the rectangular window is inferior to smooth windows such as the Hann, Hamming, or Parzen windows (3-6). In analog filter bank spectral analysis (see Sec. 5.1), the window function is determined by the shape of the filter bandpass characteristics (a wider bandwidth yields sharper time resolution). Additional time-averaging is usually introduced by the smoothing of the envelope detector following each filter.

Another consideration in frequency-domain techniques is that of pre-emphasis. The average spectrum of voiced speech falls with frequency at 6-12 dB/octave (depending on the speaker) (27). In order to decrease the dynamic range required to represent the spectrum, it is often desirable to counteract this fall-off by a fixed filter with a rising frequency response. In the digital domain, if the signal was not pre-emphasized prior to sampling, pre-emphasis is often performed by differencing, i.e.,

$$s'(n) = s(n) - s(n-1) \tag{5}$$

Differencing is roughly equivalent to differentiation in the analog domain, giving a rising 6 dB/octave characteristic (which levels off at half the sampling frequency). Other fixed digital filters may also be used for pre-emphasis.

5.1 Analog Filter Bank Spectral Analysis

Much early work on the characteristics of speech was done with spectral analysis by means of an analog filter bank, and several commercial speech recognition systems use this technique because of its real-time, simple, and inexpensive operation. Much current ASR research uses it for the same reasons.

The design of a filter bank spectrum analyzer involves questions of the number of filters, their frequency and bandwidth characteristics, and the smoothing characteristics of the envelope detectors following the bandpass filters. One early design used the Koenig scale (linear filter spacing up to 1.6 kHz, logarithmic spacing (and increasing bandwidth) above that point) and 10 ms time constant smoothing on all channels (28). More recent designs have been based on perceptual grounds, with 1/3-octave filters and smoothing inversely proportional to filter bandwidth, so that low frequency filters have relatively poor time resolution and high frequency filters have good time resolution (30-32). In the digital domain, it is of course possible to simulate a filter bank by means of the DFT (see Sec. 5.2); such an implementation may be desirable because of digital advantages (stability, flexibility), usually gained at a cost of non-real-time operation.

5.2 Discrete Fourier Transform

The discrete Fourier Transform (DFT) is the sampled data version of the Fourier transform:

$$S(k) = \sum_{n=0}^{N-1} s(n) e^{\frac{-j2\pi kn}{N}}, \quad k=0,1,2,\ldots N-1 \tag{6}$$

One interpretation of the DFT is that the signal s(n) is periodic with length N, so only N values of s(n) appear in the expression. For practical speech analysis, it is permissible to regard s(n), n=0,1,...,N-1 as a windowed portion of a much longer signal. Thus, the DFT is directly applicable to short-term spectral analysis. (The earlier discussion of the shape of the window function applies here. As written, Eq. (6) implies an undesirable rectangular window. In practice, a smooth window is used, and s(n) in Eq. (6) is replaced by s(n)w(n).) See (3-6) for an extensive discussion of the properties of the DFT.

If N is highly factorable (usually a power of 2), the DFT may be computed very efficiently (O(NlogN) computations rather than $O(N^2)$) by using a set of related algorithms known collectively as the Fast Fourier Transform (FFT) (4-6) The FFT is used routinely to do spectral analysis, autocorrelation, digital filtering, and filter bank simulation. More recently, another faster algorithm for computing the DFT for N the product of certain prime factors, has been developed by Winograd (33).

5.3 Cepstral Smoothing

A spectrum of a voiced segment of speech is shown in Fig. 1. Note that the spectrum has both a somewhat regular fine structure, due to the harmonics of the glottal source, with a fundamental frequency of about 120 Hz, and a coarse structure due to the formant frequencies. For many purposes, including ASR, it is desirable to separate these two effects, and in particular, to derive the spectral envelope or a smoothed version of the spectrum.

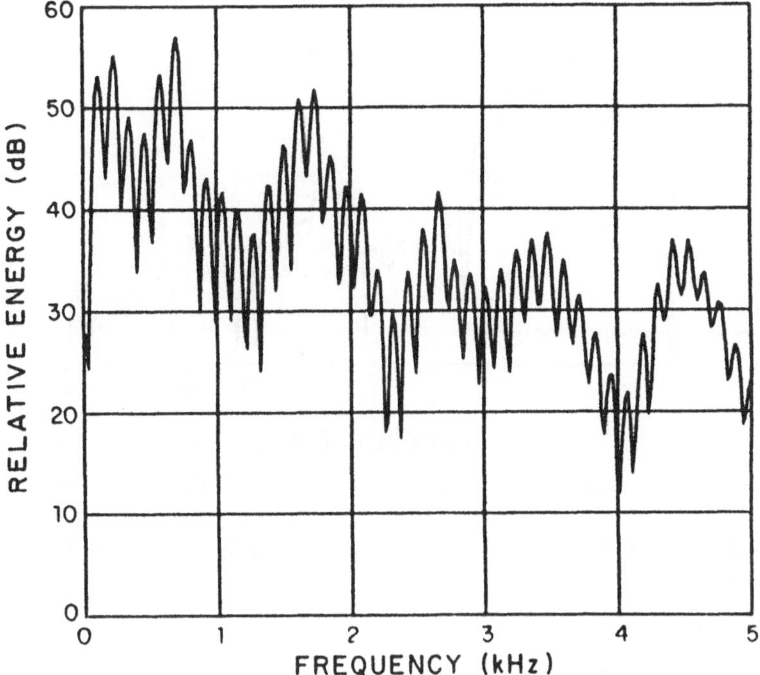

Figure 1. Spectrum of the vowel /æ/.

The cepstrum is the inverse Fourier transform of the logarithm of the power spectrum. (Think of the log spectrum of Fig. 1 as a "signal" and the cepstrum as the spectrum of that "signal".) Low-time components of the cepstrum correspond to the slowly varying part of the log spectrum (the formant structure), and high-time components to the rapidly varying part (the harmonics). Therefore, if the cepstrum is low-time windowed (keeping only the low-time component) and then transformed, a smoothed log spectrum results (3,4,34,35). In practice, these transforms are done by FFT. Figure 2 shows a vowel spectrum (different from Fig. 1) and its cepstrally smoothed counterpart; note that the smoothed version follows the coarse structure of the spectrum.

5.4 Pitch-Synchronous Spectral Analysis

The spectral analysis methods described above can be applied at arbitrary places in the signal and are often applied time-synchronously, at regular time intervals, as stated in Sec. 3.4. The fine structure in a voiced speech spectrum is due to the fact that the signal within the analysis window shows periodicity.

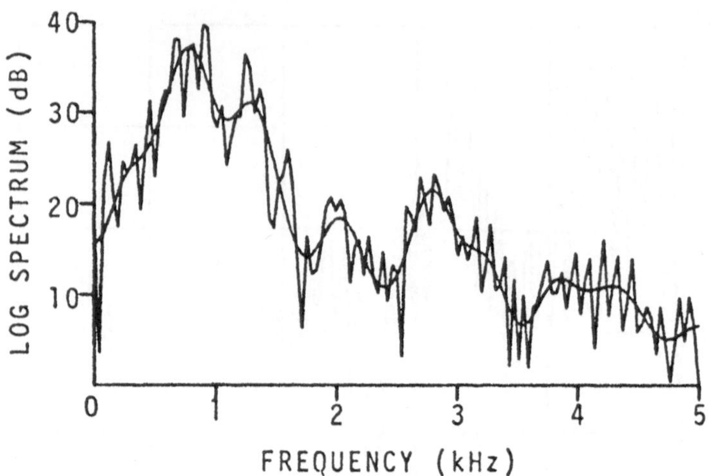

Figure 2. Spectra of the vowel /a/, showing the DFT spectrum and a cepstrally smoothed version. Note the significant peak at 2 kHz, which is not a formant (see Fig. 3).

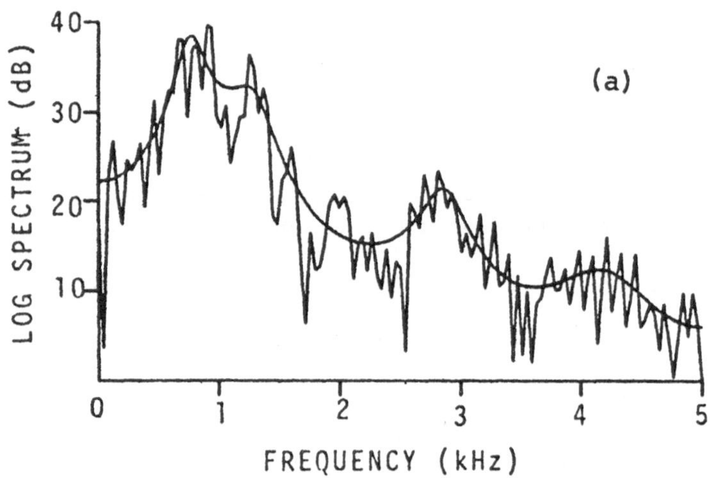

Figure 3. Spectra of the vowel /a/ (same as Fig. 2), showing the DFT spectrum and a 14-pole LP-modeled version. Note that the hump at 2 kHz is not modeled as a formant.

If instead exactly one period of the signal were analyzed, there would be no fine structure because there would be no periodicity in the signal under analysis (3,36,37). (In this case, there is no need to use a smooth window.) This is the principle of pitch-synchronous analysis. Its advantage is that no smoothing of the spectrum is necessary, but it has two disadvantages: the need for precise determination of the pitch period and the fact that the analysis frames are no longer equally spaced.

5.5 Energy in Broad Bands

In many ASR systems, a first level of phonetic segmentation is performed on the basis of a very coarse spectral analysis, such as the energies present in several broad frequency bands. Such parameters can be used for preliminary segmentation using distinctions such as sonorant/nonsonorant, continuant/obstruent, nasal/nonnasal, etc. For example, in the BBN HWIM speech understanding system (38), the following frequency bands were used:

120 - 440 Hz
640 - 2800 Hz
3400 - 5000 Hz

5.6 Linear Prediction

One of the most powerful techniques of digital speech processing is that of linear predictive (LP) analysis (3,39-41). It has become the predominant method for estimating the basic parameters of the speech signal (formants, fundamental frequency, vocal tract area function, and spectra), and it lies behind other important applications of speech processing, such as low bit-rate transmission of speech.

The basic mathematical idea behind linear prediction is that a (digital) speech signal can be approximated as a linear combination of past values of the signal, but for the purposes of this chapter, these details are less important than the fact that linear prediction permits us to model the speech signal with an all-pole model. This is important because our most useful acoustical model of the vocal tract also takes the same form (1,42), and thus from the speech signal itself we can calculate a model that directly estimates the relevant parameters of the vocal tract.

As with spectral analysis, LP analysis is applied to the speech signal on a short-time basis and thus the short-time behavior of the vocal tract is captured as it moves to form successive speech sounds. The coefficients of the LP model may be manipulated to yield excellent estimates of formant frequencies and they may be transformed via the DFT to yield a smoothed spectrum that emphasizes the vocal tract frequency response. Figure 3 shows a vowel spectrum and an LP fit to that spectrum. The frequencies and bandwidths of the LP model poles may be estimated by finding and measuring the peaks of the LP spectrum, or they may be calculated by solving for the roots of a polynomial formed by the LP coefficients.

An all-pole model is ideal for analyzing vowels, but less so for nasals, fricatives, and stops because for those classes of sounds, the acoustical description of the vocal tract contains both poles and zeros. Linear prediction is still useful in these cases, as evidenced by the high quality of LP-vocoded speech; at worst, it gives a good estimate of the spectral envelope of the signal. Modeling speech with poles and zeros is also possible, but it is a much more complex process (43), and it does not appear to have been used in ASR research.

Linear prediction has a number of interesting properties that can be exploited in the analysis of speech. Certain speech sounds have formants so close together that they form one peak, not two, in the spectrum, as shown in Fig. 4. Identifying the formant frequencies from such a spectrum presents problems. However, by multiplying the LP coefficients by a rising exponential function, the poles of the LP model are moved closer to the unit circle, sharpening their spectral peaks, as shown in Fig. 5, which permits the individual formants to be identified by spectral peaks (40,44).

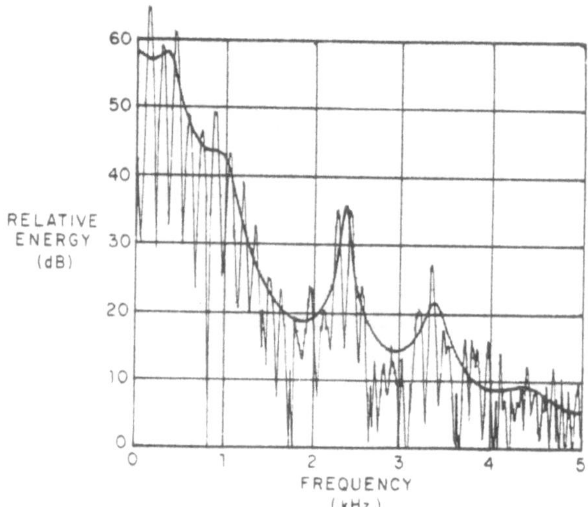

Figure 4. Spectrum of /w/, showing a 14-pole LP fit. Note that the 2nd formant at 1 kHz is not resolved as a separate peak in the LP spectrum.

Linear predictive modeling may be applied to arbitrary regions of a spectrum (45). For example, in the analysis of speech over a 10 kHz bandwidth, it may be desirable to model the 0-5 kHz region, where the important vowel formants lie, with 14 poles, and the 5-10 kHz region, where such resolution is not necessary, with a smaller number. Figure 6 shows such a two-region modeling.

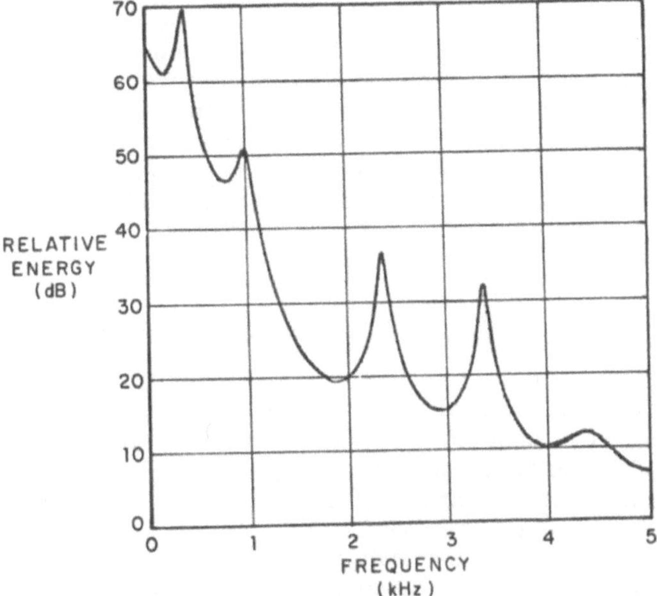

Figure 5. The same LP spectrum shown in Fig. 4, except that the LP poles have been moved closer to the unit circle to sharpen the formant peaks. Note that the 2nd formant is now visible as a clear peak.

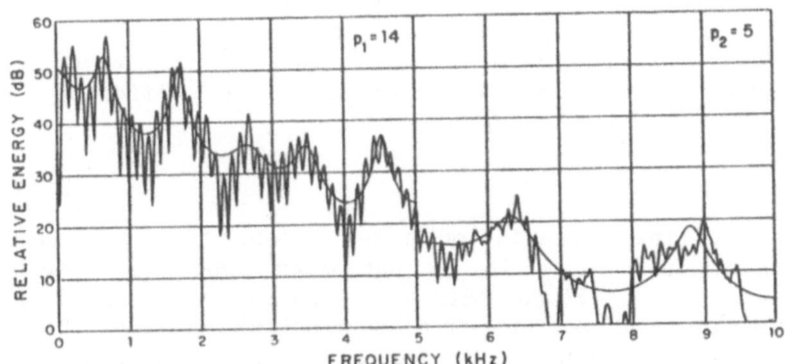

Figure 6. 0-10 kHz spectrum of /æ/, showing a 14-pole LP fit to the 0-5 kHz region and a 5-pole fit to the 5-10 kHz region.

For gross characterization of speech spectra, a 2-pole LP analysis turns out to have some useful properties because it portrays both the region of dominant energy and the degree of spread of energy in the spectrum (3,46,47).

In some ASR systems, comparisons are made between a speech interval to be classified and several (or many) reference intervals. In these cases, a distance measure between two such intervals is required. Itakura (48) developed such a distance measure of the log ratio of linear prediction residuals, derived from a maximum likelihood statistical approach. Gray and Markel (49) have investigated other LP distance measures. Such measures are used in phonetic segment classification (17) and in dynamic programming template matching (see Bridle's chapter in this volume).

5.7 Formant Tracking

Because the formants provide a fundamental description of speech sounds (42), measuring (or "tracking") their positions throughout an utterance yields important information about the vocal tract gestures that underlie it. Flanagan (50) tracked peaks in filter bank spectra, handling cases in which the formants did not form separate peaks. Bell et al. (28) described an analysis-by-synthesis method that was used in several fundamental MIT studies of articulatory/acoustic properties of speech. The advent of digital spectral analysis produced new formant tracking methods, such as that described by Schafer and Rabiner (35), which used cepstrally-smoothed spectra (Sec. 5.3) and the chirp-z transform to sharpen formant peaks.

Linear prediction has reduced the uncertainty in identifying formants that earlier techniques faced, but it has not eliminated it. Linear prediction models the entire speech signal, not just the vocal tract component of it, so not every LP conjugate pole pair is certain to model a formant. Heuristics are still necessary. LP formant tracking schemes of different types are described in (38,41,51-54).

5.8 Acoustic Tube Modeling

The linear prediction model allows the parameters of an acoustic tube model to be estimated from the speech signal (39,41,55,56). "Reasonable" area functions (i.e. vocal tract area as a function of distance from the glottis) have been obtained, but there are questions about the accuracy and uniqueness of such models due to factors such as assumptions about losses and loading in the model, the fact that LP is not an optimal model for all kinds of speech sounds, and the fact that LP is known to be poor for estimating formant bandwidths; this last is crucial for area function estimation.

The area function is an intriguing, appealing domain for characterizing speech for ASR, since it would seem to display the very articulatory gestures used by the speaker. However, it can be argued (at least for a linear prediction formulation) that it is just a transformation of the LP parameters, so it is no better than the LP model itself. Even if it were possible to estimate the area function accurately, it should be noted that the area function is (like other SSPFE parameters) a continuous function of time, in which coarticulation and phonological effects are manifest. The area function does not seem farther along the road from continuous gesture to discrete linguistic unit than do other parameters.

5.9 Fundamental Frequency Estimation

Frequency domain methods are also applied to the problem of fundamental frequency estimation. The periodic glottal excitation shows up as harmonics, the fine structure in a voiced speech spectrum. In fact, it is possible to estimate fundamental by tracking the harmonic peaks visible in a DFT spectrum (57).

Another interesting approach is the harmonic product spectrum (3,58), in which compressed versions of the spectrum are multiplied together, resulting in reinforcement of the fundamental and suppression of all other spectral features. This method is quite resistant to noise, but it requires high resolution in the DFT.

Recall from Sec. 5.3 that the fundamental shows up in the high-time portion of the cepstrum. In fact, the regularity of the harmonic structure in the spectrum produces a distinct peak in the cepstrum, which is quite useful for fundamental frequency estimation (59).

Finally, linear prediction can be used as a method of spectrally flattening the signal so that the autocorrelation method of pitch extraction (Sec. 4.4.1) can be applied (21).

6. MISCELLANEOUS ISSUES

6.1 Voiced/Unvoiced/Silence

An important characterization for all ASR systems is the division of the signal into regions of voiced speech, unvoiced speech, and "silence" or nonspeech. Simple classifications based solely on energy and the detection of fundamental frequency are only partially adequate because pitch detection is based on the detection of <u>periodicity</u>; rapid changes in articulation and glottal irregularities produce departures from periodicity without interrupting voicing. Furthermore, the nonspeech intervals may contain sufficient background noise to make energy alone insufficient for separating speech from nonspeech.

More complete V/U/S schemes make use of several parameters, each of which makes some discrimination among these classes. For example, a method described by Atal and Rabiner (60) uses five parameters:

1. Energy
2. Zero-crossing rate
3. The autocorrelation at unit sample delay
4. The 1st linear predictor coefficient
5. The energy of the linear prediction residual

The discrimination abilities of these parameters are combined in a pattern recognition paradigm requiring the use of labeled training data to characterize the three classes. Unknown speech samples are classified by the use of a 5-dimensional distance measure. Possible drawbacks of this method include its sensitivity to changes in the background noise and to changes in the loudness of the speech.

6.2 Change Parameters

In addition to knowing the values (as functions of time) of parameters such as energy, low frequency energy, and format frequencies, it is often useful to know how rapidly they are changing. Although this is an elementary calculation from the parameters themselves, it is sometimes useful to consider change-measures over intervals of 10-30 ms as elementary parameters themselves. A "spectral change" measure that sums the frequency-by-frequency difference between two successive spectra can also be useful.

6.3 Median Smoothing

Parameter tracks (such as formants or fundamental frequency) often contain "obvious" errors due to failures of algorithm heuristics to handle all special cases adequately. These errors include isolated octave or voicing errors in fundamental frequency, missing formant peaks, or spurious peaks identified as formants. Some form of automatic correction for outlying data points is desirable, but such a correction should not distort "true" abrupt transitions, such as voiced/unvoiced.

A technique that has proved useful in these applications is a running median smoothing, optionally followed by "light" linear smoothing (3,61). An L-point median corrects isolated errors of width (L-1)/2, does not smear abrupt transitions separated by (L+1)/2, and will approximately follow low-order polynomial (smooth) trends. A 3- or 5-point median is commonly used.

6.4 Speaker Differences

The speech characteristics of individual speakers differ due to differences in the size, shape, and characteristics of their vocal tracts. These differences affect SSPFE procedures in terms of expected ranges of parameters values (e.g., higher fundamental and formant frequencies for women than for men), and the SSPFE component of an ASR system may also include measurements designed to help subsequent recognition components compensate for speaker differences. These speaker-normalization measurements may include statistics of parameters (e.g., extreme values (62) or mean and standard deviations (63) of formant frequencies) for the speaker in question or a special speaker-normalizing measurement such as estimated vocal tract length (64).

7. CONCLUSION

This chapter has attempted to describe some of the considerations and techniques that go into the design of the speech signal processing and feature extraction component of an automatic speech recognition system. We have seen that the representation of the speech at this level is a multiparameter one, with each parameter portraying some linguistically relevant aspect of the intricately encoded speech signal. We also note that at this level, we are still portraying the signal as a continuously varying entity; we have not yet made the first of the continuous-to-discrete transformations that form the central task of any speech recognition system.

In this chapter, we have not presented the mathematics of the SSPFE techniques; the reader will find that in the references. Neither have we described in detail the _heuristics_ that must be employed in order for SSPFE techniques to display linguistically relevant aspects of the speech signal, but it is worth noting that this is a characteristic of SSPFE techniques such as pitch period estimation, formant tracking, etc.

SSPFE is ultimately a matter of engineering. The theories of speech production and perception help us understand the nature of the speech signal, but SSPFE is always performed in the context of some _system_, and the decisions as to the methods to use, taking into account the advantages and disadvantages of each, in a framework of the allowable complexity, cost, and processing time, are ultimately engineering decisions.

It seems to me that SSPFE can have no ultimate solution at this level. Even if we could process the speech signal so that we knew the excitation and shape of the vocal tract at each instant of time (which is all we could hope to extract from the signal), we still would not be in a position to translate directly into a phonetic or phonemic transcription. This is because, as stated above, we are still in a continuous, encoded domain. The discrete linguistic events that we seek are still farther back in the production process (or if you prefer, further ahead in the perception process).

There was a time when SSPFE methods that could give us a more precise fundamental description of the speech signal (e.g., formants) were needed. Digital signal processing has given us most of what we need in this respect, but at a price of computation time. These techniques are feasible, but not yet cheap enough and fast enough. We should not avoid such "impractical" techniques, for technological advances will soon make these esoteric techniques attainable in practical systems. Instead we must seek to increase our understanding of how the linguistic message is encoded in the speech signal, so that we shall know how to best employ this tremendous computing power to the best advantage in automatically recognizing speech.

REFERENCES

1. J. L. Flanagan, "Speech Analysis, Synthesis and Perception," 2nd ed. (Springer-Verlag, Berlin, Heidelberg, New York, 1972).
2. R. W. Schafer, J. D. Markel, "Speech Analysis" (IEEE Press, New York, 1979).
3. L. R. Rabiner, R. W. Schafer, "Digital Processing of Speech" (Prentice-Hall, Englewood Cliffs, N.J., 1978).
4. A. V. Oppenheim, R. W. Schafer, "Digital Signal Processing" (Prentice-Hall, Englewood Cliffs, N.J., 1975).
5. L. R. Rabiner, B. Gold, "Theory and Application of Digital Signal Processing" (Prentice-Hall, Englewood Cliffs, N.J., 1975).
6. A. Peled and B. Liu, "Digital Signal Processing, Theory, Design, and Implementation" (Wiley, New York, 1976).
7. P. B. Denes, E. N. Pinson, "The Speech Chain" (Anchor Press, Garden City, N.Y., 1973).
8. A. M. Liberman, F. S. Cooper, D. P. Shankweiler, M. Studdert-Kennedy, "Perception of the Speech Code," Psych. Rev., vol. 74, pp. 431-461 (1967); also in E. E. David Jr., P. B. Denes (Eds.), "Human Communication: A Unified View" (McGraw-Hill, New York, 1972), pp. 13-50.
9. N. Lindgren, "Machine Recognition of Human Language - Part II," IEEE Spectrum, vol. 2, No. 4, pp. 44-59 (1965).
10. R. Jakobson, C. G. M. Fant, M. Halle, "Preliminaries to Speech Analysis" (MIT Press, Cambridge, Mass., 1963).
11. G. Fant, "Speech Sounds and Features" (MIT Press, Cambridge, Mass., 1973).

12. N. Chomsky, M. Halle, "The Sound Pattern of English" (Harper and Row, New York, 1968).
13. D. R. Reddy, "Segmentation of Speech Sounds," J. Acoust. Soc. Am., vol. 40, pp. 307-312 (1966).
14. R. J. Niederjohn, "A Mathematical Formulation and Comparison of Zero-Crossing Analysis Techniques which have been Applied to Automatic Speech Recognition," IEEE Trans. Acoust., Speech, Signal Proc., vol. ASSP-23, pp. 373-380 (1975).
15. P. Vicens, "Aspects of Speech Recognition by Computer," Ph.D. dissertation, Stanford Univ., Stanford, Calif. (1969).
16. L. R. Rabiner, M. R. Sambur, "An Algorithm for Determining the Endpoints of Isolated Utterances," Bell System Tech. J., vol. 54, pp. 297-315 (1975).
17. CMU Computer Science Speech Group, "Speech Understanding Systems: Summary of Results of the Five-Year Research Effort at Carnegie-Mellon University," Dept. of Comp. Sci., Carnegie-Mellon Univ., Pittsburgh, Pa., Technical Report (1977).
18. J. M. Baker, "A New Time-Domain Analysis of Human Speech and Other Complex Waveforms," Ph.D. dissertation, Carnegie-Mellon Univ., Pittsburgh, Pa. (1975).
19. L. R. Rabiner, M. J. Cheng, A. E. Rosenberg, C. A. McGonegal, "A Comparative Performance Study of Several Pitch Detection Algorithms," IEEE Trans. Acoust., Speech, Signal Proc., vol. ASSP-24, pp. 399-418 (1976); also in (2).
20. M. M. Sondhi, "New Methods of Pitch Extraction," IEEE Trans. Audio Electroacoust., vol. AU-16, pp. 262-266 (1968); also in (2).
21. J. D. Markel, "The SIFT Algorithm for Fundamental Frequency Extraction," IEEE Trans. Audio Electroacoust., vol. AU-20, pp. 367-377 (1972); also in (2).
22. R. Gillmann, "A Fast Frequency Domain Pitch Algorithm," J. Acoust. Soc. Am., vol. 58, suppl. 1, p. S62 (abstract)(1975).
23. J. J. Dubnowski, R. W. Schafer, L. R. Rabiner, "Real-Time Digital Hardware Pitch Detector," IEEE Trans. Acoust., Speech, Signal Proc., vol. ASSP-24, pp. 2-8 (1976).
24. M. J. Ross, H. L. Schaffer, A. Cohen, R. Freudberg, H. J. Manley, "Average Magnitude Difference Function Pitch Extractor," IEEE Trans. Acoust., Speech, Signal Proc., vol. ASSP-22, pp. 353-362 (1974); also in (2).
25. N. J. Miller, "Pitch Detection by Data Reduction," IEEE Trans. Acoust., Speech, Signal Proc., vol. ASSP-23, pp. 72-79 (1975).
26. B. Gold, L. R. Rabiner, "Parallel Processing Techniques for Estimating Pitch Periods of Speech in the Time Domain," J. Acoust. Soc. Am., vol. 46, pp. 442-448 (1969); also in (2).
27. J. Mártony, "Studies of the Voice Source," Speech Transmission Laboratory, Royal Institute of Technology, Stockholm, QPSR 1/65, pp. 4-9 (1965).

28. C. G. Bell, H. Fujisaki, J. M. Heinz, K. N. Stevens, A. S. House, "Reduction of Speech Spectra by Analysis-by-Synthesis Techniques," J. Acoust. Soc. Am., vol. 33, pp. 1725-1736 (1961); also in (2).
29. L. C. W. Pols, "Real-Time Recognition of Spoken Words, IEEE Trans. Comput., vol. C-20, pp. 972-978 (1971).
30. D. H. Klatt, "A Digital Filter Bank for Spectral Matching," Record of the 1976 IEEE Int. Conf. on Acoust., Speech, Signal Proc., Philadelphia, Pa., pp. 537-540 (1976).
31. E. Zwicker, E. Terhardt, E. Paulus, "Automatic Speech Recognition using Psychoacoustic Models," J. Acoust. Soc. Am., vol. 65, pp. 487-498 (1979).
32. C. L. Searle, J. Z. Jacobson, S. G. Rayment, "Stop Consonant Discrimination Based on Human Audition," J. Acoust. Soc. Am., vol. 65, pp. 799-811 (1979).
33. H. F. Silverman, "An Introduction to Programming the Winograd Fourier Transform Algorithm (WFTA)," IEEE Trans. Acoust., Speech, Signal Proc., Vol ASSP-25, pp. 152-165 (1977); vol. ASSP-26, pp. 268-269 (1978); vol. ASSP-26, p. 483 (1978).
34. A. V. Oppenheim, R. W. Schafer, "Homomorphic Analysis of Speech," IEEE Trans. Audio Electroacoust., vol. AU-16, pp. 221-226 (1968); also in (2).
35. R. W. Schafer, L. R. Rabiner, "System for Automatic Formant Analysis of Voiced Speech," J. Acoust. Soc. Am., vol. 47, pp. 634-648 (1970); also in (2).
36. M. V. Mathews, J. E. Miller, E. E. David Jr., "Pitch Synchronous Analysis of Voiced Sounds," J. Acoust. Soc. Am., vol. 33, pp. 179-186 (1961); also in (2).
37. W. J. Hess, "A Pitch-Synchronous Digital Feature Extraction System for Phonemic Recognition of Speech," IEEE Trans. Acoust., Speech, Signal Proc., vol. ASSP-24, pp. 14-26 (1976).
38. W. Woods, M. Bates, G. Brown, B. Bruce, C. Cook, J. Klovstad, J. Makhoul, B. Nash-Webber, R. Schwartz, J. Wolf, V. Zue, "Speech Understanding Systems: Final Technical Progress Report," Bolt Beranek and Newman Inc., Cambridge, Mass., Report 3438, vol. II (1976).
39. B. S. Atal, S. L. Hanauer, "Speech Analysis and Synthesis by Linear Prediction of the Speech Wave," J. Acoust. Soc. Am., vol. 50, pp. 637-655 (1971); also in (2).
40. J. Makhoul, "Linear Prediction: A Tutorial Review," Proc. IEEE, vol. 63, pp. 561-580 (1975); also in (2).
41. J. D. Markel, A. H. Gray Jr., "Linear Prediction of Speech" (Springer-Verlag, Berlin, Heidelberg, New York, 1976).
42. G. Fant, "Acoustic Theory of Speech Production," 2nd ed. (Mouton, The Hague, 1970).
43. B. S. Atal, M. R. Schroeder, "Linear Prediction Analysis of Speech Based on a Pole-Zero Representation," J. Acoust Soc. Am., vol. 64, pp. 1310-1318 (1978).

44. J. Makhoul, J. Wolf, "Linear Prediction and the Spectral Analysis of Speech," Bolt Beranek and Newman Inc., Cambridge, Mass., Report 2304 (1972).
45. J. Makhoul, "Spectral Linear Prediction: Properties and Applications," IEEE Trans. Acoust., Speech, Signal Proc., vol. ASSP-23, pp. 283-296 (1975).
46. J. Makhoul, J. Wolf, "The Use of a Two-Pole Linear Prediction Model in Speech Recognition," Bolt Beranek and Newman Inc., Cambridge, Mass., Report 2357 (1973).
47. M. R. Sambur, L. R. Rabiner, "A Speaker Independent Digit Recognition System," Bell System Tech. J., vol. 54, pp. 81-102 (1975).
48. F. Itakura, "Minimum Prediction Residual Principle Applied to Speech Recognition," IEEE Trans. Acoust., Speech, Signal Proc., vol. ASSP-23, pp. 67-72 (1975).
49. A. H. Gray Jr., J. D. Markel, "Distance Measures for Speech Processing," IEEE Trans. Acoust., Speech, Signal Proc., vol. ASSP-24, pp. 380-391 (1976).
50. J. L. Flanagan, "Automatic Extraction of Formant Frequencies from Continuous Speech," J. Acoust. Soc. Am., vol. 28, pp. 110-118 (1956); also in (2).
51. J. D. Markel, "Application of a Digital Inverse Filter for Automatic Formant and F0 Analysis," IEEE Trans. Audio Electroacoust., vol. AU-21, pp. 149-153 (1973).
52. S. S. McCandless, "An Algorithm for Automatic Formant Extraction using Linear Prediction Spectra," IEEE Trans. Acoust., Speech, Signal Proc., vol. ASSP-22, pp. 135-141 (1974); also in (2).
53. S. Seneff, "Modifications to Formant Tracking Algorithm," IEEE Trans. Acoust., Speech, Signal Proc., vol. ASSP-24, pp. 192-193 (1976); also in (2).
54. R. DeMori, P. Laface, E. Piccolo, "Automatic Detection and Description of Syllabic Features in Continuous Speech," IEEE Trans. Acoust, Speech, Signal Proc., vol. ASSP-24, pp. 365-379 (1976).
55. H. Wakita, "Direct Estimation of the Vocal Tract Shape by Inverse Filtering," IEEE Trans. Audio Electroacoust., vol. AU-21, pp. 417-427 (1973); also in (2).
56. H. Wakita, "Estimation of Vocal-Tract Shapes form Acoustical Analysis of the Speech Wave: The State of the Art," IEEE Trans. Acoust., Speech, Signal Proc., vol. ASSP-27, pp. 281-285 (1979).
57. S. Seneff, "Real-Time Harmonic Pitch Detector," IEEE Trans. Acoust., Speech, Signal Proc., vol. ASSP-26, pp. 358-365 (1978).
58. M. R. Schroeder, "Period Histogram and Product Spectrum: New Methods for Fundamental Frequency Measurements," J. Acoust. Soc. Am., vol. 43, pp. 829-834 (1968).
59. A. M. Noll, "Cepstrum Pitch Determination," J. Acoust. Soc. Am., vol. 41, pp. 293-309 (1967); also in (2).

60. B. S. Atal, L. R. Rabiner, "A Pattern Recognition Approach to Voiced-Unvoiced-Silence Classification with Applications to Speech Recognition," IEEE Trans. Acoust., Speech, Signal Proc., vol. ASSP-24, pp. 201-212 (1976).
61. L. R. Rabiner, M. R. Sambur, C. E. Schmidt, "Applications of a Nonlinear Smoothing Algorithm to Speech Processing," IEEE Trans. Acoust., Speech, Signal Proc., vol. ASSP-23, pp. 552-557 (1975).
62. L. J. Gerstman, "Classification of Self-Normalized Vowels," IEEE Trans. Audio Electroacoust., vol. AU-16, pp. 73-77 (1968).
63. I. Kameny, "Automatic Acoustic-Phonetic Analysis of Vowels and Sonorants," Conference Record, 1976 IEEE Int. Conf. Acoust., Speech, Signal Proc., Philadelphia, Pa., pp. 166-169 (1976).
64. H. Wakita, "Normalization of Vowels by Vocal Tract Length and its Application to Vowel Identification," IEEE Trans. Acoust., Speech, Signal Proc., vol. ASSP-25, pp. 183-192 (1977).

PATTERN RECOGNITION TECHNIQUES FOR SPEECH RECOGNITION

J. S. Bridle

Joint Speech Research Unit
Princess Elizabeth Way
Cheltenham, Gloucestershire
England. GL52 5AJ

Abstract: This is an overview of techniques which have been developed for automatic pattern recognition, with an indication of their relevance to automatic speech recognition. The first part is concerned with data transformations, distance measures, cluster analysis and other aspects of what could be called 'classic' mathematical pattern recognition. The second part is more directly concerned with speech, and the term 'pattern recognition' is used to denote an approach to speech recognition which tries to avoid the problems of using a phoneme level of description and treats larger units such as words as patterns with a time axis.

1. INTRODUCTION

Pattern recognition is concerned primarily with the description and classification of sets of measurements taken from physical processes. Many techniques have been described. Some of them are solutions to rather abstractly posed classification problems, some are motivated primarily by simplicity of implementation and some are rooted in particular applications areas.

It is useful to classify application areas into those in which an image is naturally present, (e.g. recognising letters of the alphabet on paper) and those where the raw data really is just a set of measurements for each object to be recognised (e.g. concentrations of various chemicals in a sample). In

J. C. Simon (ed.), Spoken Language Generation and Understanding, 129-145.

speech recognition we are in an intermediate situation: The raw data is not in the form of pictures, but there are standard ways of displaying speech signals, of which the most familiar is the spectrogram.

The pattern recognition techniques described in sections 2. and 3. assume that we are dealing not with a continuous pattern like speech but with a set of multidimensional measurements. In places the techniques are illustrated by applying them to a very simple speech recognition problem: spectrum cross-sections at 10ms interval are the data points, and the task is to 'decode' the speech into a sequence of 'sound class symbols', each determined by a single spectrum shape. This example should not be taken too seriously.

2. STATISTICAL PATTERN CLASSIFICATION

In this section we assume that some 'feature extraction' process has reduced each unknown object to a set of numbers, a 'feature vector'. In a speech sound class recognition example the objects to be recognised might be spectrum cross sections, and the features could be the amount of energy in various frequency bands.

A large number of pattern classification schemes can be analysed in terms of a set of 'discriminant functions', $g_i(x)$, one for each class (4). The classification rule is to choose the class with the largest (or smallest) value of its discriminant function for the unknown input, x. We classify x as from the i'th class, c_i, if

$$g_i(x) = \max_i \; g_i(x)$$

The effect of the discriminant functions is to carve up the feature space into regions corresponding to different classifications. The boundaries between these regions, where two discriminant functions have equal values, are called discriminant surfaces. It is sometimes useful to think of pattern classification methods in terms of these regions or surfaces, indeed some methods manipulate the surfaces explicitly.

In the following paragraphs we assume that we have several correctly labelled 'training samples' and we examine the implications of various assumptions about the underlying distributions of sample points in feature space. This can lead to a particularly simple type of discriminant function in the log likelihood classifier. In certain cases this reduces again to comparing distances from an unknown point in sample space to the centroids of the distributions.

2.1 Complete knowledge of distributions for each class.

If we had the likelihood, $P(ci|x)$, of each pattern class, ci, given any observation vector, x, then the best guess for the class which generated any particular observation is simply the class which is most likely, given that observation. We classify x as from class ci if

$$P(ci|x) = \max_j P(cj|x)$$

In practice we cannot directly measure $P(cj|x)$, but using Bayes' Theorem we know that

$$P(cj|x) = P(x|cj)\, P(cj) \,/\, P(x)$$

and so we can classify x as coming from class ci if

$$P(ci)\, P(x|ci) = \max_j P(cj)\, P(x|cj)$$

If we do not know or do not want to be affected by the a-priori frequencies of occurrence of the classes this simplifies to

$$P(x|ci) = \max_j P(x|cj)$$

so

$$gi(x) = P(x|ci)$$

The problem then reduces to generating an appropriate distribution for the feature vector x for each class.

2.2 Distributions with unknown parameters.

This means that we are prepared to write down expressions for the likelihood (or probability density) of any spectrum shape, x, for each pattern class, ci, in terms of some unknown parameters. A simple example is a multivariate normal distribution for each class, with the same variance for all classes and all components of the spectrum. In that case the only significant parameters remaining to be determined would be the class means, which could be estimated even from only one example of each class.

Once we have estimated the parameters using real examples we proceed as indicated above.

2.3 General multivariate normal distributions.

If the distribution for each class is a general multivariate normal (Gaussian) distribution the parameters are the class means (or centroids), mi, and the covariance matrix for each class, Ci. In the one dimensional case each class mean is a scalar, and the covariance matrix reduces to the variance. The distribution is then

$$P(x|mi,si) = Pi(x) = \frac{1}{\sqrt{2\pi si^2}} \exp\left(-\frac{(x-mi)^2}{2si^2}\right)$$

It simplifies the algebra if we deal with the natural logarithm of the likelihood - since the logarithm is monotonic this will not affect the decision.

In one dimension this gives

$$gi(x) = -2 \text{ Ln } Pi(x) = \frac{(x-mi)^2}{si^2} + \text{Ln } 2\pi si^2$$

Note that we need the class with the minimum value of this log likelihood discriminant function.

For the multidimensional normal distribution we have

$$gi(x)=-2\text{Ln } Pi(x)=(x-mi)^T Ci^{-1} (x-mi)+\text{Ln}|Ci|+n\text{Ln}2\pi$$

For a problem of fixed dimensionality the third term can be dropped. The first term is a generalised distance involving the covariance matrix. It is called the Mahalanobis distance (4). Another way to think of the first term is as a Euclidean distance in a linearly transformed feature space in which the distribution is isotropic. This viewpoint is expanded on in the next section.

When there is both good reason to expect the covariance

matrices for different classes to be different and also enough training data to estimate them this is the appropriate method. It was used successfully by Atal and Rabiner (1) to distinguish between silence, voiced and unvoiced speech sounds in the first part of a connected digit recognition scheme. This paper can be recommended as an example of the use of several statistical pattern recognition methods in a speech recognition application.

2.4 Multivariate normal with equal covariances.

If we can assume that the covariance matrices for the different classes are the same then a considerable simplification results. The second term in the above equation is the same for all classes, so it can be dropped. We are left with

$$gi(x) = (x-mi)^T C^{-1} (x-mi)$$

Where C, the common covariance matrix, can be estimated from all the training data, after subtracting off the appropriate class means. We can factorise C into the product of its eigenvector matrix, R, and a diagonal matrix, D, of its eigenvalues

$$C=RDR^T$$

If we introduce a scaling matrix S whose diagonal elements are the inverse of the square roots of the elements of D then we can express C^{-1} as

$$C \quad = RD^{-1} R \quad = RSSR^T$$

Then

$$gi(x) = (x-mi)^T RSISR^T (x-mi)$$

$$= (SR^T (x-mi))^T (SR^T (x-mi))$$

$$= (x'-mi')^2$$

where $x'=SR^T x$ and $mi'=SR^T mi$

We have expressed gi(x) in terms of the squared Euclidean distance between a linear transformation of the data and the similarly transformed class means. This result is important practically, because it shows that a feature vector need be transformed only once before being compared with the reference 'patterns', mi'. It is also important conceptually. The effect of the rotation using R is to make the principal axes of the distributions line up with the coordinate axes. The diagonal matrix scales these axes in proportion to the standard deviation in each direction, to make each within-class distribution isotropic (spherical). The complete transformation defines a new space in which distances to class centroids can be interpreted in terms of log likelihoods. This property will be important when we consider the problem of recognising complete spoken words in terms of sequences of spectrum shapes.

Even if we believe that the covariances of the classes are different it may be worth using this method if we do not have enough data to estimate the separate covariances well. Another reason is because it involves only one linear transformation of the data instead of a separate transformation for each class. Linear transformations are rather popular, and some of the reasons are their mathematical tractability, the way they can be justified by assumptions about normal distributions, and their ease of implementation.

Whatever the form of the distributions, the above methods will define a linear transformation which makes the covariance spherically symmetrical. This will tend to turn any unimodal distributions into shapes for which the Euclidean distance makes more sense.

The linear transformation above is the first part of the method of Linear Discriminant Analysis (LDA). The second stage of LDA is a Principal Components Analysis of the transformed class centroids. This, together with the first, within-class normalisation stage, provides an ordered set of linear components of the original feature space, ordered according to their power to discriminate between classes. LDA has been used, for instance, in the problem of selecting and combining statistical features for automatic text-independent speaker recognition (8).

2.5 Multivariate normal with diagonal covariance matrices.

If we assume that for each class the variation in the features is independent, then the covariance matrices will be diagonal. It is not then neccessary to apply a rotation, and the feature axes can be scaled separately to make the variance in all directions equal, allowing squared Euclidean distance to be used as a discriminant function.

Note that if we have insufficient data to estimate a

separate covariance matrix for each class we may need to choose between assuming one common covariance matrix or separate but diagonal covariance matrices for the different classes.

2.6 Isotropic multivariate normal

The simplest case is independence and equal variance of the features. The squared Euclidean distance can be used directly, with a scale factor if the result is to be combined with other log likelihood type measures. We still have the option of assuming different variances for different classes, and this might be appropriate, for instance, if the different classes were different parts of the same word, with more variability expected in some parts than others.

2.7 Adaptive discriminant surface methods

Some pattern classification methods do not rely for their derivation on assumptions about the form of the distributions of sample points in feature space. It is possible to manipulate the discriminant surfaces directly, and there are several methods which adjust the position of (usually hyperplane) surfaces as training samples are presented. Since hyperplane discriminant surfaces appear in all the parametric cases above which involve a Euclidean distance with or without a single linear transformation of the feature space, this restriction is not too serious. Unfortunately almost all the basic adaptive hyperplane methods are designed for problems with only two classes, and extension to many classes is awkward.

2.8 Nearest neighbour methods

A second class of non-parametric pattern classification methods is nearest neighbour methods. The idea is to base the classification decision on the distance between an unknown pattern and a set of the actual training samples. The unknown is classified according to the class of the nearest labelled pattern (its nearest neighbour) or, in more sophisticated versions to the class to which most of its nearest neighbours belong (N-nearest neighbour methods).

When we have only a few training samples for each class this is the obvious thing to do, particularly if there is no obvious way of averaging the training patterns. Nearest neighbour methods have come into favour following the proof that they come close to the theoretical optimum results for arbitrary underlying distributions in the limit as the number of training samples approaches infinity (3). This result seems to have been used to justify nearest neighbour methods in practical cases when the assymtotic result is irrelevant.

3. DEALING WITH UNLABELLED TRAINING DATA

Suppose we start with a large amount of unlabelled speech data and wish to represent it compactly in a way that makes it suitable for speech recognition. Since we have no information about the classes we have to assume that the original analysis method has been chosen well, so it retains the relevant differences between different sounds while minimising the unimportant differences between sounds that we want to regard as similar. Two approaches are described below. In Principal Components Analysis we find a linear transformation that discards correlations between the original features and disturbs the position of the points as little as possible. Cluster analysis methods find a description of the data in terms of a set of clusters.

3.1 Principal Components Analysis

Assume we have a set of samples which are in the form of N-dimensional vectors. We wish to find a linear transformation to a space with less dimensions, say M, which will preserve the relative positions of the points as much as possible. The most obvious method is to discard those components of the original vectors which vary least over the data set. There will be correlations between these components, and if, as an extreme case, two components always had the same value, then we could throw one away with no loss of information. For the general case, however, the correct procedure is to rotate the distribution of points so that there is no linear correlation between the new axes, then discard the axes along which there is least variation. The sample covariance matrix contains all the relevant information about the distribution of the original data. We need to find the eigenvectors and corresponding eigenvalues of the covariance matrix, for which there are standard methods. The 'best' linear transformation to M dimensions is then given by a matrix of the eigenvectors of the covariance matrix which have the M largest eigenvalues. This transformation rotates the cloud of data points so that its principal axes lie along the new coordinate axes. The individual eigenvalues are the variance of the data along the new coordinates. As a simple example, if one eigenvector were much larger than the rest this would indicate that the variability in the data could be accounted for by one new feature only. This may not be obvious in the original N-dimensional space.

To measure the effect on the data we can transform from the reduced space back into the original space, where the new positions will lie in hyperplanes perpendicular to the discarded axes. Principal Components Analysis minimises the average squared distance between original points and their new positions. The optimality of this transformation in preserving the original information does not depend on any assumptions about the form of underlying probability density functions, or the number of training samples, but such things could well affect the usefullness of the transformation, especially on new data. Principal Components Analysis is also known as the Karhunen-Loeve transformation.

When PCA is applied to filter bank analysis of speech it is usually found that the first two or three components account for almost all the variability in the data, implying a large amount of correlation between the original components, so it achieves the objective of data reduction. What is more, the new space seems to have other interesting properties. Pols and his colleagues have shown strong relationships with formant descriptions and with perceptual similarity spaces for vowels (12), and have explored possibilities for speaker normalisation and identification of consonants from vowel-consonant transitions.

3.2 Cluster analysis

There are many automatic cluster analysis methods, and two basic types are descibed below.

Many cluster analysis algorithms work as follows. There is a set of clusters found so far, described in ways that are particular to the algorithm. Another data point is considered,and if it seems to belong to one of the clusters it goes to modify the parameters of that cluster. If it does not seem to fit then a new cluster is started for it. Depending on the particular algorithm, this process can continue for several iterations over the data, perhaps splitting or merging clusters, until a stable state is reached. It should be clear that in general the final state, and thus the way the data points are allocated to the clusters, will depend on the starting conditions and the order in which the data points are presented.

A different approach is to start by specifying a global criterion of a good clustering of any set of points, and find the best clustering. In this way the particular algorithm used to identify clusters should not affect the final answers. The problem now becomes to devise cluster criteria which lead to satifactory clusters and for which reasonably efficient algorithms can be found. An example of a computationally efficient method which is often used is the Single Link method of Jardine and Sibson (10,16). In this method a cluster is defined as a set of points all of which are within some critical distance of at least one other member of the cluster. There are efficient algorithms for determining in one operation all the partitionings of the data which result from this rule as the critical distance is varied, and this information can be displayed conveniently in a tree structure to help the investigator decide on the most appropriate critical distance. The biggest disadvantage of the single link method is that the clusters are not always meaningful. There are several related methods, with names such as N-link and hierarchical clustering. Good results have been reported recently (15) using an N-link cluster method to choose several templates to represent each word in talker-independent isolated word recognition.

Data displays, which are extremely useful in speech analysis and pattern recognition, are particularly important in cluster analysis. Before resorting to automatic cluster analysis methods for speech recognition it would probably be worthwhile to examine the data visually, perhaps two dimensions as specified by Principal Components Analysis. If we use a speech spectrum description in terms of clusters, then we ought to retain information about the relative position of the clusters.

4. RECOGNISING WORDS AS PATTERNS

The basic assumption of the whole word pattern matching approach is that different utterances of the same word by a particular talker result in similar patterns of sound. There will be variation in spectrum shape at corresponding parts of the patterns from the same word, and to deal with this the methods of section 2. apply. There will also be variations in the time scale of the patterns, and this will make it difficult to compare corresponding parts. Some of the timescale variation is no doubt subject to rules, but in most attempts at whole word matching which try to allow flexible timescales this variation is treated as if it were random.

Although it it inherently limited to fairly small vocabularies and it is difficult to accept arbitrary speakers, whole word pattern matching is the basis of the only commercially available speech recognition machines, so it needs to be taken seriously. At least we ought to know what the limitations are. Some of the techniques used in whole word pattern matching are closely related to techniques which have been used successfully in connected speech 'understanding' systems. For instance, the word verification component of HWIM is a whole word pattern matcher using a synthetic template and Harpy is based on a finite state machine model.

All but one of the commercial speech recognition machines are restricted to isolated word recognition: each word to be recognised must start and finish in silence, and the user must pause between words. In this case the usual way of dealing with the timescale variation problem is to detect the beginning and end of the word and re-sample the pattern to make it a standard length. This works well for short words, but can fail with longer, multi-syllabic words (19). Pols (14) describes an isolated word recognition system with linear time normalisation in which the spectrum matching is motivated by a log likelihood argument.

The rest of this section is concerned with approaches to whole word pattern matching which try to deal with timescale variation. We look at elastic timescale templates, time registration paths, finite state machines and spelling correction. All four approaches lead to the use of Dynamic Programming to solve an optimisation problem. The Dynamic Programming method is illustrated for the spelling correction case.

4.1 Elastic templates.

One way of thinking about time-flexible matching is to imagine the reference pattern (template) drawn on a kind of rubber sheet or on the folds of an accordion. We compare an unknown pattern with the original template and with many distorted versions, remembering only the best match. It is natural to include the amount of distortion of the timescale in our evaluation of the goodness of fit of any particular distorted version of the template with the unknown. Fishler and Elshlager (5) present such a method in general form with an efficient algorithm, illustrated with a two dimensional image recognition problem.

4.2 Time registration paths

A second view is of finding a mapping between the time axis of the template and that of an unknown pattern. This relationship can be drawn as a path in the plane whose axes are the time axes of the two patterns. Any particular 'time registration path' specifies a complete correspondence between the various parts of the two patterns. Given a path we can compute a 'distance' between the the two complete patterns. This pattern distance should include the distance (in feature space) between the corresponding parts, and perhaps a measure of the deviation of the path from linearity. Again we are interested only in the best match over all acceptable time registration paths.

It is usual for the two time axes to have been quantised by regular sampling of the features, and for the time registration paths to be restricted to be composed of segments from a limited set such as short vertical, horizontal and diagonal lines.

The time registration path idea was first used for speech recognition by Velichko and Zagoruyko (18). Itakura (9), used a somewhat different method, and Sakoe and Chiba (17) have compared several methods experimentally.

4.3 Finite state machines

A third viewpoint is obtained if each production of a word pattern is treated as if it were the result of operating a finite state machine, possibly a probabalistic finite state machine, one machine for each word. Since finite state machines are formally equivalent to finite state grammars, this approach can also be regarded as syntactic pattern recognition.

The 'word machine' is supposed to go though a sequence of

its possible states, making transitions at regular intervals. For each state transition it produces an output which is an observable item such as a spectrum shape. The actual output is drawn from a distribution whose parameters are specific to the current state. A complete specification of a word machine, in terms of all its state transition probabilities and the parameters of the output distributions for each state, is the equivalent of a template. Given such a specification and a particular observed sequence of 'outputs' we can compute the likelihood that any particular sequence of states might have occured and produced the output. This is simply the product (or the sum of logs) of the likelihoods of the supposed state transitions and the likelihoods of the outputs being produced by the states associated with each output.

We are actually after the likelihood of the most likely sequence of states. We compute this for every word machine and decide on the most likely word.

Bakis (2) used this formulation for the problem of recognising occurances of a given word in connected speech. Each observations represented 10ms of speech, and they were derived from a log power spectrum via a transformation designed by Principal Components Analysis. The output distributions for each state were modelled as multivariate normal with a diagonal covariance matrix. The states were ordered, and based on the successive 10ms observations of a prototype utterance of the word. There were three possible transitions from each state: to the next state, to the same state, and directly to the next but one. Bakis used the Viterbi Algorithm (which is a Dynamic Programming algorithm) to compute the most likely sequence of states. He used the recognised words, with the mapping on to states, to refine the parameters of the word model.

4.4 Spelling correction

A fourth approach starts by assuming that we have reduced the speech to a string of symbols. Each vocabulary word is also represented as a string of these symbols. In general the strings obtained from utterances of unknown words to be recognised will not be the same as any of the stored template strings, and we need a method of defining and then computing a measure of difference between two strings of symbols. The Levenshtein Distance or edit distance between two strings is the minimum number of insertions, deletions and substitutions needed to change one string into the other.

We shall illustrate the use of Dynamic Programming for computing the simple Levenshtein Distance, and hope that the generalisations and the application to other formulations of time-flexible pattern matching will be evident.

Let the two strings, A and B, have symbols A(1),A(2)...A(M) and B(1),B(2)...B(N). Define D(i,j) as the Levenshtein Distance between the first i symbols of A and the first j symbols of B. We require D(M,N). D(0,0) can be defined as 0.

D(i,j) can be expressed in terms of distances between shorter strings.

$$
\begin{aligned}
D(i,j) = \text{Minimum}[\quad & D(i-1,j)+1, \\
& D(i,j-1)+1, \\
& D(i-1,j-1)+0 \text{ if } A(i)=B(j) \\
\text{or } & D(i-1,j-1)+1 \text{ if } A(i)\neq B(j) \\
]&
\end{aligned}
$$

The effect of this formula can be illustrated with a simple example. Suppose our two strings are

A = ..fsi.ss.
and B = .si.kss.

We can arrange the two strings with i and j along two axes, and fill in the symbol substitution 'distances'.

		1	2	3	4	5	6	7	8	9
		.	.	f	s	i	.	s	s	.
1	.	0	0	1	1	1	0	1	1	0
2	s	1	1	1	0	1	1	0	0	1
3	i	1	1	1	1	0	1	1	1	1
4	.	0	0	1	1	1	0	1	1	0
5	k	1	1	1	1	1	1	1	1	1
6	s	1	1	1	0	1	1	0	0	1
7	s	1	1	1	0	1	1	0	0	1
8	.	0	0	1	1	1	0	1	1	0

We can compute D starting at position (0,0) and working towards (M,N) along rows or columns. The decisions can be recorded if we wish to find out the actual edit sequence(s) which give the minimum distance. Note that the coordinates for D are offset from those of the symbol numbers for the strings.

```
      1 2 3 4 5 6 7 8 9
      . . f s i . s s .
    0-1-2-3-4-5-6-7-8-9
1 . |\ \       \     \
    1 0-1-2-3-4-5-6-7-8
2 s | |\ \ \     \ \
    2 1 1-2 2-3-4-5-6-7
3 i | |\|\ \|\
    3 2 2 2 3 2-3-4-5-6
4 . | |\|\|\  |\     \
    4 3 3 3 3 3 2-3-4-5
5 k | |\|\|\|\  |\ \ \
    5 4 4 4 4 4 3 3-4-5
6 s | |\|\|\ \| |\ \
    6 5 5 5 4-5 4 3 3-4
7 s | | | |\|\  |\|\ \
    7 7 7 7 5 5 5 4 3-4
8 . |\ \ \| |\|\   | |\
    8 7 7 7 6 6 5 5 4 3
```

The Levenshtein Distance for these two strings is 3, and it is obtained by deleting one of the initial '.' symbols in A or inserting one at the start of B, deleting the 'f' from A or inserting it in B, and inserting a 'k' in A or deleting the one in B.

A Weighted Levenshtein Distance can be defined by associating different weights to the three edit operations (13) and this could be generalised to include different weights for different pairs of symbols. If we do away with symbols and store and compare spectrum shapes directly, then the replacement weights can be derived from spectrum distances, and we are close to the time-registration path formulation.

5.0 CONCLUSIONS

The above notes will not tell you how to build a speech recognition machine! They may provide some background concepts relevant to the design of some parts of a speech recognition system. Two of the most important areas left out are the question of the evaluation and selection of sets of features, and the related question of the number of parameters that can usefully be included in the design of a pattern recognition system if the amount of training data is limited. (There is a limit to the amount of information that can be extracted from a finite amount of data). Kanal (11) addressed these subjects, and there have been several other treatments since.

Real, naturally-occurring signals such as speech cannot be expected to actually fit any assumptions or models we invent. That should not stop us from first choosing methods of analysis in which our assumptions and criteria are explicit, and then inserting as many as possible of good assumptions based on experience.

6. ACKNOWLEDGEMENTS

I am grateful for illuminating discussions on these topics with colleagues, particularly Dr. M.J. Irwin and Dr. M.J. Hunt.

REFERENCES

(1) B.S. Atal and L.R. Rabiner, "A pattern-recognition approach to voiced-voiceless-silence classification with applications to speech recognition", IEEE Trans. Acoustics,Speech and Signal Processing, ASSP-24, No.3, June 1976, pp. 201-212

(2) R. Bakis, "Spoken word spotting via centisecond states", IBM Technical Disclosure Bulletin Vol.18 No.10, March 1976. pp. 3479-3481

(3) T.M. Cover and P. Hart, "The nearest neighbor decision rule", IEEE Trans. Information Theory, Vol.IT-13, pp.21-27, Jan.1967.

(4) R.O. Duda and P.E. Hart, "Pattern classification and scene analysis", New York: Wiley, 1973.

(5) M.A. Fishler and R.A.E. Elshlager, "The representation and matching of pictorial structures", IEEE Trans. Computers, Vol.C-22, No.1, pp. 67-92.

(6) G.D. Forney, "The Viterbi Algorithm", Proc. IEEE Vol.61 No.3, March 1973, pp.268-278.

(7) Y. Ho and A.K. Agrawala, "On pattern classification algorithms- introduction and survey", IEEE Trans. Automatic control, Vol.ac-13, No.6,Dec.1968, pp.676-690.

(8)M.J. Hunt, J.W. Yates and J.S. Bridle, "Automatic speaker recognition for use over communications channels." Proc. IEEE Int. Conf. on Acoustics, Speech and Signal Processing, Hartford, CT, May 1977.

(9) F.Itakura, "Minimum residual principle applied to speech recognition", IEEE Trans. Acoustics,speech and Signal Processing, Vol.ASSP-23, pp.67-72, Feb.1975.

(10) N. Jardine, "A new approach to pattern recognition", Nature, Vol.234, Dec.31, 1971 pp.526-528.

(11) L. Kanal, "Patterns in pattern recognition:1968-1974", IEEE Trans. Inform.Th., Vol.IT-20, No.6, November 1974. pp.697-722

(12) W.Klein,R.Plomp and L.C.W.Pols, "Vowel spectra,vowel spaces and vowel identification",J.Acoust.Soc.Am. 48, 999, 1970.

(13) T. Okuda, E. Tanaka and T. Kasai, "A method for the correction of garbled words based on the Levenshtein metric", IEEE Trans. computers,,Vol. C-25, No. 2, Feb. 1976, pp 172..

(14) L.C.W. Pols, "Real-time recognition of spoken words", IEEE Trans. Computers, Vol.20,No.9, pp. 972-978 1971.

(15) L.R. Rabiner, S.E. Levinson and A.E. Rosenberg, "Speaker independent recognition of isolated words using clustering techniques", J.Acoust.Soc.Am.,Vol.64, Suppl.No.1, Fall 1978, page S181, (abstract).

(16) R.Sibson, "SLINK: an optimally efficient algorithm for the single-link cluster method", Computer J., Vol.16, No.1, pp.30-34

(17) H. Sakoe and S. Chiba, "Dynamic Programming Algorithm Optimisation for Spoken Word Recognition", IEEE Trans. Acoustics,Speech and Signal Processing, Vol. ASSP-26, No.1, February 1978, pp.43-49.

(18) V.M. Velichko and N.G. Zagoruyko, "Automatic recognition of 200 words", Int.J.Man-Machine Studies, 1970,No 2, pp.223-234.

(19) G. White and R. Neely, "Speech recognition experiments using linear prediction,bandpass filtering and dynamic programming", IEEE trans. Acoustics,Speech, and Signal Processing, Vol.ASSP-24, No.2, April 1976. pp.183-188.

AUTOMATIC MAPPING OF ACOUSTIC FEATURES INTO PHONEMIC LABELS

Toshiyuki Sakai

Department of Information Science, Kyoto University

Abstract

This paper describes significant problems for automatic mapping of acoustic features into phonemic labels of the phonemic(phonetic) block in an automatic speech recognition. These problems are feature parameter, segmentation, labeling, co-articulation and speaker differences. We also discuss some general approaches for language independent problems, especially, pattern matching techniques for labeling, and describe our approach method in the LITHAN speech understanding system. Since these depend on each other, lastly, we emphasize that a system should have adaptive functions for various factors which bring out the varieties of speech.

I. INTRODUCTION

This paper describes significant problems for automatic mapping of acoustic features into phonemic labels of the phonemic (phonetic) block in an automatic speech recognition system. This block is closely related to other blocks, in particular, feature extraction and word identification(verification). Although the direct transformation from acoustic features into a word is possible in a limited spoken word recognizer without transmission loss of information caused by passing the intermediate phonetic, phonemic level in processing, the phonemic block is indispensable for a general speech understanding system with a large vocabulary set. The reason is as follows:

(1) the set of phonemes or syllables is limited in number for any set of vocabulary, therefore the rules or reference patterns used in acoustic phonemic transformation are almost independent of this kind of 'task'.

J. C. Simon (ed.), Spoken Language Generation and Understanding, 147-189.

(2) the linguistic information such as phonological rule, word juncture rule can be embedded in the phoneme sequences, from which the speaker or context independent word dictionary can be deduced.

The second assertion (2) is that we can utter any words and sentences from a limited set of phonemes [assertion (1)] and rules.

The performance of a speech understanding system in a certain task depends much on that of the word identification block. Since the latter performance is closely related to that of the phonemic block, this block may be said to be important to the system performance. Speech analysis and feature extraction level(acoustic level) have no fatal problems, because the performance has been carried out by a speech synthesis and its perception. Thus, concentrating on the phoneme block suggests the big range and depth of problems in this block. The significant items are segmentation, labeling, co-articulation and speaker differences.

Fig.1 illustrates the general schema of the acoustic-phonemic block. Firstly, input speech waves are transformed into a sequence of feature parameters in order to reduce the amount of data and to represent it by physical properties corresponding to each phoneme. Until now, many feature parameters have been extracted from speech for speech recognition (see section II). The sequence is further transformed into a sequence of symbols corresponding to linguistic units, or phonemes, although such a transformation may be unnecessary for limited spoken word recognition (see section 3.4). The most fundamental functions for this block are (1) SEGMENTATION: the interpretation of the continuously varying acoustic parameters as a sequence of discrete units which is the next higher level called linguistic structures (see section 3.1), and (2) LABELING: the characterizing operation considering the relation of phonetic equivalence between segments (see section 3.2)[1]. The results usually have errors or plural candidates (see section 3.3). This ambiguity derives from the variety of speech caused by co-articulation (see section IV) and speaker differences (see section V). The feature extraction block is said to be language-independent. On the other hand, the phonemic block is dependent on language: for example, set of phonemes (reference pattern in the Fig.1), phonological rules, or tonal/nontonal language (prosodic analysis). In this paper, we discuss some general approaches for these language independent problems (segmentaion, labeling and so on) and describe our approach method adopting the LITHAN Speech Understanding System [2-6].

II. FEATURE PARAMETERS AND THEIR PERFORMANCE

In pattern recognition, feature parameter extraction is the

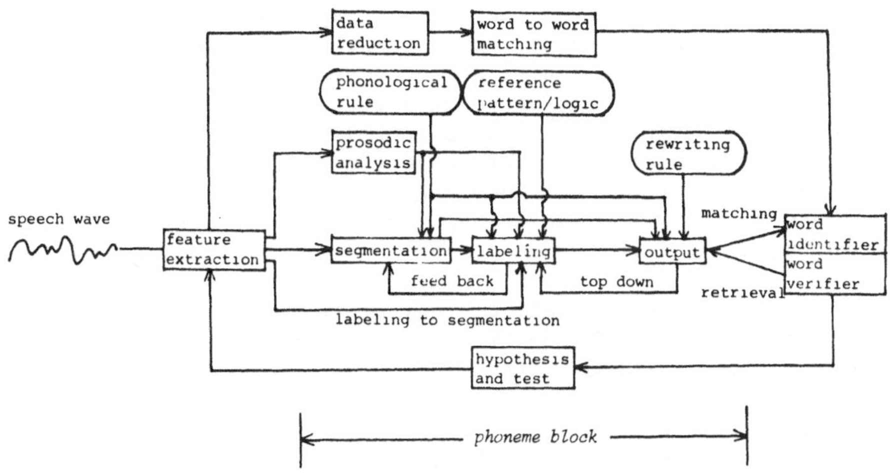

Fig. 1 Acoustic - phonemic block

Table 1 Feature parameters and their characteristics

feature	accuracy of extraction	dimension	contribution to recognition	contribution to decoding	cost & time (software)	usefulness for modeling	refer
zero-crossing	excellent	5~10	fair	poor	excellent		[illegible]
Walsh-Hadamard transform	excellent	>20	fair	fair	excellent		[illegible]
spectrum (FFT)	excellent	>20	fair	excellent	good		[illegible]
autocorrelation function	excellent	5~20	good	excellent	good		[illegible]
linear predictive coefficient	excellent	8~12	poor	excellent	good		
parcor	excellent	8~12	excellent	excellent	good		
cepstrum	excellent	8~12	excellent	excellent	fair		
smoothed power spectrum	excellent	>20	excellent	excellent	poor	good	
formant	good	3	excellent	good	poor	excellent	
vocal tract area function	fair	8~12	excellent	excellent	good poor	excellent	[illegible] [24]
position of lip, tongue, jaw	poor	3~6	good	fair	poor	excellent	[25]

most significant problem. Fortunately, the feature parameter of speech waves can be extracted almost completely by data reduction as described above. However, some flexible feature parameters having to do with phoneme recognition cannot yet be extracted completely; they are formant and anti-formant frequencies, positions of articulation, vocal tract length, area functions and so on. And also, even if the sets of parameters have the same information together, it is desirable that each component of the parameters be independent from one another from the standpoint of pattern recognition (ex. factor analysis, LPC versus parcor).

Recently, the evaluation study of features has been developed for speaker identification [7-14]. However there are few such investigations for phoneme recognition.

L.C.W.Pols et al. analyzed speech spectra obtained by an 18-channel 1/3 octave filter bank, and found that the first three formant frequencies were closely related to the three components obtained from the factor-analysis of 18 dimensions [15,16]. These results suggest that short time spectra obtained from a filter bank are superior to formant frequencies in phoneme recognition performance, because of the accuracy of feature extraction. But formant frequencies have the advantage that they correspond to the physical property of the vocal tract and therefore they can be used as a model of co-articulation (see section IV).

H.G.Goldberg compared four parametric representations for the evaluation of labeling performances; a set of energy amplitude and zero-crossing measurements from 5 octave filters (ZCC), a set of energy measurements from a 25 channel 1/3 octave filter bank (ASA), a smoothed short-time spectrum computed from the LPC filter (SPG), and the LPC coefficients themselves (ACS)[17]. The segmentation effectiveness of these 4 parametric representations was ranked as follows: SPG, ACS, ASA and ZCC. The labeling effectiveness was almost the same except for poor performance of ZCC.

P.F.Castelaz et al. performed a comparison of LPC coefficient, log area function, FFT and Zero-Crossing analysis (ZCA) technique for vowel recognition in /h-d/ context [18]. They showed that LPC algorithms yielded results inferior to selected FFT and ZCA algorithms and that the use of log area function was not shown to yield significantly different results from the use of predictive coefficients. Although the best FFT algorithm utilizing the Euclidean distance measure yielded the best overall results in a quiet environment, ZCA algorithms outperformed all of the LPC and FFT algorithms in noisy environments.

A.Ichikawa et al. [19] and G.M.White et al. [20] compared various feature parameters through spoken word recognition by using a dynamic programming technique. The recognition performance

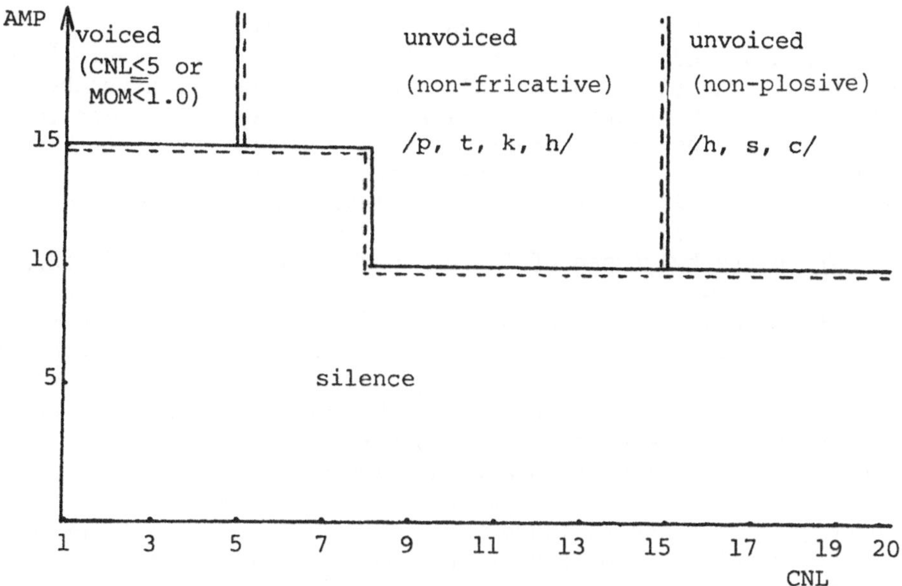

Fig. 2 Segmentation by energy level and spectral deviation

Definitions of AMP, CNL, MOM: see equations in section 3.1(a).
The central frequencies: 5-th channel=420Hz, 10-th channel=1000Hz, 15-th channel=2380Hz.

Table 2 Notation in LPC analysis of reference pattern $f(\theta)$ and input pattern $g(\theta)$

	power	correlation coefficient	prediction coefficient	/A/ parameter	normalized residual	cepstrum coefficient
$f(\theta)$	u	r_j	a_i	A_j	R_f	$c_n^{(f)}$
$g(\theta)$	v	s_j	b_i	B_j	R_g	$c_n^{(g)}$

$1 \leq i \leq p$, $-p \leq j \leq p$, p: order of AR model. $0 \leq n \leq q$, q: order of cepstrum analysis.

$$A_k = \sum_{i=|k|-p}^{p} a_i \cdot a_{i-|k|}, \qquad B_k = \sum_{i=|k|-p}^{p} b_i \cdot b_{i-|k|}$$

was ordered as follows by their experiments; smoothed logarithmic power spectrum ≒ cepstrum, partial autocorrelation coefficient (parcor), autocorrelation function and linear predictive coefficient.

Table 1 summarizes well-used parametric representations of speech sounds and their characteristics in speech processing[15-25]. As shown in the table, almost all feature parameters have the feature that the quality of speech decoding from parameters is good compared with original speech. However, it is not clearly known if these parameters are stable for speaker-factor or time-interval-factor, in other words, if these are speaker-independent or time-independent.

There are two types of feature parameters. The parameters in the upper columns are obtained from the direct transformation algorithms for speech waves (however, the smoothed power spectrum cannot be obtained perfectly). On the other hand, the parameters in the lower columns are obtained by the feature extraction algorithms from speech waves and correspond to the vocal structures of phonemes with respect to speech production. Formant frequencies are often used for the modeling of co-articulation or speaker differences (see sections IV and V), because these are regarded naturally as the physical representation of vocal geometry and have fewer dimensions. Therefore, some researchers have insisted that vocal tract area functions or positions of articulators are useful for the representaion of dynamic changes of speech or speaker differences.

III. SEGMENTATION AND LABELING

The first problem for segmentation and labeling is the definition of segment unit and phone. The most representative unit is a phoneme, because it is the smallest unit in the time axis of speech. Therefore the set is also small. However the smaller the unit is, the larger the effect of context becomes. Speaking from the view point of pattern recognition, it is best to select the shortest context-independent time unit for segmentation or labeling. Unfortunately, such a unit does not exist, because any sound is always influenced by the adjacent sounds. Although there do exist time intervals for certain sounds where these sounds are influenced by environment, the set of such units becomes enormous. This problem will be discussed in section IV.

The segmentation of syllable units may be comparatively context-independent. Therefore such methods have been tried. In this case, there are two definitions of syllabic segment: syllable segment [26] and inter-syllable segment [27,28]. The former is a linguistic unit. The latter is a speech interval between syllable centers (or nucleus) in two successive syllables. The latter method can detect the unit more accurately than the former and the acoustic features in the unit have sufficient information for the identifi-

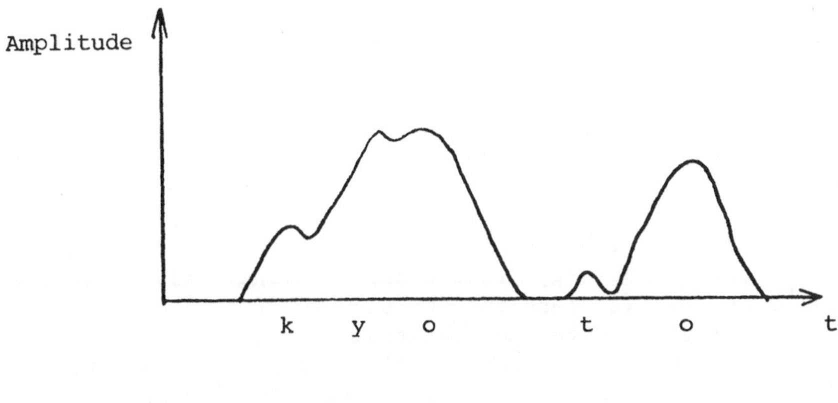

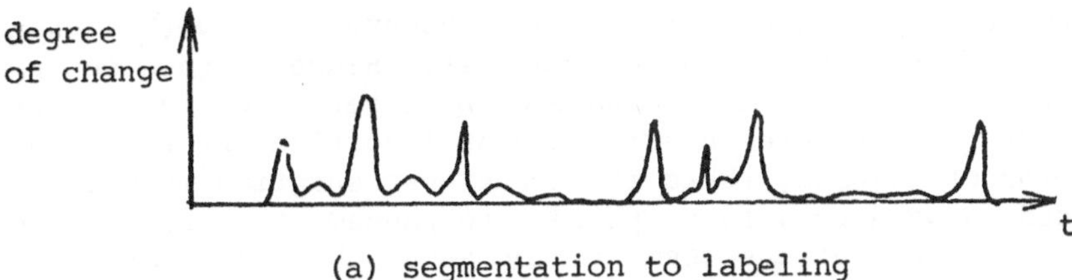

(a) segmentation to labeling

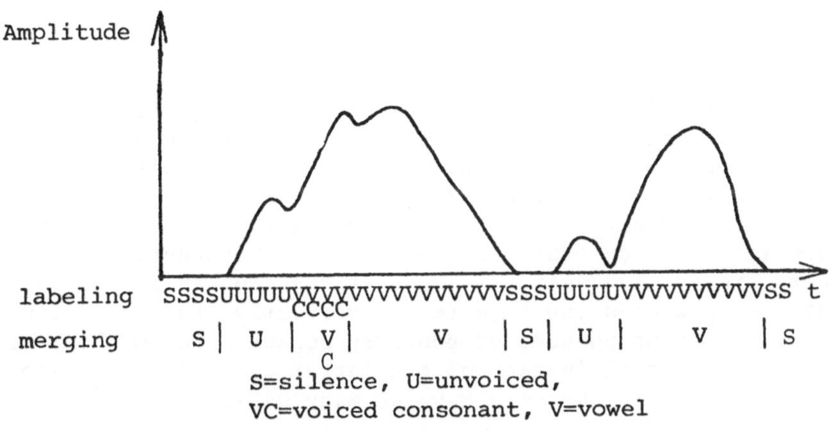

(b) labeling to segmentation (frame by frame)

Fig. 3 Direct segmentation

cation of consonants. However the set of such units becomes very large. Therefore such units may be effective in the case of pre-processing more detail segmentation or word verification (for example, testing whether there exists the phoneme /a/ in the second syllable or not).

The relationship between segmentation and labeling is very close. Therefore, these cannot be processed independently from one another and it is very difficult to treat these processes in parallel. So it is desirable to introduce a model including both segmentation and labeling processes (see section IV). There are two approaches for segmentation and labeling: (1) segmentation → labeling, and (2) labeling for each short time → merging → segmentation (label-driven segmentation)(see Fig.3). The latter method has very few omission errors. This is a great advantage, because omission errors of phonemes often cause a fatal error in word identification (see section 3.3). However this method cannot detect a transient phoneme as a unit. Since a transient phoneme cannot be recognized by acoustic characteristics in a short time interval, there are substitution or insertion errors of phonemes. On the other hand, there is another segmentation method which is based on word to word or phoneme to phoneme pattern matching. This carries out the segmentation as a sub-product of word or phoneme recognition results(see section 3.1(d)).

This section will discuss the segmentation process into a unit of phonemes.

3.1 Segmentation

The principal process of segmentation is to find out the boundary marks in time axis between different sounds. The difference of sounds depends on the language, for example, Japanese does not discriminate /l/ from /r/. This fact suggests that there exists no language-independent segmentation algorithm. Some phonemes are essentially steady state in their acoustic characteristics and others are continuously varying or transitionary in nature. If sustained phonemes can be detected perfectly, the other time domain may be transient phoneme or boundary region between phonemes. However the discrimination of the latter cases is very difficult, because transient phonemes have so many varieties.

(a) silence/voiced/unvoiced

The detection of speech sound in noise environments is very difficult, especially, the discrimination between unvoiced fricatives and noise [29]. The energy level of sound is usually used to detect "speech". On the other hand, since voiced sounds of speech are produced by vibratory action of vocal cords, in principal, they can be detected by observing the periodic repetition of speech waves or pitch period. Automatic pitch extraction is also difficult, because the repetition is pseudo periodic and there is interaction between the vocal tract and the glottal excitation. Particularly, there is the inherent difficulty in defining the exact beginning and end in voiced sounds. Therefore it requires several pitch intervals. Since a pitch extraction method is expensive, it is not desirable for a real time recognition system. But if the pitch contour is used for other processes (ex. prosodic analysis) in a speech understanding system, this method may be useful to detect voiced sounds. The "voiced" detection rate by an accurate pitch extraction method is about 97-98% [30].

Other inexpensive methods are usually used for the "voiced" detection. B.S.Atal tried a voiced-unvoiced decision method without pitch detection by using a pattern recognition approach, that is, statistical decision by using the mean and the covariance of distribution of each feature parameter (energy of speech signal, normalized prediction error, normalized autocorrelation coefficient of unit sampling time delay, first predictor coefficient of a 12-pole LPC analysis, zero-crossing rate)[31]. The effectiveness of each parameter corresponds to the order above. In the case of five parameters employed at a time, the decision rate was about 98%.

We performed the silent/voiced/unvoiced decision by using the total energy of speech signal and deviation of energy in low/high frequency ranges [2]. This is based on the fact that the spectral characteristics of glottal wave (i.e. voiced) has the slope of about -12dB/oct and therefore the energy of lower frequencies is superior to that of higher frequencies in voiced sounds. The speech wave was analyzed by a 20 channel 1/4 octave filter bank (center frequencies: 210Hz-5660Hz), and sampled at every 10ms interval and digitized into 10 bits. The outputs were regarded as a time series of vectors consisting of 20 components $(x_1, \ldots, x_{20})$. Energy level (AMP) and parameters of spectral deviation (CNL and MOM) were defined as follows.

$$\mathrm{AMP} = \left(\sum_{i=1}^{20} x_i \right)^{1/2}$$

CNL is the minimum channel number which satisfies the next condition. $\sum_{i=1}^{j} x_i \geq \mathrm{AMP}/2$

$$\mathrm{MOM} = (x_{18} + 2x_{19} + 3x_{20})/\mathrm{AMP}$$

The decision rules are shown in Fig.2. This algorithm takes sounds with weak energy (eg. /h/,/s/) into consideration. The correct decision rate was more than 95%.

(b) successive vowels

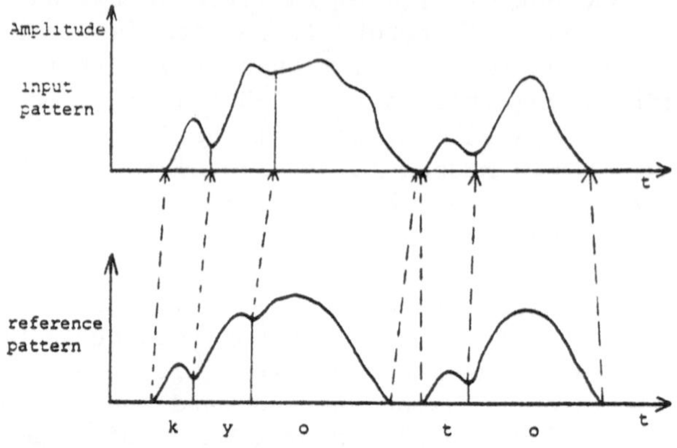

(a) pattern matching of word to word

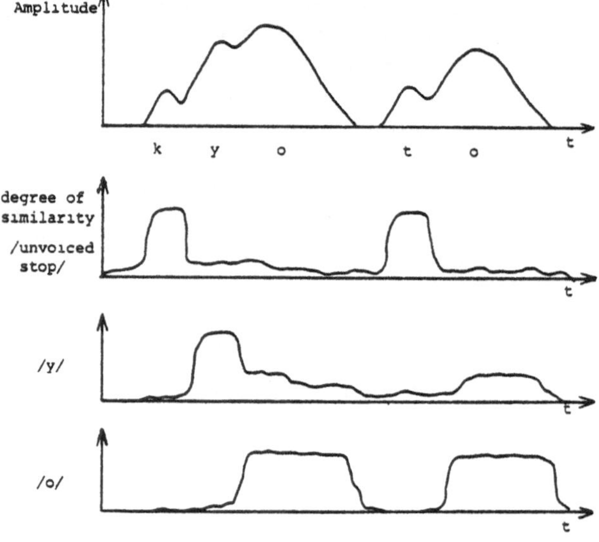

(b) pattern matching of phoneme to phoneme

Fig. 4 Indirect segmentation

Since the acoustic characteristics of successive vowels vary continuously, segmentation is also difficult. For example the characteristics in glide of diphthong /ai/ is very similar to that of /e/. The segmentation in voiced parts is usually performed by using the measurements of change of feature parameters, such as amplitude, energy and spectrum. However, the degree of change differs from context to context. Thus, the segmentation process of successive vowels must have the following functions: (1) automatic adaptation to threshold value and (2) introduction of segment duration factor. For example, a temporary sound /e/ appearing in the context /ai/ may be rejected by reason of its short interval.

(c) consonant

Voiced sounds except for vowels can be regarded as voiced consonants. These parts can also be detected by the existence of concave of energy level in time axis. However the degree of change in energy level is different from consonant to consonant or context to context. The change is not a cue to the detection or recognition of voiced consonants for human perception[32](see Table 6). This fact suggests that the best recognition process might be different for machine and human. In Japanese, since successive consonants do not appear with some exceptions, there exist some systems which do not detect consonants actively 27 . In English, the segmentation of successive consonants is necessary. In such a case, syntactic approaches, including the rewriting rules are effective [28],[33-35].

(d) indirect segmentation

In an automatic spoken word recognition system, a word to word pattern matching method is often used. In this case, the speech signal is continuously matched with a set of reference patterns for words. Whenever a match is achieved, the presence of phonemes and the location are indirectly detected as shown in Fig.4 (a). This technique is used for making phonemic reference patterns in the case of one/many speakers and also in modifying reference patterns in learning process (see section 3.2 (b)).

Instead of word to word matching, phoneme to phoneme matching is considered [36]. Although this approach cannot segment the speech wave into a continuous and non-overlapping phoneme sequence, it makes possible a parallel processing. Recently, T. Nakajima et al. proposed a concept of phoneme filtering on the basis of this approach [37]. Such an approach has been tried for the detection/identification of a special phoneme [38].

3.2 Labeling

There are many methods for labeling the segment into one of phonemic categories, namely:

(1) undeterministic method
  (a) pattern matching method [2,27,39]
  (b) probabilistic or statistical/stochastic method [2,16,40]
(2) (semi-)deterministic method
  (c) syntactic method [35]
  (d) logical or decision tree method (including heuristic method) [38,41,42]

These methods have been also used for spoken word recognition. Fig.5 illustrates the relationship among them. Pattern matching methods and probabilistic methods are based on the comparison between a reference pattern and an unknown input pattern by using a concept of distance. Here the reference pattern for probabilistic method is given by statistics, for pattern matching, by a template. Such methods are weak for detecting complex feature cues, however owing to the global features, the results obtained are more stable than syntactic or logical methods. A non-linear pattern matching, which is used for dynamic patterns, corresponds to a stochastic method. The output of this decision is an analog quantity.

On the other hand, syntactic or logical methods employ yes or no decisions. However an application of such a decision is very difficult for patterns with variety like speech, while it is also difficult to make syntactic rules or design for automata accepting specific symbol strings. Thus, recently, a 'soft' automaton and 'fuzzy logic' have been proposed for pattern recognition. Thus, the differences among them are gradually decreasing and the reference pattern is usually represented by a network or graph. In this section, we discuss the pattern matching method and the probabilistic method (the syntactic method will be described in another presentation).

(a) distance measurement

At first, it is necessary to define the distance between a reference pattern and an input pattern for a pattern matching method. K.Shikano and M.Kohda compared theoretically and experimentally the properties and performances of the following four distance measures in LPC spectral matching [43]; the likelihood ratio measure (or linear predictive residual)[44], normalized residual power measure [27,45], COSH measure [46], and LPC cepstrum distance [46]. Let $f(\theta)$ and $g(\theta)$ be a reference spectral envelope and an input spectral envelope obtained by LPC, respectively. In this case, $f(\theta)$ and $g(\theta)$ are represented by the following equations (refer to Table 2).

$$f(\theta) = \frac{uR_f}{2\pi} \cdot \frac{1}{|1+a_1e^{j\theta}+\cdots+a_pe^{jp\theta}|^2} = \frac{uR_f}{2\pi} \cdot \frac{1}{\sum_{k=-p}^{p} A_k e^{jk\theta}}$$

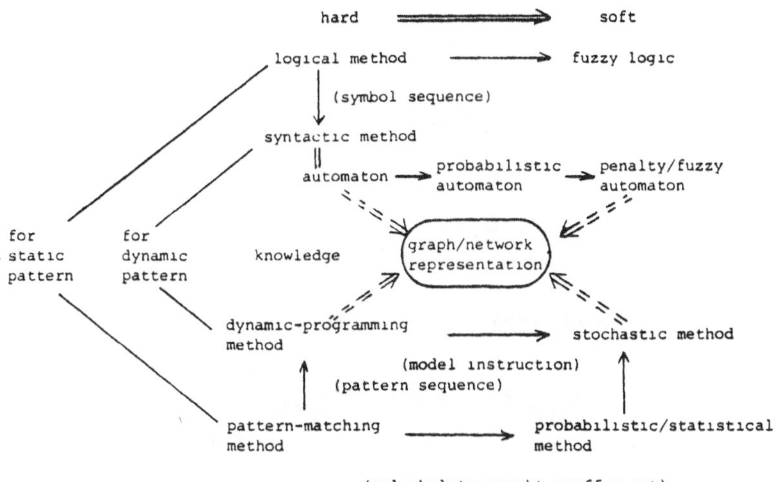

Fig. 5 Relationship among labeling techniques

Table 3 Results of vowel recognition by various classification methods

method	city-block distance	Euclid distance	linear discriminant function	quadratic discriminant function
speaker-independent (common speaker)	86.3%	87.4%	91.8%	93.0%
speaker-dependent (individual speaker)	91.8%	92.2%	95.0%	95.0%

$$g(\theta) = \frac{vR_g}{2\pi} \cdot \frac{1}{|1+b_1e^{j\theta}+\cdots+b_pe^{jp\theta}|^2} = \frac{vR_g}{2\pi} \cdot \frac{1}{\sum_{k=-p}^{p} B_k e^{jk\theta}}$$

These above distances are defined as follows.

(1) Likelihood ratio measure (LRM)

$$LRM = \frac{1}{2\pi}\int_{-\pi}^{\pi}\left\{\log\frac{f(\theta)}{g(\theta)} + \frac{g(\theta)}{f(\theta)} - 1\right\}d\theta = \log\frac{uR_f}{vR_g} + \frac{v}{uR_f}\cdot\sum_{k=p}^{-p} A_k \cdot s_k - 1$$

(2) Normalized residual (NR)

$$NR = \min_{\sigma^2}\frac{1}{2\pi}\int_{-\pi}^{\pi}\left\{\log\frac{\sigma^2 f(\theta)}{g(\theta)} + \frac{g(\theta)}{\sigma^2 f(\theta)} - 1\right\}d\theta = \sum_{k=p}^{-p} A_k s_k$$

(3) COSH measure (COSH)

$$COSH = \frac{1}{2\pi}\int_{-\pi}^{\pi}\left\{\log\frac{f(\theta)}{g(\theta)} + \frac{g(\theta)}{f(\theta)} - 1\right\}d\theta + \frac{1}{2\pi}\int_{-\pi}^{\pi}\left\{\log\frac{g(\theta)}{f(\theta)} + \frac{f(\theta)}{g(\theta)} - 1\right\}d\theta$$

$$= \frac{v}{uR_f}\cdot\sum_{k=p}^{-p} A_k \cdot s_k + \frac{u}{vR_g}\cdot\sum_{k=p}^{-p} B_k \cdot r_k - 2$$

(4) LPC cepstrum measure (CEPSTRUM)

$$c^{(f)} = \log\frac{uR_f}{2\pi}, \quad c_n^{(f)} = -a_n - \sum_{k=1}^{n-1}\frac{n-k}{n} a_k c_{n-k}^{(f)}, \quad c_n^{(f)} = c_{-n}^{(f)}$$

$$c^{(g)} = \log\frac{vR_g}{2\pi}, \quad c_n^{(g)} = -b_n - \sum_{k=1}^{n-1}\frac{n-k}{n} b_k c_{n-k}^{(g)}, \quad c_n^{(g)} = c_{-n}^{(g)}$$

$$CEPSTRUM = \sqrt{\sum_{i=-\infty}^{\infty}(c_i^{(f)} - c_i^{(g)})^2} = \frac{1}{2\pi}\int_{-\pi}^{\pi}\left(\log\frac{f(\theta)}{g(\theta)}\right)^2 d\theta$$

Apparently, LRM and NR do not satisfy the properties of distance (symmetry). But NR, in particular, is not an expensive calculation. They compared the various conditions; (1) no modification, (2) power normalization (u=v), (3) residual power normalization ($uR_f = vR_g$), (4) u and v adaptation for minimization of above distance. The effectiveness was ordered as follows; concerning distance COSH, CEPSTRUM, LRM, NR and for condition (3) ≒ (4), (2), (1). However, there were no significant differences among them.

H.F.Siverman and N.R.Dixon compared the following speech spectral classification methods [47]. They used the power spectrum obtained by FFT as feature parameters.

(1) Maximum direction cosine method (DCOS)

$$d_j = \frac{\frac{1}{N}\sum_{i=1}^{N} x_i \cdot P_{ij}}{\left(\frac{1}{N}\cdot\sum_{i=1}^{N} x_i^2\right)^{1/2} \cdot \left(\frac{1}{N}\cdot\sum_{i=1}^{N} P_{ij}^2\right)^{1/2}}$$

where P_{ij} is the i-th frequency component of the j-th pattern and x_i is the i-th frequency component of a candidate spectrum and j is the class identity name.

(2) Generalized minimum distance method

$$c_j(l, a_1, a_2) = \left[\frac{1}{N}\cdot\sum_{i=1}^{N} \left| x_i - P_{ij} - f(|\overline{x} - \overline{P}_j|)\cdot(\overline{x} - \overline{P}_{ij})\right|^{l}\right]^{1/l}$$

where the symbol '-' denotes the mean over the elements, $l = \{1/3$ $1/2, 1, 2, 3\}$ and

$$f(y) = \begin{cases} 0 & y > a_2 \\ \dfrac{y - a_2}{a_2 - a_1} & a_1 \leq y \leq a_2 \\ 1 & y < a_1 \end{cases}$$

where $[a_1, a_2]$ is set to [0,0](=MDO, Minimum distance with no mean correction), [0,Ymax](=MDL, Minimum distance with linear mean correction), [Ymax,Ymax](=MDF, Minimum distance with full mean correction). The effectiveness was as follows; MDL, DCOS, MDF, MDO for distance and 2 ≒ 3, 1, 1/2, 1/3, for l.

We also compared several discrimination methods by using the power spectrum obtained from a filter bank [2]. The results are shown in Table 3. Speech materials employed were meaningful words in VCV contexts, spoken by 10 male adults. The quadratic discriminant function based on Bayes decision rules has the most powerful performance in vowel recognition. But this method consumes rather much computational time in comparison with other algorithms.

On spectral distance, we think that the logarithmic transformation is better than the absolute value, because we prefer the relative intensity between frequency components of speech sound spectrum to the instantaneous amplitude; and the auditory sense for the intensity of speech sound is proportional to the logarithm of the intensity itself.

(b) creation of reference pattern

First, we must answer the following questions for the creation of reference patterns.
(1) how many templates should be created for each phoneme.
(2) how to extract samples for the creation from speech sounds.
(3) how to create the reference pattern from samples within a class.

For a logical method, T.Nakajima et al. made up about 4000 logical decision rules for vowel recognition [48]. Many samples and much effort are necessary for making such logical rules. For a probabilistic method, the reference function is also made from the statistics of many reference samples. The statistics are usually obtained by regarding the distribution of feature paramaters as a normal distribution. In this case, the mean and covariance matrix are calculated from these samples [2,15].

On the other hand, for a pattern matching method, many samples for making reference patterns are comparatively unnecessary. The reference pattern is usually made of the arithmetic mean of each component of some samples in the case where they have the same number of components. The dynamic programming technique can be used for the mean-value operation between two samples with different length or different number of components [49]. The dynamic programming technique is also used when the extraction of samples is necessary in continuous speech(see Fig.3)[45].

K.Tanaka proposed a clustering technique to make the reference patterns [50]. Since the categories obtained by such a clustering technique application may be different linguistic categories and are different from speaker to speaker, it cannot be used positively in the field which includes linguistic information or rewriting rules for unspecified speakers. Therefore, it is desirable to cluster each phoneme into categories. M.Kohda et al. classified in advance vowels /a/, /i/, and /u/ into two categories, respectively [45]. However, we had the following questions.
(1) whether each vowel is always divided into two classes for any speaker or not.
(2) whether the way of division of vowel in a context is the same for all speakers or not.
(3) If we classify spectral patterns of each vowel or a phoneme in a word dictionary into two classes, we must investigate characteristics of phonemes in all contexts. This is very difficult.

Therefore we proposed a clustering technique which makes two reference patterns for each vowel from one reference pattern by using learning samples, obtained from input speech [51].

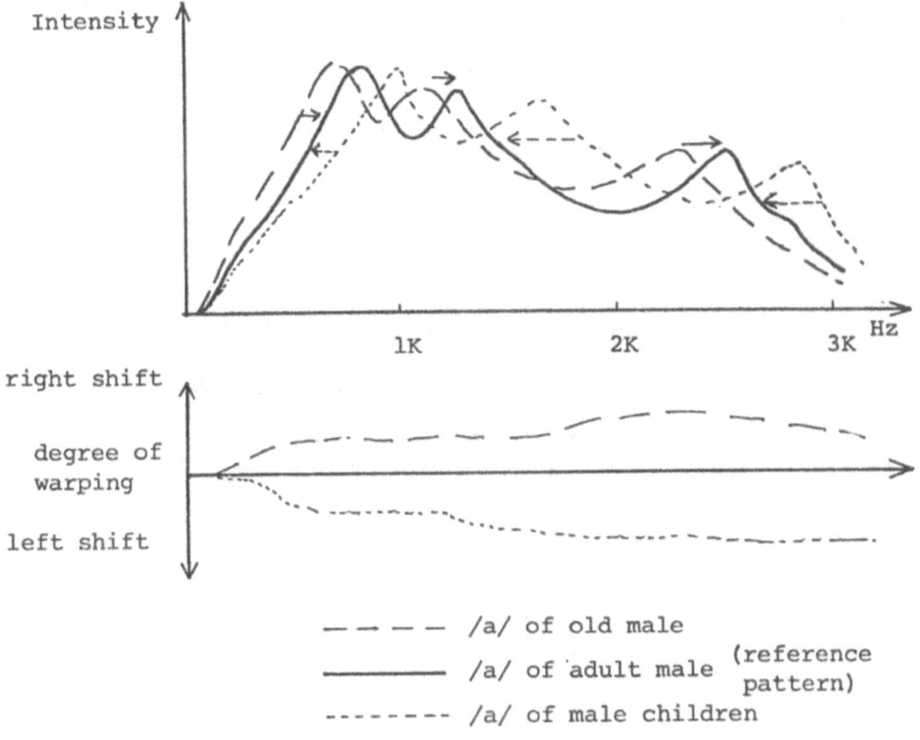

Fig. 6 Frequency warping by dynamic programming

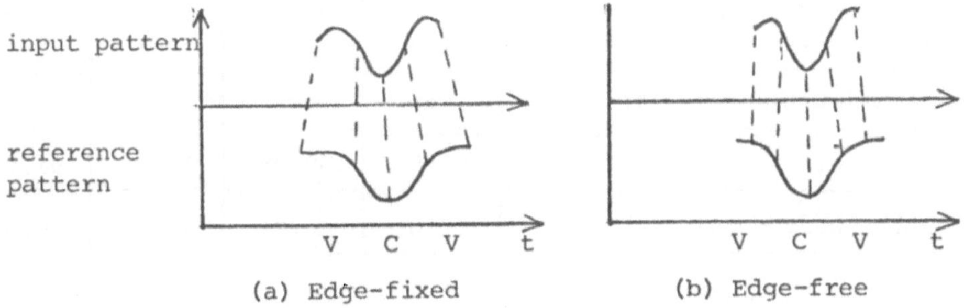

Fig. 7 Edge-fixed and Edge-free DP matching

Table 4 An example of phoneme lattice and evaluation of various output forms of phoneme lattice.

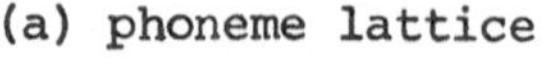

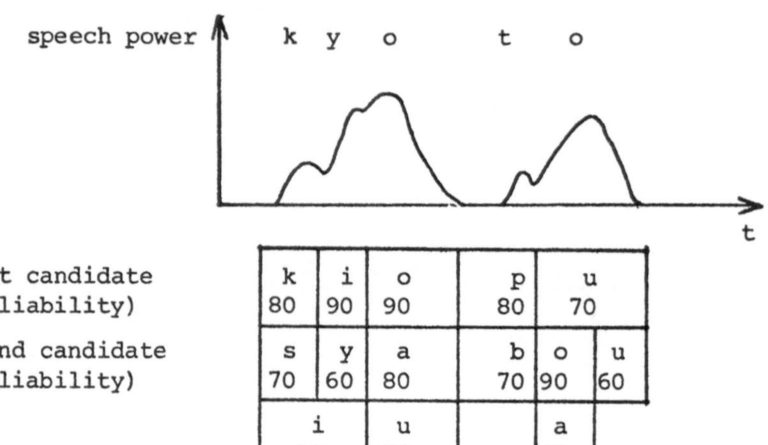

(b) experimental results of arithmetic expressions

information included in phoneme lattice	information used			
first candidate of phoneme	yes	yes	yes	yes
second candidate of phoneme	yes	yes	no	yes
reliability of candidates	yes	no	no	yes
duration time of segment	yes	yes	yes	no
word recognition rate	93.5%	89.0%	91.3%	91.8%
sentence recognition rate	60.0%	40.0%	53.0%	58.0%

(c) pattern matching by dynamic programming

Dynamic programming is an extremely useful technique for achieving nonlinear time warping (adjustment) to match patterns between two sequences with different length [52,53]. Although dynamic programming algorithms make good use of the properties of speech sounds such as continuous, uni-directional and nonlinear warping in time, it may overlook a subtle variation of local significant features. Such a phenomenon is an inherent disadvantage of pattern matching or probabilistic methods. Therefore, recently, some slope constraints have been introduced into the warping function [2,54]. This technique is useful not only in word recognition but also phoneme recognition. K.Tanaka [55] and H.Matsumoto and H.Wakita [56] proposed a phoneme recognition method for unspecified speakers based on a frequency warped spectral matching by using dynamic programming. Although their methods still have the disadvantage of restrictions of the warping function, such an approach may be useful to normalize a frequency domain for unspecified speakers (see Fig.6). In this case, the warping functions must be constrained so that they take account of linear scaling for the differences in vocal-tract length and of slight non-linear differences. For example, in Fig.6, they should be designed to be located only at one side, either left or right. From the view point of pattern recognition, it is desirable to warp for three dimensions in parallel; time, frequency and intensity.

R.Nakatsu and M.Kohda also used the dynamic programming technique for the recognition of consonants [27]. Since consonants are transient sounds, these cannot be recognized perfectly by using only the pattern at a special time. And also, the pattern of consonant is affected by adjacent vowels. Therefore, they used the vowel-consonant-vowel(VCV) unit for the recognition. The VCV syllable of input speech is represented by a succession of autocorrelation function, s_1, $s_2 \dots s_m$, where i denotes the i-th frame. On the other hand, the reference pattern of VCV syllable is represented by a succession of the maximum likelihood spectral parameters A_1, $A_2 \dots A_n$. The similarity between these two successions is obtained by

$$L = \max_{f} \sum_{i=1}^{m} l(i,f(i))$$

where $l(i,j) = -\log(\sum_k s_{ik} A_{jk})$ (remember NR distance), and

$$f(1)=1,\ f(m)=n,\ f(i)-f(i-1)=0,\ 1 \text{ or } 2.$$

These conditions indicate that the start point and the end point of the VCV syllable segment coincide with those of the reference pattern. This method is called edge-fixed DP matching (see Fig.7

(a)). However, VCV syllable segments extracted from connected speech seldom contain steady parts of vowels. On the other hand, VCV syllable reference patterns, which are made up from isolated VCV syllable speech, usually contain steady vowel parts. Therefore, to make VCV syllable segment corresponding to a part of a reference pattern, the above conditions were modified as follows. $1 \leq f(1)$ $f(m) \leq n$. Such a method is called edge-free DP matching, word spotting [57] or direct matching [5]. Fig.7(b) illustrates this example.

3.3 Output forms and errors in phoneme recognition

Errors are unavoidable with segmentation and labeling. To avoid segmentation errors, the output of segmentation is often represented by the form of a segment lattice which permits plural candidates and overlapping [58]. Since the phoneme recognition results are also not always correct, the outputs usually contain plural candidates even for a segment. Thus, the output of labeling also consists of a lattice, called 'phoneme lattice'. There are two types in the segment or phoneme lattice, whether each segment or phoneme has a value of reliability (or score, likelihood, confidence degree)[2,27] or not [58]. Of course, the more candidates there are, the higher the number of correct phonemes in the lattice, though the recognition results also become more ambiguous.

K.Shikano and M.Kohda proposed an estimation method of substantial phoneme recognition rate from a given phoneme lattice without reliability based on the mutual information from view point of communication theory [59]. However, the estimation in the case with reliability is very difficult. We cannot know exactly the optimum number of candidates in the lattice of the case with reliability. Of course, it depends on the phoneme recognition performance of a system. We evaluated various output forms of phoneme lattice on the basis of the word and sentence recognition rate of arithmetic expressions [2]. The results are shown in Table 4(b). Table 4(a) shows an example of a phoneme lattice with reliability. From these results we found that the phoneme lattice with reliability is superior to that without reliability in our phoneme recognition system.

Further we investigated the significance of segmentation and labeling on the basis of pseudo phoneme block, that is, a pseudo phoneme sequence which is generated by a random generator according to given substitution, omission and insertion probabilities for every phoneme [60]. Table 5 shows the simulation results of 100 Japanese city names recognition. The last column denoted by '*' shows the recognition rate when the system knew that the segmentation was perfectly correct. Although the results may slightly depend on our word recognition method [2,5], we obtained the following conclusions.

Table 5 Evaluations of segmentation and labeling errors by simulation.

segmentation		recognition rate in case of Japanese 100 city name task				
omission	insertion					
0%	0%	94.2%	95.7%	94.6%	94.3%	98.4%
10	0	72.2	76.1	74.9	71.7	77.3
0	10	93.1	97.1	96.3	94.7	98.1
0	20	94.8	96.8	94.9	95.5	98.2
10	10	73.8	75.6	77.4	74.9	80.2
10	20	73.5	75.5	74.8	73.2	78.4
* 0	* 0	96.9	98.3	98.7	98.2	99.8
vowel		80	90	80	80	90
voiced consonant		40	40	60	40	60
unvoiced consonant		80	80	80	90	90

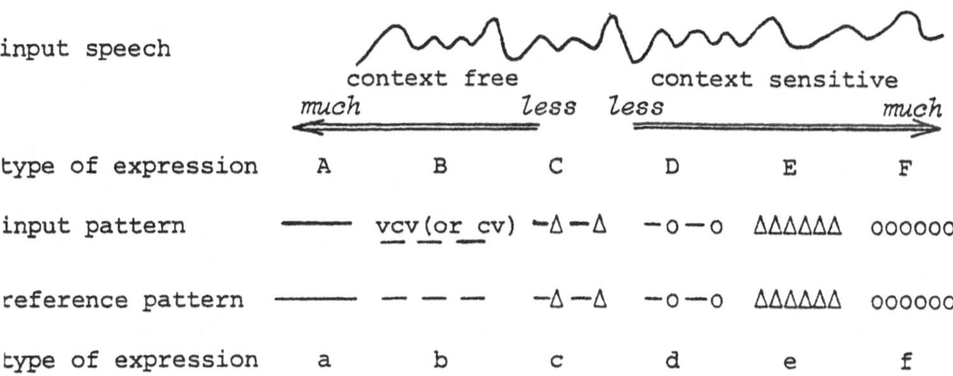

Fig. 8 Kinds of input pattern and reference pattern on matching

— a time series expressed by feature parameters every frame
Δ a segment expressed by feature parameters
o a phoneme corresponding to a segment
vcv denotes vowel-consonant-vowel syllable

(1) The omission error in segmentation often becomes fatal on word recognition.
(2) The correct rate of labeling is not sensitive for recognition of about one hundred words.
(3) The insertion error is not sensitive to word recognition. This fact suggests that it is not useful to attempt to merge successive segments.

3.4 Acoustic-Phonemic Processing without Phoneme Recognition

This is the process without phoneme recognition on an acoustic-phonemic level. It is the reduction process of feature parameters in order to reduce the amount of data. Since this output comes from the input of a word identification block, of course, it should not abandon the linguistic information. The sustained sounds can be represented by the central or average sound but transient sounds also cannot be reduced. Therefore this operation of reduction process, for time domain, is considered as a nonlinear sampling technique.

We can classify the acoustic representation of a word into six types as shown in Fig.8 [2]. We discussed the mapping procedure from the time sequence of acoustic features (type A) to the sequence of phonemes (type F) in preceding sections. In spoken word recognition, the representation of type A has been often used because we need not worry about co-articulation. However, since the amount of data for a large size vocabulary becomes enormous, this type should not be adopted for a speech understanding system. Thus, acoustic representation by a sequence of syllable units was considered [27]. Since this type (B) is independent of a vocabulary set, it is useful for languages with a small number of syllables like Japanese. However these reference patterns (types A and B) are very expensive, for the learning of reference patterns.

Types C and D are acoustic representations in which only the sustained sounds are reduced to few frames by central or phonemic categories (type D). Type E is a nonlinear sampling of a sequence of acoustic features and is obtained from the data reduction of transient sounds corresponding the type C.

All these types are mapped into a word by matching with a given word represented by various types (a-f).

IV. CO-ARTICULATION

Speech sounds which are produced through the vocal apparatus are influenced by context, as the apparatus moves continuously to save momentum. This phenomenon is called co-articulation, and it makes automatic recognition very difficult, in particular segmentation and labeling. In this section, we discuss various aspects of this problem.

(a) perception

The phenomenon of co-articulation has been investigated through perceptual experiments [61-64]. H.Kuwahara and H.Sakai investigated the perceptual properties of vowels segmented from isolated monosyllables and connected speech, as well as those of monosyllables and larger units segmented from connected speech [64]. They obtained the following results:

(1) The perception of a vowel in connected speech was found to be seriously impaired by the removal of its environment. The identification rate of vowels in clearly pronounced connected speech ranged from 52% to 70%, when taken out of their environments and presented in isolation.
(2) When monosyllabic segments were taken out of connected speech and presented in isolation, only 42% were identified correctly. Because of the existence of the preceding consonantal environment, however, the rate for the vowels in this case was improved up to about 80%. In the case of bisyllabic segments, the rate of correct identification was about 70%.

We investigated the perception of seven voiced consonants taken out from VCV contexts [32]. The experimental methods and results are summarized in Fig.9 and Table 6, respectively. From these results, we could conclude at least the following two facts: (1) more linguistic information of voiced consonants is contained in a glide than in the central part of the consonant, and (2) even if the glide is taken out of a VCV utterance, the intelligibility is beyond 60-70%. The second fact is very important information for us, because we can treat the central part more easily than the glide. The classification rate, 60%, of voiced consonants is sufficient for an automatic spoken word recognizer or speech understanding system (refer to Table 5). This fact was also confirmed from the classification experiment by machine, which was made for comparison between machine and human listeners. The reference pattern of a central part for each voiced consonant was calculated from speech materials from ten male adults. The results is shown in the last column of Table 6. From these experiments we concluded that the precise detection of voiced consonants is more important than the development of the classification technique.

(b) analysis

Recently, the direct observations of articulatory movements have been possible [65,66] and the production models are being tried out on the basis of these observations [67]. Observations of co-articulation for extracted acoustic features have also been investigated [68-72]. K.N.Stevens and A.S.House showed that there

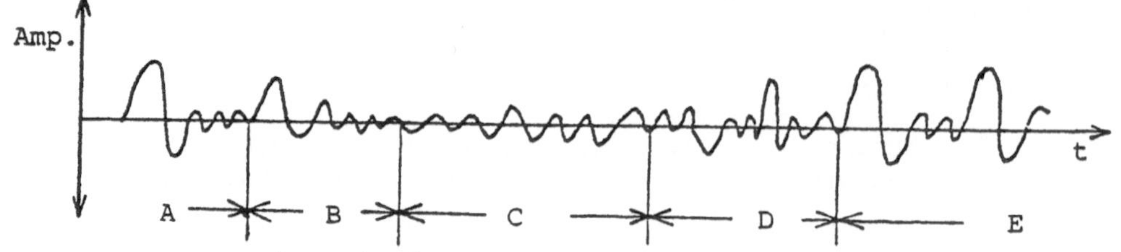

A: stationary part of preceding vowel
B: on-glide
C: central part of voiced consonant
D: off-glide
E: stationary part of following vowel
S: silence

Fig. 9 Segmentation of VCV utterance

Table 6 Identification rate of voiced consonants taken out of VCV utterances.

(a)∿(f) : by perceptual experiment
(g) : by machine recognition

method	phoneme	m	n	ŋ	b	d	r	z	average
a	A + B + C + D + E	99%	100%	100%	100%	98%	99%	88%	98%
b	*A + B + C + D + E	99	100	100	93	93	100	92	97
c	A + C + E	100	100	94	87	7	100	0	70
d	A + C + C + E	100	93	80	93	40	100	0	76
e	A + B + D + E	100	100	100	100	87	100	13	86
f	A + B + S + D + E	100	100	100	100	100	87	7	85
g	machine recognition	76	76	60	80	73	96	96	80

* after normalization of speech power

are systematic shifts in vowel formant frequencies depending on the place of articulation of the consonant [68].

S.E.G.Öhman investigated the formant transitions in vowel - stop consonant - vowel utterances spectrographically, and found that because the terminal frequencies of the formants in VCV utterances depend not only on the consonants but on the entire vowel context, the stop consonant loci are, therefore, not unique [69].

We analyzed the phenomenon of co-articulation by using the multi-variate statistical method [73]. Nine consonants were uttered in two-syllable $C_1V_1C_2V_2$ (consonant-vowel-consonant-vowel) words. Using /p,t,k,b,d,g,m,n,ŋ/ for C_1, C_2 and /a,e,o/ for V_1, V_2, an adult man uttered each word of all combinations, three times at random. These utterances were analyzed by a 20-channel 1/4 octave filter bank. The analyzed spectra were represented by the following linear model of multivariate analysis of variance for four-factor design with repeated measurements.

$$
\begin{aligned}
X_{ijklm} &= K && \text{general level}\\
&+ \alpha_i + \beta_j + \gamma_k + \delta_l && \text{main effect}\\
&+ \varepsilon_{ij} + \xi_{ik} + \eta_{il} + \theta_{jk} + \lambda_{jl} + \mu_{ki} && \text{two-factor interaction}\\
&+ \nu_{ijk} + \rho_{ijl} + \sigma_{ikl} + \tau_{jkl} && \text{three-factor interaction}\\
&+ \phi_{ijkl} && \text{four-factor interaction}\\
&+ \chi_{ijklm} && \text{residual}
\end{aligned}
$$

where, $1 \leq i \leq 9$ (for C_1), $1 \leq j \leq 3$ (for V_1), $1 \leq k \leq 9$ (for C_2), $1 \leq l \leq 3$ (for V_2). The normalized value of these factors at nine time points are shown in Fig.10. From these results, we obtained the following conclusions:

(1) In the stationary parts of each phoneme, the main effect of the phoneme is maximum.

(2) The interaction between two continuous phonemes is, of course, smaller than each main effect. The main effect of adjacent phonemes is larger than that of such interactions.

(3) Influences of the preceding and following vowels of V_1 and V_2 on C_2 are almost the same.

(c) modeling

The phenomenon of co-articulation is very complex, because the beginning and ending times of each vocal apparatus are different. The research of modeling is still at the stage of trial and error. All proposed methods have used the formant transition. H.Kasuya et al. proposed the following experimental formula for normalization of the second formant [74]. Let F_{2i}, F_{2m}, and F_{2f} be the second

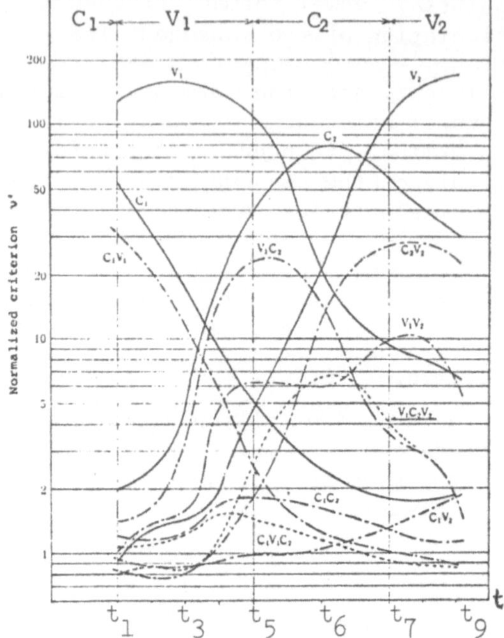

Fig. 10 Multivariate analysis of variance for four-factor design in $C_1V_1C_2V_2$ utterances.

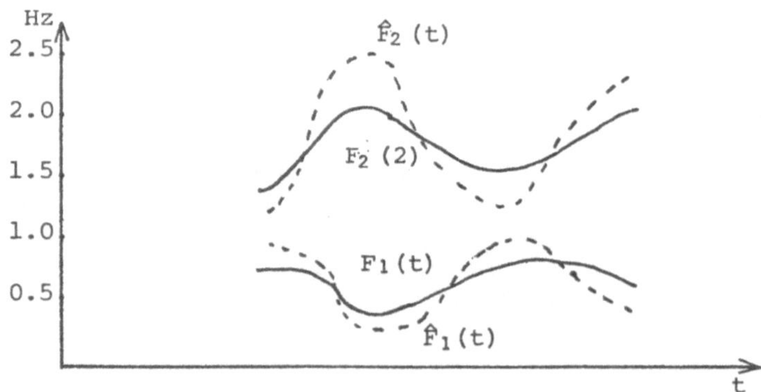

Fig. 11 First and second formant loci of /ioio/. Solid lines and dotted lines represent the original and the modified formant loci, respectively.

formant frequencies at the initial point, middle point and final point in a vowel part, respectively. They normalized the formant by the following formula.

$$F_2 = F_{2m} - K\cdot(F_{2i} + F_{2f}) + C$$

where, K and C are constant.

H.Kuwahara and H.Sakai generalized the above method [75]. Let $F_i(t)$, i=1,2, be the i-th formant frequency at time t. The normalized formant frequency was estimated by

$$\hat{F}_i(t) = F_i(t)\left\{1 + \int_{-T}^{T} w(t)dt\right\} - \int_{-T}^{T} w(t)\cdot F_i(t-t)dt$$

where, w(t) is the weighting function of Gauss function. Fig.11 shows an example of this normalization.

H.Fujisaki et al.[76] and S.Itahashi et al.[77] proposed the critically damped second order linear model of co-articulation as shown in Fig.12. The transfer function and step response are given by

$$H(s) = r^2/(s^2 + 2rs + r^2), \qquad F(t) = 1 - (1 + rt)\cdot e^{-rt}$$

Let $F_{n,1}, \ldots F_{n,m}$ be the target n-th formant frequencies which are activated at time $t_1, \ldots t_m$, respectively and U is a unit function. Then, the input of this linear system is

$$C_n = F_{n,1}U(t-\tau_1) + \sum_{j=2}^{m} (F_{n,j} - F_{n,j-1})\cdot U(t-\tau_j) \qquad t \geq \tau_1$$

and the response of this system is given by

$$F_n(t) = F_{n,1} + \sum_{j=2}^{m} (F_{n,j} - F_{n,j-1})\cdot[1-\{1+\gamma(t-\tau_i)\} \times \exp\{-\gamma(t-\tau_j)\}]\cdot U(t-\tau_j)$$

H.Fujisaki and Y.Sato obtained τ_i and $F_{n,j}$ from the observation values of input speech by using the above model and analysis by synthesis technique. Although this method can perform both segmentation and recognition at the same time, it requires a certain amount of computation time.

We proposed a new real-time oriented algorithm which normalized the effect of co-articulation at the stage of matching between a just recognized phoneme sequence and a phoneme sequence in a word dictionary [5]. Let us consider a matching model such as is illustrated in Fig.13. Let us define the similarity between symbols of m-angle and n-angle as $100 - 20\cdot|m-n|$. In this example, the

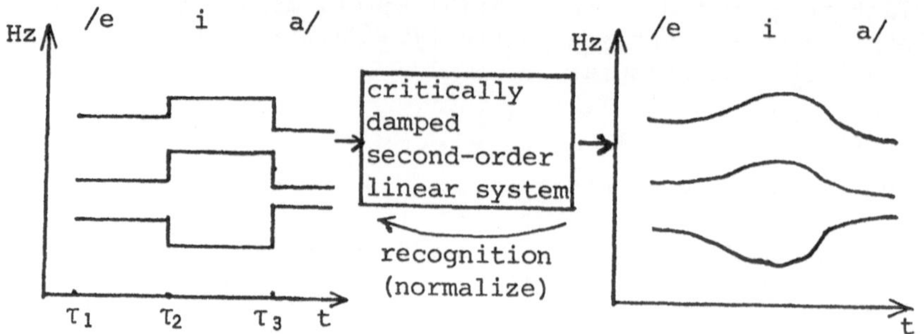

Fig. 12 Conversion of idealized target values into actual values of formant frequency by a model of co-articulatory process.

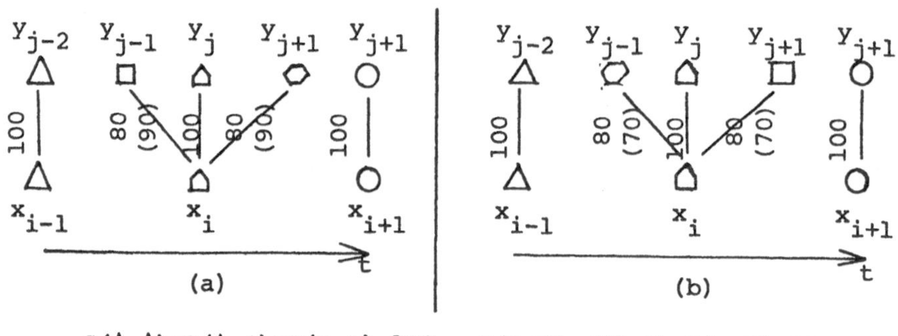

S(Δ,Δ)=S(⌂,⌂)=S(o,o)=100, S(⌂,□)=S(⌂,⬠)=S(o,⬠)=80,

S(Δ,⬠)=S(⌂,o)=60, S(Δ,□)=S(o,□)=40.

Fig. 13 Graphic model of normalization of co-articulation at the stage of matching between two phoneme sequences. The scores in parentheses denote scores after normalization.
x: reference phoneme, y: input phoneme(candidate)

Table 7 Average of feature parameters of isolated vowels for each group.

group	No. of persons	sex	age (years)	height (cm)	pitch (Hz)	formant	average frequencies (in Hz) of five vowels a	i	u	e	o
M1	20	male	10.3	137.4	270	F1	1064	419	510	729	776
						F2	1626	2183	1336	2258	1503
						F3	2789	3363	2495	3143	3055
M2	20	male	22.6	169.8	146	F1	796	337	397	578	605
						F2	1290	2112	1142	1847	1258
						F3	2547	3114	2311	2642	2841
M3	20	male	53.0	165.6	142	F1	688	326	398	489	554
						F2	1149	2168	968	1754	1320
						F3	2202	3108	2169	2635	2548
F1	20	female	10.5	140.2	268	F1	928	432	512	673	708
						F2	1530	1935	1237	2018	1252
						F3	2309	3389	2111	3058	2672
F2	20	female	19.0	156.8	247	F1	1004	371	420	625	728
						F2	1595	2304	1379	2328	1339
						F3	2881	3288	2846	3125	2867
F3	20	female	51.1	153.6	205	F1	866	359	419	550	696
						F2	1546	2429	1149	2144	1588
						F3	2503	3348	2834	3003	2771

similarity obtained from matching for the case of (a) is the same as in the case of (b), if the co-articulation (or context) is neglected. In Fig.13, if we can regard the association between x_i and $[y_{j-i}, y_j, y_{j+i}]$ as a valid association, we could assume that y_{j-i} (or y_j) is a transient segment between x_{i-1} and x_i (or x_i and x_{i+1}). In this case, the following inequalities might be satisfied, because the transient segment represents an intermediate phoneme between two successive phonemes in the word.

$$S(x_{i-1}, y_{j-1}) > S(x_{i-1}, x_i), \qquad S(x_{i+1}, y_{j+1}) > S(x_{i+1}, x_i)$$

where, $s(x_i, y_j)$ denotes the similarity between x_i and y_j. Therefore, we consider that the normalization of co-articulation can be performed by the following modification of the similarity.

$$S(x_i,y_{j-1}) = S(x_i,y_{j-1}) + K[S(x_{i-1},y_{j-1}) - S(x_{i-1},x_i)]$$

$$S(x_i,y_{j+1}) = S(x_i,y_{j+1}) + K[S(x_{i+1},y_{j+1}) - S(x_{i+1},x_i)]$$

where, K is constant. For example, if K is 0.5, $S(x_i,y_{j-1})$ in the case of (a) and (b) of Fig.13 become 90 and 70, respectively.

V. SPEAKER DIFFERENCES

The feature parameters for phonemes vary from speaker to speaker. These variations derive from the differences in vocal apparatus and dialect of each speaker and these lead to difficulties in recognizing the speech of unspecified speakers. The researches on speaker identification try to extract these variations. On the other hand, the researches on speech (phoneme) recognition try to eliminate them. However, it is also necessary to do such operations to know the true nature of speaker differences. In this section, we discuss this problem from various approaches.

(a) perception and analysis

K.Tabata et al.[78] and Ito et al.[79] investigated the factor of speaker differences by separating the characteristics of speech into those of vocal tract and glottis wave, through perceptual experiments. They found that the vocal tract contains more characteristics of the speaker than the glottis wave. T.Nakajima et al. proposed that the characteristics of glottis could be eliminated by the adaptive inverse filter of the 2nd and 3rd order [24].

We analyzed 1800 isolated vowels of 120 persons [80]. The main results are shown in Table 7. There are two types of speaker differences: the differences of an age and sex group (inter-group differences) and the differences in the same group (intra-group differences). Of course, although we assumed that the former is greater than the latter, it was confirmed by the results of acoustic analysis. Therefore, firstly it is necessary to eliminate the inter-group differences for vowel recognition. Further, we experimented with vowel recognition by using linear discriminant functions which were derived from the statistics of the formant frequencies, parcor coefficients and reflection coefficients of 60 persons (10 persons from each group). When the group of a speaker was unknown, the recognition rate was about 77%, but when it was known, about 98% for training samples. On the other hand, for test samples of unknown 60 speakers, the results were about 74% and 87%, respectively.

And also we evaluated the speaker-factor in V_1CV_2 utterances by using a multivariate analysis of variance as described in the previous section, where /a,i,u,e,o/ for V_1 and V_2, /m,n,ŋ/ for C and the number of speakers was 10 [73]. The result is shown in Fig.14. We found the following main results.

(1) The main effect of the speaker-factor is relatively large at all time points, and especially larger than that of the consonant-factor at the stationary part of the nasal consonant.
(2) The interaction between the speaker-factor and the vowel-factor is larger than the main effect of the consonant-factor at the stationary part of the vowel. However it is smaller than the interaction between the consonant-factor and the vowel-factor at the boundary between the consonant and the vowel, where the important information about the consonant seems to exist.

The second is consistent with the experiment results of R.Nakatsu and M.Kohda [81], that is, that the recognition of consonants by using VCV patterns of a specific person was not as sensitive for many speakers. However, S.Saito and S.Furui reported that there was personal information in dynamic characteristics of speech spectra [82].

It is well-known that even the feature parameters of the same speaker vary from time to time. H.Matsumoto and T.Nimura analyzed the VCV utterances by taking the time-factor (over 259 days) into consideration through the same multivariate variance method [83]. From their analysis, it was found that the time-factor was as large as the interaction between speaker-factor and phoneme-factor. And also M.Kohda and S.Saito reported that the influence of the time interval between reference samples and test samples is not significant in the case of less than one month [84].

(b) normalization of speaker differences

Two approaches have been investigated in order to solve the problem of speaker differences: normalization and learning. The fundamental idea of the normalization approach [85-88] is based on

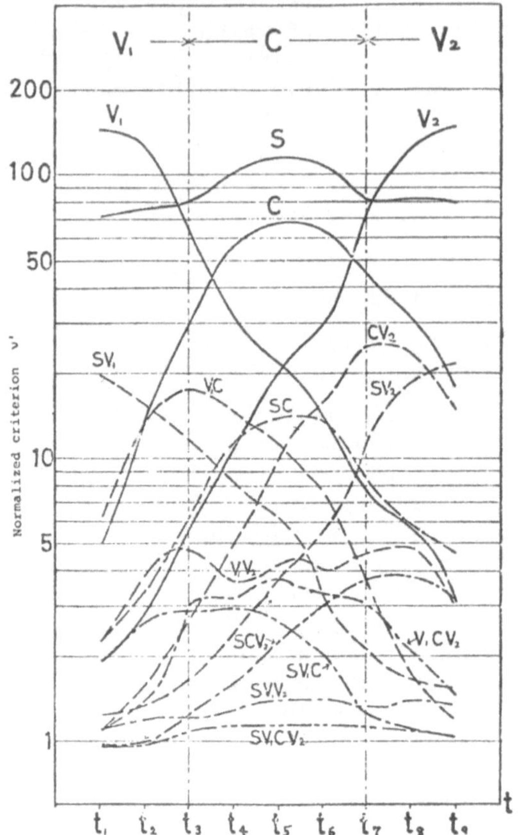

Fig. 14 Multivariate analysis of variance for four-factor design in VCV utterances (S: speaker-factor).

the assumption that there exist certain invariant relationships. Almost all methods hypothesize that the vocal-tract configurations of the speakers are similar to each other and differ only in length.

H.Fujisaki et al. transformed the first three formant frequencies (F_1, F_2, F_3) to the polar coordinates (r,θ,ϕ), that is [86],

$$r = (F_1^2 + F_2^2 + F_3^2)^{1/2}$$

$$\theta = \tan^{-1}(F_1/F_2)$$

$$\phi = \tan^{-1}\{(F_1^2 + F_2^2)^{1/2}/F_3\}$$

Thus, the above hypothesis leads to the normalized coodinates(θ,ϕ), because only r depends on the vocal-tract length.

H.Wakita normalized directly the observed formants and bandwidths to the standard values by using the estimated vocal tract length (l), as follows [87].

$$\hat{F}_i = F_i \cdot l/l_s \qquad \hat{B}_i = B_i \cdot l/l_s$$

where, l_s is the standard vocal-tract length.

However such a hypothesis is not always satisfactory because vocal geometry is different among men, women and children. It is known, for example, that women and children have shorter pharynges in relation to their oral cavities [88]. Furthermore we found that the relationships among formants are not constant even among men and that it varies from age to age (see Table 7).

M.R.Sambur and L.R.Rabiner proposed a speaker-adaptive recognition method [89]. In this technique for self-normalization of parameters, many of the most significant thresholds were obtained from measurements made directly on the speech sample being recognized. This makes possible a speaker-or time-independent system, but it requires at least one sample of each spoken word.

H.Matsumoto and H.Wakita have been trying a speaker-independent spectral matching method by using a dynamic programming technique as described in section 3.2(b).

(c) learning of speaker differences

Although the normalization approach is very theoretical, the current study still has to solve many problems relating to speech recognition. On the other hand, the learning approach [45,90-93] is a reliable and practical method in the current art of speech recognition.

Speaker-dependent reference patterns are usually made from a few learning samples. However many samples are necessary to learn the reference patterns of all phonemes. Therefore, M.Kohda and S.Saito proposed a selective procedure which selected the fewest optimum samples for learning a vocabulary set [90]. The spectra of voiced consonants vary not only from speaker to speaker, but also with contexts. In order to learn them speedily, we used the learned spectra of vowels, because it is very difficult to learn the spectra by a few samples of voiced consonants [91]. Further, S.Furui extended our method as follows [92]. When the feature parameters of phonemes are represented by the vector with N components, unknown reference is estimated by

$$\hat{x}_i = (r/\sum_{j\in X_L} w_{ij})\cdot \sum_{j\in X_L} w_{ij}\cdot \phi_{ij}\cdot x_j + (1-r)y_i$$

$$i\in X_L\ , \quad 0\leq r\leq 1$$

where, X = set of phonemes

X_L = set of learned phonemes

ϕ_{ij} = NxN matrix for the estimation of parameters of the phoneme i from the parameters of the phoneme j.

$w_{ij} = \frac{1}{N}\sum_{l=1}^{N} R_{ijl}$ weight coefficient.

R_{ijl} = multiple correlation in the case that the l-th component of the phoneme i is estimated by the parameters of the phoneme j.

$$y_i = \begin{cases} x_i & : i \in X_L \\ \bar{x}_i & : i \in X - X_L \end{cases}$$

$\bar{x}_i$ = average feature parameters of the phoneme i for every speaker.

B.T.Lowerre [93] and we [91] have tried a non-supervised learning method which does not directly require learning samples. This approach has also the advantages that it can trace the spectral time-variation caused by the environments or exhaustion of speaker and can be used together with other learning methods.

VI. CONCLUSION

In this paper, we described and discussed significant problems of phoneme recognition. Because each problem, segmentation, labeling, co-articulation and speaker differences, is not simple and cannot be solved by a simple model, we must treat the different phenomena caused by various factors simultaneously for each problem.

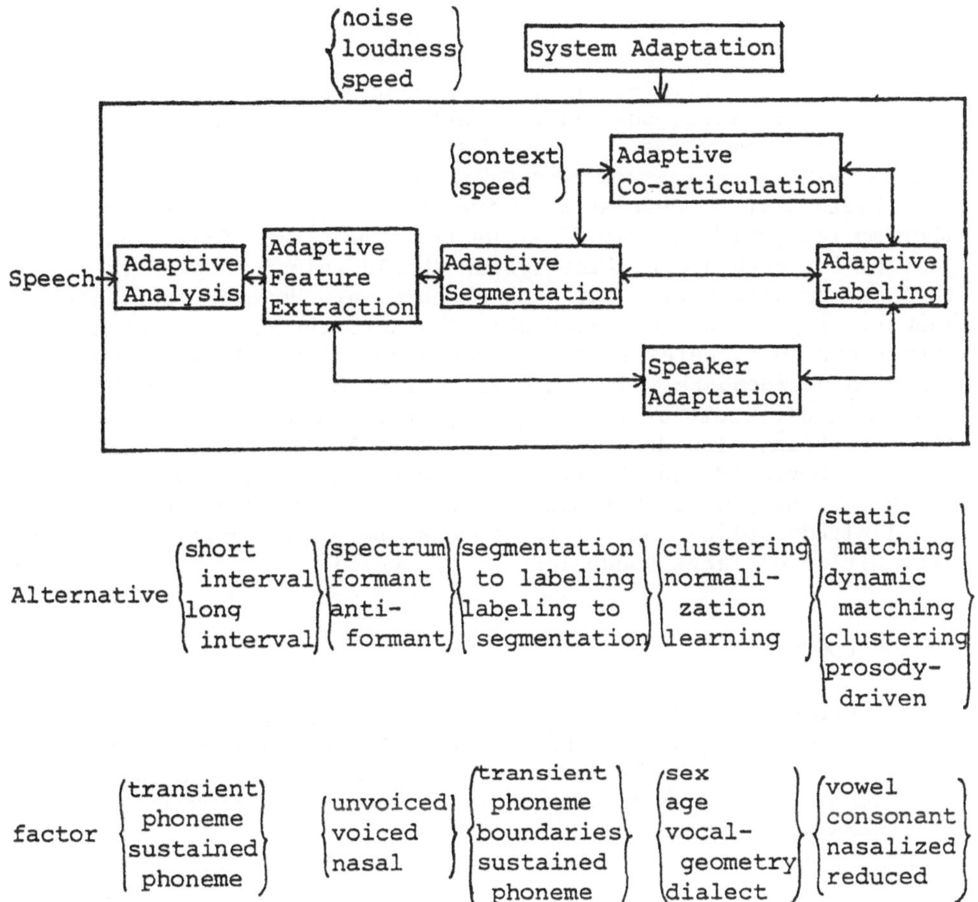

Fig. 15 An adaptive model for phoneme recognition

Therefore a recognition system must have a switching function to select suitable functions for each case at every stage.

We can classify the varieties of speech into four types. The first is the variety of phonemes themselves such as voiced/unvoiced, sustained sound/transient sound, nasal/non-nasal and so on. The second is the variety of sounds caused by the manner of speech production such as speed of utterance, loudness, prosody and so on. The third is the process of speech production such as stress/reduced, nasalization, so-called co-articulation and so on. The last is speaker differences.

Lastly, we propose an adaptive system [94] for automatic phoneme recognition. Fig.15 shows the model, and alternatives and factors for each block. The system should select automatically the optimum alternative for each phenomenon. However it cannot know readily by only one step which factors the observed phenomena correspond to. Therefore such a hierarchical model must have the function of feedback. Such a hierarchical model may make the problems more difficult by dividing them into sub-problems, because these problems also depend on one other. But it has the advantage that a sub-problem can be treated more easily than a large problem. We think, the pattern matching method on the basis of word to word has limited application for continuous speech recognition, speech recognition of large vocabulary or unspecified speakers.

ACKNOWLEDGEMENTS

I would like to express many thanks to Dr. Sei-ichi Nakagawa who has been of such great help when preparing this manuscript, and I also wish to thank Dr. Yasuhisa Niimi and Dr. Nakagawa with whom we had a discussion concerning the contents of my lecture in the NATO ASI programme.

References

[1] D.R.Broad and J.E.Shoup: Concepts for acoustic phonetic recognition, in Speech Recognition, ed. R.Reddy, pp.243-274, Academic Press (1975).

[2] S.Nakagawa: A machine understanding system for spoken Japanese sentences, Ph.D thesis, Kyoto University (1976).

[3] T.Sakai and S.Nakagawa: A speech understanding system of simple Japanese sentences in a task domain, IECEJ Trans. Vol-60E, No.1, pp.13-20(1977).

[4] T.Sakai ans S.Nakagawa: Speech understanding system - LITHAN - and some applications, Proceedings of the 3rd IJCPR, pp.621-625 (1976).

[5] S.Nakagawa and T.Sakai: A word recognition method from a classified phoneme string in the LITHAN speech understanding system, Conference Record of ICASSP, pp.726-730(1978).

[6] S.Nakagawa and T.Sakai: On parsing direction and tree search in the LITHAN speech understanding system, ASA and ASJ Joint Meeting, JASA, Vol.64S, No.1 (1978).

[7] J.J.Wolf: Efficient acoustic parameters for speaker recognition, JASA, Vol.51, No.6, pp.2044-2056(1972).

[8] B.S.Atal: Effectiveness of linear predictive characteristics of the speech wave for automatic speaker identification and verification, JASA, Vol.55, No.6, pp.1304-1312 (1974).

[9] A.E.Rosenberg and M.R.Sambur: New technique for automatic speaker verification, IEEE Trans. Vol.ASSP-23, No.2, pp.169-176 (1975).

[10] M.R.Sambur: Selection of Acoustic Features for speaker identification, IEEE Trans. Vol.ASSP-23, No.2, pp.176-182 (1975).

[11] J.D.Markel, B.T.Oshika and A.H.Gray: Long-term averaging for speaker recognition, IEEE Trans. Vol.ASSP-25, No.4, pp.330-337 (1977).

[12] E.Bung et al.: Statistical techniques for automatic speaker recognition, Conference Record of ICASSP, pp.772-775(1977).

[13] R.S.Cheung and B.A.Eisenstein: Feature selection via dynamic programming for text-independent speaker identification, IEEE Trans. Vol.ASSP-26, No.5, pp.397-403(1978).

[14] Y.Grenier: Speaker identification from linear prediction, Proceedings of the 4-th IJCPR, pp.1019-1021(1978).

[15] W.Klein, R.Plomp and L.C.W.Pols: Vowel spectra, vowel spaces, and vowel identification, JASA, Vol.48, No.4, pp.999-1009(1970).

[16] L.C.W.Pols, H.R.C.Tromp and R.Plomp: Frequency analysis of Dutch vowels from 50 male speakers, JASA, Vol.53, No.4, pp.1093-1101(1973).

[17] H.G.Goldberg: Segmentation and labeling of speech: a comparative performance evaluation, Ph.D thesis, Carnegie-Mellon University (1975).

[18] P.F.Castelaz and R.J.Niederjohn: A comparison of linear prediction, FFT, zero-crossing analysis techniques for vowel recognition, Conference record of ICASSP, pp.541-545(1978).

[19] A.Ichikawa, Y.Nakano and K.Nakata: Evaluation of various parameter sets in spoken digits recognition, IEEE Trans. Vol.AU-21, No.3, pp.202-209(1973).

[20] G.M.White and R.B.Neely: Speech recognition experiments with linear prediction, bandpass filtering and dynamic programming, IEEE Trans, Vol.ASSP-24, No.2, pp.183-188(1976).

[21] H.A.Barger and K.R.Rao: A comparison study of phonemic recognition by discrete orthogonal transforms, Conference Record of ICASSP, pp.553-556(1978).

[22] S.Chiba, M.Watari and T.Watanabe: A speaker-independent word recognition system, Proceedings of the 4-th IJCPR, pp.995-999 (1978).

[23] H.Kasuya and H.Wakita: Speech segmentation and feature normalization based on area functions, Conference Record of ICASSP, pp.29-32(1976).

[24] T.Nakajima et al.: Estimation of vocal tract area function by adaptive reconvolution and adaptive speech analysis system, ASJ Trans. Vol.31, No.3, pp.157-166(1978, in Japanese).

[25] K.Shirai and H.Honda: Feature extraction for speech recognition based on articulatory model, Proceedings of the 4-th IJCPR, pp.1064-1068 (1978).

[26] P.Mermelstein: Automatic segmentation of speech into syllabic units, JASA, Vol.53, No.4, pp.880-883(1975).

[27] R.Nakatsu and M.Kohda: Speech recognition of connected words, Proceedings of the 4-th IJCPR, pp.1009-1011(1978).

[28] H.Kasuya and H.Wakita: On segmentation of continuous speech, Technical report on speech of ASJ, S78-10 (1978, in Japanese).

[29] L.R.Rabiner and M.R.Sambur: An algorithm for determing the endpoints of isolated utterances, Bell Sys. Tech. J. Vol.54, pp.297-315(1975).

[30] L.R.Rabiner, et al.: A comparative performance study of several pitch detection algorithms, IEEE Trans. Vol.ASSP-24, No.5, pp.399-418(1976).

[31] B.S.Atal and L.R.Rabiner: A pattern recognition approach to voiced-unvoiced-silence classification with applications to speech recognition, IEEE Trans. Vol.ASSP-24, No.3, pp.201-212 (1976).

[32] S.Nakagawa and T.Sakai: Some properties of Japanese sounds through perceptual experiments and spectral analyses, Studia Phonologica XI, pp.48-64(1977).

[33] W.A.Lea, M.F.Madress and T.E.Skinner: A prosodical-guided speech understanding strategy, IEEE symposium on speech recognition, pp. 38-44 (1974).

[34] P.Mermelstein: The syntax of acoustic segments, Conference Record of ICASSP, pp.29-32(1976).

[35] R. Demori, P.Laface and E.Piccolo: Automatic detection and description of syllabic features in continuous speech, IEEE Trans. Vol.ASSP-24, No.5, pp.365-379(1976).

[36] K.W.Otten: Approaches to the machine recognition of conversational speech, in Advances in Computers, ed. M.Yovits, pp.127-163, Academic Press(1971).

[37] T.Nakajima and T.Suzuki: Application of the articulatory feature vowel system to continuous speech, Record of Joint Meeting of ASJ, 2-2-5, Oct. 1978(in Japanese).

[38] P.Mermelstein: On detecting nasals in continuous speech, JASA, Vol.61, No.2, pp.581-587(1977).

[39] N.R.Dixon and H.F.Silverman: A general language-operated direction implementation system (GLODIS): its application to continuous speech recognition, IEEE Trans. Vol.ASSP-24, No.2, pp.137-162(1976).

[40] F.Jelinek: Continuous speech recognition by statistical methods, Proceedings of the IEEE, Vol.64, No.4, pp.532-556 (1976).

[41] D.R.Reddy: Computer recognition of connected speech, JASA, Vol.42, pp.329-347(1967).

[42] C.J.Weinstein et al.: A system for acoustic-phonetic analysis of continuous speech, IEEE Trans, Vol.ASSP-23, No.1, pp.54-67 (1975)

[43] K.Shikano and M.Kohda: On the LPC distance measures for vowel recognition in continuous utterances, Technical report on speech of ASJ, S78-19(1978, in Japanese).

[44] F.Itakura: Minimum prediction residual principle applied to speech recognition, IEEE Trans. Vol.ASSP-23, No.1, pp.67-72 (1975).

[45] M.Kohda, S.Hashimoto and S.Saito: Spoken digit mechanical recognition system, IECEJ Trans. Vol.55-D, No.3, pp.186-193 (1972, in Japanese).

[46] A.H.Gray and J.D.Markel: Distance measures for speech processing, IEEE Trans. Vol.ASSP-24, No.5, pp.380-391(1976).

[47] H.F.Siverman and N.R.Dixon: A comparison of several speech-spectra classification methods, IEEE Trans, No.4, pp.289-298(1976).

[48] T.Nakajima and T.Suzuki:Study on variation of vowel tract shapes in continuous speech and vowel discrimination experiment based on articulatory feature extraction, Technical report on speech of ASJ, S77-42 (1977, in Japanese).

[49] Y.Niimi: A method for forming universal reference patterns in an isolated word recognition system, Proceedings of the 4-th IJCPR, pp.1022-1032(1978).

[50] K.Tanaka: A standard category pattern making method with application to phoneme recognition, Proceedings of the 4-th IJCPR, pp.1030-1032(1978).

[51] S.Nakagawa and T.Sakai: A real time spoken word recognition system in a large vocabulary with learning capability of speaker differences, Proceedings of the 4-th IJCPR, pp.985-989 (1978).

[52] V.M.Velichko and N.G.Zagoruiko: Automatic recognition of 200 words, Int.J.Man-Machine Studies, Vol.2, pp.223-234 (1970).

[53] H.Sakoe and S.Chiba: A dynamic programming approach to continuous speech recognition, Report. 7-th ICA, 20-c-13(1971).

[54] H.Sakoe and S.Chiba:Dynamic programming algorithm optimization for spoken word recognition, IEEE Trans. Vol.ASSP-26, No.1, pp.43-49(1978).

[55] K.Tanaka: A dynamic processing approach to extraction and categorization of phonemic information, Conference Record of ICASSP, pp.5-8(1976).

[56] H.Matsumoto and H.Wakita: Vowel normalization by frequency warping, ASA and ASJ Joint Meeting, JASA, Vol.64S, No.1 (1978).

[57] R.W.Christiansen and C.K.Rushforth: Detecting and locating key words in continuous speech using predictive coding, IEEE Trans. Vol.ASSP-25, No.5, pp.361-367(1977).

[58] W.A.Woods: Motivation and overview of BBN SPEECHLIS: an experimented prototype for speech understanding research, IEEE Trans. Vol.ASSP-23, No.1, pp.2-10 (1975).

[59] K.Shikano and M.Kohda: An estimation system of phoneme recognition rate of phoneme lattice, Record of Joint Meeting of ASJ, 3-1-17, Oct. 1977 (in Japanese).

[60] H.Mizukami: Influence of phoneme recognition ability on word recognition rate, Graduation thesis, Dept. of Inform. Science, Kyoto University (1979, in Japanese).

[61] Y.Takeuchi: Perceptual study of segmented Japanese monosyllables, Studia Phonologica I, pp.70-85(1961, in Japanese).

[62] S.E.G.Öhman: Perception of segment of VCCV utterances, JASA, Vol.40, No.5, pp.979-988(1966).

[63] W.A.Grimm: Perception of segments of English spoken consonant vowel syllables, JASA, Vol.40, No.5, pp.1454-1461(1966).

[64] H.Kuwahara and H.Sakai: Perception of vowels and C-V syllables segmented from connected speech, ASJ Trans. Vol.28, No.5, pp.225-234(1972, in Japanese).

[65] T.Gray: Articulatory movements in VCV sequences, JASA, Vol.62, No.1, pp.183-193(1977).

[66] S.Kiritani and H.Hirose: Correlation analysis of the temporal patterns of articulatory movement and EMG, ASA and ASJ Joint Meeting, JASA, Vol.64S, No.1(1978).

[67] S.Sekimoto and S.Kiritani: Parameter description of tongue point movements in the production of Japanese vowels, ASA and ASJ Joint Meeting, JASA, Vol.64S, No.1(1978).

[68] K.N.Stevens and A.S.House: Perturbation of vowel articulations by consonantal context: an acoustical study, J. Speech Hearing Res. Vol.6, pp.111-128(1963).

[69] S.E.G.Örman: Coarticulation in VCV utterances: spectrographic measurements, JASA, Vol.39, No.1, pp.151-168 (1966).

[70] K.N.Stevens, A.S.House and A.P.Poul: Acoustical description of syllabic nuclei: an interpretation in terms of a dynamic model of articulation, JASA, Vol.40, No.1, pp.123-132(1966).

[71] K.M.N.Menon, P.J.Jensen and D.Dew: Acoustic properties of certain VCC utterances, JASA, Vol.46, No.2, pp.449-457(1970).

[72] D.J.Broad and R.H.Fertig: Formant-frequency trajectories in selected CVC-syllable nuclei, JASA, Vol.47, No.6, pp.1572-1582(1970).

[73] K.Tabata and T.Sakai: Evaluation of the Speaker-factor in Japanese VCV utterances, IECEJ Trans. Vol.60E, No.6, pp.284-289(1977).

[74] H.Kasuya, H.Suzuki and K.Kido: On properties of formant frequencies of vowels in meaningless words composed of three mores, Technical report on Electric Acoustics of IECEJ, EA68-13 (1968, in Japanese).

[75] H.Kuwahara and H.Sakai: Normalization of coarticulation effect for a sequence of vowels in connected speech, ASJ Trans. Vol.29, No.2, pp.91-99(1973, in Japanese).

[76] Y.Saito and H.Fujisaki: Formulation of the process of coarticulation in terms of formant frequencies and its application to automatic speech recognition, ASJ Trans. Vol.34, No.3, pp.177-185(1978, in Japanese).

[77] S.Itahashi and S.Yokoyama: Formant trajectory tracking and its approximation by second order linear system, Record of Joint Meeting of ASJ, 2-1-11, May, 1973(in Japanese).

[78] K.Tabata, A.Kamei and Y.Ohno: Hearing evaluation of speaker factor in vowel utterances, Record of Joint Meeting of ASJ, 1-5-11, Apr.1977 (in Japanese)

[79] K.Ito and S.Saito: Analysis of talker information of speech wave, Record of Joint Meeting of ASJ, 2-1-3, Oct. 1977 (in Japanese).

[80] H.Shirakata: Changes in feature parameters of Japanese vowels by age and sex of speakers, and recognition of vowels, Master thesis, Dept. of Inform. Science, Kyoto University(1979, in Japanese).

[81] F.Nakatsu and M.Kohda: On the performance of the acoustic processor in the on-line conversational speech recognition system, Record of Joint Meeting of ASJ, 4-2-7, Apr. 1977 (in Japanese).

[82] S.Saito and S.Furui: Personal information in dynamic characteristics of speech spectra, Proceedings of the 4-th IJCPR, pp.1014-1018(1978).

[83] H.Matsumoto and T.Nimura: Text-independent speaker identification using canonical discriminant analysis, the effect of speaker-factor, phoneme x speaker factor, and temporal variation factor, Technical report on Electronics and Acoustics of IECEJ, EA77-33(1977, in Japanese).

[84] M.Kohda and S.Saito: Influence of long-term variations of learning and unknown samples on recognition rate of spoken digits, Record of Joint Meeting of ASJ, 1-3-23, Oct. 1973 (in Japanese).

[85] L.J.Gerstman: Classification of self-normalized vowels, IEEE Trans. Vol.AU-16, pp.78-80 (1968).

[86] H.Fujisaki, N.Nakamura and K.Yoshimoto: Normalization and recognition of sustained Japanese vowels, ASJ Trans. Vol.26, No.3, pp.152-153 (1970).

[87] H.Wakita: Normalization of vowels by vocal-tract length and its application to vowel identification, IEEE Trans. Vol. ASSP-25, No.2, pp.183-192 (1977).

[88] G.Fant: Speech sounds and features, M.I.T. Press (1973).

[89] M.R.Sambur and L.R.Rabiner: A speaker-independent digit recognition system, BELL S.T.J., Vol.54, pp81-102 (1975).

[90] S.Saito and M.Kohda: Spoken word recognition using the restricted number of learnig samples, Conference Record of ICASSP, pp.229-232 (1976).

[91] S.Nakagawa and T.Sakai: A real time spoken word recognition system with various learning capabilities of the speaker differences, IECEJ Trans. Vol.61-D, No.6. pp.395-402 (1978, in Japanese).

[92] S.Furui: An efficient learning method for spoken word recognition, Technical report on speech of ASJ, S77-43 (1977, in Japanese).

[93] B.T.Lowerre: Dynamic speaker adaption in the HARPY speech recognition system, Conference Record of ICASSP, pp.788-790 (1977).

[94] T.Sakai: Adaptive system of pattern recognition, in Methodologies of Pattern Recognition, ed. S.Watanabe, pp.457-480, Academic Press, (1969).

AUTOMATIC PHONEME RECOGNITION IN CONTINUOUS SPEECH: A SYNTACTIC APPROACH

Renato De Mori
Istituto di Scienze dell'Informazione
Corso Massimo d'Azeglio, 42
I-10125 - Torino (Italy)

ABSTRACT. A model for assigning phonetic and phonemic labels to speech segments is presented. The system executes fuzzy algorithms that assign degrees of worthiness to structured interpretations of syllabic segments extracted from the signal of a spoken sentence. The knowledge source is a series of syntactic rules whose syntactic categories are phonetic and phonemic features detected by a precategorical and a categorical classification of speech sounds. Rules inferred from experiments and results for male and female voices are presented.

1. INTRODUCTION

Considerable effort has been made in the last few years to develop systems capable of understanding connected speech. Recent reviews of the problems involved in designing such systems and of the solutions proposed have been made by Reddy /1-2/, Martin /3/, Jelinek /4/, Klatt /5/, Wolf /6/, Reddy et al. /7/, Walker et al. /8/, Woods et al. /9/, De Mori /10/.

The purpose of this paper is to propose a method for the interpretation of speech patterns and to give some experimental results of its application. The interpretation of speech patterns involves the emission of hypotheses concerning possible phonemic transcriptions of syllable segments automatically extracted from a numerical representation of energy-frequency-time obtained by short-term spectral analysis of a spoken sentence.

J. C Simon (ed.), Spoken Language Generation and Understanding, 191-220.

Each hypothesis is evaluated and a degree of trustworthiness is assigned to it in such a way that it can be further processed for generating and coherently evaluating hypotheses about the words /11/, the syntactic structure and the semantics of the spoken sentence /12/.

The interpretation of speech patterns is difficult because any vagueness may affect the features extracted from the acoustic data and there is a degree of imprecision in the relations between acoustic features and their phonetic or phonemic interpretation.

For example, a sonorant intervocalic consonant may be characterized by a marked dip in the time evolutions of the signal energy, and in the energy in a frequency band from 3 to 5 kHz; furthermore, for such consonants, the low frequency energy (roughly below 1 kHz) is much higher than the high frequency energy (above 5 kHz). Given a specific acoustic pattern of an intervocalic consonant with its adjiacent vowels, a speech understanding system trying to interpret the pattern needs to take into account the fact that the consonant may be sonorant. This possibility can be evaluated numerically following the theory of possibility proposed by Zadeh /13/. To evaluate this possibility the vagueness inherent in the terms "marked dip" and "much higher energy" has to be numerically represented by "degrees of compatibility" between the statements and the acoustic pattern. This can be done by applying Zadeh's theory of possibility. To get the interpretation it has to be established how to combine the degrees of plausibility (or membership) of a marked dip in signal energy, in the 3-5 kHz energy and the degree of plausibility with which the low frequency energy is much higher than the high frequency energy in order to obtain the possibility of the interpretation: "the consonant is sonorant".

In a previous paper /14/, a method for obtaining acoustic patterns was proposed, together with a language for their description and syntax-directed procedures for the recognition of the vowel positions and for the segmentation of continuous speech into pseudo-syllable-segments (PSS). In this paper, a method based on fuzzy restrictions is proposed for extraxting acoustic features (such as "marked dip") from the description of acoustic patterns. A method based on fuzzy relations is also proposed for relating acoustic features with phonetic and phonemic interpretations. Finally, the use of restrictions and relations to compute the possibility of a hypothesis (a phonetic or a phonemic interpretation of a speech pattern) is illustrated.

The concepts of fuzzy restrictions and fuzzy relations are due to Zadeh /15-16/ and will be briefly recalled in Section 2 of this paper. The algorithms used for evaluating the possibility of a hypothesis will be referred to as fuzzy algorithms.

Recently, fuzzy set theory has been applied for decision-making in vowel and speaker recognition /17/. In such an approach, an unknown vowel is represented by a vector of spectral measurements $X=x_1, x_2, \ldots, x_n, \ldots, x_N$ and a membership function $\mu_j(X)$ is associated with pattern X for the j-th class. $\mu_j(X)$ is defined as a decreasing function of the Euclidean distance $d(X, R_j)$ between X and R_j, the reference vector of the j-th class. The decision is made by assigning X to the class j for which $\mu_j(X)$ is maximum. The application of fuzzy sets in such a case is just a way of evaluating the closeness of points in a Euclidean space.

The approach proposed in this paper and the rules described in Sect. 2 are an attempt to formalize the intuitive logic used by a phonetician. Membership functions are defined as a measure of the plausibility of an interpretation of a speech pattern.

The advantages of such an approach lie in its flexibility. Each relation and each restriction can be established to represent as closely as possible the knowledge attained in research on phonetics and to optimize the recognition performances.

The results were limited to one type of PSS, namely vowel-consonant-vowel (VCV) pseudosyllables, but they refer to every possible coarticulation of such syllables in the Italian language; the syllables were extracted from continuous speech without any limitation and the results were obtained by applying the same rules for different (male and female) speakers.

The fuzzy algorithms presented in this paper are used in a speech understanding system (SUS) organized with several levels of knowledge sources (KS_s). Each KS consists of syntactic rules relating an item of a given level, represented by a syntactic category, with items of lower levels. There are many levels between the acoustic and the lexical ones.

The first levels just above the acoustic level have phonetic features as syntactic categories. Features that will be introduced in Sect. 2, like vocalic-nonvocalic, sonorant-nonsonorant, etc., are extracted by a precategorical classification wirh a procedure that does not require any knowledge of the context.

Other categories, like liquid and nasal, are introduced to generate phonetic transcriptions with procedures that are context-dependent. Finally, phonemes belonging to the hypothesized phonetic classes are attached to the corresponding speech intervals, using algorithms discussed in Sect. 3.

Fuzzy algorithms are used whenever a KS is invoked to generate an interpretation (a hypothesis). The hypothesis is a syntactic category H related by rules to items previously hypotheseized and belonging to lower levels of interpretation. Each rule used is associated with a fuzzy relation and each lower level item used by the rule has been previously associated with a possbility of being a correct interpretation of a speech pattern p. To use H as an interpretation of p as well it is necessary to evaluate the possibility that H represents p correctly. This possibility is computed by composing the fuzzy relation between H and its constituents with the possibility of the consituents themselves.

Details on the composition of fuzzy relations and restrictions can be found in papers by Zadeh /18, 22/.

The possibilities of the hypotheses are used for scheduling the activation of the KSs following the system that controls the process of understanding. Problems of knowledge representation and control strategies would make this paper too long and diverse. Non-homogeneous parts will be considered in other papers; a preliminary and concise presentation of them can be found in /32/.

The reasons for using possibilities instead of probabilities or purely heuristic methods for scoring hypotheses can be summarized as follows.

The theory of possibility, in contrast with heuristic approaches, offers algorithms for composing hypotheses evaluations which are consistent with axioms in a well-developed theory; the theory of possibility, rather than the theory of probability, relates to the perception of degrees of evidence instead of degrees of likelihood or frequency. The aim of the approach proposed in this paper is to express the evidence of a hypothesis concerning a speech pattern with the possibility of having high evidence measurements for clearly interpretable patterns even if the patterns and the features considered are scarcely probable. In any case, using possibilities does not prevent us from using statistics in the estimation of membership functions. But this estimation does not necessarily require such a large number of experiments as the estimation of a probability density. The use of phonetic and phonemic features as syntactic categories of a fuzzy grammar is a new idea allowing one to incorporate in a knowledge source the vagueness with which the relations between acoustic and phonetic or phonemic features are known.

Using such rules, degrees of evidence of speech interpretation can be obtained and a change of a single rule may have the same effect of updating a large number of prototypes in a more conventional parametric approach.

2. ALGORITHMS FOR PRECATEGORICAL CLASSIFICATION

Generalities

Precategorical classification consists primarily of the assignment of phonetic features to non-vocalic segments. The main purpose of such a classification is for segmentation of continuous speech into presudosyllable segments (PSS) and for driving a context-dependent extraction of more detailed features.

The spectra are then processed in order to obtain some "gross spectral features":

- S : the total energy of a spectrum,
- B : the energy in the 200-900 Hz band,
- F : the energy in the 5-10 kHz band,
- A : the energy in the 3-5 kHz band,
- R_V: the ratio between B and F.

The gross features are described by a language presented in /14/ and the description in terms of these features, denoted DGF, is used for precategorical classification, denoted PC; PC is then used for segmenting continuous speech into syllabic units by a syntactic recognition algorithm introduced in /14/.

The results of the precategorical classification are used for driving the extraction of detailed spectral features such as formant evolutions, characteristics of frication noise, burst, etc. following an approach presented in /14/. Such features are then described by a language presented in /14/ and a linguistic description DDF of the acoustic patterns is given.

Precategorical classification has also been postulated in human perception (see Stevens /20/ for a discussion on this item).

The approach proposed in this paper is based on the detection of phonetic categories related to some of the distinctive features proposed by Hughes and Hemdal /21/. The tree of Fig. 1 summarizes a scheme for precategorical classification that has been implemented, giving satisfactory results with few ambiguities. A node represents a feature that can be hypothesized only if the feature corresponding to the father has been previously hypothesized. Thus, for example, the feature "affricate" can be hypothesized only for nonsonorant consonants. Let us call such features "Precategorical Classification Features" (PCF).

The phonetic features shown in the tree in Fig. 1 are syntactic categories related to the lower level description DGF of the acoustic features by a "branching questionnaire" of the type proposed by Zadeh in a recent paper /22/. Thus PCF is assigned to a speech segment after answering a composite classificational question $Q \hat{=} B$, where the bodies of the component questions Q_j (j=1, 2, ..., N) are fuzzy sets $B_1, B_2, \ldots, B_n$ involved in an analytic representation of N /22/. The fuzzy sets $B_1, B_2, \ldots, B_n$ are linguistic variables defined over the range of acoustic measurements that phoneticians have found useful for characterizing distinctive features.

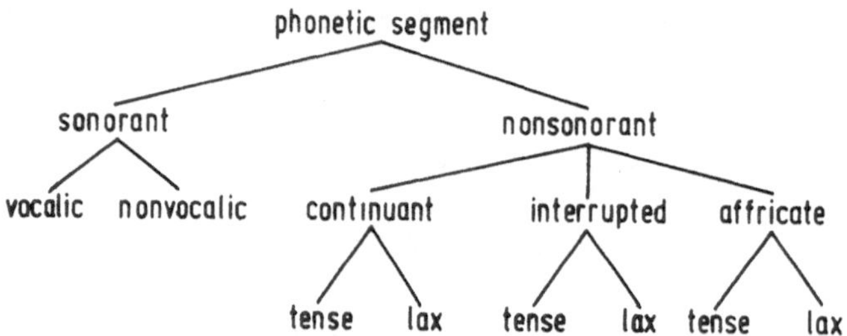

Fig. 1. Tree of the features detectable at the level of precategorical classification using context-independent algorithms.

Let Q_i (i=1, 2, ..., K) be the composite question related to the i-th PCF to be hypothesized. The answers to component questions Q_{ij} of the composite question Q_i are fuzzy linguistic variables (e.g. high, medium, low, more or less high, more or less low) whose membership functions are computed from diagrams stored in the phonetic source of knowledge of the SUS and are defined over the universe of an acoustic measurement. The answers to the composite question Q_i are related by syntactic rules to the answers to the component questions Q_{ij}; thus each answer to Q_i can be associated with a membership function whose value is computed under the control of semantic rules from the values of the membership functions associated with each answer to the component questions.

This use of fuzzy algorithms models to some extent the fact

that most of the acoustic-phonetic properties of speech sounds are only known with a degree of vagueness, e.g.: the signal energy is high for vowels, nonsonorant consonants have high frequency components, in unvoiced stops there is an interval of silence followed by some noise.

A fuzzy linguistic variable, representing a judgement that can be expressed after the inspection of some acoustic parameters, is defined by a fuzzy restriction /15/:

$$R(X, u)$$

where u is a generic value of an acoustic parameter and X is the subjective judgement. Abbreviating R(X, u) to R(X), the degree to which the assignment equation:

$$X = u : R(X) \tag{1}$$

is satisfied, still has to be established. Denoting such a degree as the compatibility C(u) of u with R(X), by definition /15/ it follows that:

$$C(u) = \mu_{R(X)}(u) \tag{2}$$

where $\mu_{R(X)}(u)$ is a membership function defining R(X) as a fuzzy subset of u.

For the sake of simplicity, the membership functions will be labelled by abbreviations of adjectives of phonetic features.

Fig. 2 shows the compatibility function of the fuzzy variable X:: = < high consonant durations > over the universe of the possible values of u which represents time in this example.

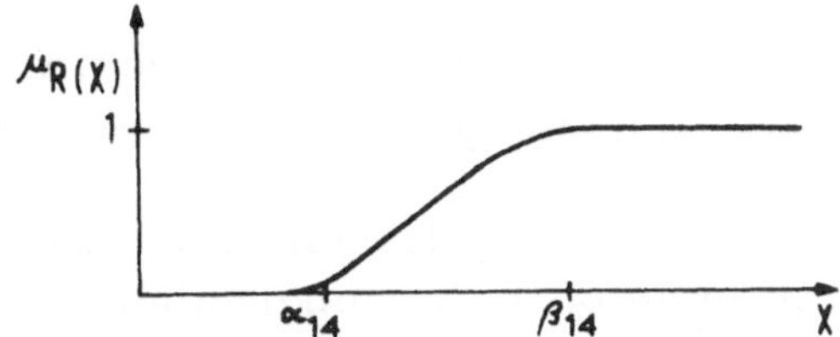

Fig. 2. Compatibility function of the fuzzy variable x::=<high consonant duration>

The fuzzy variables form a basis for a composite question Q. An answer to Q may be interpreted as a specification of the grade of evidence of a phonetic feature in a speech segment. This grade is a function of the grades of membership of the speech segment with every fuzzy variable.

The membership functions associated with each restriction were established subjectively after inspection of the distribution of acoustic measurements made on a large number of sound samples.

Let Σ be the alphabet of the linguistic variables that may be answered to the component questions of the "questionnaire" used for characterizing the phonetic feature (FF). Obviously $X \in \Sigma$. Let $\underline{p}$ be an acoustic pattern represented by its description in terms of acoustic features. We are interested in the following possibility: Poss $\{$(FF) is in $\underline{p}\}$. Let $D \in \Sigma^*$ be a string of linguistic variables; the string D identifies a composite fuzzy restriction $R_{\underline{p}}(D)$ defined on some of the acoustic attributes of $\underline{p}$. The KS invoked for generating or verifying the hypothesis that (FF) is in $\underline{p}$ is a fuzzy naming relation R(D, (FF)).

Fuzzy naming relations have been introduced by Zadeh /23/ and are represented in the approach proposed in this paper by simple fuzzy grammars. Fuzzy grammars, also introduced by Zadeh /23/, are similar to stochastic grammars (see Fu /24/ for details); their rewriting rules are associated with a degree of plausibility. In our case this degree of plausibility represents the imprecision of the knowledge. Examples of fuzzy grammars will be given below and in the successive sections. The possibility that a feature FF is present in a pattern $\underline{p}$ is obtained by the following composition:

$$R_{\underline{p}}(FF) = R_{\underline{p}}(D) \circ R(D, (FF))$$

and is computed as follows:

$$\text{Poss}\ \{(FF) \text{ is in } \underline{p}\} = \operatorname*{Sup}_{D \in \Sigma^*} \{\mu_{\underline{p}}(D) \wedge \mu((FF), D)\} \tag{3}$$

where $\wedge$ is the min operator, $\mu_{\underline{p}}(D)$ is the compatibility of the string D with the pattern $\underline{p}$ and $\mu((FF), D)$ is the compatibility of the name (FF) with the string D.

The intersection of two fuzzy linguistic variables has a membership equal to minimum of the membership of the variables. In this way the intersection is the largest fuzzy set which is contained in both the fuzzy linguistic variables because its membership is smaller than or equal to the membership of the variables for every point of the space where the variables are defined.

The use of linguistic variables and fuzzy algorithms is an attempt to represent acoustic-phonetic knowledge, which is mostly expressed in a non-numerical form, by a formalism suitable for computer systems. Acoustic-phonetic features are then processed in a logical, non-linear, hierarchical system of concurrent processes; such a system is certainly more suitable for application in SUS than the crude mathematical approach of measuring distances in an N-dimensional space. Furthermore, the degrees of consistency of a hypothesis do not need to be known with great accuracy and should reflect a subjective, heuristic evaluation of the help the hypothesis can offer in grasping the meaning of the spoken sentence.

The hypothesis assignment consists basically of table look--up and the evaluation of min-max functions, making the program execution rapid; this aspect, together with flexibility in modifying the strategy as well as the sources of knowledge adds motivations for using the proposed method in SUS design.

Sonorant-Nonsonorant classification

The classification of a consonant as sonorant or nonsonorant is performed after the detection of vocalic intervals (see /14/ for details) and is obtained as an answer to a composite question:

Q_1 = is the interval corresponding to the acoustic pattern p non sonorant ?

The question Q_1 has six component questions:

- Q_{11} = how high is R_V ?
- Q_{12} = how low is the minimum dip in R_V with respect to the values of R_V in the preceding and following vowels ?
- Q_{13} = what is the minimum value of S with respect to the value assumed by S on the silences ?
- Q_{14} = what is the duration of the signal dip ?
- Q_{15} = what is the minimum dip in the signal ?
- Q_{16} = what is the maximum dip in R_V ?

Each question admits two possible answers: low or high. These answers will be indicated as:

l_{1i}, h_{1i}

where the subscripts 1i (i=1, 2, . . . , 6) refer to the answer of the 1i-th question; it is assumed that:

$$l_{ji} = \overline{h}_{ji} \qquad (4)$$

The answer to such questions are fuzzy linguistic variables defined over the ranges in which the parameters they refer to may vary. Such parameters are defined as follows:

- $u_{11} = R_v$
- $u_{12} = \min (R_{vp} - R_{vc}; R_{vf} - R_{vc})$
- $u_{13} = \min_{S \in dip} (S - S_{sil})$
- u_{14} = consonant duration
- $u_{15} = \min (S_p - S_c; S_f - S_c)$
- $u_{16} = \max (R_{vp} - R_{vc}; R_{vf} - R_{vc})$,

where the subscript p refers to the detected vowel preceding the interval in which a nonsonorant feature is being sought; the subscript f refers to the detected vowel following the interval, the subscript c refers to the consonant interval and S_{sil} is the level of the signal energy in the silence.

The membership functions were defined after considering the range of the above parameters for each consonant in every context assigning values different from 1 to $\mu_{h_{1i}}$ and $\mu_{l_{1i}}$ (i=1, 2, ..., 6) only in the range where sonorant and nonsonorant sounds may coexist.

Notice that $\mu_{l_{1i}}$ is for $\mu_{low}(u_{1i})$ and $\mu_{h_{1i}}$ is for $\mu_{high}(u_{1i})$ furthermore $\mu_{l_{1i}} = 1 - \mu_{h_{1i}}$. (5)

The fuzzy restrictions obtained from the experiments and corresponding to the linguistic values high have the diagram shown in Fig. 2; the break points, indicated as α_{14} and β_{14} in Fig. 2 are the bounds of the interval where the memberships are neither 0 nor 1. In general, the break points for the variable h_{ji} will be indicated as α_{ji} and β_{ji}; these parameters assume different values depending on the questions, as listed in Tab. I.

TABLE I

Variable	α_{ji}	β_{ji}	Dimension
u_{11}	-15	7	dB
u_{12}	- 3	21	dB
u_{13}	2	20	dB
u_{14}	50	130	msec
u_{15}	6	23	dB
u_{16}	0	24	dB
u_{21}	10	16	dB
u_{22}	3	14	dB
u_{31}	-30	-20	dB
u_{32}	12	18	dB
u_{33}	80	130	msec
u_{34}	20	70	msec
u_{35}	20	30	dB

The answers to atomic questions are related to the values of hypotheses about phonetic features by fuzzy rewriting rules

/23/. These rules are the usual syntactic rules involved in the definition of phrase grammars. Furthermore, a degree of "grammaticality" and a semantic rule are associated with each syntactic rule. The grammaticality is an a-priori evaluation of the plausibility of the syntactic rule. The semantic rule associated with a syntactic one is used for computing the value of the hypothesis expressed by the left-side member of the syntactic rule as a function of the membership functions of the right--side components.

A set of fuzzy rules has been inferred for defining the syntactic categories sonorant and nonsonorant that correspond to the answers to Q_1. The inference method is described elsewere /25/. The inferred rules are given by the following:

Π_1

$\langle\text{nonsonorant}\rangle :: \overset{1}{=} \langle\text{dip in S}\rangle$ (P11)

$\langle\text{nonsonorant}\rangle :: \overset{1}{=} \langle\text{dip in } R_V\rangle$ (P12)

$\langle\text{dip in S}\rangle :: \overset{0.85}{=} l_{13} p_S$ (P13)

$\langle\text{dip in S}\rangle :: \overset{0.9}{=} h_{14} p_S$ (P14)

$\langle\text{dip in S}\rangle :: \overset{1}{=} h_{15} p_S$ (P15)

$\langle\text{dip in S}\rangle :: \overset{1}{=} h_{14} l_{13} p_S$ (P16)

The notation $\langle\text{dip in S}\rangle :: \overset{0.85}{=} l_{13}$ means that l_{13} (a low difference between the minimum of S and the silence level) allows us to generate the hypothesis that there is a dip in S with 0.85 plausibility.

The semantic rule associated with the syntactic definition of $\langle\text{dip in S}\rangle$ is derived from (3) as follows:

$$\mu_{\langle\text{dip in S}\rangle} = p_S \wedge \{(0.85 \wedge \mu_{l_{13}}) \vee (0.9 \wedge \mu_{h_{14}}) \vee \vee h_{15} \vee (\mu_{l_{13}} \wedge \mu_{h_{14}})\} \quad ; \tag{6}$$

p_S is a boolean variable defined as follows:

$$p_S = \begin{cases} 1 & \text{if } (S - S_{sil}) \leq 8 \text{ dB} \\ 0 & \text{if } (S - S_{sil}) > 8 \text{ dB} \end{cases}$$

$\langle\text{dip in } R_V\rangle :: \overset{0.6}{=} l_{11}$ (P19)

$\langle\text{dip in } R_V\rangle :: \overset{0.9}{=} h_{12}$ (P110)

$\langle\text{dip in } R_V\rangle :: \overset{0.9}{=} h_{16}$ (P111)

$\langle\text{dip in } R_V\rangle :: \overset{1}{=} l_{11} h_{12}$ (P112)

$\langle\text{dip in } R_V\rangle :: \overset{1}{=} l_{11} h_{16}$ (P113)

$\langle \text{dip in } R_V \rangle \quad :: \overset{1}{=} h_{12} h_{16}$ (P114)

The semantic rule associated with the definition of ⟨dip in R_V⟩ is:

$$\mu_{\langle \text{dip } R_V \rangle} = (0.6 \wedge \mu_{l_{11}}) \vee (0.9 \wedge \mu_{h_{12}}) \vee (0.9 \wedge \mu_{h_{16}}) \vee (\mu_{l_{11}} \wedge \mu_{h_{12}}) \vee (\mu_{l_{11}} \wedge \mu_{h_{16}}) \vee (\mu_{h_{12}} \wedge \mu_{h_{16}}) \,. \tag{7}$$

The symbols $\vee$ and $\wedge$ represent the <u>max</u> and the <u>min</u> operators respectively.

The definition of the syntactic category ⟨sonorant⟩ could be simply established as: ⟨sonorant⟩ = $\overline{\langle \text{nonsonorant} \rangle}$.

Nevertheless, it has been found that this phonetic feature is better characterized in terms of fuzzy rules involving the answers to component questions that chiefly characterize the sonorant consonants. These elements are the complements of the elements appearing in the definition of nonsonorant.

π_2

⟨sonorant⟩	$:: \overset{0.85}{=} \langle \text{SRV} \rangle$	(P21)
⟨sonorant⟩	$:: \overset{0.9}{=} \langle \text{SS} \rangle$	(P22)
⟨sonorant⟩	$:: \overset{1}{=} \langle \text{SRV} \rangle \langle \text{SS} \rangle$	(P23)
⟨SRV⟩	$:: \overset{0.9}{=} h_{11}$	(P24)
⟨SRV⟩	$:: \overset{0.}{=} l_{12}$	(P25)
⟨SRV⟩	$:: \overset{1}{=} l_{16}$	(P26)
⟨SRV⟩	$:: \overset{1}{=} h_{11} \; l_{12}$	(P27)
⟨SS⟩	$:: \overset{1}{=} l_{14}$	(P28)
⟨SS⟩	$:: \overset{1}{=} l_{15} \, p_S$	(P29)

The semantic rules associated with the definitions of ⟨SRV⟩ and ⟨SS⟩ are given as follows:

$$\mu_{\langle \text{SRV} \rangle} = (0.9 \wedge \mu_{h_{11}}) \vee (0.9 \wedge \mu_{l_{12}}) \vee \mu_{l_{16}} \vee (\mu_{l_{12}} \wedge \mu_{h_{11}}) \tag{8}$$

$$\mu_{\langle \text{SS} \rangle} = \mu_{l_{14}} \vee (p_S \wedge \mu_{l_{15}}) \,. \tag{9}$$

The measure of the hypotheses ⟨sonorant⟩ and ⟨nonsonorant⟩ are computed using the (3), considering ⟨dip in S⟩, ⟨dip in R_V⟩, ⟨SRV⟩, ⟨SS⟩ as descriptions of the acoustic pattern <u>p</u> and the grammaticalities of π_1 and π_2 as memberships of the fuzzy relations between the left side and the right side phrases of the rules. One gets:

$$\text{Poss}\{\langle \text{nonsonor.} \rangle \text{ is in } \underline{p}\} = \mu_{\langle \text{dip in S} \rangle} \vee \mu_{\langle \text{dip in } R_V \rangle} \tag{10}$$

$$\text{Poss}\{\langle \text{sonorant} \rangle \text{ is in } \underline{p}\} = (\mu_{\langle \text{SRV} \rangle} \wedge 0.85) \vee (\mu_{\langle \text{SS} \rangle} \wedge 0.9) \vee (\mu_{\langle \text{SRV} \rangle} \wedge \mu_{\langle \text{SS} \rangle}) \tag{11}$$

Classification of the other features

For a further classification of nonsonorant sounds, a composite question Q_3 was introduced having the following components:

- Q_{31} = how high is R_V ?
- Q_{32} = what is the maximum value of S in the interval where R_V is less than -12 dB ?
- Q_{33} = what is the duration of dip in the signal amplitude between the two vowels ?
- Q_{34} = what is the duration of the interval where R_V is less than -12 dB ?

The answers to these questions are:

l_{3i}, h_{3i} (i=1, 2, 3, 4).

The answers to Q_3 are linguistic variables defined on the following parameters:

- $u_{31} = u_{11}$
- $u_{32} = S_M$, maximum of the signal in the interval where R_V is less than -12 dB
- u_{33} = duration of dip in S
- u_{34} = duration of the interval for which R_V is less than -12 dB
- $u_{35} = u_{15}$.

The feature ⟨nonsonorant-continuant⟩ is defined as follows:

$$\langle \text{nonsonorant-continuant} \rangle = h_{32} \cdot h_{34} \cdot l_{33} . \qquad (12)$$

Among the nonsonorants that are not continuant, a further subdivision between interrupted and affricate is made by considering the duration and the relative positions of the amplitude dip and the dip in R_V. A detailed definition of the questions pertaining to the features interrupted and affricate is omitted here for the sake of brevity.

A further specification of the nonsonorant sounds is made by assuming the feature "tense" or "lax" in accordance with the following rules:

$\Pi 4$

$\langle \text{tense} \rangle :: \stackrel{1}{=} l_{31}$

$\langle \text{tense} \rangle :: \stackrel{1}{=}$ interrupted $\cdot\, l_{31}$

$\langle \text{tense} \rangle :: \stackrel{1}{=}$ interrupted $\cdot\, h_{33}$

$$\langle\text{tense}\rangle \;::\overset{1}{\underset{1}{=}}\; \text{interrupted} \cdot h_{35}$$
$$\langle\text{lax}\rangle \;::=\; \overline{\text{tense}}$$

$$\mu_{\langle\text{tense}\rangle} = (\mu_{l_{31}} \wedge \mu_{h_{34}}) \vee \tag{13}$$

$$\vee\, \mu_{\langle\text{interrupted}\rangle} \wedge (\mu_{l_{31}} \vee \mu_{h_{33}} \vee \mu_{h_{35}})\;; \tag{14}$$

$$\mu_{\langle\text{lax}\rangle} = 1 - \mu_{\langle\text{tense}\rangle}$$

3. PHONEME CLASSIFICATION

Phoneme labelling of pseudo-syllables is performed by a fuzzy algorithm consisting of a branching questionnaire. The emission of a hypothesis about the presence of a phoneme is the answer to a composite question associated with a membership function.

Phoneme classification is a context-dependent operation, except for non-reduced vowels that can be recognized with no knowledge of the context. Vowels belong to sonorant segments, for which formants are tracked. Emission of hypotheses about vowels is primarily based on the analysis of the plot of F2 (the second formant frequency) versus F1 (the first formant frequency) for the sonorant portion of the PSS. The graph drawn in the F1-F2 plane by the pronounciation of a syllable is described by a language whose detail definition is given in /14/.

Formant tracking (see /29/ for details) is performed on a pseudo-syllable segment by looking for formant arcs made by concatenation of spectral peaks having high energy. Particular care has been taken in order to guarantee as much as possible agreement between FFT and LPC spectra and to control formant tracking by transition rules.

The primitives of the language are basically stable zones (SZ) and lines (LN) (see /31/ for definitions). Stable zones correspond to the stationary portions of the speech waveform. It is assumed that nonreduced vowels are always represented by stable zones, with a duration of more than 40 msecs.

Vowel classification is based on the detection of stable zones in speech intervals that have been previously hypothesized as vocalic segments. Some SZs in segments that have not been previously classified as "vocalic" may also cause this assignment if some conditions involving relations with gross spectral features are satisfied. This assignment is controlled by rules that are omitted here for the sake of brevity.

The vocalic SZs are then labelled by one or more vowel symbols, and a membership function is assigned to each label. This operation uses a definition of vowel loci as fuzzy sets in the F1-F2 plane. Each SZ has the coordinates of its center of gravity as attributes. The coordinates identify a point in the F1-F2 plane that may belong to one or more vowel loci with certain membership functions.

Vowel loci, defined as fuzzy sets, have an assignment of membership functions that can be made by a subjective adjustment of probability densities estimated after a long period of learning. The resulting assignment is speaker-dependent.

In the approach presented in this paper, a vocalic hypothesis v_i is defined by the following rule:

$$\pi_5$$

$$v_i \ ::= \langle \text{vocalic} \rangle \langle SZ_v \rangle \ v_i(F_1, F_2) \ ;$$

where $\langle$ vocalic $\rangle$ is a phonetic feature attached during precategorical classification, $\langle SZ_v \rangle$ represents the fact that a vocalic stable zone has been found inside the vocalic segment and $v_i(F_1, F_2)$ represents the fact that the center of gravity of SZ_v is in the locus of the vowel v_i.

A semantic rule is associated with π_5 to evaluate the vocalic hypothesis v_i:

$$\mu_{v_i} = \mu_{\langle \text{vocalic} \rangle} \wedge \mu_{\langle SZ_v \rangle} \wedge \mu_{v_i(F_1, F_2)} \tag{15}$$

$\mu_{\langle \text{vocalic} \rangle}$ is a result of the precategorical classification, $\mu_{v_i(F_1, F_2)}$ is the degree of membership of the center of gravity of SZ_v in the fuzzy set which is the locus of v_i in the F1-F2 plane, and $\mu_{\langle SZ_v \rangle}$ is a measure, defined by heuristic rules, of the "stationarity" of the spectra of the vocalic segment.

Classification of sonorant sounds into phonemes requires a preliminary recognition of the features liquid or nasal. This operation as well as the phonemic transcription is controlled by rules that may be context-dependent. Such rules are functions of answers to the questions of a branching questionnaire.

After experiments on liquid and nasal intervocalic consonants it has been found that, if the algorithm used for vowels finds consistent paths with no discontinuities joining the formant lines of the two vowels, it generally tracks the right formants.

This result, together with the constraint that the first formant is allowed to have only a downward shift in its frequency in terval, made it possible to track correctly the formants for all the liquid and for some of the nasal consonants. Such a correctness was confirmed in many cases by synthesis experiments.

For some nasal consonants, particularly in the context of two front vowels, formant tracking appears to be very difficult (especially for the second formant of / m /) even if FFT and up to 16-pole LPC spectra are used together. This is due to the complex interaction of the anti-resonances of the vocal tract and the resonances of the pharingal and nasal tract.

Such difficulties have been successfully avoided by controlling formant tracking with rules for those cases for which there is not a single, consistent path in the consonant region that joins, with no discontinuities, the formant lines of the vowels.

Liquid-nasal classification

The liquid-nasal classification is a process of generation of hypotheses about the consonant sounds classified as sonorants with high membership function value in the precategorical classification. The decision within the sonorant consonant class is the answer to the composite question:

$$Q_L \hat{=} B_L$$

The answer to Q_L is obtained by a relation on the answer to the component questions Q_{Li} $(i=1, 2, \ldots, K_L)$ the answer to the component questions Q_{Li} are fuzzy linguistic variables h_{Li}, m_{Li}, l_{Li} defined over the continuous intervals of acoustic parameters defined by the previous and the following parameters:

- $F_i(nT)$: the n-th sample of the i-th formant frequency (T is a sampling period = 10 msec)
- $A_i(nT)$: the n-th sample of the energy associated with the i-th formant
- $D_{12} = A_1(n^*T) - \mathrm{Max}\{A_2(n^*T), A_3(n^*T)\}$; where n^*T is the time interval where $A_2(nT)$ has the absolute minimum in the syllable
- $D_{13} = A_2(n^*T) - A_3(n^*T)$.

m_{Li} is a fuzzy set having membership equal to 1 when both μ_{hL_i} and μ_{lL_i} are less than 0.7, equal to zero when μ_{hL_i} or μ_{lL_i} are equal to 1; it assumes values between zero and one in the other intervals.

After experiments on hundreds of consonants in VCV syllables extracted from sentences pronounced by a single male talker, a set of rules has been derived. This set of rules is considered as a starting assumption for an inference procedure that will be developped in a recognition stage. The inference will have the purpose of refining the definition of the fuzzy sets associated to each component question. Furthermore it will better specify the relations between atomic answers and the answers to the composite question and the semantic (relations between membership functions) associated with such relations.

For the sake of simplicity, the intervals for which the membership functions assume value 1 and 0 are identified by the boundary values l_{ji} and h_{ji} that are listed in Tables II, III, together with their associated acoustic variable x_{ji}. The parameters h_{Li}, l_{Li} and x_{Li} for the composite question Q_L are summarized in Table II; $l_{L8} = l_{L9} = 0$ means that the variables l_{L8} and l_{L9} have membership function equal to 1 only when the corresponding durations are 0.

TABLE II

Question	x_{Li}	h_{Li}	l_{Li}	Dimension
Q_{L1}	D_{12}	15	4	dB
Q_{L2}	D_{12}	15	14	dB
Q_{L3}	D_{12}	12	10	dB
Q_{L4}	D_{12}	7	6	dB
Q_{L5}	D_{12}	12	9	dB
Q_{L6}	D_{12}	10	5	dB
Q_{L7}	D_{23}	1	-6	dB
Q_{L8}	Dur(dip(S))20 (Dur(dip(R)=0)		0	msec
Q_{L9}	Dur(dip(S))20		0	msec

TABLE III

Question	x_{LQ_i}	h_{LQ_i}	l_{LQ_i}	Dimension
Q_{LQ1}	Dur(dip(s))40		20	msec
Q_{LQ2}	D_{23}	2	0	dB
Q_{LQ3}	Dur(conson)50		30	msec
Q_{LQ4}	F_{1c}	560	400	Hz
Q_{LQ5}	F_{2c}	1300	300	Hz

TABLE IV

Question	x_{Ni}	h_{Ni}	l_{Ni}	Dimension
Q_{N1}	D_{12}	23	17	dB
Q_{N2}	D_{12}	23	15	dB
Q_{N3}	F_{12}	1100	950	Hz
Q_{N4}	F_{22}	2000	1500	Hz
Q_{N5}	F_{21}	1700	1400	Hz
Q_{N6}	D_{12}	20	13	dB
Q_{N7}	D_{12}	20	15	dB
Q_{N8}	F_{12}	1700	1600	Hz
Q_{N9}	D_{12}	26	13	dB
Q_{N10}	F_{22}	2120	2000	Hz
Q_{N11}	F_{21}	1360	1150	Hz
Q_{N12}	F_{22}	1700	1500	Hz
Q_{N13}	F_{21}	1600	1400	Hz
Q_{N14}	F_{22}	2000	1600	Hz
Q_{N15}	F_{21}	900	800	Hz
Q_{N16}	F_{22}	1900	1400	Hz
Q_{N17}	F_{21}	1200	1000	Hz
Q_{N18}	F_{22}	1320	1280	Hz
Q_{N19}	F_{22}	1460	1260	Hz
Q_{N20}	F_{22}	1200	1000	Hz
Q_{N21}	Max(F_2)	1200	800	Hz
Q_{N22}	Max(F_2)	1240	900	Hz
Q_{N23}	Max(F_2)	1520	1160	Hz
Q_{N24}	D_{12}	26	15	dB
Q_{N25}	D_{12}	15	9	dB
Q_{N26}	F_{22}	1700	1400	Hz

It has been found possible to classify correctly with membership equal to 1 all the training set using regular expressions as relations between the answers to the atomic questions and the answers to the composite question. Some of the terms of this relations are valid only if some associated predicates are true. These predicates refer to the vocalic context of the sonorant sound and are represented by brackets containing the vowel before the consonant followed by the vowel after the consonant; an asterix represents a "don't care" condition. The fact that the relation derived shows a context dependency in the distinction between liquids and nasals means that context dependency is relative to the particular acoustic parameters used. Nevertheless, the experiments carried out for deriving rules in general cases make appear unlikely that the distinction liquid/nasal is feasible with context-independent procedure, even if this is possible for limited protocols.

The expressions obtained for the answers to the question Q_L are the following:

$$\begin{aligned} \text{LIQUID} = {} & l_{L1}+m_{L1}\,[u,*]l_{L2}+[e,*](l_{L3}+m_{L3}h_{L7})+ \\ & +[i,*](m_{L8}+h_{L8}(l_{L4}+m_{L4}h_{L7}))+[o,*](l_{L5}+ \\ & +m_{L5}h_{L7})+[a,*](l_{L6}+m_{L6}(l_{L9}+m_{L9}h_{L7}))) \end{aligned} \tag{16}$$

$$\begin{aligned} \text{NASAL} = {} & h_{L1}+m_{L1}([e,*](h_{L3}+m_{L3}l_{L7})+[i,*]h_{L8}(h_{L4}+ \\ & +m_{L4}l_{L7})+[o,*](h_{L5}+m_{L5}l_{L7})+[a,*]h_{L9}(h_{L6}+ \\ & +m_{L6}l_{L7}) \end{aligned} \tag{17}$$

The membership functions can be computed with min-max operations.

Applications to the classification of liquids

The classification of liquids mainly concerns "l" and "r" because the liquid "gl" (like in the Spanish word Sevilla) is easily distinguished by a syntax controlled procedure that recognize typical evolutions in the F1-F2 graph. The acoustic parameters involved in this classification are dur (dip (S)), D_{23}, dur (cons), i.e. the duration of the consonant tract, F_{1c} and F_{2c} that are respectively the first and the second formant frequencies corresponding to the minimum of the second formant energy.

The question Q_{LQ} for the classification of liquids has component questions and fuzzy set definitions according with Table III; the expressions derived are the following:

$$\text{"LIQUID l"} = h_{LQ1} + m_{LQ1}(l_{LQ2} + h_{LQ3} + m_{LQ3}(l_{LQ4}h_{LQ5})) \quad (18)$$

$$\text{"LIQUID r"} = l_{LQ1} + m_{LQ1}(l_{LQ3} + h_{LQ4}) \quad (19)$$

Application to the detailed classification of nasals

Parameter of the atomic questions

The classification of nasals is obtained as an answer to the composite question Q_N. This answer is obtained by a relation on the answers to the component questions Q_{Ni} $(i=1, 2, \ldots, K_N)$; the answers to the component questions Q_{Ni} are fuzzy linguistic variables h_{Ni}, l_{Ni} defined by Table IV on the basis of the acoustic parameters previously defined and the following ones:

- F_{21} : the lowest value of $F_2(nT)$ at the time corresponding to the beginning of the downward shift of the first formant frequency of the nasal consonant;
- F_{22} : the value of $F_2(nT)$ at the end of the downward shift of the first formant frequency of the nasal consonant.

Detection of F_{21} and F_{22} is very easy because the formant evolutions in the F1-F2 plane are described by a language / 9 / that contains lines as primitives.

The recognition rules

The recognition rules for the classification of nasals were inferred subjectively with the goals of avoiding a wrong answer with degree of worthiness equal to one and trying to minimize the number of times the wrong answer assumes a degree of worthiness higher than the right answer.

The vocalic contexts are represented by places of articulation and the break points at the end of the intervals where $(l_{Ni}=1)\&(h_{Ni}=0)$ or $(h_{Ni}=1)\&(l_{Ni}=0)$ are reported in Table IV with column heading h_{Ni} and l_{Ni} respectively. In Table IV, the type of parameters on which the questions are asked is reported with column heading p_{Ni}.

The places of articulation have been determined on the basis of the values assumed by the second formant on the stationary portion of the vocalic segment according to the following definitions:

$1900 \text{ Hz} < F_2$ front (f),

$1300\text{ Hz} < F_2 < 1900\text{ Hz}$ central (c),

$F_2 < 1300\text{ Hz}$ back (b).

An analytic representation of the inferred rules is the following:

$$\begin{aligned}\langle m\rangle = {} & [ff](l_{N1}+l_{N2})+[fc](l_{N3}+l_{N4}+l_{N5})+[fb](l_{N6}+l_{N7}+l_{N8})+ \\ & +[cf](l_{N9}+l_{N10})+[cc](l_{N11}+l_{N12})+[cb](l_{N13}+l_{N14})+ \\ & +[bf](l_{N15}+l_{N16}+l_{N17})+[bc](l_{N18}+l_{N19}+l_{N20})+ \\ & +[bb](l_{N21}+l_{N22})\end{aligned} \qquad (20)$$

$$\begin{aligned}\langle n\rangle = {} & [ff](h_{N1}+h_{N2})+[fc](h_{N3}+h_{N4}+h_{N5})+[fb](h_{N6}+h_{N7}+h_{N8})+ \\ & +[cf](h_{N9}+h_{N10})+[cc](h_{N11}+h_{N12})+[cb](h_{N13}+h_{N14})+ \\ & +[bf](h_{N15}+h_{N16}+h_{N17})+[bc](h_{N18}+h_{N19}+h_{N20})+ \\ & +[bb](h_{N21}+h_{N22})\end{aligned} \qquad (21)$$

Classification of stop-sounds

After analyzing all the possible-vowel stop consonant-vowel pseudo-syllables extracted from continuous speech and conside ring data from four speakers, a set of rules have been inferred for the classification of stop sounds.

The context dependency of these rules are expressed by the places of articulation (front/central/back) of the preceding and the following vowel. The recognition rules are based on formant transitions and a description of the burst spectra. In order to have speaker-independent rules, no absolute measurements were made, rather relative measurements were considered.

The values F_{2e}, F_{3e} of the second and third formant are ex tracted at the end of the transition and the analogous values F_{2b} F_{3b} are extracted at the beginning of the transition. In order to reduce the effect of speaker dependences the ratios $RF_e=F_{3e}/F_{2e}$ and $RF_b=F_{3b}/F_{2b}$ were computed and stop sound hypotheses were defined as fuzzy sets over the universes of such ratios. For example, Fig. 3 shows the fuzzy sets $H^b{}_{fc}(g)$, $H^b{}_{fc}(d)$ and $H^b{}_{fc}(b)$ defined over the universe of F_{3b}/F_{2b}; the indices fc define the vocalic context: front-central, the subfix b represents the beginning of the transition, the parentheses contain the phonemes for which the hypotheses hold. Burst spectra are obtained by averaging the spectra of the intervals in which R_v

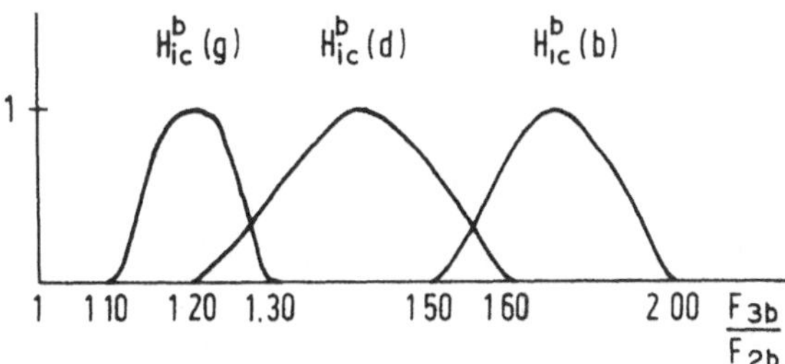

Fig.3. Fuzzy sets for the recognition of stop consonants

is less than -12 dB. The average spectra are then described by a sequence of pairs. The pairs are ordered according to the intensities of energy concentrations (energoids) in the spectrum; each pair corresponds to an energoid of the spectrum whose energy is higher than a fixed percentage (70%) of the absolute energy maximum of the spectrum. Each pair contains a symbol representing the frequency interval in which the gravity center of energoid falls; Tab. V shows the definitions of the symbols. The second element of the pair is a number expressing the ratio between the energy of the energoid and the energy in the 0.2 - 1 kHz interval; this makes spectral descriptions independent from the intensity of the speech waveform.

TABLE V

Frequency bounds (kHz)	Symbol
1 ÷ 2	A
2 ÷ 2.5	B
2.5 ÷ 3	C
3 ÷ 3.5	D
3.5 ÷ 4	E
4 ÷ 4.5	G
4.5 ÷ 5	H
5 ÷ 6	K
6 ÷ 7	L
7 ÷ 8	M
8 ÷ 10	N

The syntax for the recognition of stop consonants is given in the following. For the sake of brevity, the definition of fuzzy sets are omitted and for the sake of simplicity the universe on which the fuzzy linguistic variables are defined, are explicitly mentioned between parentheses. For example, low R_B means a low ratio between the energy of the energoid coded by B and the energy in the 0.2 - 1 kHz band. The symbol ✱ means that no other symbols have to be detected. The symbols belonging to the alphabet for the description of the spectra are the only variables that are not fuzzy; if they are not detected in the burst spectrum, their value is zero and the corresponding ratio of energies is not computed. The context of the consonant, for example "front-back" is indicated by the initials and is a pair of fuzzy variables; for each variable the membership is computed in the formants space. The rules for the classification of stop sounds are expressed analytically by regular expressions for the sake of simplicity; low R means that none of the ratios of energies reaches a given threshold.

$$
\begin{aligned}
\langle b\rangle ::= {} & [fb](H^{e}_{fb}(b)+H^{e}_{fb}(bd)(B \text{ low } R_{B}+K \text{ low } R_{K}+A(P \text{ low } R_{P}+ \\
& +L \text{ low } R_{L}))+[cc](H^{e}_{cc}(bd)+H^{e}_{cc}(bd)) \text{low } R + \\
& +[cf](H^{e}_{cf}(bd)) \text{low } R+[cb](H^{e}_{cb}(bd)) \text{low } R+[fc]H^{b}_{fc}(b)+ \\
& +[ff](D+F+K \text{ low } R_{K})+[bb]H^{e}_{bb}(b)(C+K \text{ low } R_{K})+[bc]H^{e}_{bc}(b)\cdot \\
& \cdot(K \text{ low } R_{K}+E \text{ low } R_{E}+B)+[bf]H^{b}_{bf}(b)(\text{low } R+B+H \text{ low } R_{H})
\end{aligned}
$$

$$
\begin{aligned}
\langle d\rangle ::= {} & [fb]H^{e}_{fb}(bd)(K \text{ high } R+EM+AF+AK+EN)+[cc](H^{e}_{cc}(bd)+ \\
& + H^{b}_{cc}(bd))(A \text{ high } R_{A}\ B \text{ low } R_{B} +K \text{ high } R_{K})+ \\
& +[cf]H^{e}_{cf}(bd)(AC+K \text{ high } R_{K}\ F \text{ high } R_{F})+[cb]H^{e}_{cb}(bd)\cdot \\
& \cdot(D \text{ high } R_{D}+C \text{ high } R_{C}+E \text{ high } R_{E})+[fc]H^{b}_{fc}(d)(K \text{ high } R_{K}+ \\
& +B \text{ high } R_{B})+[ff](K \text{ high } R_{K}\ A+EM+EN+DC+DE)+ \\
& +[bb]H^{e}_{bb}(d) +[bc]H^{e}_{bc}(d)
\end{aligned}
\tag{22}
$$

$$
\begin{aligned}
\langle g\rangle ::= {} & [fb]H^{e}_{fb}(g)(K \text{ high } R_{K}+D \text{ high } R_{D}\ N+E \text{ high } R_{E}\ N \text{ high } R_{N})+ \\
& +[cc](H^{b}_{cc}(g)+H^{e}_{cc}(g))(A \text{ high } R_{A}\ B \text{ high } R_{B}\ D \text{ high } R_{D})+ \\
& +[cf]H^{e}_{cf}(g)(H \text{ high } R_{H}+D \text{ high } R_{D})+[cb]H^{e}_{cb}(g)(A \text{ high } R_{A}+ \\
& +F \text{ high } R_{F})+[fc]H^{b}_{fc}(g)(K \text{ high } R_{K}+E \text{ high } R_{E}\ L \text{ high } R_{L})+ \\
& +[ff](K \text{ high } R_{K} * +MD+FN)+[bb]H^{e}_{bb}(g)(MN+LN)+ \\
& +[bc]H^{e}_{bc}(g)(C \text{ high } R_{C}+M \text{ high } R_{M})+[bf]H^{b}_{bf}(g)(K \text{ high } R_{K}+ \\
& +L \text{ high } R_{L})
\end{aligned}
$$

The rules presented here are subject to refinements during successive stages of inference. Firthermore, they show how the context helps in solving many ambiguities and that informations about formant transitions are useful only when context dependencies are taken into account.

Similar rules have been inferred for the tense interrupted <p>, <t>, <k>; they are omitted for the sake of brevity.

4. EXPERIMENTAL RESULTS

The experimental result of the emission of hypotheses about the features sonorant and nonsonorant for four talkers, three male and one female, using the syntactic rules π_1 and π_2

TABLE VI
Summary of experimental results (4 speakers)

feature	over all error rate
sonorant/nonsonorant (grammar = 1)	8%
sonorant/nonsonorant (variable grammar)	5%
continuant/interrupted/affricate	7%
tense/lax	6%
nasals	6%

with all the grammaticalities set equal to one, show an average error rate of 8%. The results refer to syllables extracted from spoken sentences of various and unconstrained syntactic structure. A hundred syllables were analyzed for each talker. Although it has been found that adapting some of the α_{1i} and α_{1i} to the talker and letting all the grammaticalities be equal to one, leads to some improvement, a better result has been obtained by leaving α_{1i} and α_{1i} unchanged and assigning grammaticalities lower than one to some rules. The assignment shown in the rules π_1 and π_2 gave an average error rate of 5% for about 400 samples pronounced by the four talkers.

The distribution of the errors shows no remarkable differences among the talkers, suggesting that the features considered are talker-independent.

The results of the classification continuant/interrupted/affricate for nonsonorant sounds give an average error of 7% and the results of the classification tense/lax a 6%. For the classification tense/lax the performançe has been improved by using more complex fuzzy rules; details concerning these are omitted here for the sake of brevity.

For the classification of nasals, an overall error rate of 6% is achieved in the recognition of nasals obtained for 200 utterances by 4 male speakers. None of these errors corresponds to the assignment of unit degree of worthiness to a wrong hypothesis but only to the fact that a wrong interpretation obtained a degree of worthiness higher than the right one.

A summary of the results is shown in tab. VI.

REFERENCES

/1/ - D. Raj Reddy, ed., "Speech Recognition: Invited Papers of the IEEE Symp.", New York: Academic Press, 1975.

/2/ - D. Raj Reddy, "Speech Recognition by Machine: A Review", Proc. IEEE, vol. 64, No. 4, Apr. 1976, p. 501-531.

/3/ - T.B. Martin, "Practical applications of voice input to machines", Proc. IEEE, vol. 64, No. 4, Apr. 1976, p. 487-501.

/4/ - F. Jelinek, "Continuous Speech Recognition by Statistical Methods", Proc. IEEE, vol. 64, No. 4, Apr. 1976, p. 532-556.

/5/ - D.H. Klatt, "Review of the ARPA speech understanding project", Journ. Acoust. Soc. of America, vol. 62, Dec. 1977, p. 1345-1366.

/6/ - J.J. Wolf, "Speech Recognition and Understanding" in Digital Pattern Recognition, K.S. Fu ed., New York: Springer, 1976.

/7/ - D. Raj Reddy ed., "Speech Understanding System - Summary of results of the five-year research effort at Carnegie Mellon University" Pittsburgh, Pa. 1977.

/8/ - D.E. Walker ed., "SRI Speech Understanding Systems - Final report, Stanford Research International", Menlo Park, Ca., 1977.

/9/ - W.A. Woods et al., "Speech Understanding Systems - Final Report, Bolt, Beranek and Newman Inc.", Cambridge, Mass., 1976, Rept. No. 3438.

/10/ - R. De Mori, "Recent advances in Speech Recognition", Invited paper Proc. 4th International Joint Conference on Pattern Recognition, Kyoto, Oct. 1978, p. 106-124.

/11/ - R. De Mori, P. Torasso, "Lexical Classification in a Speech Understanding System Using Fuzzy Relations", Proc. IEEE-ASSP Conference, Philadelphia, 1976, p. 565-568.

/12/ - M. Coppo, L. Saitta, "Semantic support for a Speech Understanding System based on fuzzy relations", Proc. IEEE Conference on Cybernetics and Society, Washington D.C., 1976, p. 520-525.

/13/ - L.A. Zadeh, "Fuzzy sets as a basis for a theory of possibility", Fuzzy Sets and Systems, vol. 1, p. 3-28, 1978.

/14/ - R. De Mori, P. Laface, E. Piccolo, "Automatic detection and description of syllabic features in continuous speech", IEEE Trans. on Acoustic Speech and Signal Processing, vol. ASSP-24, Oct. 1976, p. 365-379.

/15/ - L.A. Zadeh, "The concept of a linguistic variable and its application to approximate reasoning - II", Information Sciences, 8, 1975, p. 301-357.

/16/ - L.A. Zadeh, "Similarity Relations and fuzzy Ordering", Information Sciences, vol. 3, 1971, p. 177-200.

/17/ - S.K. Pal, D.D. Majumder, "Fuzzy set and Decisionmaking approaches in vowel and Speaker Recognition", IEEE Trans. on Systems, Man and Cybernetics, Aug. 1977, p. 625-629.

/18/ - L.A. Zadeh, "The concept of a linguistic variable and its application to approximate reasoning - III", Information Sciences, vol. 9, 1975, p. 43-80.

/19/ - R. De Mori, S. Rivoira, A. Serra, "A special purpose computer for digital signal processing", IEEE Trans. on Computers, vol. C-24, Dec. 1975, p. 1202-1211.

/20/ - K.N. Stevens, "Potential role of property detectors in the perception of consonants", M.I.T.-R.L.E., Quarterly Progress Report No. 110, July 1973, p.155-168.

/21/ - G.W. Hughes, J.F. Hemdal, "Speech Analysis", Rept. AFCRL-65-681 (P13552), Purdue University, 1965.

/22/ - L.A. Zadeh, "A fuzzy algorithmic approach to the definition of complex or imprecise concepts", International Journal on Man-Machine Studies, 1976, 8, p. 249-291.

/23/ - L.A. Zadeh, "Fuzzy languages and their relation to human and machine intelligence", Proc. Int. Conference on Man and Computer, Bordeaux (France), ed. by S. Kargle Basel, Munchen, New York 1972, p. 130-165.

/24/ - K.S. Fu, "Syntactic methods in pattern recognition", Academic Press, 1974.

/25/ - R. De Mori, L. Saitta, "Automatic Learning of naming relations over finite languages", ISI Internal Report, Univ. of Turin, 1978.

/26/ - R. De Mori, P. Laface, P. Torasso, "Automatic classification of liquids and nasals in continuous speech", Proc. IEEE - ICASSP 77 Conference, Hartford (Conn.), p.644--647.

/27/ - R. De Mori, R. Gubrinowicz, P. Laface, "Inference of a Knowledge source for the Recognition of nasals in Continuous Speech", to be published on IEEE Trans. on Acoustic, Speech and Signal Processing.

/28/ - S.E.G. Öhman, "Coarticulation in VCV utterances: Spectrographic Measurements", Journal Acoust.Soc.Amer., vol. 39, N. 1, p. 151-168, 1966.

/29/ - P. Laface, "Ambiguities in feature extraction from continuous speech", Proc. 4th International Joint Conferen ce on Pattern Recognition, Kyoto (Japan), Nov. 1978, p. 1056-1058.

/30/ - C.J. Weinstein, S.S. McCandless, L.F. Mondshein, V. F. Mondsehin, V.W. Zue, "A system for acoustic-phonetic analysis of continuous speech", IEEE Trans. on Acoustics, Speech and Signal Processing, vol. ASSP-23, Feb. 1975, p. 54-67.

/31/ - R. De Mori, "A descriptive technique for automatic speech recognition", IEEE Trans. on Audio and Electroacoustics, vol. AU-21, Apr. 1973, p. 89-100.

/32/ - R. De Mori, P. Laface, "Representation of phonetic and phonemic knowledge in a speech understanding system", Proc. AISB/GI Conference on Artificial Intelligence, Hamburg (FRG), July 1978, p. 201-203.

/33/ - R. De Mori, "Syntactic recognition of speech patterns", in Syntactic Pattern Recognition, Applications, ed. K.S. Fu, Springer Verlag, 1977.

/34/ - R. De Mori, "Design for a syntax-controlled acoustic classifier", Information Processing 74, ed. by J.L. Rosenfeld, North-Holland 1974, p. 753-757.

/35/ - R. De Mori, S. Rivoira, A. Serra, "A speech understanding system with learning capability", Proc. Fourth Int. Joint Conference on Artificial Intelligence, Tbilisi (USSR) Sept. 1975, p. 468-475.

/36/ - P. Mermelstein, "A. phonetic-context controlled strategy for segmentation and phonetic labelling speech", IEEE Transactions on Acoustics Speech and Signal Processing, vol. ASSP-23, Feb. 1975, p. 79-82.

/37/ - A.M. Liberman, "The grammars of speech and language", Cognitive Psychology I, Oct. 1970, p. 301-323.

/38/ - M. Studdert-Kennedy, "Speech perception" Status Report No. SR 39-40, Haskins Laboratories, New Haven (Conn.), July/Dec. 1974, p. 1-52.

APPENDIX

Some examples

The following values are assumed by the parameters u_{1i} (i=1, 2, ..., 6) for the consonant /n/ of a pseudosyllable /oni/:

u_{11} = 15.5 dB
u_{12} = -20 dB
u_{13} = 10.4 dB
u_{14} = 100 msec
u_{15} = 7.45 dB
u_{16} = 4.5 dB

Using the diagram of Fig. 2 with the break points given by Tab. I and remembering the definitions given by (4) and (5) one gets:

$\mu_{h_{11}} = 1$, $\mu_{l_{11}} = 0$,
$\mu_{h_{12}} = 0$, $\mu_{l_{12}} = 1$,
$\mu_{h_{13}} = 0.55$, $\mu_{l_{13}} = 0.45$,
$\mu_{h_{14}} = 0.7$, $\mu_{l_{14}} = 0.3$,
$\mu_{h_{15}} = 0.1$, $\mu_{l_{15}} = 0.9$,
$\mu_{h_{16}} = 0.2$, $\mu_{l_{16}} = 0.8$.

Applying (6), (7), (8), (9) one gets:

$$\mu_{\langle \text{dip in S} \rangle} = \text{Min}\left\{0, \left[\text{Max}(\text{Min}(0.85, 0.45), \text{Min}(0.9, 0.7), \text{Min}(0.45, 0.7), 0.1\right]\right\} = 0$$

$$\mu_{\langle \text{dip in } R_V \rangle} = \text{Max}\left\{\text{Min}(0.6, 0), \text{Min}(0.9, 0), \text{Min}(0.9, 0.2), \text{Min}(0, 0), \text{Min}(0, 0.2), \text{Min}(0, 0.2)\right\} = \text{Max}(0, 0, 0.2, 0, 0, 0) = 0.2$$

$$\mu_{\langle \text{SRV} \rangle} = \text{Max}\left\{\text{Min}(0.9, 1), \text{Min}(0.9, 1), \text{Min}(1, 1), 0.8\right\} = \text{Max}(0.9, 0.9, 1, 0.8) = 1$$

$$\mu_{\langle \text{SS} \rangle} = \text{Max}(0.3, 0.9) = 0.9$$

Applying (10) and (11) one gets:

$$\text{Poss}\left\{\langle\text{sonorant}\rangle \text{ is in } \underline{p}\right\} = \text{Max}\left\{\text{Min}(0.85, 1), \text{Min}(0.9, 0.9), \text{Min}(1, 0.9)\right\} = \text{Max}(0.85, 0.9, 0.9) = 0.9$$

Poss{⟨nonsonorant⟩} is in $\underline{p}$ = Max(0, 0.2) = 0.2

Thus the hypothesis that the consonant is sonorant is more consistent.

Using the same approach as for the preceding example, for the pseudo syllable /edi/, the following values have been measured:

$u_{11} = -13$ dB
$u_{12} = 0$ dB
$u_{13} = 0$ dB
$u_{14} = 100$ msec
$u_{15} = 16.7$ dB
$u_{16} = 5.6$ dB

$\mu_{h_{11}} = 0.1$, $\mu_{l_{11}} = 0.9$,
$\mu_{h_{12}} = 0.8$, $\mu_{l_{12}} = 0.2$,
$\mu_{h_{13}} = 0$, $\mu_{l_{13}} = 1$,
$\mu_{h_{14}} = 0.7$, $\mu_{l_{14}} = 0.3$,
$\mu_{h_{15}} = 0.6$, $\mu_{l_{15}} = 0.4$,
$\mu_{h_{16}} = 0.4$, $\mu_{l_{16}} = 0.6$.

$$\mu_{\langle\text{dip in S}\rangle} = \text{Min}\left\{1, \left[\text{Max}(\text{Min}(0.85, 1), \text{Min}(0.9, 0.7), \text{Min}(1, 0.7), 0.6)\right]\right\} = 0.85$$

$$\mu_{\langle\text{dip in }R_V\rangle} = \text{Max}\left\{\text{Min}(0.6, 0.9), \text{Min}(0.9, 0.8), \text{Min}(0.9, 0.4), \text{Min}(0.9, 0.8), \text{Min}(0.9, 0.4), \text{Min}(0.8, 0.4)\right\} = \\ = \text{Max}(0.6, 0.8, 0.4, 0.8, 0.4, 0.4) = 0.8$$

$$\mu_{\langle\text{SRV}\rangle} = \text{Max}\left\{\text{Min}(0.9, 0.1), \text{Min}(0.9, 0.2), \text{Min}(0.2, 0.1), 0.6\right\} = 0.6$$

$$\mu_{\langle\text{SS}\rangle} = \text{Max}(0.3, 0.4) = 0.4$$

Poss{⟨nonsonorant⟩ is in $\underline{p}$} = Max(0.85, 0.8) = 0.85

Poss{⟨sonorant⟩ is in $\underline{p}$} = Max{Min(0.85, 0.6), Min(0.9, 0.4), Min(0.6, 0.4)} = 0.6

Thus the most consistent hypothesis is that the consonant is nonsonorant.

For the classification tense/lax, the following examples can be considered:

<u>pseudo-syllable /edi/</u>

measurements:

$u_{31} = -13$ dB
$u_{33} = 100$ msec
$u_{35} = 16.7$ dB

memberships:

$\mu_{l_{31}} = 0.05$
$\mu_{h_{33}} = 0.4$
$\mu_{h_{35}} = 0$
$\mu_{\langle tense \rangle} = 0.4$
$\mu_{\langle lax \rangle} = 0.6$

<u>pseudo-syllable /epi/</u>

measurements:

$u_{31} = -11.5$ dB
$u_{33} = 160$ msec
$u_{35} = 26.7$ dB

memberships:

$\mu_{l_{31}} = 0$
$\mu_{h_{33}} = 1$
$\mu_{h_{35}} = 0.67$
$\mu_{\langle tense \rangle} = 1$
$\mu_{\langle lax \rangle} = 0$

<u>pseudo-syllable /ozo/</u>

measurements:

$u_{31} = -23$ dB
$u_{32} = 28$ dB
$u_{33} = 30$ msec
$u_{34} = 90$ msec
$u_{35} = 3$ dB

memberships:

$\mu_{h_{32}} = 1$
$\mu_{h_{34}} = 1$
$\mu_{l_{33}} = 1$
$\mu_{\langle \text{nonsonorant-continuant} \rangle} = 1$
$\mu_{l_{31}} = 0.3$
$\mu_{\langle \text{tense} \rangle} = 0.3$
$\mu_{\langle \text{lax} \rangle} = 0.7$

<u>pseudo-syllable /ifi/</u>

measurements:

$u_{31} = -32$ dB
$u_{32} = 40$ dB
$u_{33} = 40$ msec
$u_{34} = 140$ msec
$u_{35} = 0$ dB

memberships:

$\mu_{h_{32}} = 1$
$\mu_{h_{34}} = 1$
$\mu_{l_{33}} = 1$
$\mu_{\langle \text{nonsonorant-continuant} \rangle} = 1$
$\mu_{l_{31}} = \mu_{\langle \text{tense} \rangle} = 1$
$\mu_{\langle \text{lax} \rangle} = 0$

SYNTACTIC APPROACH TO PATTERN RECOGNITION+

K. S. Fu

School of Electrical Engineering
Purdue University
W. Lafayette, Indiana 47907
U.S.A.

Abstract. Syntactic approach to pattern recognition is introduced. Special topics discussed include primitive selection and pattern grammars, syntactic recognition and error-correcting parsing, and clustering analysis for syntactic patterns.

I. Introduction

The many different mathematical techniques used to solve pattern recognition problems may be grouped into two general approaches [1,2]. They are the decision-theoretic (or discriminant) approach and the syntactic (or structural) approach [3]. In the decision-theoretic approach, a set of characteristic measurements, called features, are extracted from the patterns. Each pattern is represented by a feature vector, and the recognition of each pattern is usually made by partitioning the feature space. On the other hand, in the syntactic approach, each pattern is expressed as a composition of its components, called subpatterns and pattern primitives. This approach draws an analogy between the structure of patterns and the syntax of a language. The recognition of each pattern is usually made by parsing the pattern structure according to a given set of syntax rules. In this paper, we briefly review the recent progress in syntactic pattern recognition and some of its applications.

A block diagram of a syntactic pattern recognition system is shown in Fig. 1. We divide the block diagram into the recogni-

+This work was supported by the National Science Foundation Grant ENG 78-16970, and the U.S. - Italy Cooperative Science Program.

J. C. Simon (ed.), Spoken Language Generation and Understanding, 221-251.

tion part and the analysis part, where the recognition part consists of preprocessing, primitive extraction (including segmentation and recognition of primitives and relations), and syntax (or structural) analysis, and the analysis part includes primitive selection and grammatical (or structural) inference.

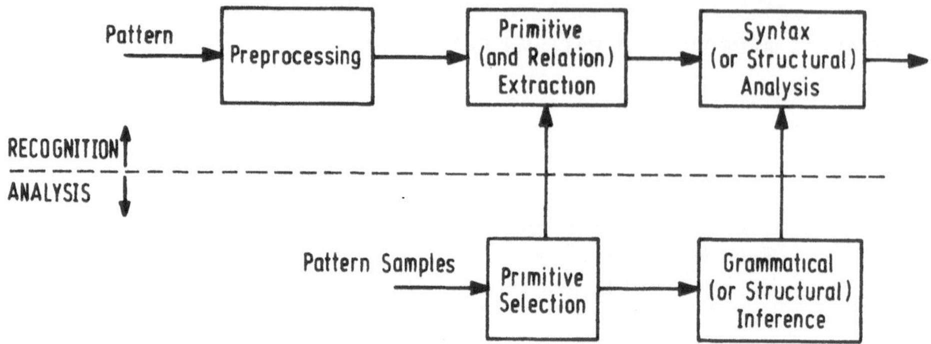

Fig.1. Block diagram of a syntactic pattern recognition system

In syntactic methods, a pattern is represented by a sentence in a language which is specified by a grammar. The language which provides the structural description of patterns, in terms of a set of pattern primitives and their composition relations, is sometimes called the "pattern description language." The rules governing the composition of primitives into patterns are specified by the so-called "pattern grammar." An alternative representation of the structural information of a pattern is to use a "relational graph," of which the nodes represent the subpatterns and the branches represent the relations between subpatterns.

Following the notations used in [6], the definition of grammars and languages is briefly reviewed.

Definition 1. A grammar is a 4-tuple

$G = (N, \Sigma, P, S)$ where

(1) N is a finite set of nonterminal symbols

(2) Σ is a finite set of terminal symbols disjoint from N.

(3) P is a finite subset of

$$(N \cup \Sigma)^* N (N \cup \Sigma)^* \times (N \cup \Sigma)^*$$

An element (α, β) in P will be written $\alpha \rightarrow \beta$ and called a production.

(4) S is a distinguished symbol in N called the start symbol.

Definition 2. The language generated by a grammar G, denoted L(G), is the set of sentences generated by G. Thus,

$$L(G) = \{\omega | \omega \text{ is in } \Sigma^*, \text{ and } S \Rightarrow$$

where a relation => on $(N \cup \Sigma)^*$ is defined as follows: if $\alpha\beta\gamma$ is in $(N \cup \Sigma)^*$ and $\beta \rightarrow \delta$ is a production rule in P then $\alpha\beta\gamma \Rightarrow \alpha\delta\gamma$, and $\overset{*}{\Rightarrow}$ denotes the reflexive and transitive closure of =>.

If each production in P is of the form $A \rightarrow \alpha$, where A is in N and α is in $(N \cup \Sigma)^*$ then the grammar G is a context-free grammars. The set of languages generated by context-free grammars is called context-free languages.

Figure 2 gives an illustrative example for the description of the boundary of a submedian chromosome. The hierarchical structural description is shown in Fig. 2(a), and the grammar generating submedian chromosome boundaries is given in Fig. 2(b). A more detailed example on the application of syntactic approach to spoken word recognition is described in Appendix.

II. Primitive Selection and Pattern Grammars

Since pattern primitives are the basic components of a pattern, presumably they are easy to recognize. Unfortunately, this is not necessarily the case in some practical applications. For example, strokes are considered good primitives for script handwriting, and so are phonemes for continuous speech, however, neither strokes nor phonemes can easily be extracted by machine. A compromise between its use as a basic part of the pattern and its easiness for recognition is often required in the process of selecting pattern primitives.

There is no general solution for the primitive selection problem at this time [3-5]. For line patterns or patterns described by boundaries or skeletons, line segments are often suggested as primitives. A straight line segment could be characterized by the locations of its beginning (tail) and end (head), its length, and/or slope. Similarly, a curve segment might be described in terms of its head and tail and its curvature. The information characterizing the primitives can be considered as their associated semantic information or as features used for primitive recognition. Through the structual description and the semantic specification of a pattern, the semantic information associated with its subpatterns or the pattern itself can then be determined. For pattern description in terms of regions, half-planes have been proposed as primitives. Shape and texture measurements are often used for the description of regions.

After pattern primitives are selected, the next step is the construction of a grammar (or grammars) which will generate a language (or languages) to describe the patterns under study. It is known that increased descriptive power of a language is paid for in terms of increased complexity of the syntax analysis system (recognizer or acceptor). Finite-state automata are capable or recognizing finite-state languages although the

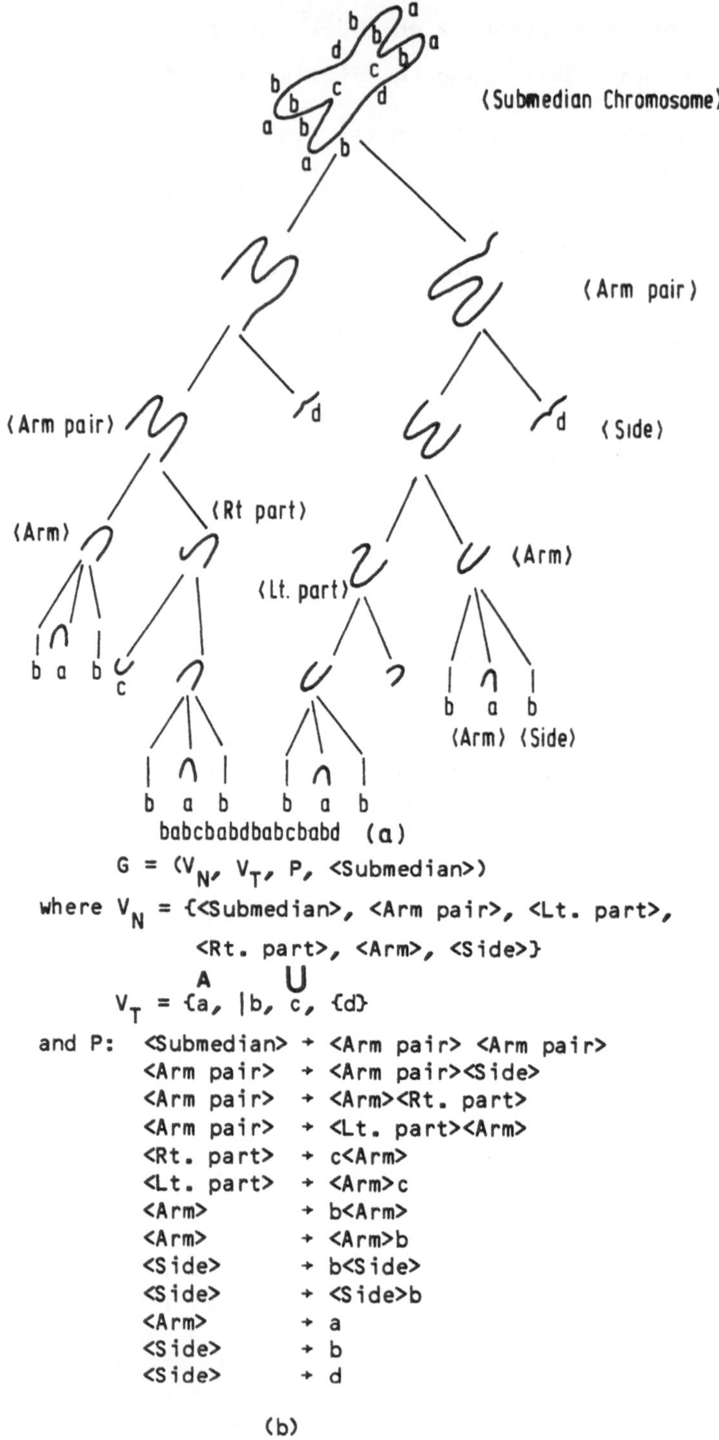

Figure 2. Syntactic Representation of Submedian Chromosome

descriptive power of finite-state languages is also known to be weaker than that of context-free and context-sensitive languages. On the other hand, nonfinite, nondeterministic procedures are required, in general, to recognize languages generated by context-free and context-sensitive grammars. The selection of a particular grammar for pattern description is affected by the primitives selected, and by the tradeof between the grammar's descriptive power and analysis efficiency.

If the primitives selected are very simple, more complex grammars may have to be used for pattern description. On the other hand, a use of sophisticated primitives may result in rather simple grammars for pattern description, which in turn will result in fast recognition algorithms. The interplay between the complexities of primitives and of pattern grammars is certainly very important in the design of a syntactic pattern recognition system. Recently, attributed grammars, of which each production rule has its associated attribute (or semantic) rules, have been used for syntactic pattern recognition [19]. Context-free programmed grammars, which maintain the simplicity of context-free grammars but can generate context-sensitive languages, have recently been suggested for pattern description [3].

A number of special languages have been proposed for the description of patterns such as English and Chinese characters, chromosome images, spark chamber pictures, two-dimensional mathematics, chemical structures, spoken words, and fingerprint patterns [3,7]. For the purpose of effectively describing high dimensional patterns, high dimensional grammars such as web grammars, graph grammars, tree grammars, and shape grammars have been used for syntactic pattern recognition [3,8,9].

Ideally speaking, it would be nice to have a grammatical (or structural) inference machine which would infer a grammar from a given set of patterns. Unfortunately, not many convenient grammatical inference algorithms are presently available for this purpose [10-16]. Nevertheless, recent literatures have indicated that some simple grammatical inference algorithms have already been applied to syntactic pattern recognition, particularly through man-machine interaction [17-20].

III. Syntactic Pattern Recognition Using Stochastic Languages

In some practical applications, a certain amount of uncertainty exists in the process under study. For example, due to the presence of noise and variation in the pattern measurements, segmentation error and primitive extraction error may occur, causing ambiguities in the pattern description languages. In order to describe noisy and distorted patterns under ambiguous situations, the use of stochastic languages has been suggested [3]. With probabilities associated with grammar rules, a stochastic grammar generates sentences with a probability distribution. The probability distribution of the sentences can be used to model the noisy situations.

A stochastic grammar is a four-tuple $G_s = (V_N, V_T, P_s, S)$ where P_s is a finite set of stochastic productions. For a stochastic context-free grammar, a production in P_s is of the form

$$A_i \xrightarrow{P_{ij}} \alpha_j, \; A_i \in V_N, \; \alpha_j \in (V_N \cup V_T)^*$$

where P_{ij} is called the production probability. The probability of generating a string x, called the string probability p(x), is the product of all production probabilities associated with the productions used in the generation of x. The language generated by a stochastic grammar consists of the strings generated by the grammar and their associated string probabilities.

By associating probabilities with the strings, we can impose a probabilistic structure on the language to describe noisy patterns. The probability distribution characterizing the patterns in a class can be interpreted as the probability distribution associated with the strings in a language. Thus, statistical decision rules can be applied to the classification of a pattern under ambiguous situations (for example, use the maximum-likelihood or Bayes decision rule). A block diagram of such a recognition system using maximum-likelihood decision rule is shown in Figure 3. For a given stochastic finite-state grammar G_s, we can construct a stochastic finite-state automaton to recognize only the language $L(G_s)$ [3]. For stochastic context-free languages, stochastic syntax analysis procedures are in general required. Because of the availability of the information about production probabilities, the speed of syntactic analysis can be improved through the use of this information [3]. Of course, in practice, the production probabilities will have to be inferred from the observation of relatively large numbers of pattern samples [3]. When the imprecision and uncertainty involving in the pattern description can be modeled by using the fuzzy set theory, the use of fuzzy languages for syntactic pattern recognition has recently been suggested [21].

Other approaches for the recognition of distorted or noisy patterns using syntactic methods include the use of transformational grammar [22,23] and approximation [24], and the application of error-correcting parsing techniques [25-29].

IV. Syntactic Recognition

Conceptually, the simplest form of recognition is probably "template-matching." The sentence describing an input pattern is matched against sentences representing each prototype or reference pattern. Based on a selected "matching" or "similarity" criterion, the input pattern is classified in the same class as the prototype pattern which is the "best" to match the input.

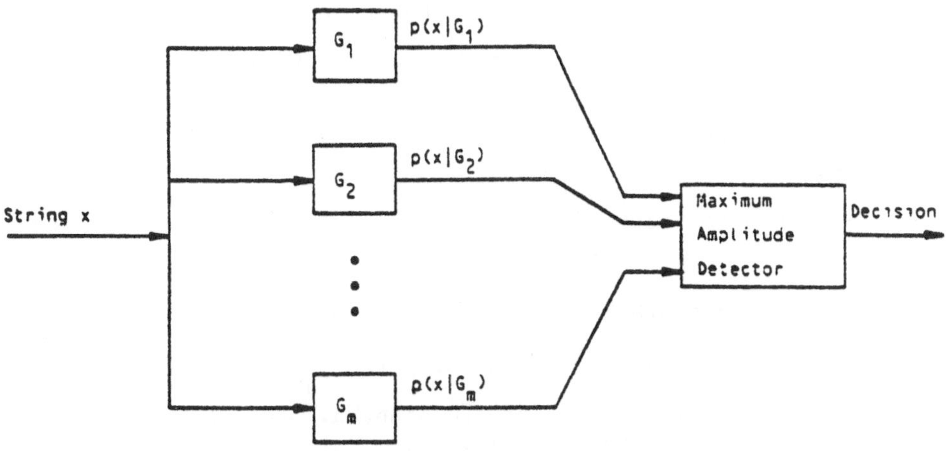

Fig. 3. Maximum-Likelihood Syntactic Recognition System

The structural information is not recovered. If a complete pattern description is required for recognition, a parsing or syntax analysis is necessary. In between the two extreme situations, there are a number of intermediate approaches. For example, a series of tests can be designed to test the occurrence or nonoccurrence of certain subpatterns (or primitives) or certain combinations of them. The result of the tests, through a table lookup, a decision tree, or a logical operation, is used for a classification decision. Recently, the use of discriminant grammars has been proposed for the classification of syntactic patterns [30].

There are many parsing algorithms proposed for context-free languages [6]. For the class of general context-free languages, Earley's parsing algorithm is briefly described in this section.

<u>Algorithm 0</u>. Early's parsing algorithm [6]

<u>Input</u>: A context-free grammar $G = (N,\Sigma,P,S)$ and an input string $x = a_1a_2\cdots a_n$ in Σ^*.

<u>Output</u>: The parse lists $I_0, I_1, \cdots, I_n$

<u>Method</u>:

(A) Construction of I_0

<u>Step 1</u>. If $S \to \alpha$ is in P, add $[S\to\cdot\alpha,0]$ to I_0. Perform Step 2 and Step 3 until no new items can be added to I_0.

<u>Step 2</u>. If $[B\to\gamma\cdot,0]$ is on I_0, add $[A\to\alpha B\cdot\beta,0]$ for all $[A\to\alpha\cdot B\beta,0]$ on I_0.

<u>Step 3</u>. Suppose that $[A\to\alpha\cdot B\beta,0]$ is an item in I_0. Add to I_0, for all productions in P of the form $B\to\gamma$, the item $[B\to\cdot\gamma,0]$ provided this item is not already in I_0.

(B) Construction of I_j from $I_0, I_i, \cdots, I_{j-1}$

<u>Step 4</u>. For each $[B\to\alpha\cdot a\beta,i]$ in I_{j-1} such that $a=a_j$, add $[B\to\alpha a\cdot\beta,i]$ to I_j.

Perform Step 5 and Step 6 until no new item can be added.

<u>Step 5</u>. Let $[A\to\alpha\cdot,i]$ be an item in I_j. Examine I_i for items of the form $[B\to\alpha\cdot A\beta,k]$. For each one found, we add $[B\to\alpha A\cdot\beta,k]$ to I_j.

<u>Step 6</u>. Let $[A\to\alpha\cdot B\beta,i]$ be an item in I_j. For all $B\to\gamma$ in P, add $[B\to\cdot\gamma,j]$ to I_j.

Note that x is in L(G) if and only if there is some item of the form $\{S\to\alpha\cdot,0]$ in I_n. An algorithm is available to recover the pattern structure or the parse of x from the parse lists [6]. It can be shown that the time complexity of Algorithm 0 is in general $O(n^3)$, that is, Algorithm 0 can be executed in $O(n^3)$ suitably defined elementary operations when the input is of length n. If the underlying grammar G is unambiguous, the time complexity reduces to $O(n^2)$.

A parsing procedure for recognition is, in general, nondeterministic and, hence, is regarded as computationally inefficient. Efficient parsing could be achieved by using special classes of languages such as finite state and deterministic languages for pattern description. The tradeoff here between the descriptive power of the pattern grammar and its parsing efficiency is very much like that between the feature space selected and the classifier's discrimination power in a decision-theoretic recognition system. Special parsers using sequential procedures or other heuristic means for efficiency improvement in syntactic pattern recognition have recently been constructed [25,31-33].

V. Error-Correcting Parsing

In practical applications, pattern distortion and measurement noise often exist. Pattern segmentation errors and misrecognitions of primitives (and relations) and/or subpatterns will lead to erroneous or noisy sentences rejected by the grammar characterizing its class. Recently, the use of an error-correcting parser as a recognizer of noisy and distorted patterns has been proposed [25-29]. In the use of an error-correcting parser as a recognizer, the pattern grammar is first expanded to include all the possible errors into its productions. The original grammar is transformed into a covering grammar that generates not only the correct sentences, but also all the possible erroneous sentences. For string grammars, three types of error - substitution, deletion and insertion - are considered. Misrecognition of primitives (and relations) are regarded as substitution errors, and segmentation errors as deletion and insertion errors.

The distance between two strings is defined in terms of the minimum number of error transformations used to derive one from the other by Aho and Peterson [34]. When the error transformations are defined in terms of substitution, deletion and insertion errors, the distance measurement coincides with the definition of Levenshtein metric [35]. For a given input string y and a given grammar G, a minimum-distance error-correcting parser (MDECP) is an algorithm that searches for a sentence z in L(G) such that the distance between z and y, d(z,y) is the minimum among the distances between all the sentences in L(G) and y. The algorithm also generates the value of d(z,y). We simply define this value to be the distance between L(G) and y and denote it as $d_1(L(G),y)$.

When a given grammar is a context-free grammar (CFG), its MDECP can be implemented by modifying Earley's parsing algorithm. We also extend the definition of the distance between L(G) and y, $d_1(L(G),y)$, to the definition of $d_K(L(G),y)$, the average distance between y and the K sentences in L(G) that are the nearest to y. The computation of $d_K(L(G),y)$ can be implemented by further modification of the algorithm of MDECP.

<u>Definition 3</u>. For two strings. $x, y \in \Sigma^*$, we can define a transformation $T: \Sigma^* \to \Sigma^*$ such that $y \in T(x)$. The following three transformations are introduced:

(1) substitution error transformation

$$\omega_1 a \omega_2 \overset{T_S}{\vdash\!\!\text{-----}} \omega_1 b \omega_2, \text{ for all } a, b \in \Sigma, a \neq b,$$

(2) deletion error transformation

$$\omega_1 a \omega_2 \overset{T_D}{\vdash\!\!\text{-----}} \omega_1 \omega_2, \text{ for all } a \in \Sigma$$

(3) insertion error transformation

$$\omega_1 \omega_2 \overset{T_I}{\vdash\!\!\text{-----}} \omega_1 a \omega_2, \text{ for all } a \in \Sigma$$

where $\omega_1, \omega_2 \in \Sigma^*$

Definition 4. The distance between two strings $x, y \in \Sigma^*$, $d^L(x,y)$, is defined as the smallest number of transformations required to derive y from x.

Example 1. Given a sentence x = cbabdbb and a sentence y = cbbabbdb, then

X = cbabdbb

$$\overset{T_S}{\vdash\!\!\text{-----}}\ \text{cbabbbb}\ \overset{T_S}{\vdash\!\!\text{-----}}\ \text{cbabbdb}\ \overset{T_I}{\vdash\!\!\text{-----}}\ \text{cbbabbdb} = y$$

The minimum number of transformations required to transform x to y is three, thus, $d^L(x,y) = 3$.

The metric defined in Definition 4 gives exactly the Levenshtein distance between two strings [35]. A weighted Levenshtein distance can be defined by assigning nonnegative numbers σ, γ and δ to transformations T_S, T_D and T_I respectively. Let $x, y \in \Sigma^*$ be two strings, and let J be a sequence of transformations used to derive y from x, then the weighted Levenshtein distance between x and y, denoted as $d_W(x,y)$ is

$$d_W(x,y) = \min_j \{\sigma\cdot k_j+\gamma\cdot m_j+\delta\cdot n_j\} \tag{1}$$

where k_j, m_j and n_j are the number of substitution, deletion, and insertion error transformations respectively in J.

We shall propose a weighted metric that would reflect the difference of the same type of error made on different terminals. Let the weights associated with error transformations on terminal a in a string $\omega_1 a\omega_2$ where $a \in \Sigma$, ω_1 and $\omega_2 \in \Sigma^*$, be defined as follows:

(1) $\omega_1 a\omega_2 \overset{T_S,\ S(a,b)}{\vdash\!\!\text{-----------}} \omega_1 b\omega_2$ for $b \in \Sigma$, $b \neq a$, where S(a,b) is the cost of substituting a for b. Let S(a,a) = 0.

(2) $\omega_1 a\omega_2 \overset{T_D,\ D(a)}{\vdash\!\!\text{-----------}} \omega_1\omega_2$ where D(a) is the cost of deleting a from $\omega_1 a\omega_2$.

(3) $\omega_1 a\omega_2 \overset{T_I,\ I(a,b)}{\vdash\!\!\text{-----------}} \omega_1 ba\omega_2$ for $b \in \Sigma$, where I(a,b) is the cost of inserting b in front of a.

We further define the weight of inserting a terminal b at the end of a string x to be,

$$(4) \quad x \overset{T_{I'},\ I'(b)}{\vdash\!\!\text{----------}} xb, \text{ for } b \in \Sigma.$$

Let $x, y \in \Sigma^*$ be two strings, and J be a sequence of transformations used to derive y from x. Let $|J|$ be defined as the sum of the weights associated with transformations in J, then the weighted distance between x and y, $d^W(x,y)$ is defined as

$$d^W(x,y) = \min_J \{|J|\} \tag{2}$$

Let L(G) be a given language and y be a given sentence the essence of minimum-distance error-correcting parsing is to search for a sentence x in L(G) that satisfied the minimum distance criterion as follow

$$d(x,y) = \min_z \{d(z,y) | z \in L(G)\} \tag{3}$$

We note that the minimum-distance correction of y is y itself if $y \in L(G)$.

We shall extend the minimum-distance ECP proposed by Aho and Peterson [34] to all three types of metric; L, w, and W. In [34], the procedure for constructing an ECP starts with the modification of a given grammar G by adding the three types of error transformations in the form of production rules, called error productions. The grammar G is now expanded to G' such that L(G') includes not only L(G), but all possible sentences with the three types of errors. The parser constructed according to G' with a provision added to count the number of error productions used in a derivation is the error-correcting parser for G. For a given sentence y, the ECP will generate a parse, Π, which consists of the smallest number of error productions. A sentence x in L(G) that satisfies the minimum-distance criterion (measured by using Levenshtein distance) can be generated from Π by eliminating error productions. With some modifications, this minimum-distance ECP can easily be extended to the three metrics proposed. We first give the algorithm of constructing an expanded grammar, in which the nonnegative numbers associated with error-productions are the weights associated with their corresponding error transformations with respect to the metric used.

<u>Algorithm 1</u>. Construction of expanded grammar
<u>Input</u>: A CFG $G = (N,\Sigma,P,S)$
<u>Output</u>: A CFG $G' = (N',\Sigma',P',S')$ where P' is a set of weighted productions.
<u>Method</u>:
<u>Step 1</u>. $N' = N \cup \{S'\} \cup \{E_a | a \in \Sigma\}$, Σ' Σ

Step 2. If $A \to \alpha_0 b_1\alpha_1 b_2 \ldots b_m\alpha_m$, $m \geq 0$ is a production in P such that $\alpha_i \in N^*$ and $b_i \in \Sigma$, then add $A \to \alpha_0 E_{b_1} \alpha_1 E_{b_2} \ldots E_{b_m} \alpha_m$, 0 to P', where each E_{b_i} is a new non-terminal, $E_{b_i} \in N'$ and 0 is the weight associated with this production.

Step 3. Add the following productions to P'.

Production Rule	L	w	weight W (metric)	
(a) $S' \to S$	0	0	0	
(b) $S' \to Sa$	1	δ	$I'(a)$	for all $a\in\Sigma'$
(c) $E_a \to a$	0	0	0	for all $a\in\Sigma$
(d) $E_a \to b$	1	σ	$S(a,b)$	for all $a\in\Sigma$, $b\in\Sigma'$ and $b \neq a$
(e) $E_a \to \lambda$	1	γ	$D(a)$	for all $a\in\Sigma$
(f) $E_a \to bE_a$	1	δ	$I(a,b)$	for all $a\in\Sigma$, $b\in\Sigma'$

In Algorithm 1 the production rules added in Step 3(b), 3(d), 3(e) and 3(f) are called error productions. Each error production corresponds to one type of error transformation on a particular symbol in Σ. Therefore, the distance measured in terms of error transformations can be measured by error productions used in a derivation. The parser is a modified Earley's parsing algorithm with a provision added to accumulate the weights associated with productions used in a derivation. The algorithm is as follows.

Algorithm 2. Minimum-distance error-correcting parsing algorithm

Input: An expanded grammar $G' = (N',\Sigma',P',S')$ and an input string $y = b_1b_2\ldots b_m$ in Σ'^*

Output: $I_0, I_1 \ldots I_m$ the parse list for y, and d(x,y) where x is the minimum-distance correction of y.

Method:

Step 1. Set j = 0. Then add $[E\to\cdot S',0,0]$ to I_j.

<u>Step 2</u>. If $[A\to\alpha\cdot B\beta,i,\xi]$ is in I_j, and $B\to\gamma$, η is a production rule in P' then add item $[B\to\cdot\gamma,j,0]$ to I_j.

<u>Step 3</u>. If $[A\to\alpha\cdot,i,\xi]$ is in I_j and $[B\to\beta\cdot A\gamma,k,\xi]$ is in I_i, and if no item of the form $[B\to\beta A\cdot\gamma,k,\phi]$ can be found in I_j, then add an item $[B\to\beta A\cdot\gamma,k,\eta+\xi+\zeta)$ to I_j where ζ is the weight associated with production $A\to\alpha$. If $[B\to\beta A\cdot\gamma,k,\phi]$ is already in I_j, then replace ϕ by $\eta+\xi+\zeta$ if $\phi > \eta+\xi L\zeta$.

<u>Step 4</u>. If $j = m$ go to Step 6, otherwise $j = j+1$.

<u>Step 5</u>. For each item in I_{j-1} of the form $[A\to\alpha\cdot b_j\beta,i,\xi]$ add item $[A\to\alpha b_j\cdot\beta,i,\xi]$ to I_j, go to Step 2.

<u>Step 6</u>. If item $[E\to S',0,\xi]$ is in I_m. Then $d(x,y) = \xi$, where x is the minimum-distance correction of y, exit.

In Algorithm 2, the string x, which is the minimum-distance correction of y, can be derived from the parse of y by eliminating all the error productions. The extraction of the parse of y is the same as that described in Earley's algorithm.

We shall define the distance between a string and a given language based on any one of the three metrics as follows.

<u>Definition 5</u>. Let y be a sentence, and L(G) be a given language, the distance between L(G) and y, $d_K(L(G),y)$, where K is a given positive integer, is;

$$d_K(L(G),y) = \min\{\sum_{i=1}^{K} \frac{1}{K} d(z_i,y) \mid z_i \in L(G)\} \quad (4)$$

In particular, if K = 1, then

$$d_1(L(G),y) = \min\{d(z,y) \mid z \in L(G)\} \quad (5)$$

is the distance between y and its minimum-distance correction in L(G).

As the distance between a string (a syntactic pattern) and a language (a set of syntactic patterns) is defined, a minimum-distance decision rule can be stated as follows: suppose that there are two classes of patterns, C_1 and C_2 characterized by grammar G_1 and G_2 respectively. For a given syntactic pattern y with unknown classification, decide $y \in \begin{matrix} c_1 \\ c_2 \end{matrix}$ if $d(L(G_1),y) \lessgtr d(L(G_2),y)$.

When errors can be specified probabilistically or the pattern grammar is stochastic, the maximum-likelihood and Bayes error-correcting parsers have been proposed [25,28]. A special case of using a finite-state grammar with two types of error probabilistically specified has been applied to spoken word recognition [50]. It has been demonstrated that using a stochastic error-correcting technique, a 98% accuracy can be achieved on spoken word recognition even when the accuracy of primitive recognition (phoneme labeling) is 77% [50].

The sequential parsing procedure suggested by Persoon and Fu [33] has been applied to error-correcting parser to reduce the parsing time [25]. By sacrificing a small amount of error-correcting power (that is, allowing a small error in parsing), a parsing could be terminated much earlier before a complete sentence is scanned. The trade-off between the parsing time and the error committed can be easily demonstrated. In addition, error-correcting parsing for transition network grammars [28] and tree grammars [29] has also been studied. For tree grammars, five types of error - substitution, deletion, stretch, branch, and split - are considered. The original tree pattern grammar is expanded by including the five types of error transformation rule. The tree automaton constructed according to the expanded tree grammar and the minimum-distance criterion is called an error-correcting tree automaton (ECTA). When only substitution errors are considered, the structure of the tree to be analyzed remains unchanged. Such an error-correcting tree automaton is called a "structure-preserved error-correcting tree automton" (SPECTA) [36]. Another approach to reduce the parsing time is the use of parallel processing [37].

VI. Clustering Analysis for Syntactic Patterns

In statistical pattern recognition, a pattern is represented by a vector, called a feature vector. The similarity between two patterns can often be expressed by a distance, or more generally speaking, a metric in the feature space. Cluster analysis can be performed on a set of patterns on the basis of a selected similarity measure [38]. In syntactic pattern recognition a similarity measure between two syntactic patterns must include the similarity of both their structures and primitives. In Section V, we have described several distance measures for strings, which leads to the study of clustering analysis for syntactic patterns. The conventional clustering methods, such as, the minimum spanning tree, the nearest (or K-nearest) neighbor classification rule and the method of clustering centers can be extended to syntactic patterns.

The studies described in Section 6.1 are mainly on the pattern-to-pattern basis [39]. An input sentence (a pattern) is compared with sentences in a formed cluster, one by one, or with the representative (cluster center) of the cluster. In Section 6.2 we shall use the distance measure between a sentence and a language [40]. The proposed clustering procedure is combined with a grammatical inference procedure and an error-correcting parsing technique. The idea is to model the formed cluster by inferring a grammar, which implicitly characterize the structural identity of the cluster. The language generated by the grammar may be larger than the set consisting of the members of the cluster, and includes some possible similar patterns due to the recursive nature of grammar. Then the distance between an input sentence (a syntactic pattern) and a language (a group of syntactic pattern) is computed by using a ECP. The recognition is based on the nearest neighbor rule.

6.1 Sentence-to-Sentence Clustering Algorithms

(A) A Nearest Neighbor Classification Rule

Suppose that C_1 and C_2 are two pattern sets, represented by sentences $X_1 = \{x_1^1, x_2^1, \ldots x_{n_1}^1\}$ and $X_2 = \{x_1^2, x_2^2, \ldots, s_{n_2}^2\}$ respectively. For an unknown syntactic pattern y, decide that y is in the same class as C_1 if

$$\min_j d(x_j^1, y) < \min_\ell d(x_\ell^2, y);$$

and y is in class C_2 if

$$\min_j d(x_j^1, y) > \min_\ell d(x_\ell^2, y). \qquad (4.1)$$

In order to determine $\min_j d(x_j^i, y)$, for some i, the distance between y and every element in the set X_i have to be computed individually. The string-to-string correction algorithm proposed in [39] yields exactly the distance between two strings defined in Definition 4.

The following algorithm computes the weighted distance $d^W(x,y)$ between two strings [39].

Algorithm 3.

Input: i) Two strings $x = a_1a_2\ldots a_n$ and $y = b_1b_2\ldots b_m$.

ii) A table of weights associated with error transformations on a_i's and b_j's.

Output: The weighted distance $d^W(x,y)$.

Method:

Step 1: Set $\delta(0,0) = 0$.

Step 2: Do i = 1,n.

$\delta(i,o) = \delta(i-1,0) + D(a_i)$

Step 3: Do j = 1,m.
$\delta(0,j) = \delta(0,j-1) + I(a_1,b_j)$

Step 4: Do i = 1,n.
Do j = 1,m.
$\varepsilon_1 = \delta(i-1,j-1) + S(a_i,b_j)$
$\varepsilon_2 = \delta(i-1,j) + D(a_i)$
$\varepsilon_3 = \delta(i,j-1) + I(a_{i+1},b_j)$ if $i < n$ or
$\varepsilon_3 = 8(i,j-1) + I'(b_j)$ if $i = n$.
$\delta(i,j) = \mathrm{Min}\{\varepsilon_1,\varepsilon_2,\varepsilon_3\}$

Step 5: $d^W(x,y) = \delta(n,m)$ exit.

The nearest neighbor classification rule can be easily extended to the K-nearest neighbor rule. Let $\tilde{X}_i = \{\tilde{x}_1^i,\tilde{x}_2^i,\ldots\tilde{x}_{n_i}^i\}$ be a reordered set of X_i such that $d(\tilde{x}_j^i,y) \leq d(\tilde{x}_\ell^i,y)$ iff $j < \ell$, for all $1 \leq j, \ell \leq n_i$, then

$$\text{decide } y\varepsilon \begin{matrix} C_1 \\ C_2 \end{matrix} \text{ if } \sum_{j=1}^{K} \frac{1}{K} d(\tilde{x}_j^1,y) \begin{matrix} < \\ > \end{matrix} \sum_{j=1}^{K} \frac{1}{K} d(\tilde{x}_j^2,y) \quad (6)$$

We shall describe a clustering procedure in which, the classification of an input pattern is based on the nearest (or K-nearest) neighbor rule.

Algorithm 4.
Input: A set of samples $X = \{x_1,x_2,\ldots,x_n\}$ and a design parameter, or threshold, t.
Output: A partition of X into m clusters, $C_1,C_2,\ldots,C_m$.
Method:
Step 1. Assign x_1 to C_1, j=1, m=1.
Step 2. Increase j by one. If $D = \min_\ell d(x_\ell^i,x_j)$ is the minimum, $1 \leq i \leq m$, and
(i) $D \leq t$, then assign x_j to C_i
(ii) $D > t$, then initiate a new cluster for x_j, and increase m by one.
Step 3. Repeat Step 2 until every string in X has been put in a cluster.

Note that, in Algorithm 4, a design parameter is required. A commonly used clustering procedure is to construct a minimum spanning tree. Each node on the minimum spanning tree represents an element in the sample set X. Then partition the tree. Actually, when the distances between all of the pairs, $d(x_i,x_j)$, x_i, $x_j \in X$, are available, the algorithm for constructing minimum spanning tree is the same as that where X is a set of feature vectors in the statistical pattern recognition [2,38].

(B) The Cluster Center Techniques

Let us define a β-metric for a sentence x_j^i in cluster C_i as follow,

$$\beta_j^i = \frac{1}{n_i} \sum_{\ell=1}^{n_i} d(x_j^i, x_\ell^i) \tag{7}$$

Then, x_j^i is the cluster center of C_i, if $\beta_j^i = \min_\ell \{\beta_\ell^i | 1 < \ell < n_i\}$, x_j^i is also called the representative of C_i, denoted A_i.

The following clustering algorithm is given in [2,38].

<u>Algorithm 5.</u>
<u>Input:</u> A sample set $X = \{x_1,x_2,...,x_n\}$.
<u>Output:</u> A partition of X into m clusters
<u>Method:</u>
<u>Step 1.</u> Let m elements of X, chosen at random, be the "representatives" of the m clusters. Let them be called $A_1, A_2,...,A_m$.
<u>Step 2.</u> For all i, $x_i \in X$ is assigned to cluster j, iff $d(A_j,x_i)$ is minimum.
<u>Step 3.</u> For all j, a new mean A_j is computed. A_j is the new representative of cluster j.
<u>Step 4.</u> If no A_j has changed, stop. Otherwise, go to Step 2.

6.2 A Proposed Nearest Neighbor Syntactic Recognition Rule

With the distance between a sentence and a language defined in Section V, we can construct a syntactic recognizer using the nearest (or K-nearest) neighbor rule. Suppose that we are given two classes of patterns characterized by grammar G_1 and G_2, respectively. For an unknown syntactic pattern y, decide that y is in the same class as $L(G_1)$ if

$$d(L(G_1),y) < d(L(G_2),y)$$

and decide that y is in the same class as $L(G_2)$ if,

$$d(L(G_2),y) < d(L(G_1),y) \tag{8}$$

The distance $d(L(G_i),y)$ can be determined by a minimum-distance ECP constructed for G_i. Consequently, a grammatical inference procedure is required to infer a grammar for each class of pattern samples. Since the parser also gives the structural description of y, the syntactic recognizer gives both the classification and description of y as its output. We shall summarize the procedure in the following algorithm.

<u>Algorithm 6</u>.
<u>Input</u>: m sets of syntactic pattern samples

$$X_1 = \{x_1^1, x_2^1, \ldots, x_{n_1}^1\}, \ldots, X_m = \{x_1^m, x_2^m, \ldots, x_{n_m}^m\}$$

and a pattern y with unknown classification.

<u>Output</u>: The classification and structural description of y.

<u>Method</u>:

<u>Step 1</u>. Infer m grammars $G_1, G_2, \ldots, G_m$, from $X_1, X_2, \ldots, X_m$, respectively.

<u>Step 2</u>. Construct minimum-distance ECP's, $E_1, E_2, \ldots, E_m$ for $G_1, G_2, \ldots, G_m$, respectively.

<u>Step 3</u>. Calculate $d(L(G_k),y)$ for all i = 1,...,m. Determine ℓ such that

$$d(L(G_\ell),y) = \min_k d(L(G_k),y)$$

y is then classified as from class ℓ. In the meantime, the structural description of x can be obtained from E_ℓ.

Using the distance defined in Section V as a similarity measure between a syntactic pattern and a set of syntactic patterns, we can perform a cluster analysis to syntactic patterns. The procedure again involves error-correcting parsing and grammatical inference. In contrast to the nearest neighbor rule in Algorithm 6 which uses a supervised inference procedure, the procedure described in this section is basically non-supervised. When the syntactic pattern samples are observed sequentially, a grammar can be easily infered for the sample observed at each stage of the clustering procedure. We propose the following clustering procedure for syntactic patterns:

Algorithm 7.

Input: A set of syntactic pattern samples $X = \{x_1, x_2, \ldots, x_n\}$ where x_i is a string of terminals or primitives. A threshold t.

Output: The assignment of x_i, $i = 1, \ldots, n$ to m clusters and the grammar $G^{(k)}$, $k = 1, \ldots, m$, characterizing each cluster.

Method:

Step 1. Input the first sample x_1, infer a grammar $G_1^{(1)}$ from x_1, $L(G_1^{(1)}) \supseteq \{x_1\}$.

Step 2. Construct an error-correcting parser $E_1^{(1)}$ for $G_1^{(1)}$

Step 3. Input the second sample x_2, use $E_1^{(1)}$ to determine whether or not x_2 is similar to x_1 by comparing the distance between $L(G_1^{(1)})$ and x_2, i.e., $d(x_2, L(G_1^{(1)}))$, with a threshold t.

(i) If $d(x_2, L(G_1^{(1)})) < t$, x_1 and x_2 are put into the same cluster (Cluster 1). Infer a grammar $G_2^{(1)}$ from $\{x_1, x_2\}$.

(ii) If $d(x_2, L(G_1^{(1)})) \geq t$, initiate a new cluster for x_2 (Cluster 2) and infer a new grammar $G_1^{(2)}$ from x_2. In this case, there are two clusters characterized by $G_1^{(1)}$ and $G_1^{(2)}$, respectively.

Step 4. Repeat Step 2, construct error-correcting parsers for $G_2^{(1)}$ or $G_1^{(2)}$ depending upon $d(x_2, L(G_1^{(1)})) <$ t or $d(x_2, L(G_1^{(1)})) \geq t$, respectively.

Step 5. Repeat Step 3 for a new sample. Until all the pattern samples are observed, we have m clusters characterized by $G_{n_1}^{(1)}$, $G_{n_2}^{(2)}, \ldots,$ $G_{n_m}^{(m)}$, respectively.

The parsers (non-error-correcting) constructed according to $G_{n_1}^{(1)}, G_{n_2}^{(2)}, \ldots, G_{n_m}^{(m)}$ could then form a syntactic recognizer directly for the m-class recognition problem.

The threshold t is a design parameter. It can be determined from a set of pattern samples with known classifications. For example, if we know that the sample x_i is from Class 1 characterized by $G^{(1)}$ and the sample x_j is from Class 2 characterized by $G^{(2)}$, then $t < d(x_i, x_j)$. Or, more generally speaking,

$$t < \text{Min} \{d(L(G^{(2)}), x_i) \; , \; d(L(G^{(1)}), x_j)\} \tag{9}$$

For m classes characterized by $G^{(1)}, G^{(2)}, \ldots, G^{(m)}$, respectively, we can choose

$$t < \underset{k,\ell}{\text{Min}} \{d(L(G^{(\ell)}), x^{(k)})\}, \; k \neq \ell \tag{10}$$

where $x^{(k)}$ is a pattern sample known from Class k and $L(G^{(\ell)})$ is the grammar characterizing Class ℓ ($\ell \neq k$). If the above required information is not available, an appropriate value of t will have to be determined on an experimental basis until a certain stopping criterion is satisfied (for example, with a known number of clusters).

VII. Concluding Remarks

We have briefly reviewed some recent advances in the area of syntactic pattern recognition. Due to noise and distortions in real world patterns, syntactic approach to pattern recognition was regarded earlier as only effective in handling abstract and artificial patterns. However, with the recent development of distance or similarity measures between syntactic patterns and error-correcting parsing procedures, the flexibility of syntactic methods has been greatly expanded. Errors occurring at the lower-level processing of a pattern (segmentation and primitive recognition) could be compensated at the higher level using structural information. Using a distance or similarity measure, nearest-neighbor and k-nearest-neighbor classification rules can be easily applied to syntactic patterns. Furthermore, with a distance or similarity measure, a clustering procedure can be applied to syntactic patterns. Such a nonsupervised learning procedure can also be very useful for grammatical inference in syntactic pattern recognition [20,40].

It has been noticed from the recent advances that semantic information has been used more and more with the syntax rules in characterizing patterns. Quite often, semantic information involving spatial information can be expressed syntacticaly such as attributed grammars, and relational trees and graphs

[3,11,14,19,41,42]. Parsing efficiency has become a concern in syntactic recognition. Special grammars and parallel parsing algorithms have been suggested for speeding up the parsing time. Structural information of a pattern can also be used as a guide in the segmentation process through the syntactic approach [19,43,44]. Syntactic representation of patterns such as hierarchical trees and relational graphs should also be very useful for database organization. Several recent publications have already shown such a trend [45-48].

In some applications, both the decision-theoretic and the syntactic methods may be used. One possibility is to use a decision-theoretic method for the recognition of primitives, and then to use a syntactic method for the recognition of subpattern and pattern itself. The second possibility is the use of stochastic grammars with which the syntactic recognition is made in the decision-theoretic sense (maximum-likelihood or Bayes). The term "mixed or combined approach" is often used to denote such an approach.

APPENDIX

Syntactic Recognition of Spoken Words

The application of syntactic approach to the recognition of sounds, words, and continuous speech has recently received increasing attention [51-54]. The method used by De Mori [51] for recognition of spoken digits is summarized in this appendix. Zero crossings are used to characterize each segment of the incoming speech signal [55,56]. A careful analysis of the sequences of zero-crossing intervals from many spoken words indicated that intervals of 20-msec length provide meaningful short time statistics of the intervals. Inspection of these statistics leads to the conclusion that the zero-crossing intervals in each 20-msec segment can be classified into a few groups. The number of the zero-crossing intervals classified into the groups during a segmentation interval can then be used as features of the speech segment. Consequently, the segmentation of each pattern (speech signal) is based on 20-msec time intervals and the range of the zero-crossing intervals of the outputs of the low-pass filter (LPF) and high-pass filter (HPF), respectively, is subdivided according to Table A.1. A simple block diagram of the preprocessing part of the recognition system is shown in Fig. A.1.

For the nth segment of 20-msec length, compute

$$R_L(nT) = \{R_{L1}(nT),\ R_{L2}(nT),\ldots,R_{L7}(nT)\}$$

$$R_H(nT) = \{R_{H1}(nT),\ldots,R_{H4}(nT)\}, \quad T = 20 \text{ msec}$$

where $R_{ki}(nT)$ is the number of zero-crossing intervals from the kth filter

TABLE A.1
Specifications for the Groups of Zero-Crossing Intervals[a]

	7 groups for LPF[b]	4 groups for HPF[b]
t^0	7.0	0.9
t_1	3.0	0.6
t_2	1.6	0.4
t_3	1.2	0.3
t_4	1.0	0.1
t_5	0.8	
t_6	0.6	
t_7	0.4	

[a] An incoming zero-crossing interval of duration t from the output of the low-pass filter (or the high-pass filter) is assigned to the group (L,i) (or (H,i)).

[b] Units msec.

(k = L for LPF, k = H for HPF) assigned to the group (L,i) or (H,i) during the nth segmentation period. From this informaton, then calculate the following parameters:

$$B_L(nT) = \frac{\sum_{i-1}^{7} (i-1)R_{Li}(nT)}{\sum_{i=1}^{7} R_{Li}(nT)} \quad \text{(quantized into 60 levels)}$$

$$B_H(nT) = \frac{\sum_{i=1}^{4} (i-1)R_{Hi}(nT)}{\sum_{i=1}^{4} R_{Hi}(nT)} \quad \text{(quantized into 30 levels)}$$

Thus, each spoken word can be represented pictorially on the $B_L - B_H$ plane. Figure A.2 shows such a plot for the Italian spoken digit "nine" (NOVE).

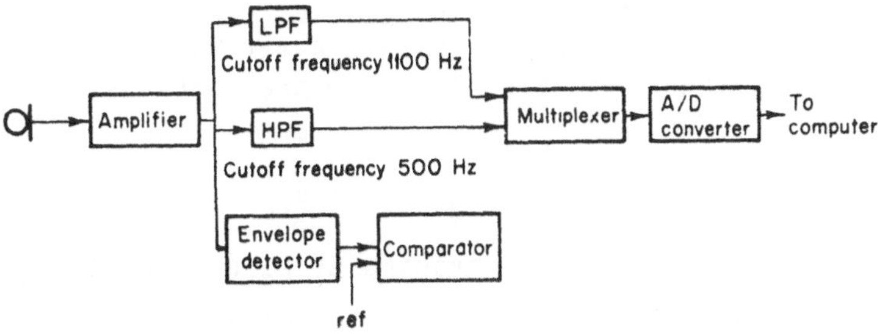

Fig. A.1. Preprocessing system.

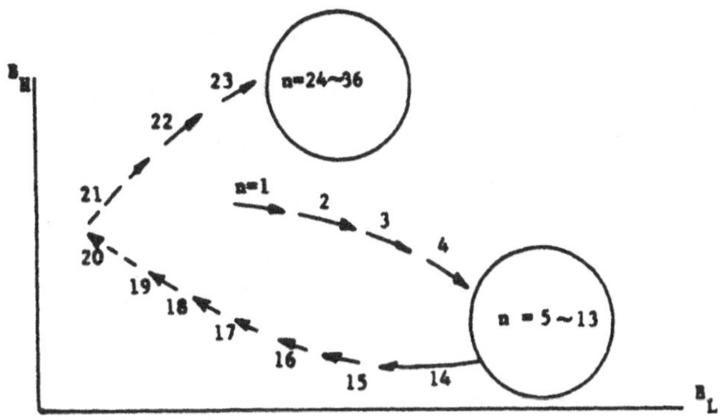

Fig. A.2. $B_L - B_H$ representation of NOVE.

A relatively dense set of points is generated by the pronunciation of a vowel or a semivowel.

(A) Syntactic Representation of a Spoken Word (in B_L-B_H Plane)

<u>Primitives</u> (<u>Terminals</u>)

(1) Silence interval (SL) - a sequence of points in the $B_L - B_H$ plane whose coordinates are all zeros following at least one point whose coordinates are both nonzero. These points are labeled as

$$N \cdot p_1 \cdot p_2$$

where N is the symbol used for (SL), and p_1 and p_2 denote the starting position and the duration of the (SL), respectively. For example, N • 1010 • 11 represents an (SL) starting from the tenth segmentation interval after the beginning of the word for a duration of three successive segments.

(2) Stable zone (SZ) - a set of points representing successive segments of the word lies within a relatively small region and the number of points is higher than the specified threshold. The (SZ) is described by

$$S \cdot s_1 \cdot s_2 \cdot s_3 \cdot s_4$$

where s_1 and s_2 have the same meaning as p_1 and p_2 for (SL), s_3 and s_4 are the coordinates of the center of gravity of the (SZ).

(3) Lines (LN) - nonstationary portions of the acoustics waveform approximated by straight-line segments. The (LN) is described by

$$L \cdot l_1 \cdot l_2 \cdot l_3 \cdot l_4$$

where l_1 and l_2 have the same meaning as p_1 and p_2 for (SL), l_3 is the length of the line segment, and l_4 is the slope of the line represented in octal code.

Nonterminals

(1) Silence fragment (NF) - an (SL) with the following description:

$$N \cdot n_1 \cdot n_2 \cdot 0 \cdot 0 \cdot 0 \cdot 0 \cdot 0$$

where n_1 and n_2 have the same meaning as p_1 and p_2. In terms of a production rule, (NF) is defined as

$$(NF) \rightarrow (SL)$$

(2) O fragment (OF) - a stable zone which cannot be composed by other primitives. Its description is

$$0 \cdot o_1 \cdot o_2 \cdot o \cdot o \cdot o \cdot o_6 \cdot o_7$$

where o_1, o_2, o_6, and o_7 have the same meaning as s_1, s_2, s_3, and s_4, respectively. The production is

$$(OF) \rightarrow (SZ)$$

For example, a vowel between two silences usually leads to an (OF).

(3) I fragment (IF) - a picture ending with an (SZ) preceded by an (LN), an (SZ), or an (SZ) followed by an (LN). In production form,

$$(IF) \rightarrow \gamma(SZ)$$

where $\gamma \to$ (LN), or $\gamma \to$ (SZ)(LN). The (IF) description is

$$I \cdot i_1 \cdot i_2 \cdot i_3 \cdot i_4 \cdot i_5 \cdot i_6 \cdot i_7$$

where i_1 and i_2 have the same meaning as o_1 and o_2, and i_6 and i_7 are the coordinates of the center of gravity of the last (SZ). i_3 is a composition code defined by the following table:

(IF) composition	i_3 Code
(LN)(SZ)	01
(SZ)(SZ)	11
(SZ)(LN)(SZ)	101

If i_3 = 11, i_5 represents the slope (in octal code) of the line segment joining the two (SZ)'s, and i_4 denotes the length of the line segment joining the centers of gravity of the two (SZ)'s divided by 2^3.

(4) V fragment (VF) - a picture composed by two lines and terminated by an (SZ); that is,

$$(VF) \to \gamma^2 (SZ)$$

A (VF) description is

$$V \cdot v_1 \cdot v_2 \cdot v_3 \cdot v_4 \cdot v_5 \cdot v_6 \cdot v_7$$

where v_1, v_2, v_6, and v_7 are similar to i_1, i_2, i_6, and i_7 of the (IF), v_3 is the composition code with a value 1 for (SZ) and 0 for (LN). v_5 is the sequence of slopes of the line segments each one of which is expressed in octal code. v_4 is the sequence of lengths of the line segments divided by 2^3.

(5) Z fragment (ZF) - a picture with three lines.

$$(ZF) \to \gamma^3 (SZ)$$

A (ZF) description is

$$Z \cdot z_1 \cdot z_2 \cdot z_3 \cdot z_4 \cdot z_5 \cdot z_6 \cdot z_7$$

where $z_1,\ldots,z_7$ have the same meaning as $v_1,\ldots,v_7$ of a (VF).

Grammar: $G = (V_N, V_T, P, S)$, where

V_N = {(NF),(OF),(IF),(VF),(ZF),γ,S,ZERO,ONE,TWO,

THREE, FOUR, FIVE, SIX, SEVEN, EIGHT, NINE}

V_T = {(SL), (LN), (SZ)}

P:

S → ZERO, S → ONE, S → TWO, S → THREE

S → FOUR, S → FIVE, S → SIX, S → SEVEN

S → EIGHT, S → NINE,

(NF) → (SL), (OF) → (SZ), (IF) → γ(SZ),

(VF) → γ^2(SZ), (ZF) → γ^3(SZ),

γ → (LN), γ → (SZ), γ → (SZ)(LN)

ZERO → (IF) $\{0 \le i_4 \le 1, 6 \le i_5 \le 7, 32 \le i_6 \le 48, 0 \le i_7 \le 16\}$

ZERO → (VF) $\{00 \le v_4 \le 22, 26 \le v_5 \le 47, 32 \le v_6 \le 48, 0 \le v_7 \le 16\}$

ZERO → (ZF) $\{000 \le z_4 \le 111, 026 \le z_5 \le 047, 32 \le z_6 \le 48, 0 \le z_7 \le 16\}$

ONE → (VF) $\{00 \le v_4 \le 22, 06 \le v_5 \le 27, 32 \le v_6 \le 48, 0 \le v_7 \le 17\}$

ONE → (ZF) $\{000 \le z_4 \le 222, 072 \le z_5 \le 274, 32 \le z_6 \le 48, 0 \le z_7 \le 16\}$

TWO → (IF) $\{1 \le i_4 \le 2, 1 \le i_5 \le 2, 32 \le i_6 \le 40, 18 \le i_7 \le 24\}$

TWO → (VF) $\{00 \le v_4 \le 22, 01 \le v_5 \le 13, 32 \le v_6 \le 40, 16 \le v_7 \le 24\}$

THREE → (OF) $\{24 \le o_6 \le 40, 16 \le o_7 \le 24\}$

THREE → (IF) $\{0 \le i_4 \le 1, 3 \le i_5 \le 3, 24 \le i_6 \le 40, 16 \le i_7 \le 24\}$

FOUR → (OF) $\{24 \le o_6 \le 56, 16 \le o_7 \le 24\}$
(NF)(IF) $\{0 \le i_4 \le 1, 6 \le i_5 \le 7, 32 \le i_6 \le 48, 08 \le i_7 \le 24\}$

FOUR → (IF) $\{1 \le i_4 \le 2, 2 \le i_5 \le 3, 40 \le i_6 \le 56, 16 \le i_7 \le 24\}$,
(NF)(OF) $\{32 \le o_6 \le 48, 0 \le o_7 \le 16\}$

FOUR → (VF) $\{11 \le v_4 \le 22, 03 \le v_5 \le 03, 40 \le v_6 \le 56, 16 \le v_7 \le 24\}$
(NF)(OF) $\{32 \le o_6 \le 48, 0 \le o_7 \le 16\}$

FIVE → (VF) $\{41 \le v_4 \le 42, 44 \le v_5 \le 44, 8 \le v_6 \le 32, 24 \le v_7 \le 30\}$
(NF)(OF) $\{32 \le o_6 \le 40, 16 \le o_7 \le 24\}$

FIVE → (ZF) $\{411 \le z_4 \le 622, 404 \le z_5 \le 444, 8 \le z_6 \le 32, 24 \le z_7 \le 30\}$
(NF)(OF) $\{32 \le o_6 \le 40, 16 \le o_7 \le 24\}$

FIVE → (ZF) $\{411 \le z_4 \le 622, 404 \le z_5 \le 444, 8 \le z_6 \le 32, 24 \le z_7 \le 30\}$
(NF)(IF) $\{0 \le i_4 \le 0, 0 \le i_5 \le 0, 32 \le i_6 \le 40, 16 \le i_7 \le 24\}$

SIX → (VF) $\{11 \le v_4 \le 42, 42 \le v_5 \le 53, 16 \le v_6 \le 32, 24 \le v_7 \le 30\}$

SIX → (ZF) $\{111 \le z_4 \le 422, 423 \le z_5 \le 533, 16 \le z_6 \le 32, 24 \le z_7 \le 30\}$

SEVEN → (IF) $\{3 \le i_4 \le 7, 4 \le i_5 \le 5, 32 \le i_6 \le 40, 18 \le i_7 \le 24\}$
(NF)(OF) $\{32 \le o_6 \le 40, 16 \le o_7 \le 24\}$

SEVEN → (VF) $\{12 \le v_4 \le 33, 43 \le v_5 \le 43, 0 \le v_6 \le 16, 16 \le v_7 \le 30\}$
(NF)(OF) $\{32 \le o_6 \le 40, 16 \le o_7 \le 24\}$

EIGHT → (OF) $\{32 \le o_6 \le 48, 0 \le o_7 \le 16\}$
(NF)(OF) $\{32 \le o_6 \le 48, 0 \le o_7 \le 16\}$

EIGHT → (IF) $\{0 \le i_4 \le 1, 2 \le i_5 \le 3, 32 \le i_6 \le 48, 0 \le i_7 \le 16\}$
(NF)(OF) $\{32 \le o_6 \le 48, 0 \le o_7 \le 16\}$

NINE → (ZF) $\{321 \le z_4 \le 543, 720 \le z_5 \le 730, 32 \le z_6 \le 40, 16 \le z_7 \le 24\}$

NINE → (VF) $\{52 \le v_4 \le 54, 74 \le v_5 \le 74, 32 \le v_6 \le 40, 0 \le v_7 \le 16\}$

NINE → (VF) $\{21 \le v_4 \le 42, 20 \le v_5 \le 30, 32 \le v_6 \le 40, 16 \le v_7 \le 24\}$

(B) Recognition: A bottom-up parsing, implemented in terms of two push-down transducers, is used.

REFERENCES

1. K. S. Fu and A. Rosenfeld, "Pattern Recognition and Image Processing", IEEE Trans. on Computers, Vol. C-25, No. 12, 1976.

2. K. S. Fu, Digital Pattern Recognition, Springer-Verlag, 1976.

3. K. S. Fu, Syntactic Methods in Pattern Recognition, Academic Press, 1974.

4. K. Hanakata, "Feature Selection and Extraction for Decision Theoretic Approach and Structural Approach" in Pattern Recognition-Theory and Application, ed. by K. S. Fu and A. B. Whinston, Noordhoff International Publishing Co. Leyden, The Netherlands, 1977.

5. C. H. Chen, "On Statistical and Structural Feature Extraction" in Pattern Recognition and Artificial Intelligence, ed. by C. H. Chen, Academic Press, 1976.

6. A. V. Aho and J. D. Ullman, The Theory of Parsing, Translation, and Compiling, Vol. 1, Prentice-Hall, 1972.

7. K. S. Fu, Syntactic Pattern Recognition Applications, Springer-Verlag, 1977.

8. J. Gips, Shape Grammars and Their Use, Birkhauser Verlag, Basel and Stuttgart, 1975.

9. K. S. Fu, "Tree Languages and Syntactic Pattern Recognition" in Pattern Recognition and Artificial Intelligence, ed. by C. H. Chen, Academic Press, 1976.

10. K. S. Fu and T. L. Booth, "Grammatical Inference-Introduction and Survey", IEEE Trans. on Systems, Man and Cybernetics, Vol. SMC-5, Jan. and July 1975.

11. S. M. Chou and K. S. Fu, "Inference for Transition Network Grammars", Proc. Third International Joint Conference on Pattern Recognition, Nov. 8-11, 1976, Coronado, Calif., U.S.A.

12. G. B. Porter, "Grammatical Inference Based on Pattern Recognition", Proc. Third International Joint Conference on Pattern Recognition, Nov. 8-11, 1976, Coronado, Calif., U.S.A.

13. L. Miclet, "Inference of Regular Expressions", Proc. Third International Joint Conference on Pattern Recognition, Nov. 8-11, 1976, Coronado, Calif., U.S.A.

14. J. M. Brayer and K. S. Fu, "Some Multidimensional Grammar Inference Methods" in Pattern Recognition and Artificial Intelligence, ed. by C. H. Chen, Academic Press, 1976.

15. J. M. Brayer and K. S. Fu, "A Note on the k-tail Method of Tree Grammar Inference", IEEE Trans. on Systems, Man and Cybernetics, Vol. SMC-7, No. 4, April 1977, pp. 293-299.

16. A. Barrero and R. C. Gonzalez, "A Tree Traversal Algorithm for the Inference of Tree Grammars", Proc. 1977 IEEE Computer Society Conference on Pattern Recognition and Image Processing, June 6-8, Troy, N.Y.

17. H. C. Lee and K. S. Fu, "A Syntactic Pattern Recognition System with Learning Capability", Proc. Fourth International Symposium on Computer and Information Sciences (COINS-72), Dec. 14-16, 1972, Bal Harbour, Florida.

18. J. Keng and K. S. Fu, "A System of Computerized Automatic Pattern Recognition for Remote Sensing", Proc. 1977 International Computer Symposium, Dec. 27-29, Taipei, Taiwan.

19. K. C. You and K. S. Fu, "Syntactic Shape Recognition Using Attributed Grammars", IEEE Trans. on Systems, Man, and Cybernetics, Vol. SMC-9, No. 6, June 1979.

20. S. Y. Lu and K. S. Fu, "Stochastic Tree Grammar Inference for Texture Synthesis and Discrimination", Computer Graphics and Image Processing, Vol. 7, No. 5, Oct. 1978.

21. R. DeMori, P. Laface and M. Sardella, "Use of Fuzzy Algorithms for Phonetic and Phonemic Labelling of Continuous Speech", Instituto di Scienze dell' Informazione, Universita di Torino, Italy.

22. A. K. Joshi, "Remarks on Some Aspects of Language Structure and Their Relevance to Pattern Analysis", Pattern Recognition, Vol. 5, No. 4, 1973.

23. B. K. Bhargava and K. S. Fu, "Transformation and Inference of Tree Grammars for Syntactic Pattern Recognition", Proc. 1974 IEEE International Conference on Cybernetics and Society, Oct., Dallas, Texas.

24. T. Pavlidis, "Syntactic Pattern Recognition on the Basis of Functional Approximation" in Pattern Recognition and Artificial Intelligence, ed. by C. H. Chen, Academic Press, 1976.

25. S. Y. Lu and K. S. Fu, "Stochastic Error-Correcting Syntax Analysis for Recognition of Noisy Patterns", IEEE Trans. on Computers, Vol. C-26, No. 12, Dec. 1977, pp. 1268-1276.

26. L. W. Fung and K. S. Fu, "Stochastic Syntactic Decoding for Pattern Classification", IEEE Trans. on Computers, Vol. C-24, July 1975.

27. M. G. Thomason and R. C. Gonzalez, "Error Detection and Classification in Syntactic Pattern Structures", IEEE Trans. on Computers, Vol. C-24, 1975.

28. K. S. Fu, "Error-Correcting Parsing for Syntactic Pattern Recognition", in Data Structure, Computer Graphics and Pattern Recognition, ed. by A. Klinger, et. al., Academic Press, 1977.

29. S. Y. Lu and K. S. Fu, "Error-Correcting Tree Automata for Syntactic Pattern Recognition", IEEE Trans. on Computers, Vol. C-27, Nov. 1978.

30. C. Page and A. Filipski, "Discriminant Grammars, An Alternative to Parsing for Pattern Classification", Proc. 1977 IEEE Workshop on Picture Data Description and Management, April 20-22, Chicago, Ill.

31. T. Pavlidis, "Syntactic Feature Extraction for Shape Recognition", Proc. 3rd International Joint Conf. on Pattern Recognition, Nov. 8-11, 1976, Coronado, Calif., pp. 95-99.

32. G. Stockman, L. N. Kanel and M. C. Kyle, "Structural Pattern Recognition of Carotid Pulse Waves Using a General Waveform Parsing System", Comm. ACM, Vol. 19, No. 12, Dec. 1976, pp. 688-695.

33. E. Persoon and K. S. Fu, "Sequential Classification of Strings Generated by SCFG's", International Journal of Computers and Information Sciences, Vol. 4, Sept. 1975.

34. A. V. Aho and T. G. Peterson, "A Minimum Distance Error-Correcting Parser for Context-Free Languages", SIAM Journal on Computing, Vol. 4, Dec. 1972.

35. V. I. Levenshtein, "Binary Codes Capable of Correcting Deletions, Insertions and Reversals," Sov. Phys. Dokl., Vol. 10, Feb. 1966.

36. S. Y. Lu and K. S. Fu, "Structure-Preserved Error-Correcting Tree Automata for Syntactic Pattern Recognition", Proc. 1976 IEEE Conference on Decision and Control, Dec. 1-3, Clearwater, Florida.

37. N. S. Chang and K. S. Fu, "Parallel Parsing of Tree Languages", Proc. 1978 IEEE Computer Society Conference on Pattern Recognition and Image Processing, May 31 - June 2, Chicago, Ill.

38. R. O. Duda and P. E. Hart, Pattern Classification and Scene Analysis, Wiley, 1972.

39. S. Y. Lu and K. S. Fu, "A Sentence-to-Sentence Clustering Procedure for Pattern Analysis", IEEE Trans. on Systems, Man and Cybernetics, Vol. SMC-8, No. 5, May 1978.

40. K. S. Fu and S. Y. Lu, "A Clustering Procedure for Syntactic Patterns", IEEE Trans. on Systems, Man and Cybernetics, Vol. SMC-7, No. 10, Oct. 1977, pp. 734-742.

41. R. W. Ehrich and J. P. Foith, "Representation of Random Waveforms by Relational Trees", IEEE Trans. on Computers, Vol. C-25, July 1976, pp. 725-736.

42. P. V. Sankar and A. Rosenfeld, "Hierarchical Representation of Waveforms", TR-615, Computer Science Center, University of Maryland, College Park, Md. 20742, Dec. 1977.

43. J. Keng and K. S. Fu, "A Syntax-Directed Method for Land-Use Classification of LANDSAT Images", Proc. Symposium on Current Mathematical Problems in Image Science, Nov. 10-12, 1976, Monterey, Calif.

44. S. Tsuji and R. Fujiwana, "Linguistic Segmentation of Scenes into Regions", Proc. Second International Joint Conference on Pattern Recognition, August 13-15, 1974, Copenhagen, Denmark.

45. T. Kunii, S. Weyle and J. M. Tenenbaum, "A Relational Data Base Scheme for Describing Complex Pictures with Color and Texture", Proc. Second International Joint Conference on Pattern Recognition, August 13-15, 1974, Copenhagen, Denmark.

46. R. H. Bonczek and A. B. Whinston, "Picture Processing and Automatic Data Base Design", Computer Graphics and Image Processing, Vol. 5, No. 4, Dec. 1976.

47. R. L. Kashyap, "Pattern Recognition and Data Base", Proc. 1977 IEEE Computer Society Conference on Pattern Recognition and Image Processing, June 6-8, Troy, N.Y.

48. S. K. Chang, "Syntactic Description of Pictures for Efficient Storage Retrieval in a Pictorial Data Base", Proc. 1977 IEEE Computer Society Conference on Pattern Recognition and Image Processing, June 6-8, Troy, N.Y.

49. S. Y. Lu, "A Tree-to-Tree Distance and Its Application to Cluster Analysis", IEEE Trans. on Pattern Analysis and Machine Intelligence, Vol. PAMI-1, April 1979.

50. R. L. Kashyap, "Syntactic Decision Rules for Recognition of Spoken Words and Phrases Using a Stochastic Automaton," IEEE Trans. on Pattern Analysis and Machine Intelligence, Vol. PAMI-1, April 1979.

51. R. De Mori, "A desriptive technique for automatic speech recognition," IEEE Trans. Audio Electroacoustics, AU-21, 89-100, 1972.

52. R. Newman, K. S. Fu, and K. P. Li, "A syntactic approach to the recognition of liquids and glides," Proc. Conf. Speech Commun. Process, Newton, Massachusetts, April 24-26, 1972.

53. W. A. Lea, "An approach to syntactic recognition without phonemics," Proc. Conf. Speech Commun. Process, Newton, Massachusetts, April 24-26, 1972.

54. A. Kurematsu, M. Takeda, and S. Inoue, "A Method of Pattern Recognition Using Rewriting Rules," Tech. Note No. 81, Res. and Develop. Lab., Kokusai Denshiu Denwa Co. Ltd., Tokyo, Japan, June 1971.

55. M. R. Ito and R. W. Donaldson, "Zero-crossing measurements for analysis and recognition of speech sounds," IEEE Trans. Audio Electroacoustic, AU-19, 235-242, 1971.

56. G. D. Ewing and J. F. Taylor, "Computer recognition of speech using zero-crossing information," IEEE Trans. Audio Electroacoustics, AU-17, 37-40, 1969.

57. [illegible], "[illegible] for [illegible] acoustic [illegible]," IEEE Trans. Audio Electroacoustics, [illegible] 1973.

58. R. [illegible], "A [illegible] recognition of [illegible] and [illegible]," [illegible] Pattern Recognition [illegible] Techniques [illegible]

59. [illegible], "An approach to [illegible] recognition [illegible]," [illegible] Speech [illegible] Recognition [illegible]

60. [illegible] using [illegible] Acoustic [illegible]

61. [illegible] for analysis and recognition of speech sounds," IEEE Trans. [illegible]

62. [illegible] speech [illegible]

PHONEME RECOGNITION USING A COCHLEAR MODEL *

P. ALINAT

THOMSON-CSF, DASM, Cagnes-sur-Mer, France

Abstract : In this paper the following points are briefly described about a system for syntactic recognition of phonemes :

- The analyser : filter bank with parameters based on the human cochlea
- The results obtained from this analyser concerning the classification rules for vowels, fricative and stop consonants
- The organisation of a system for syntactic recognition of phonemes using features as base level
- Some results obtained with this recognition system.

* This work has been supported by "Direction des Recherches, Etudes et Techniques"

J. C. Simon (ed.), Spoken Language Generation and Understanding, 253-262.

1. INTRODUCTION.

In a traditional approach to speech study the following levels may be taken into account : features (for example, existence and position of a formant) - phonemes - words - phrases - sentences. It is well known that the feature for successive phonemes are intermixed in the sound stream (co-articulation) [1]. Similar effect is present at the phoneme and word levels.

It can be said that every speech recognition or understanding system is decision - theoretic (discriminant) at a certain level and syntactic above this level. For isolated word recognition, this level is currently the word but for multispeaker connected speech recognition, it is often lower : phonemes or features. See for instance, in these proceedings, the papers of R. De Mori [2] and T. Sakai [3]. There are advantages to using low base level. It is better for connected speech because the junction between words can be taken into account. Secondly, there is less calculation to be done for a very large working vocabulary (a specific number of phonemes is opposed to an infinite number of sentences). Lastly, it seems that the use of the feature level allows for a better speaker independence.

Based on the above mentionned context, the following points will be briefly described in this paper about a system for syntactic recognition of phonemes :

- The analyser (filter bank with parameters based on the human cochlea
- The results obtained from this analyser concerning the classification rules for vowels, fricative and stop consonants
- The organisation of the system for syntactic recognition of phonemes using features as base level
- Some results of this recognition system.

2. ANALYSER.

The analyser is composed of a bank of 96 band-pass filters (followed by detection-integration stages) [4]. The central frequencies of the filters are based on the human cochlea [5,6] and their frequency resolutions slightly inferior to those given by Zwicker et al [7]. Figure 1 gives examples of the output of the analyser. This analyser is a very rough cochlea model because the filters are without non-linearities and couplings between them. In spite of this, certain very simple phoneme classification rules have been obtained through the use of this analyser.

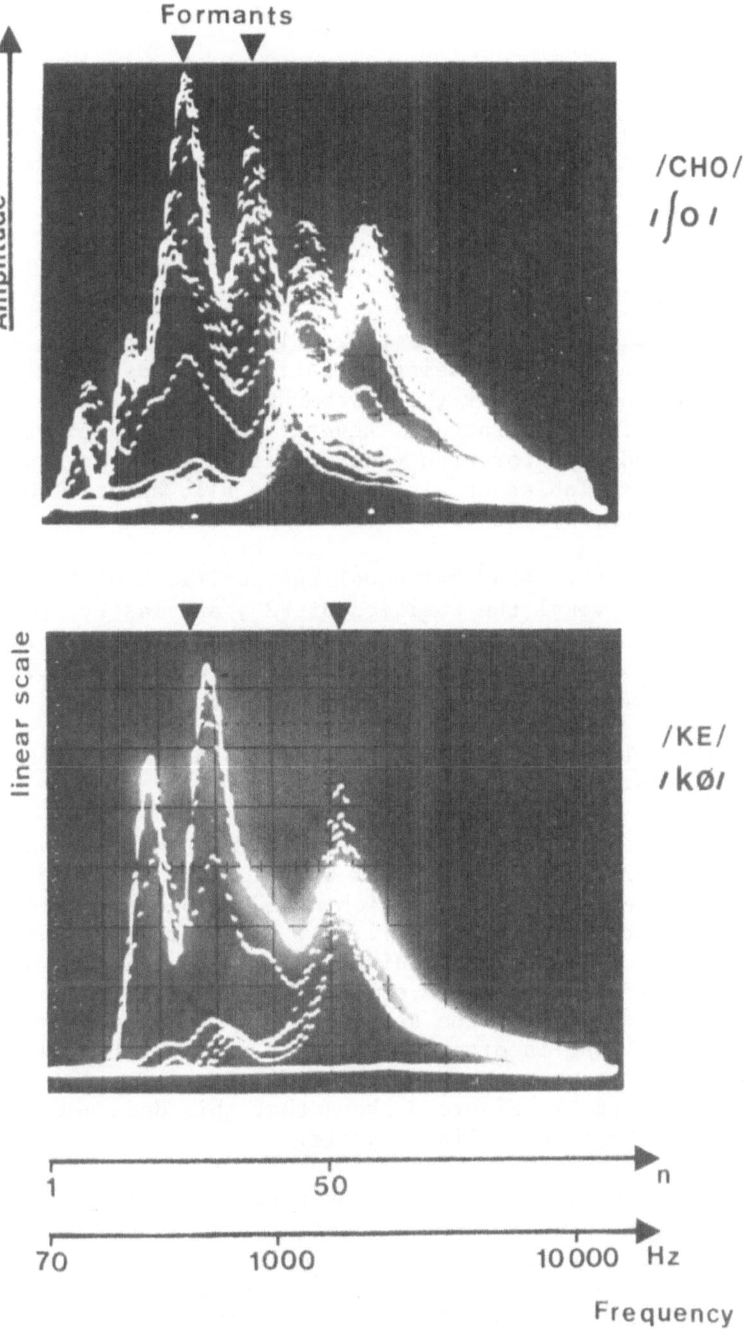

Figure 1 - Examples of outputs of the analyser for 2 isolated syllables with a curve each 8 ms (n is the filter number)

3. CLASSIFICATION RULES FOR VOWELS AND FRICATIVE CONSONANTS.

The purpose of the study was to find rules to distinguish one oral vowel from another, as well as one nasal vowel or one fricative consonant from another. This work was limited to the French language. The utterance of each phoneme depends on three factors :

- effects of coarticulation
- the speaker himself
- stress.

This third factor must be carefully eliminated from the data when classification rules are being established. To do so, clearly pronounced isolated CV or VC syllables were used. Taking into account the first two factors, several speakers (5 males, 2 females) were used, and for each vowel or fricative consonant about 10 different syllables were used. All in all, about 1150 vowels and 520 fricatives were studied.

Figure 2 gives for each oral vowel the positions of F1 and F2 ; for each nasal vowel the position of F2 ; and lastly, for the fricative consonnant the position of only one formant. The methods used in estimating the positions of the formants are given in reference [4].

From the results so obtained it was possible to establish the following classification rules:

Each oral vowel is characterized by 2 formant zones : one for F1 and one for F2 ; each nasal vowel by 1 formant zone for F2 ; and each fricative (except /f,v/ which have not been studied here) by one single formant zone as well. Only 7 % of the formants observed fall outside of the corresponding zones. Note that the open-closed vowel pairs (for instance /e,ɛ/ or /ɸ,œ/) differ only in the F1 zone whereas they share the same F2 zone. It can also be observed that the same zone is often common to 2 ou 3 vowels and that the cochlea scale reckons the length of F1 as being more or less equal to that of F2. Figure 3 shows that this does not at all hold true when reckoned on a linear scale.

These results prove the special interest for this type of scale which has been known for some time (for example the Mel scale). In fact, the rules obtained are very simple ; especially simpler than the methods using normalization by vocal-tract length for instance [8]. Another advantage of these rules is that they are easy to modify so that the effects of coarticulation may be taking into account : the zones are simply modified in function with the neighbouring phonemes and their stress.

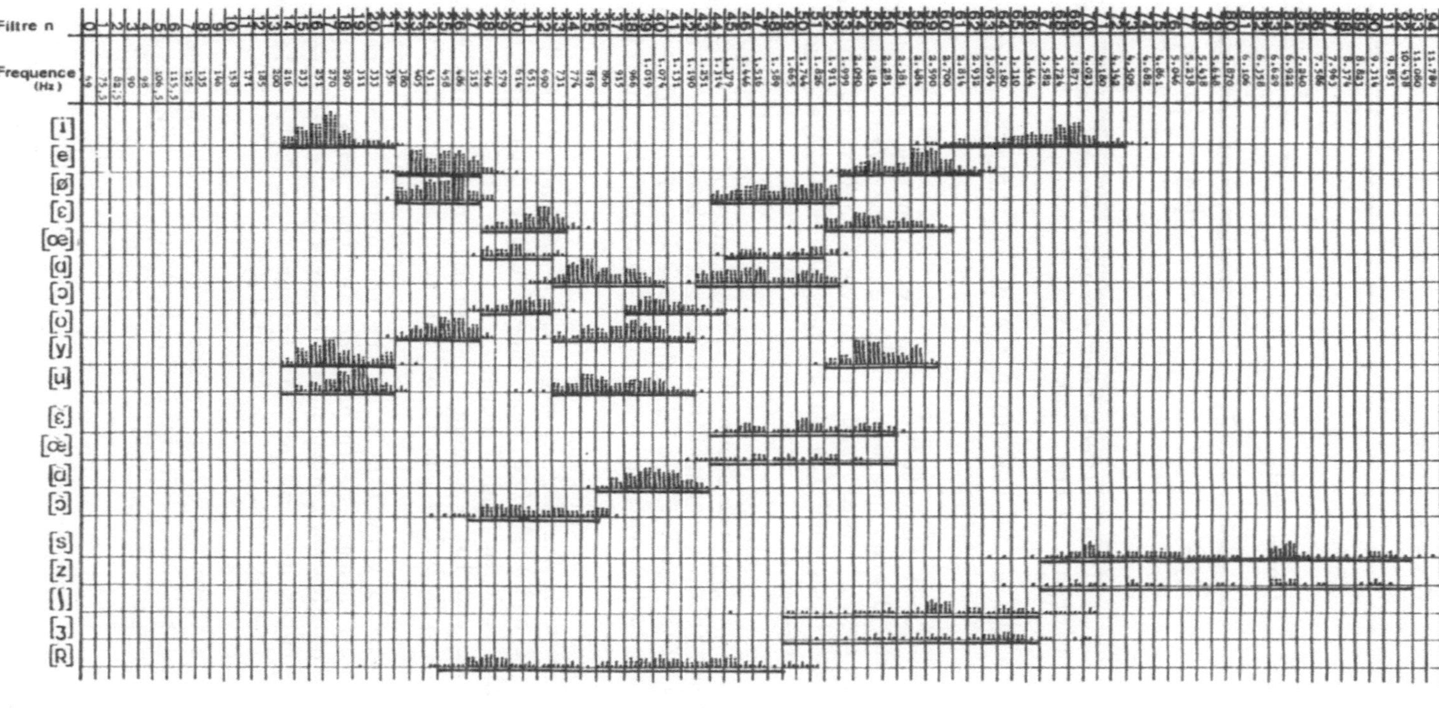

Figure 2 - F1 and F2 Formant positions on the cochlea scale for vowels and fricative consonants. Each point corresponds to one observation (speakers : 5 males, 2 females)

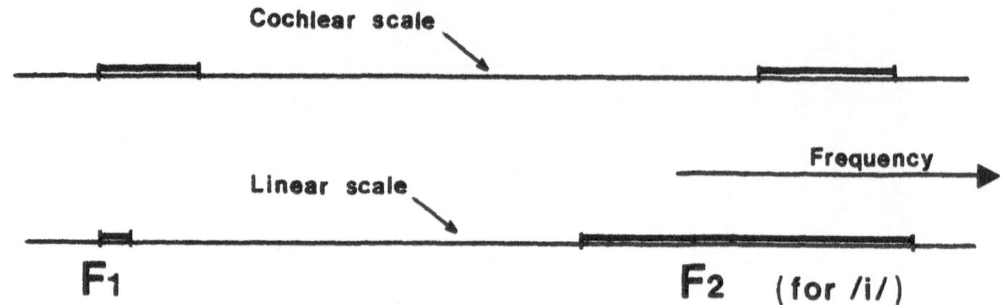

Figure 3 - Comparison between cochlear and linear scale (same point at 70 and 12000 Hz)

4. CLASSIFICATION RULES FOR THE PLACE OF ARTICULATION IN STOP CONSONANTS.

Here the purpose of the study was to set down the rules used to distinguish the 3 stop consonant classes : /pb/-/td/-/kg/. For the same reasons as for vowels and fricatives, only isolated spoken syllables have been used (24 male and 10 female speakers, all in all 1836 syllables).

La Rivière et al [9] has shown the importance of the burst for this problem. Thus, rules pertaining to the burst or to the onset of the stop (this latter used in the absence of a burst) have been sought after. Both burst and onset were analysed by the filter bank.

Briefly the rules obtained are as follows :
/k,g/ : narrow peak near F2 of the adjacent vowel
/t,d/ : diffuse and high frequency spectrum, or from time to time a narrow peak in high frequency
/p,b/ : diffuse and low frequency spectrum, or from time to time a narrow peak below the F2 of the adjacent vowel.

These rules depend very little on the following vowel, that is to say by only one parameter. In particular, they make no use of the trajectories of the formants during their transition into the vowel. These rules are not very different from the results more recently obtained by Stevens et al [11].

A program localizing the burst or the onset and using these rules has led to the following confusion matrix (concerning isolated CV syllables).

Spoken ⟶	/p or b/	/t or d/	/k or g/
/p,b/	93 %	10	4
/t,d/	3	86	4
/k,g/	4	4	92

5. PHONEME RECOGNITION.

The system is composed of : the analyser (filter bank) - a parameter extractor (independent of the context) : Amplitude, vocal source informations, positions of the formants etc... - and finally the phoneme recognition processor.

In this last stage the operations are as follows :

- The localization of vowels is made by using amplitude, vocal source information (voiced but non-fricative sound) and stability of formants. This leads to the notion of syllables more noise resistant than phonemes.

- For each vowel thus localized the following parameters are estimated : length, amplitude, nasalization, F1 and F2 positions. Then the degree of accentuation can be known.

- Consonants are regrouped into general classes : fricative, stop, semi-vowels, R, MN or L, which are processed in this order. For each general class, if need be, the consonants are localized between vowels and this by following some rules touching succession and coarticulation.

For each consonant thus localized, some parameters are estimated, so that it can be differentiated from the other consonants in the same general class and characterized by its own particular stress.

At this stage, speech has been converted into a phonemic string, with each phoneme labelled according to its general class name and characterized by a certain number of parameters. Figure-4 and Table I give examples of results obtained for the isolated word /ʃokola/. Note that certain rules of coarticulation (for instance, displacement of formant positions depending on adjacent phonemes) will be taken into account further on in the processing.

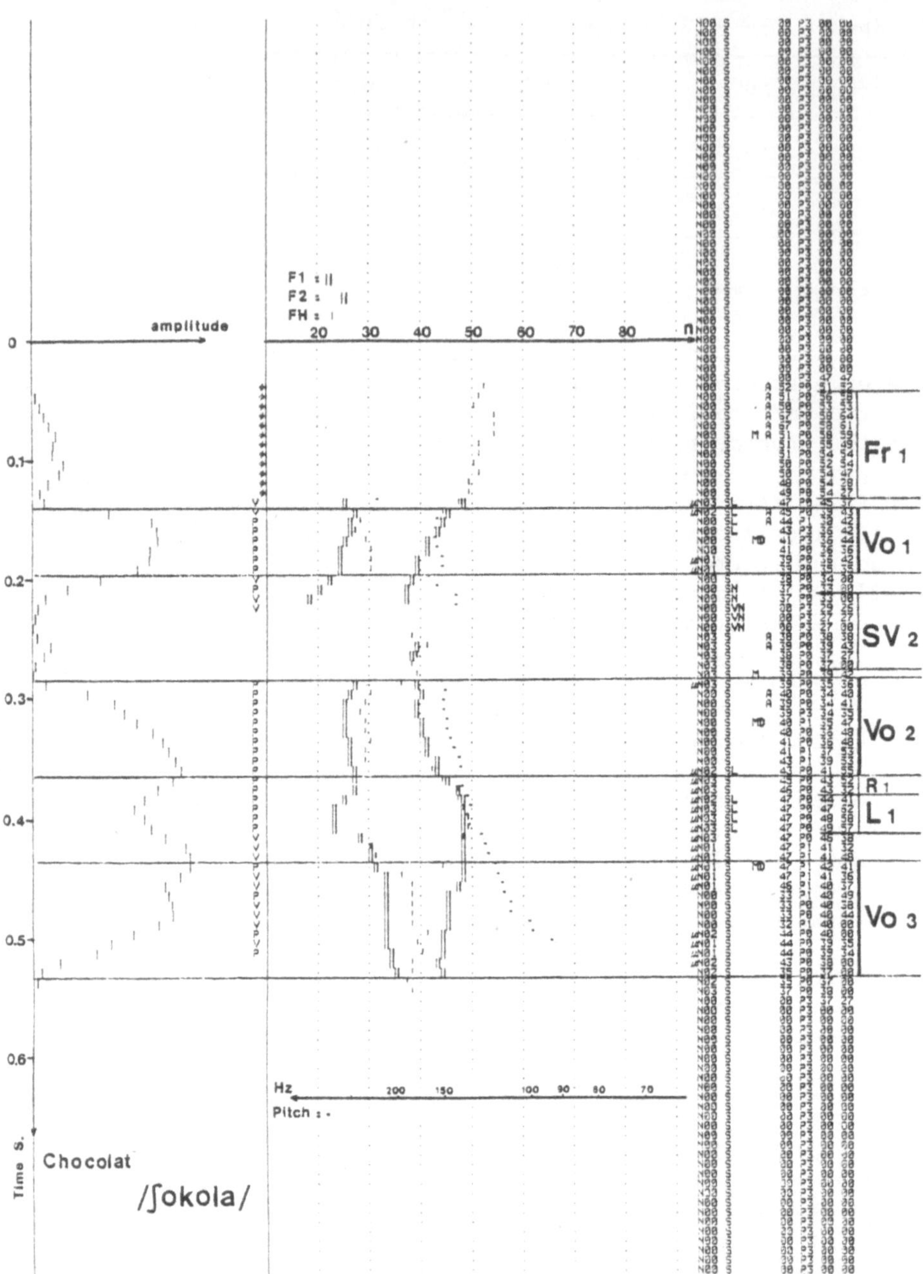

Figure 4 - Example of system output : Parameters extracted each 8 ms (a line each 8 ms) and, in the right column, recognized phonemic string. The parameters estimated for each phonemes are given in Table I.

TABLE I - Parameters of the phonemes for the word of Figure 4

Vowel	Ampl.	Length	F1	F2	NAS
Vo 1	183	5	25	41	30
Vo 2	214	6	25	40	13
Vo 3	220	8	33	45	18
Fricative consonant	**Ampl.**	**Length**	**Voiced**	**FH**	
Fr 1	61	8	No	52	
Stop consonant	**Ampl.**	**Length**	**Voiced**	**P,T,K**	
SV 2	11	6	yes	K	
/R/	**Length**	**F1**	**F2**		
R1	4	26	44		
/L,M,N/	**Ampl.**	**Length**	**L,M,N**	**F1**	**F2**
L1	155	4	L	23	47

5. RESULTS AND CONCLUSION.

The phoneme recognition system was tested with isolated words uttered clearly and slowly. 200 words were used (10 male speakers x 20 words taken from a 100 word vocabulary). The results of this testing might be summarized as follows :

Extra phonemes : 15 % (mainly R, MN, L)
Missing phonemes : 1 % (more if prononciation were quicker)
Errors about the formants of oral vowels : 2 %
nasal vowels :18 %
Fricative consonants unvoiced : 100 % detected
voiced : 65 % detected as fricative
35 % detected as MN, L or voiced stop
Stop consonants : 96 % detected with the following confusion matrix :

Spoken	P or B	T or D	K or G
PB	77 %	7	16
TD	25	70	5
KG	8	7	85

As a conclusion, the results are satisfying. Some details however need to be improved (for example, voiced fricative detection). Finally, the phonemic string could be used in the recognition of short phrases or sentences with simple syntax.

REFERENCES

[1] A.M. Liberman, F.S. Cooper, D.P. Shankweiler and M. Studdert Kennedy, "Perception of the speech code" Psychological Review vol 74, n° 6, pp 431-461, Nov 1967

[2] R. De Mori "Automatic phoneme recognition in continuous speech : a syntactic approach" NATO ASI on Spoken Language Generation and Understanding, Bonas June 26-July 7 1979

[3] T. Sakai "Automatic mapping of acoustic features into phonemic labels" NATO ASI on Spoken Language Generation and Understanding, Bonas June 26-July 7 1979

[4] P. Alinat "Etude des phonèmes de la langue française au moyen d'une cochlée artificielle. Application à la reconnaissance de la parole" Rev. Techn. Thomson-CSF vol 7 n° 1 mars 1975

[5] G. Von Bekesy "Experiments in hearing" Mc Graw Hill, New York, 1960

[6] D.D. Greenwood "Critical bandwidth and the frequency coordinates of the basilar membrane" J. Acoust. Soc. Amer. vol 33 1961, pp 1344-1356

[7] E. Zwicker, G. Flottorp and S.S. Stevens "Critical band width in loudness summation" J. Acoust. Soc. Amer vol 29 n° 5, 1957 pp. 548-557

[8] H. Wakita "Normalization of vowels by vocal-tract length and its application to vowel identification" IEEE Trans. Acoustics Speech and Signal Processing, vol ASSP-25, pp.183-192 April 1977

[9] C. La Rivière, H. Winitz and E. Herriman "Vocalic transitions in the perception of voiceless initial stops" J. Acoustc. Soc. Amer. vol 57, n° 2, febr. 1975, pp. 470-475

[10] P. Alinat "Etude du trait permettant de distinguer entre les 3 classes de consonnes explosives PB, TD, KG" 9ème Journées d'Etudes sur la Parole, Lannion 1978

[11] K.N. Stevens and S.E. Blumstein "Invariant cues for place of articulation in stop consonants" J. Acoust. Soc. Amer. 64 (5) Nov. 1978, pp. 1358-1368.

PITCH DETERMINATION OF SPEECH SIGNALS - A SURVEY

Wolfgang J. Hess

Lehrstuhl für Datenverarbeitung
Technische Universität München
Postfach 202420
8000 München 2, W. Germany

Abstract. - *In this paper, the various pitch determination methods and algorithms (PDAs) are grouped into two major classes: time-domain PDAs and short-term analysis PDAs. The short-term analysis PDAs leave the signal domain by a short-term transformation. They supply a sequence of average pitch estimates from consecutive frames. The individual algorithm is characterized by the short-term transform it applies. The time-domain methods, on the other hand, track the signal period by period. Extraction and isolation of the fundamental harmonic, and investigation of the temporal signal structure are the two extremes between which most of these PDAs are found. After a short review of these principles, the paper finally discusses different application-oriented aspects, i.e. the role of the PDA in phonetics, education, phoniatrics, and speech communication systems.*

1. INTRODUCTION

Pitch (i.e.fundamental frequency F_0 and fundamental period duration T_0) takes on a key position in the acoustic speech signal. The prosodic information of an utterance is predominantly determined by this parameter. The ear is by an order of magnitude more sensitive to changes of fundamental frequency than to changes of other speech signal parameters (Flanagan 1972; Flanagan and Saslow 1958). The quality of vocoded speech is essentially influenced by the quality and faultlessness of the pitch measurement (Gold 1977). Hence, the importance of this parameter claims for good and reliable measurement methods.

J. C. Simon (ed.), Spoken Language Generation and Understanding, 263-278.

Literally hundreds of methods for pitch determination have been developed. This article will give a short unified survey of some of the prevailing principles. As to the realization, we will not distinguish between an analog or digital hardware device and an algorithmic solution: they are all regarded as pitch determination algorithms (PDAs). Nor will we distinguish as to what is actually measured: the fundamental frequency F_0, the period duration T_0, or individual signal periods. F_0 and T_0 are interconnected by the relation

$$F_0 = 1/T_0 \qquad (1)$$

In case individual periods are measured, T_0 will be momentarily defined as the elapsed time from the beginning of one period until the beginning of the next one.

Pitch determination is one subproblem of voice source parameter determination. The other big subproblem is "where the periods are". This task, also referred to as "voiced/unvoiced discrimination", will not be discussed here. In most PDAs it is actually done apart from the real pitch determination. So we will presuppose that the voiced/unvoiced decision has been done already, and that the PDA has only voiced signals left to be analyzed.

In the following, we will first categorize the different methods of pitch determination. After that, the various categories will be briefly discussed. Finally, we will shortly review possible application areas and the various requirements and presumptions associated therewith.

2. CLASSIFICATION AND CATEGORIZATION OF PDAs.

A PDA can be subdivided into three steps of processing: (a) the preprocessor, (b) the basic extractor, and (c) the postprocessor (McKinney 1965). The basic extractor does the real measurement task: it converts the input signal into a series of pitch estimates. The task of the preprocessor is data reduction in order to facilitate the operation of the basic extractor. The postprocessor operates in a more application-oriented way: error correction, smoothing of the pitch contour, and graphic display are some of its possible tasks.

The existing PDA principles can be split up into two gross categories when the input signal of the basic extractor is taken as a criterion. If this signal has the same time base as the original speech signal, the PDA operates in time domain. In all other cases, the time domain has been left somewhere in the preprocessor. Since the speech signal is time-variant, this cannot be done other than by a short-term transformation. Accordingly, we have the two PDA categories: (a) time-domain PDAs, and (b) short-term analysis PDAs. The schematic overview of fig. 1 already

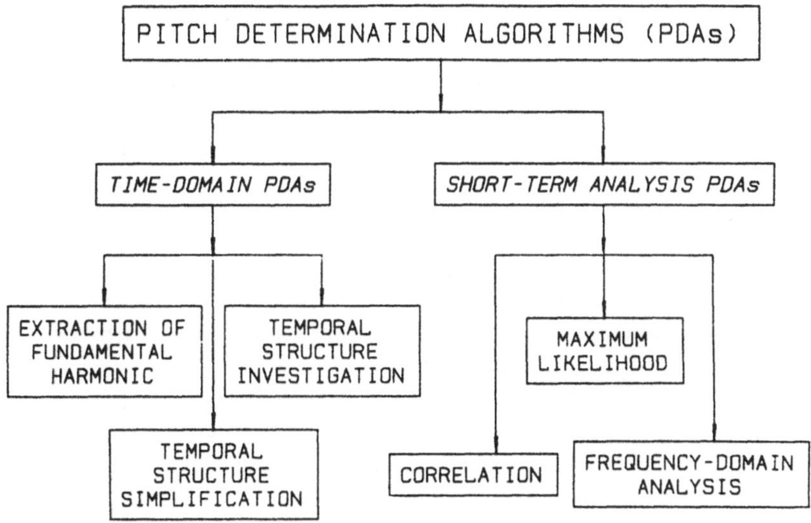

Fig. 1 - Principles of pitch determination.

adds some aspects to this general classification that will be discussed in more detail in the following sections.

3. SHORT-TERM ANALYSIS PDAs.

3.1 Principle of short-term analysis. - In all short-term analysis (STA) PDAs, a short-term transformation is performed in the preprocessor step. For this, the speech signal is split up into a series of frames (fig. 2).A frame is obtained by taking a limited number of consecutive samples from a given starting point, K, to the ending point, (K+N-1). The frame length, N, is chosen short enough so that the parameter(s) to be measured can be assumed approximately constant within the frame.On the other hand, N must be large enough to guarantee that the parameter remains measurable. For most STA PDAs this means that a frame contains at least two or three complete periods. The frame interval, on the other hand, is selected in such a way that all significant parameter changes are documented in the measurements; we must remember that the sequence of consecutive parameter measurements forms a sampled signal which obeys the sampling theorem, as well as the signal itself. The short-term transformation,so to speak, is now intended to behave like a concave mirror which focuses all the scattered information on pitch,as far as available within the frame, into one single peak in the spectral domain. This peak is then determined by a peak detector, which is the usual implementation of the basic extractor in these PDAs. Hence, the output signal of the basic extractor is a sequence of average

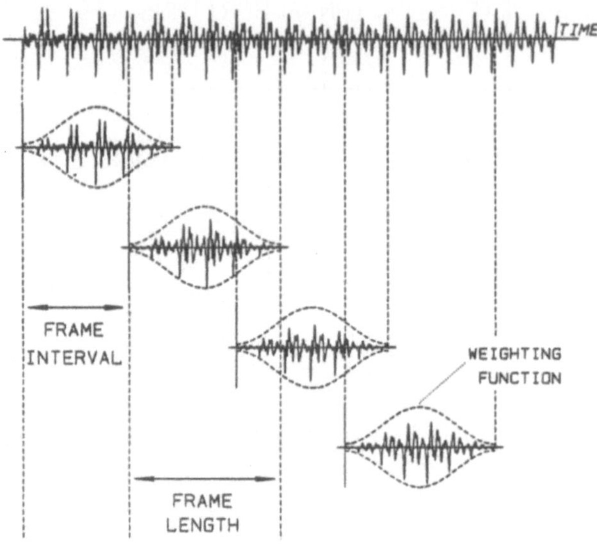

Fig. 2 - Principle of short-term analysis.

pitch estimates, not a period-by-period tracking of the signal. The short-term transform causes the phase relations between spectral domain and original signal to be lost; at the same time, however, the algorithm loses much of its sensitivity to phase distortions and signal degradation; in particular, it gets insensitive to low-frequency band limitation, as it occurs in the telephone channel. Unfortunately, the increased reliability of the algorithm is accompanied by an increased computing effort (which is at least one order of magnitude higher than for a time-domain PDA). Most of this effort goes into the numerical calculation of the transform. Besides the search for reliability, the search for a fast implementation has therefore been an important issue in the design of STA PDAs, although the recent technological developments towards high speed low-cost digital hardware decrease the relevance of this aspect.

3.2 The individual PDAs. - Not all the known spectral transforms show the desired focusing effect. The ones which do are all in some way related to the power spectrum: correlation techniques, frequency-domain techniques, and the maximum likelihood approach (fig. 3). Among the correlation techniques, we find the well known autocorrelation function (ACF) which became successful in pitch determination when it was combined with center clipping of the signal (Sondhi 1968, Galand et al. 1976, Rabiner 1977). The computation of the ACF is significantly speeded up when additi-

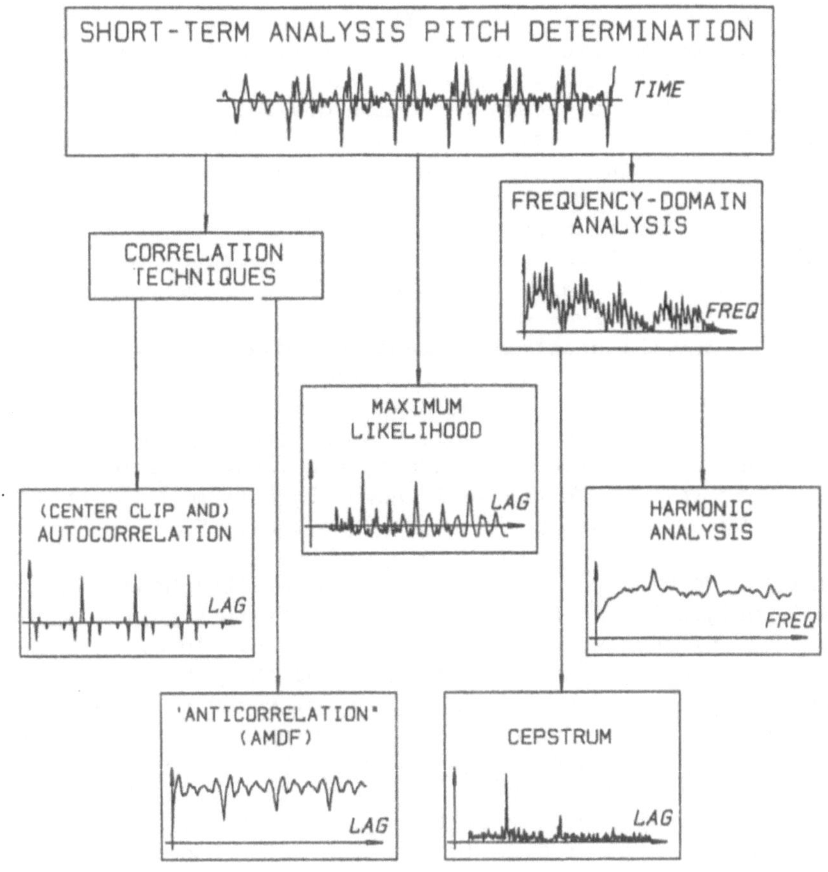

Fig. 3 - Short-term analysis pitch determination

onal peak clipping is performed (Dubnowski et al. 1976). The counterpart of the ACF is given by the "anticorrelation", i.e. the average magnitude difference function (AMDF):

$$\mathrm{AMDF}\ (k) = \sum_N |\ x(n) - x(n+k)\ | \qquad (2)$$

When the lag (delay time) k equals the period duration T_0, the AMDF will show a strong minimum. Since the AMDF needs no multiplication, it is very fast. In addition, the definition of the AMDF does not presuppose quasistationarity and can thus cope with very short frames of one pitch period or even less (Moorer 1974, Ross et al. 1974). The AMDF principle has also been successfully applied to the LPC residual (Un and Yang 1977).

The frequency-domain methods are also split up into two groups. Direct determination of F_0 as the location of a peak in the power spectrum is unreliable. It is thus preferred to investigate the harmonic structure of the signal. One way to do this is spectral compression,which exploits the fact that the fundamental frequency is the greatest common devider of all harmonics. The power spectrum is compressed along the frequency axis by a factor of two, three etc. and added to the original power spectrum.This operation gives a peak at F_0 resulting from the coherent additive contribution of all higher harmonics (Schroeder 1968, Noll 1970, Terhardt 1979). A corresponding technique has been developed in the time domain using a filter bank (Miller 1970). An alternative method which applies a majority decision principle for the spectral distance of adjacent harmonics has been proposed by Seneff (1978).The second frequency-domain technique leads back into the time domain. Instead of transforming the power spectrum itself (which would lead to the ACF), however, the inverse transform acts on the logarithmic power spectrum. This results in the well known "cepstrum" (Noll 1967), which shows a distinct peak at the "quefrency" T_0.

Finally, we have to mention the maximum likelihood approach. This is originally a mathematical procedure to separate a periodic signal of unknown period duration T_0 (Noll 1970) from Gaussian noise within a finite signal.Since neither the speech signal is periodic nor the background noise (plus the aperiodic components of speech) can be expected as Gaussian, this approach has to be slightly modified in order to work in a PDA (Wise et al. 1976, Friedman 1977, 1978). This PDA as well as the spectral compression method are extremely noise-resistive. Even at a signal-to-noise ratio of 0 dB (where the speech gets unintelligible) some of these PDAs can still operate.

In summary, short-term analysis PDAs provide a sequence of average pitch estimates rather than a measurement of individual periods.They are not much sensitive to phase distortions and to absence of the fundamental harmonic. On the other hand,the computing effort is relatively high.

4. TIME-DOMAIN PITCH DETERMINATION METHODS.

This group of PDAs is less homogenous than the short-term analysis methods. One possibility to split them up is according to the way how the burden of data reduction is distributed among the preprocessor and the basic extractor. Doing this, we find most time-domain PDAs between two extremes (fig. 4):

(1) The whole burden is imposed on the preprocessor. In the extreme case,only the waveform of the first harmonic is offered to the basic extractor.

(2) The whole burden is imposed on the basic extractor, which then has to cope with the whole complexity of the temporal signal structure. In the extreme case, the preprocessor is

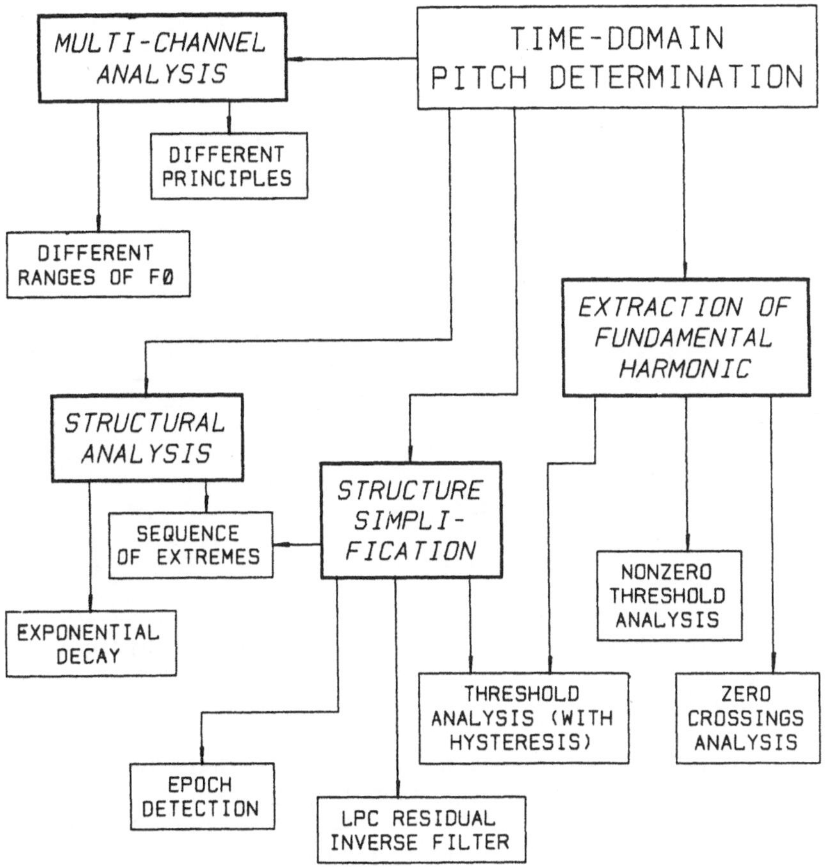

Fig. 4 - Principles of time-domain PDAs.

totally omitted.

The time-domain PDA is able to track the signal period by period. At the output of the basic extractor we find a sequence of period boundaries (pitch "markers"). Since the local information on pitch has to be taken from each period individually, time-domain PDAs usually are more sensitive to local signal degradations and thus less reliable than the majority of their short-term analysis counterparts. On the other hand, time-domain PDAs are at least in principle able to operate correctly when the signal itself is irregular due to temporary voice perturbation.

<u>4.1 Structural Analysis.</u> - The pitch period is the truncated response of the vocal tract to an individual glottal impulse. Since the vocal tract behaves like a lossy linear system, its impulse response consists of a sum of exponentially damped os-

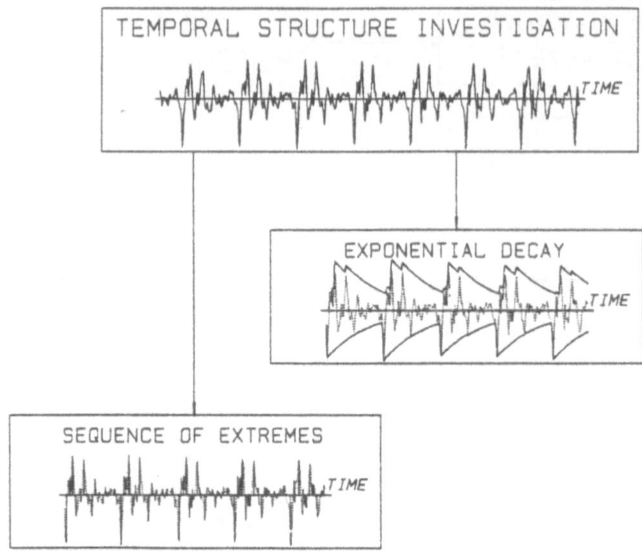

Fig. 5 - Time-domain PDAs: temporal structure investigation.

cillations. It is therefore to be expected that the magnitude of the significant peaks in the signal is greater at the beginning of the period than versus the end (fig. 5). Appropriate investigation of the signal peaks (maxima and/or minima) will thus lead to an indication of periodicity.

There are problems associated with this approach,however. First, the frequencies of the dominant damped waveforms are determined by the local formant pattern and may change abruptly. Second, the damping of the formants,particularly of a low first formant, is often quite weak and can be overrun by temporary changes of the signal level. Third, if the signal is phase distorted, different formants may be excited at different points in time.These problems are surmountable, but they lead to relatively complicated algorithmic solutions which have to regard a great variety of temporal structures. Since most of the program instructions are decisions, however, these PDAs run relatively fast. The usual way to carry out the analysis is the following (Reddy 1967, Miller 1974, Baudry and Dupeyrat 1977, Tucker and Bates 1978):

(1) Do a moderate low-pass filtering to remove the influence of higher formants.
(2) Determine all the local maxima and minima.
(3) Exclude those extremes which are found insignificant until one significant point per period is left.
(4) Reject obviously incorrect points by local correction.

For restricted applications, these PDAs can be remarkably simple (Gold and Rabiner 1969, Tillmann 1979).

A different solution has been realized in the analog domain (Dolansky 1955, Filip 1969, Winckel 1963, Zurcher 1977). The envelope of the period is modeled by a diode-resistance-capacitance circuit with a short rise time constant and a comparatively long decay time constant. This circuit emphasizes the principal peaks of the signal and suppresses all the others. Unfortunately, its performance strongly depends on the proper adjustment of the decay time constant. For that reason, this relatively simple device works well only for a restricted range of F_0 (about 2 octaves). A manual range switch or the preset of an expected value is required if a wider range of F_0 is to be analyzed.

4.2 Fundamental Harmonic Detection. - F_0 can be detected in the signal via the waveform of the fundamental harmonic. This harmonic is extracted from the signal by extensive low-pass filtering in the preprocessor. The basic extractor can then be relatively simple. Fig. 6 shows the principle of three basic extractors: the zero crossings basic extractor (ZXABE) as the simplest device, the nonzero threshold basic extractor (TABE), and, finally, the TABE with hysteresis. The ZXABE sets a marker whenever

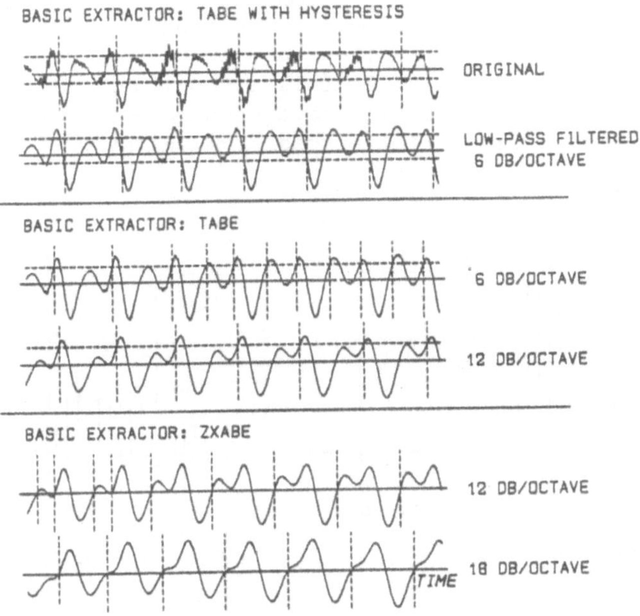

Fig. 6 - Basic extractors for fundamental harmonic extraction.

the zero axis is crossed with a defined polarity. This requires that the input waveform has two and only two zero crossings per period. The TABE sets a marker whenever a given nonzero threshold is exceeded. The TABE with hysteresis acts like the normal TABE except that the marker is not set before a second (lower) threshold is crossed in opposite direction. This more elaborate device requires a lesser degree of low-pass filtering in the preprocessor.

The requirement of extensive low-pass filtering is one of two severe weak points of this otherwise fast and simple principle. For the ZXABE, an attenuation of 18 dB/octave is necessary over the range of F_0 to be determined (McKinney 1965, Erb 1972). Accordingly, the amplitude of the signal at the basic extractor will vary by more than 50 dB due to the variations of F_0 alone. This dynamic range, increased by the intrinsic dynamic range of the signal (at least another 30 dB) is too much for the PDA to work correctly. The application of a ZXABE thus limits the possible range of F_0 and makes a manual switch or preset necessary. For the TABE, the problem is not so acute, but the fact that the threshold must be adapted to the overall signal amplitude complicates the design of the PDA. The second weak point is that this principle is a priori restricted to signals where the first harmonic is present. Low-frequency band limitation, as it occurs in telephone channels, brings these PDAs into failure.

4.3 Enhancement and reconstruction of the first harmonic. - If the principle of first harmonic extraction is to be applied to band-limited signals, the first harmonic must be enhanced or reconstructed. One way to do this is nonlinear distortion. In that respect, many proposals have been made from the beginning on (e.g. Grützmacher and Lottermoser 1937,Risberg et al. 1960);

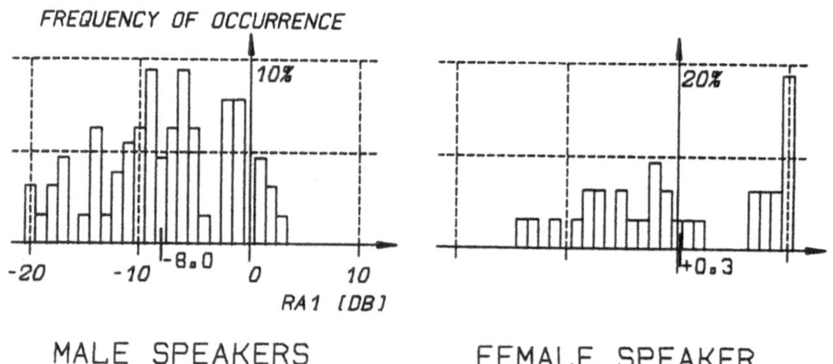

Fig. 7 - Relative amplitude RA1 of the first harmonic for undistorted signals.

much of this work has been reviewed by McKinney (1965). There is no uniform opinion, however, found in the literature as to which nonlinear function (NLF) does this job in an optimal way. To further clarify this issue, the following experiment was carried out (Hess 1979): Forty-eight isolated periods from 3 speakers (one female, two male) representing most of the vowels and sustained voiced consonants, were nonlinearly distorted by several NLFs: odd ones, even ones (such as full-wave rectification), "mixed" ones (neither even nor odd, e.g. half-wave rectification), and, finally, rectified SSB modulation (Risberg et al. 1960). The effect of these NLFs was then described measuring the relative amplitude RA1 of the first harmonic

$$\mathrm{RA1} := 20 \log \frac{A(1)}{\max\,(A(i))}\ ; \quad i = 2(1) \ldots \text{ while } f(i) \text{ below } 1200\text{ Hz} \tag{3}$$

Fig. 7 shows the histogram of this parameter for the undistorted signal. There is considerable discrepancy between the data for the male and the female speakers. The results of the experiment are shown in fig. 8. In this figure, RA1 after distortion is

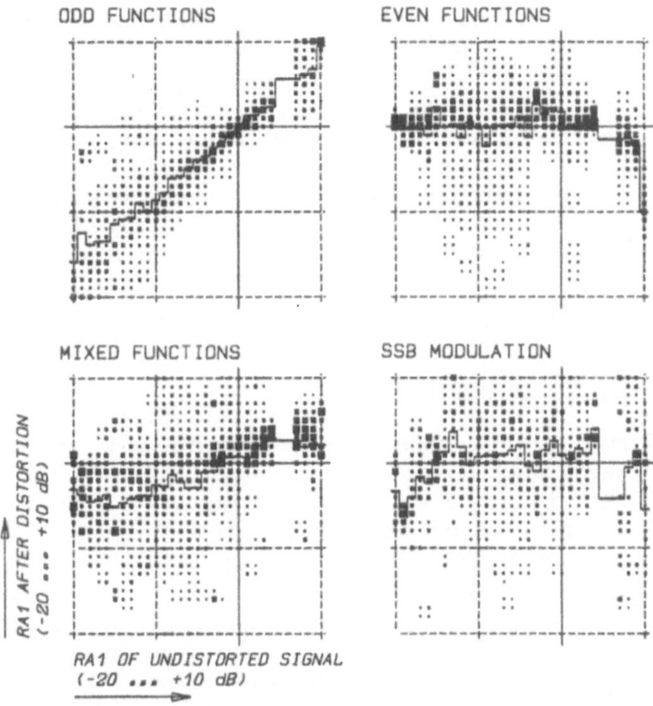

Fig. 8 - The effect of nonlinear distortion on the relative amplitude RA1 of the first harmonic.

displayed over RA1 of the undistorted signal. The odd functions, in general, show little effect. The even functions as well as the rectified SSBM greatly enhance the first harmonic when it is weak, but they attenuate it when it is dominant, as it occurs in the female voice. The mixed NLFs are not optimal. Thus we can state that no single NLF is able to enhance the first harmonic of the signal in an optimal way.

4.4 Multi-channel approaches. - Except for the algorithmic investigation of the temporal structure, most simple time-domain PDAs are restricted with respect to the range of F_0 or the type of signal to be processed. One way to increase the range or the reliability of these PDAs is to implement several of them in parallel and to perform some decision as to which one has the "correct" output. The partial PDAs may then be identical in design and process a partial range of F_0 each (Léon and Martin 1970, Dibbern 1972); or, on the other hand, they may apply different principles. The PDA by Fant (1959), for instance, uses two, the one by Hess (1979) three nonlinear functions to enhance the first harmonic in different ways. The selection criteria in order to find the most likely channel are defined by a certain channel hierarchy (Léon and Martin 1970), by a regularity check (Hess 1979), or by syntactic rules (de Mori et al. 1977). The selection is continuously checked so that the PDA can change its choice at any time.

In summary, time-domain PDAs are mostly simpler and faster than short-term analysis PDAs. On the other hand, they are less reliable. Structural analysis algorithms are sensitive to phase distortions and short spurious signals, such as switching clicks. Many of the simpler PDAs can only process a limited range of F_0. The presence of the first harmonic is often required. A way out of this dilemma is the use of multi-channel PDAs where the advantages of several principles can be combined.

5. EVALUATION AND APPLICATION.

If the performance of a measuring instrument is to be evaluated, one usually needs another instrument with at least the same accuracy. If this is not available, objective criteria to check and adjust the behavior of the new device are required. In pitch determination, both these bases of comparison are difficult to generate or totally missing.There is no PDA which operates without errors, and there is such a variety of signals and marginal conditions that one cannot test any situation in advance. Thus it becomes understandable that designers of new PDAs have seldom provided detailled data on the performance of their algorithms. One of the rare studies which evaluates the performance of PDAs in detail was done by Rabiner et al. (1976) and McGonegal et al. (1977) in connection with a speech communication system. A reference pitch contour was obtained interactively using the semi-automatic PDA by McGonegal et al. (1975). Seven PDAs (two time--domain, five short-term analysis) were involved in this investigation. The main results read as follows:

(1) No PDA actually works without errors,even under good recording conditions. All kinds of errors can and do occur, although there are significant differences in reliability between individual PDAs.

(2) The subjective evaluation does not match the preference of the objective evaluation,i.e. none of the objective criteria (number of gross errors, noisiness of contour, voiced/unvoiced errors) actually correlates well with the subjective scale of preference. That means, the question is still open as to which type of errors in pitch determination are the really annoying ones for the human ear.

The area of speech communication systems is one of the important applications of PDAs. There are other application areas, however, which must not be neglected: (1) phonetics and linguistics (including musicology), i.e. the measurement of pitch contours as carriers of prosodic, phonetic, and musical information; (2) education: training aids for the deaf and teaching aids for foreign languages; and (3) the application in medicine as a help in voice pathology and phoniatrics. Each of these applications has a different profile of requirements. Table 1 tries to compile and to weight some of these claims. Due to lack of space, we

Table 1 - Profile of requirements for PDAs depending on the respective application

	SPC	*PHO*	*EDU*	*MED*
Accuracy	2	1	5	3
Quality and facility of display	5	1	1	1
Range of F_0	2	4	4	2
Robustness against signal degradation	1	3	5	5
Manual preset of range unnecessary	1	3	5	5
Cost of implementation	3	4	1	3
Real time performance	2	4	1	5
Ease of Operation	5	2	2	3

SPC *Speech communication* *PHO* *Phonetics/linguistics*
EDU *Teaching aids* *MED* *Medicine (phoniatrics)*

1 *mandatory* *2* *important* *3* *not negligible*
4 *marginal* *5* *insignificant*

cannot discuss this issue in great detail. But some important points may be stated.For all applications except speech communication,the display is a crucial question. For phonetics, on-line performance is insignificant since the phonetician usually can wait for the results. The student of a foreign language needs the results of his utterance at once; on the other hand, he doesn't need them very accurately. The hardest requirements, however, are given in the speech communication area since the work is done automatically, and since the human ear is a most critical judge. Hence, most of the existing PDAs have been developed for this application, and most of the evaluations and reviews - including this one - have been made from this point of view. For this application, there are still open questions: none of all these principles works perfectly well for any situation, and it seems hard to believe that a new, revolutionary principle can completely solve the problem. The combination of several known principles with complementary behavior, together with a sophisticated control and selection logic, however, might form a new concept which then permits research in pitch determination to advance one further step towards a general solution.

BIBLIOGRAPHY.

BAUDRY M., DUPEYRAT B. (1976): Analyse du signal vocal - Utilisation des extrema du signal et de leurs amplitudes - détection du fondamental et recherche des formants. Proc.Journ. Etud.Parole (Nancy), vol. 7, pp. 247...257.

DE MORI R., LAFACE P., MAKHONINE V.A., MEZZALAMA M. (1977): A syntactic procedure for the recognition of glottal pulses in continuous speech. Pattern Recognition,vol. 9, pp. 181...189

DIBBERN U. (1972): Grundfrequenzmessung bei der menschlichen Sprache. Fortschritte der Akustik. Bericht von der Gemeinschaftstagung Stuttgart, pp. 345...348

DOLANSKY L.O. (1955): An instantaneous pitch-period indicator. J.Acoust.Soc.Am., vol. 27, pp. 67...72

DUBNOWSKI J.J., SCHAFER R.W., RABINER L.R. (1976): Real-time digital hardware pitch detector. IEEE-T-ASSP, vol. 24, pp. 2...8

ERB H.J. (1972.2): Untersuchung und Vergleich von Verfahren zur Erkennung der Sprachgrundfrequenz. Darmstadt: TH Darmstadt; Res.Rept. 41, Inst. f. Übertragungstechnik, 41 pp.

FANT C.G.M. (1958): Modern Instruments and Methods for Acoustic Studies of Speech. Proc. 8th Congress of Linguistics, Oslo, pp. 282...358

FILIP M. (1969): Envelope periodicity detection. J.Acoust.Soc. Am., vol. 45, pp. 719...732

FLANAGAN J.L. (1972.1): Speech analysis, synthesis, and perception. Berlin/New York: Springer.

FLANAGAN J.L., SASLOW M.G. (1958): Pitch discrimination for synthetic vowels. J.Acoust.Soc.Am., vol. 30, pp. 435...442

FRIEDMAN D.H. (1977): Pseudo-maximum-likelihood pitch extraction. IEEE-T-ASSP, vol. 25, pp. 213...221

FRIEDMAN D.H. (1978): Multidimensional pseudo-maximum-likelihood pitch estimation. IEEE-T-ASSP, vol. 26, pp. 185...196

GALAND C., ESTEBAN D., DUBUS F. (1976): Détection de la mélodie par autocorrélation non linéaire. Proc.Journ.Etud.Parole (Nancy), vol. 7, pp. 333...345.

GOLD B. (1977): Digital Speech Networks. Proc.IEEE, vol. 65, pp. 1636...1658.

GOLD B., RABINER L.R. (1969): Parallel processing techniques for estimating pitch periods of speech in the time domain. J.Acoust.Soc.Am., vol. 46, pp. 442...448.

GRÜTZMACHER M., LOTTERMOSER W. (1937): Über ein Verfahren zur trägheitsfreien Aufzeichnung von Melodiekurven. Akustische Zeitschrift, vol. 2, pp. 242...248

HESS W.J. (1976): A pitch-synchronous digital feature extraction system for phonemic recognition of speech. IEEE-T-ASSP, vol. 24, pp. 14...25.

HESS W.J. (1979): Time-domain pitch period extraction of speech signals using three nonlinear digital filters. Proc.IEEE ICASSP-79, Washington DC, pp. 773...776

LÉON P., MARTIN PH. (1969): Prolégomènes à l'étude des structures intonatives. Paris, Montréal: Dídier; Studia Phonetica, No. 2, 225 pp.

McGONEGAL C.A., RABINER L.R., ROSENBERG A.E. (1975): A semiautomatic pitch detector (SAPD). IEEE-T-ASSP, vol. 23, pp. 570...574

McGONEGAL C.A.,RABINER L.R., ROSENBERG A.E. (1977): A subjective evaluation of pitch detection methods using LPC synthesized speech. IEEE-T-ASSP, vol. 25, pp. 221...229

McKINNEY N.P. (1965): Laryngeal frequency analysis for linguistic research. Ann Arbor MI: Communic.Sciences Lab., Univ. of Michigan; Res.Rept. No. 14, 340 pp.

MILLER N.J. (1975): Pitch detection by data reduction. IEEE-T-ASSP,vol. 23, pp. 72...79

MILLER R.L. (1970): Performance characteristics of an experimental harmonic identification pitch extraction system (HIPEX). J.Acoust.Soc.Am., vol. 47, pp. 1593...1601

MOORER J.A. (1974): The optimum comb method of pitch period analysis of continuous digitized speech. IEEE-T-ASSP, vol. 22, pp. 330...338

NOLL A.M. (1967): Cepstrum Pitch Determination.J.Acoust.Soc.Am., vol. 41, pp. 293...309.

NOLL A.M. (1970): Pitch determination of human speech by the harmonic product spectrum, the harmonic sum spectrum, and a maximum likelihood estimate.Microwave Inst.Conf.Proc. - Symposium on computer processing in communications, april 1969, vol. 19, pp. 779...797

RABINER L.R. (1977): On the use of autocorrelation analysis for pitch detection. IEEE-T-ASSP, vol. 25, pp. 24...33

RABINER L.R., CHENG M.J., ROSENBERG A.E., McGONEGAL A. (1976): A comparative study of several pitch detection algorithms. IEEE-T-ASSP, vol. 24, pp. 399...413

REDDY D.R. (1967): Pitch period determination of speech sounds. Comm.ACM, vol. 10, pp. 343...348

RISBERG A., MÖLLER A., FUJISAKI H. (1960): Voice Fundamental Frequency Tracking. STL-QPSR, No. 1, pp. 3...5

ROSS M.J., SHAFFER H.L., COHEN A., FREUDBERG R., MANLEY H.J. (1974): Average magnitude difference function pitch extractor. IEEE-T-ASSP, vol. 22, pp. 353...361

SCHROEDER M.R. (1968): Period histogram and product spectrum: New methods for fundamental-frequency measurement.J.Acoust. Soc.Am.,vol. 43, pp. 829...834.

SENEFF S. (1978): Real-time harmonic pitch detector.IEEE-T-ASSP, vol. 26, pp. 358...364

SONDHI M.M. (1968): New methods of pitch extraction. IEEE-T-AU, vol. 16, pp. 262...266

TERHARDT E. (1979): Calculating virtual pitch. Hearing Research, vol. 1.

TILLMANN H.G. (1978): Bestimmung der Stimmperiode in der Zeitfunktion des digitalen Sprachschallsignals. Forschungsberichte des IPSK, München, vol. 9, pp. 207...213

TUCKER W.H., BATES R.T.H. (1978): A pitch estimation algorithm for speech and music. IEEE-T-ASSP, vol. 26, pp. 597...604

UN C.K., YANG S.C. (1977): A Pitch Extraction Algorithm Based on LPC Inverse Filtering and AMDF. IEEE-T-ASSP, vol. 25, pp. 565...572

WINCKEL F. (1963): Tonhöhenextraktor zur Messung und Steuerung von Stimme und Sprache. Arch.Ohren-Nasen-Kehlkopfheilkunde, vol. 182, pp. 651...655.

WISE J.D., CAPRIO J.R., PARKS T.W. (1976): Maximum likelihood pitch estimation. IEEE-T-ASSP, vol. 24, pp. 418...423

ZURCHER J.F. (1977): La mesure du fondamental par la détection de crêtes. Techniques employés, résultats. Proc.Journ. Etud.Parole (Aix-en-P.), vol. 8, pp. 119...126.

SPEECH ANALYSIS USING SYNTACTIC METHODS AND A PITCH SYNCHRONOUS FORMANT DETECTOR ON THE DIRECT SIGNAL

M. BAUDRY* and B. DUPEYRAT**

*Institute de Programmation - Université Paris VI,
4 Place Jussieu 75005 Paris, France
**CEN-Saclay - SES/SIR - Bât. 51 - B.P. n° 2
F-91190 Gif-sur-Yvette, France

ABSTRACT

The acoustical parameters are progressively extracted from the sampled signal up to the phonemes by use of syntactic methods. The rewriting rules, which are context-sensitive, are controled by the parameters that are evaluated on the signal itself. This leads the syntactic analysis to a main semantic point.

These techniques are applied in time domain to the speech signal. Some preprocessing is done in order to code the amplitudes of the (positive and negative) peaks and the elapsed time between them.

These temporal parameters allow a very easy and efficient coding of the speech signal and enable us to implement an isolated word recognition system.

The syntactic technics are applied at all the levels of the acoustic analysis of the signal. We obtain a pre-classification of the phonemes in 4 classes, prosodic informations and some acoustic features for the identification step. A "formant extractor", which makes use of the peaks and is synchronous with the pitch period, gives complementary informations for the identification step .

The phonetic recognition results are given for various classes of french phonemes. These results are achieved on a mini-computer operating in real time.

J. C. Simon (ed.), Spoken Language Generation and Understanding, 279-292.

INTRODUCTION

Speech waveform is directly analysed in the amplitude-time domain. The only informations transmitted to the computer are the position in time and the amplitude of the extrema of the signal. (Fig. 1) The sampling rate is 10 kHz, and the amplitude coding is carried out on 64 levels. The program runs on a mini-computer MULTI 20-06 INTERTECHNIQUE with 32 K bytes memory size. (BAUDRY M. (1)).

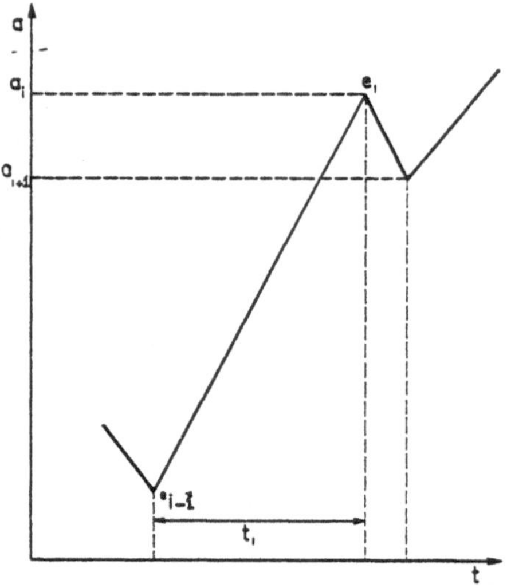

Fig. 1 Definition of an extremum e_i and its two attributes : amplitude a_i and duration t_i

We have implemented an isolated word recognition system using a INTEL-8086 16 bit microprocessor with 14 K bytes memory.

SEGMENTATION USING SYNTACTIC METHODS

The observation of the signal properties, done while manually segmenting it, led us to use particular rewriting rules.

Properties of the rewriting rules :

A set of attributes (numerical values : amplitudes, durations, etc...) is associated to the language elements. The element properties change according to the values of the attributes. We will consider them as the semantic information associated to these elements. Using the semantic information, the rewriting rules are valisated by the verification of predicates that operate on these attributes. Usually, the rules are context-sensitive. In order to easily compute the attributes, we use bottom-up analysis.

The real time processing of the signal leads us
- firstly to perform the syntactic analysis from left to right, i.e. each time a new terminal element is acquired ;
- secondly to build a non ambiguous syntactic tree. Hence, the set of rules is built in such a way that, once the rules are ordered, the first rule that can be applied is used systematically. In other words, the tree is built from the bottom as slowly as possible towards the axiom. One can notice that, fornow, the rules are amnually inferred.

Segmentation of the speech signal :

The acoustic analysis is achieved in a hierarchical way and gives 4 main steps. Syntactic technics are applied to each of the steps.

- detection of silent parts in the speech signal, mainly the silence preceding a burst,
- detection of the voiced/unvoiced acoustic feature,
- detection of each pitch period in the voiced sounds,
- distinction between vowels and consonnants by following the evolution of the maximal amplitude of each D.C.V.

This analysis allows to obtain some others acoustic features as friction,burst, voice onset time in the plosives. We get also prosodic informations as the fondamental frequency, duration and amplitude of the phonemes but these informations are not used at the present time.

The acquisition and the analysis of the signal is realised in real-time, then it has been possible to verified the validity of the rules with a large number of utterances and for different speakers.

Example : V.O.T. and Pitch Period Detection for a plosive.

Let AP and AN be the positive and the negative arch respectivily delimited by zero-crossings of the signal (Fig. 2 et 3).

Every arch has two attributes : its duration t_i and its maximal amplitude a_i (Fig. 1)

A pitch period CV is defined by its beginning DCV and its duration (Fig. 2).

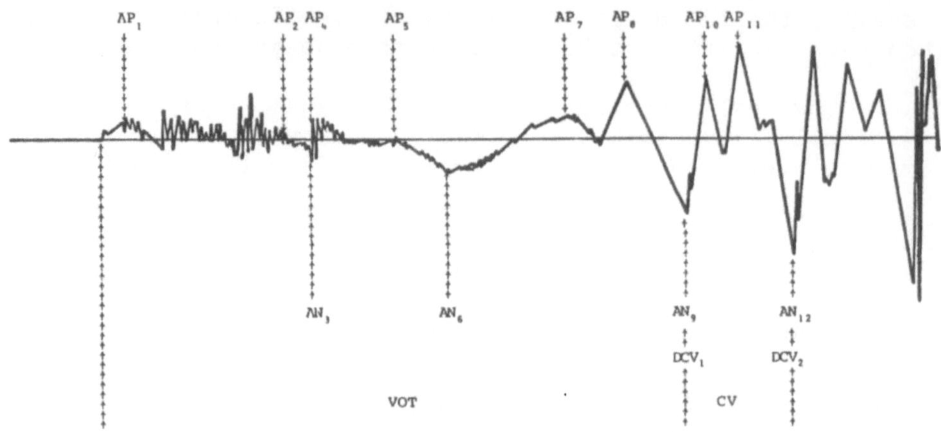

Fig. 2: AP positive arch between two zero-crossings
AN negative arch between two zero-crossings.
Sound /k/

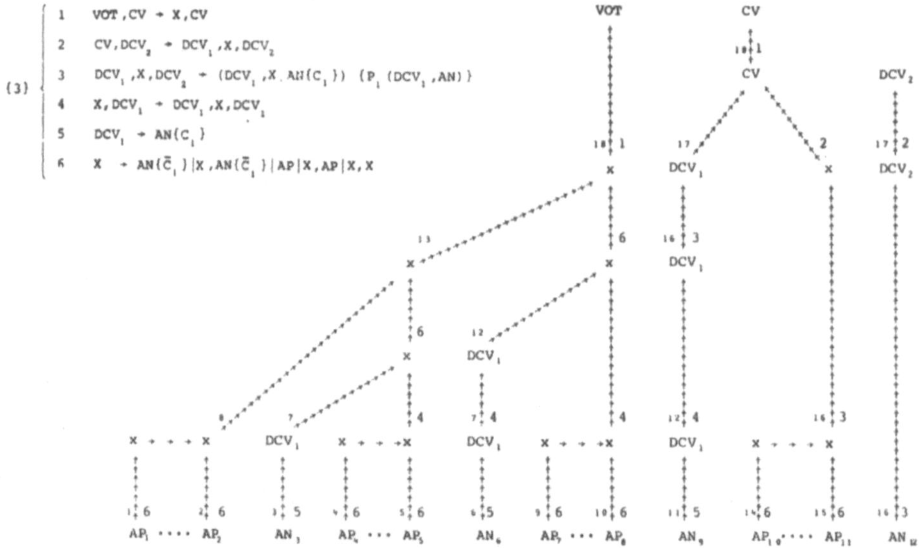

Fig. 3: This syntactic tree of the sound /k/ (displayed on the Fig. 1) is built using the set of rewriting rules 3. To the left of the arrows we find the order in which the rules are applied, and to the right the number of the rules.

At the beginning of an utterance, the rapid variations of the pitch period and the plosive perturbation cause uneasiness in the DCV detection. An example of the sound /kɛ/ is shown on figure 2. The burst is on arch AN 6 (and AP 7) preceded by friction ; the arch AN 6 should not be taken into account as a DCV, otherwise the distance between arches AN 6 and AN 9 would provide a wrong pitch period value.

The visual observation of an important set of bursts has led us to the set of rules {3} . We give thereafter somme explanations on this rules. Figure 3 shows the building of the syntactic tree corresponding to the signal in figure 2. To the left of the arrows we find the order in which the rules are applied, and to the right, the number of the rule.

The condition C 1 is defined to use thresholds relative to t_i and a_i. The predicate P 1 uses relations between the t_i and the a_i of the two arches.

X is any string of positive and negative arches that do not satisfy condition C 1.

DCV 1 is a possible arch of DCV that is to be validated by rule 3.

Rule 4 corresponds to the dropping of a false DCV 1 and the choice of a new DCV 1.

If this analysis fails, we make the hypothesis that the segment analysed is not really voiced and the voiced-unvoiced decision is to be reexamined.

Results.

The preclassification is performed with a small error rate (less than 5 %) in a V.C.V. context. Successive vowels and semi-vowels may be cumulated, and a vowel may be cut in two pieces if its articulation is unstable. Among voiced consonant clusters, the groups /mn/ as well as /p/ or /b/ followed by /l/ or /r/ are seldom separated.

PHONEME-ANALYSIS USING A PITCH SYNCHRONOUS FORMANT DETECTOR

The analysis of the segmented and preclassified phonemes is done using the frequency and energy values of the formants.

The algorithm used to extract the formant information is based on the following somple idea. (Fig. 4)

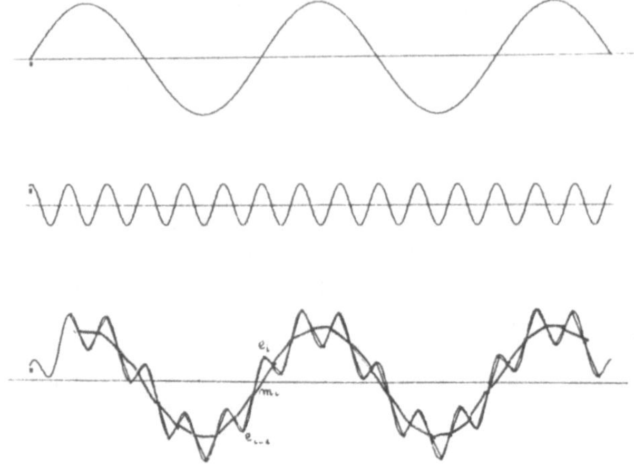

$$s_1 = a_1 \sin 2\pi f_1 t$$
$$s_2 = a_2 \sin (2\pi f_2 t + \varphi)$$
$$s = s_1 + s_2$$

Fig. 4 : The set of points m_i gives a sampling of s_1

- Let the signal s be the sum of the two sinusoïds s_1 and s_2, a_1 and a_2 being their respective amplitudes and f_1 and f_2 their respective frequencies. (We assume $f_1 < f_2$)

- Let us code the signal by the extrema e_i and let us join the successive extrema by the segments (e_{i-1}, e_i).

- Let m_i be the middle of the segment (e_{i-1}, e_i).

One can show that if the condition :

$$\frac{f_2}{f_1} < \frac{a_1}{a_2} < \left(\frac{f_2}{f_1}\right)^2 \qquad (1)$$

does not hold, then the set of the m_i's is a sampling of the signal s_1.

The condition (1) is obtained by studying the discontinuities of the number of the zero-crossings of the signal s - derivative.

Let us note that the m_i's and the e_i's are in equal number. Therefore if the e_i sample holds the Shannon condition, the m_i's too.

The limits computed for pure sinusoïds are approximately the same as those computed for damped sinusoïds.

Description of the formant detection algorithm.

The following algorithm uses the extrema coding and looks for the low frequencies of the signal by elimination of the oscillations dues to the highest frequencies.

It is composed of two parts and is recursive.

- let e_i be the ith extremum of the signal to be analysed. Il is defined by the amplitude a_i and the time interval t_i between e_{i-1} and e_i (Fig. 1)

1) Elimination of the highest frequency component.

a) We compute the histogram H of the t_i's in the considered time domain (basically a pitch period). (Fig. 5 and 6)

Fig. 5 : This is an example of sound /ε/ where the signal s is coded using the extrema.

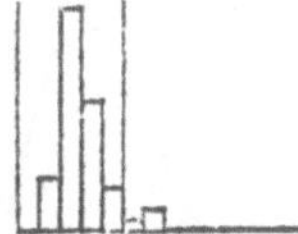

Fig. 6 : Histogram of t_i and computation of the threshold s_t on the signal s.

b) We determine the threshold s_t that corresponds to the first minimum following the first significant maximum of the histogram H.

We assert then that all the extrema, duration of which is less than or equal to s_t correspond to the highest frequency component.

c) We compute D the average value of the t_i's durations when $t_i < s_t$ on the considered time domain. We can estimate from D the value of this highest frequency to eliminate

$$f = \frac{1}{2D}$$

d) We compute A, the absolute average of the differencies of successive amplitudes of the segments (e_{i-1}, e_i) when $t_i \leqslant s_t$. We consider that the value of A varies as the energy of the highest frequency component to eliminate. (Fig. 7)

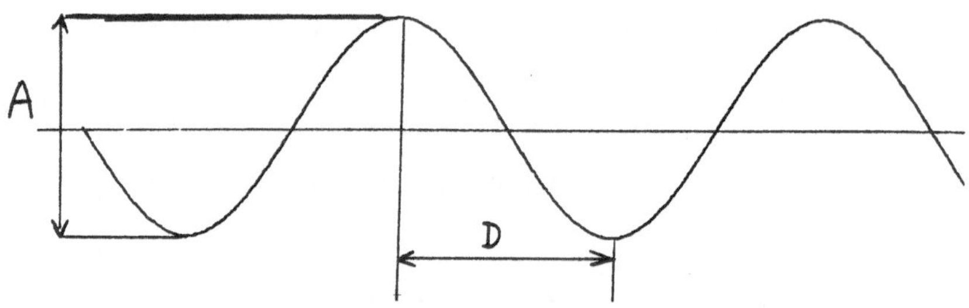

Fig. 7 : Meaning of the values of D and A for a sinusoïd.

e) We build the signal S_f obtained by substraction between s and the eliminated component.

If $t_i > s_t$, the sample e_i is preserved otherwise it is replaced by the middle of the segment (e_{i-1}, e_i).

We call S_f the filtered signal considering the analogy with the result of a low-pass filtering. (Fig. 8)

Fig. 8 : Signal S_f obtained by substracting to S the most high frequency component.

Therefore a new processing of S_f is needed.

2) <u>Recognition of the new extrema of the S_f signal.</u> After the processing explained in 1_e, all the new samples are not extrema. We observe now oscillations corresponding to frequencies greather than $\frac{1}{2\ s_t}$, due to the amplitude digitisation.

a) We compute all the extrema of the S_f signal.

Let γ segments be the segments verifying $t_i \leqslant s_t$ and δ segments be the others. (Fig. 9)

Fig. 9 : Computation of the segments γ and δ

b) Let n be the number of consecutive γ segments. Let t_γ the sum of the corresponding t_i's. Let $t_{1\delta}$ and $t_{2\delta}$ be the adjacent δ segments.

We observe two cases :

If n is odd, we eliminate the intermediate γ segments: to this effect, $t_{1\delta}$ is replaced by $t_{1\delta} + t_\gamma + t_{2\delta}$ and $a_{1\delta}$ by $a_{2\delta}$ (Fig. 10)

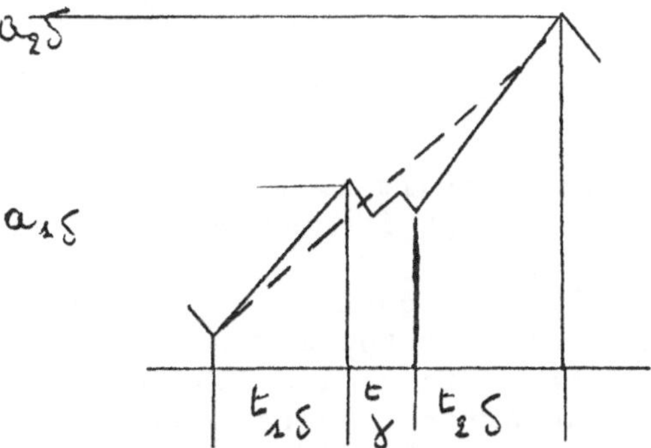

Fig. 10 : Example with n odd

If n is even, $t_{1\delta}$ is replaced by $t_{1\delta} + \frac{t_\gamma}{2}$, $a_{1\delta}$ is preserved. $t_{2\delta}$ is replaced by $t_{2\delta} + \frac{t_\gamma}{2}$ and $a_{2\delta}$ is also preserved. (Fig. 11)

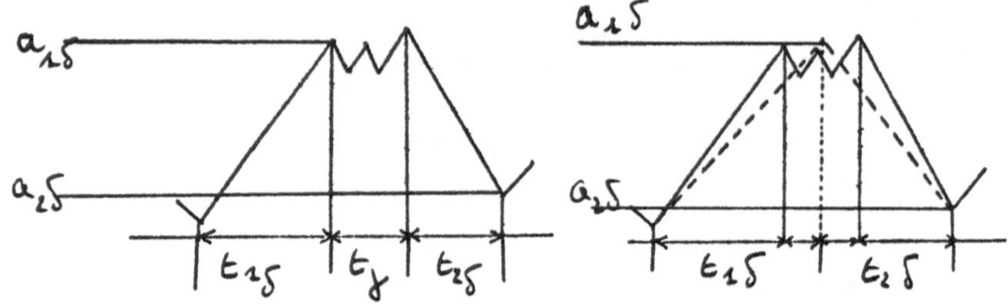

Fig. 11 : Example with n even

After this second step S_f no longer contains any segments of duration less than s_t and the signal is now coded with extrema. (Fig. 12)

s_t becomes now possible to run again this algorithm on the new signal and so on until the lowest frequency (fundamental frequency) is reached.

3) <u>Utilisation</u>.
The algorithm was tested under the following conditions. Sampling rate 10 KHz, number of amplitude levels : 64.

Fig. 12 : Result of the filtering of the most high frequency of the signal s

At this sampling frequency the hardware system that computes the extrema from the samples of the signal does not provide a correct discrimination of the 2nd and 3th formant because the error is very important for the frequencies greather than 1 500 Hz.

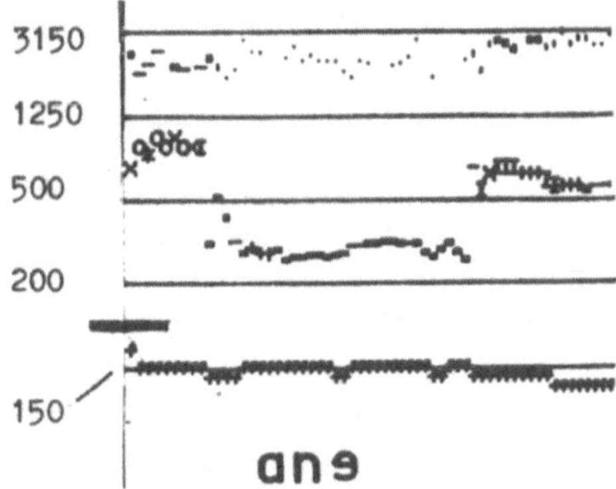

Fig. 13 : Graphic representation of the results of the formant extractor algorithm.
Comment : The scale is different for the fondamental frequency.

Because we don't have a pre emphasis at the input of our system, the energy of the 3th formant is generaly week compared with the 2nd.

We have studied a software program that provides a better computation of the duration between the extrema. This gives out a better discrimination of the 2nd and 3th formant.

The results of the algorithm may be represented on a spectrogram with the time as x, the logarithm of the frequency computed for each pitch period as y and the coefficient A (energy) represented by the degree of darkness. (Fig. 13)

Identification :

We use the results of this algorithm to identify the vowels and to follow the formant transitions.

For the vowels, we choose one pitch period in the stable part. We compute the frequencies and the normalised energies of the formants. This values are stored during the learning phase. We compute then a weighted distance to recognise the vowels.

The formant values found for the vowels agree with values found through pitch synchronous F F T.

The recognition step is done with a different material than that of the learning procedure.

Nasals are identified by the same method as vowels.

Work is still in progress concerning consonant identification. To analyse the transitions we look at the formant evolution. The results obtained for the vowels are good if the phonemes are correctly segmented.

Conclusion :

The proposed algorithm is much simpler than the algorithms commonly used to compute the formant frequency and energy. We should remark that we don't have direct access to the formant bandwidth.

RESULTS OF THE PHONETIC RECOGNITION

These results concern phonemes that are correctly segmented.

Vowels : for on speaker, the right phoneme is among the first three candidates in 98 % of the cases.

Consonants : the rate of recognition for voiced and unvoiced fricatives is about 80 %.

For the other consonants, we use secondary phonetic features since the results of automatic transition analysis are not yet available. Voiced nasal plosives are correctly segmented in 80 % of the cases. Unvoiced plosives are preclassified in 95 % of the cases when in the middle of a word.

There are two classes of unvoiced plosives : /p/ and /t,k/ separated at 60 %. Liquids : the rate of recognition is 80 % for /1/, 50 % for /r/ when the segmentation is right.

ISOLATED WORD RECONGITION SYSTEM

1 - Methodology :

The recognition of an isolated word begins with the discrimination between the signal and the silence which is in fact the ambient noise. For that purpose we apply an algorithm defined in BAUDRY (1) whose parameters are derived from the energy and duration of the signal. A word isolated in this way is then split into n windows of equal length ; n must be such that there is at least one window per phoneme for each word.

Using time intervals between zero-crossings of the signal and/or its derivative, we build a particular histogram.

Let us consider H_d the histogram of the t_i (Fig. 1) and H the histogram of the intervals between two zero-crossings of the signal, duration of AP and AN (Fig. 2). In fact H and H_d are built with 8 classes which are not linearly spaced.

Then we construct the histogram H_h on one hand with the 5 first classes of H_d which are more representatives of the high frequency components, and on the other hand with the 3 last classes of H which are more representatives of the low frequency components of the signal (for more details see BAUDRY (1)).

So each word is coded in a very easy and efficient way into n histograms H_h.

We implemented two kinds of normalization :

- linear normalization : this one normalizes the duration of the entire word.

- nonlinear normalization : there we compare two words using a dynamic programming technique. Processing can start as soon as the word is being entered but is more intricate and time-consuming.

There is no speaker adaptation, so the system is basically monospeaker (which does not prevent considering many speakers, each one possessing his/her own set of word templates).

2 - Application :

The signal is sampled at 8 kHz with 16 amplitude levels. We chose n = 16 for the number of windows. When the speaker is cooperative, the two kinds of normalization have quite similar results : about 97 % of good recognition. So we chose the first kind which is faster and more simple, for own INTEL-8086 micro-processor application.

A set of 32 word templates requires using a 4k bytes memory, 8k bytes are needed for signal acquisition and less than 2k for the program.

Answer is given less than half a second after the end of the speech.

CONCLUSION

The syntactic methods described above allows a simple formalisation of the visual observations on the signal. These methods are such that the whole program runs in 3 times the real time on a medium mini-computer. Finally, we have applied these methods to lexical search (BAUDRY, DUPEYRAT, LEVY (3)), particularly by implementing an automatic inference of the rewriting rules.

REFERENCES :

(1) BAUDRY M. - Etude du signal vocal dans sa représentation Amplitude-Temps. Algorithmes de segmentation et de reconnaissance de la parole. Thèse d'Etat - 15/06/78 - Université Paris VI.

(2) SIMON J.C., DUPEYRAT B. - Analyse rapide des composants formantiques d'un signal de parole. Note C.R. Acd. Sc. Paris - t 284 (2/05/77) Série A page 1089 à 1092.

(3) BAUDRY M. DUPEYRAT B. LEVY R - La recherche lexicale en reconnaissance de la parole. Traitement par une méthode d'analyse syntaxique utilisant une inférence automatique. 10ème Journées d'Etude sur la Parole - GRENOBLE 1979.

VARIABILITY OF VOWEL FORMANT FREQUENCY IN DIFFERENT SPEECH STYLES

Gérard CHOLLET

Institut de Phonétique, CNRS LA 26I, AIX EN PROVENCE

IASCP, University of Florida, GAINESVILLE

A phonetician can optimize his measurements of acoustic parameters of speech using ISASS (an Interactive Speech Analysis Synthesis System) with respect to some criterion of his choice. This system has been used to measure formant frequency of vowels in logatoms, isolated words, and connected discourse. The statistical analysis of these measurements quantifies the decreasing effects of coarticulation as speech articulation gets more constrained (from normal conversation to readings of sentences and isolated words). A framework for a descriptive model of the effects of stress, syllable type, articulation habits, intentions of the speaker on syllable nuclei is outlined, and its implications for speech understanding and synthesis research are discussed.

J. C. Simon (ed.), Spoken Language Generation and Understanding, 293-308.

INTRODUCTION

Many speech communication engineers have experimented with distress the variability of acoustic speech parameters. This variability is responsible in part for the unsatisfactory results of speech understanding systems and for the unnaturalness of rule driven speech synthesizers. An explanatory model of this variability seems highly desirable but requires a better understanding of the articulatory processes.

A descriptive model at the acoustic level would be adequate for automatic speech understanding, transmission, and synthesis purposes as well as for speaker identification. This is the goal of this research.

We believe such a model should consider speech as a non-random process and use random variable only to reflect our failure to explain some of the variability, to take into account the inaccuracy of our measurements (for example due to difficulty in separating noise from speech), or to describe more clearly a macro-phenomenom (in a similar way, statistical thermodynamics is an adequate macro-description of molecular cinetics).
Let us hypothesize some of the potential sources of variability of acoustic parameters and point out that many of these are quite difficult to quantify .

If we restrict our discussion to <u>intra-speaker</u> variability, we must consider at least the physiological and psychological state of the locutor and the intended purpose of the speech signal (style). In fact, a speaker adjusts his articulation effort in such a way that his message is just intelligible for his auditors (Principle of least effort, ZIPF, 1949).
This is why spontaneous utterances are different from laboratory ellicited speech in many (acoustic) respects. On the other hand, an explanation of <u>inter-speaker</u> variability must take into account physiological differences (length and shape of the vocal tract, physiology of the vocal folds,...), articulatory habits of the locutor (dialect, style, speaking rate, intensity, FO tessitura, phonation type, ...), and certainly psychological differences (alertness, mood,...). Many of these sources of variability are difficult to control even in a laboratory environment where the microphone, the instrumentation, the task... could be stressful to the subjects.
Anyway, it seems almost impossible to record truly natural speech in the laboratory. A close approximation is achieved with the following technique (suggested by O. METTAS) : A speaker is seated across from the researcher at a table on which there is a microphone. A recording technician and his equipment are also quite visible. A traditional recording session takes place. The recorder is then turned off, the technician leaves, the experimenter puts away his papers, and researcher and subject chat for a while.

Actually, the microphone is connected to another tape recorder which is out of sight and still running.

SPEECH CORPUSES

We have focussed our attention on surreptitiosly recorded speech samples of the Paris corpus (MALECOT, 1972) which consists of 50 half-hour conversations with upper middle class Parisian speakers. MALECOT used a microphone up his sleeve, an FM emiter under his jacket, and an FM receiver and tape recorder in an attache-case. The speakers were only recorded under these conditions and therefore this corpus is only adequate to study the intra and interspeaker variability for this speech style.

The observed variability of vowel formant frequency will be discussed in this paper. It was found so much larger than that reported in the literature that a controlled corpus was recorded and analysed. For this purpose, two speakers were recorded under the following conditions :

I) natural conversations (surreptitious recordings)
2) conversation in front of a microphone
3) a talk given in front of an audience
4) reading of a long passage
5) reading isolated sentences
6) reading a list of 700 words
7) articulating isolated vowels.

Conditions 5,6 and 7 have been included to provide comparative data with results previously reported by a number of other investigators. Furthermore identical speech analysis algorithms have been used for all conditions, a guarantee for coherence.

We would like to contrast this methodology to another used by P. LADEFOGED and his colleageres (1976) to compare different speech styles. They used the following technique :
Desired word samples were elicited without the experimenter actually saying them but by guiding the conversation and asking for rhymes. Then the subjects were told the words and instructed to use them in sentences. Then they were asked to repeat sentences uttered by the experimenter. Finally, they were given a list of isolated words to read aloud.
We would like to point out that such an elicitation is a variation of laboratory conditions rather that a variation of style as it is usually understood. Even so, some significant variations could have been seen but dit not emerge from the small amount of data that were collected. The investigators concluded that style of speech has little effect on vowel articulation. We would like to demonstrate here what effect speech style has on vowel articulation.

SIGNAL ANALYSIS

The sound spectrograph was found very inadequate to perform the acoustic analysis of large corpuses. Firstly it needs to be readjusted quite frequenly. Secondly, it is quite slow. Thirdly, the accuracy of the measurements is limited and dependant on the experimenter phonetic knowledge. What is needed is an automatic, algorithmic technique of acoustic analysis (CHOLLET, 1976) and some means to verify the parameters measured (by synthesis). Modern digital signal processing techniques can provide these means and have been used heavily in the design of ISASS (an Interactive Speech Analysis synthesis System).

The speech signal is filtered, sampled, quantized and stored in computer mass memory (disk or tape). It is then segmented automatically in "breath" groups (speech segments separated by pauses of at least 300 msec). The time amplitude waveform of each breath group is displayed on a graphic terminal and played back (D.A.) simultaneously. A phonetic transcription is entered by the operator at this point and linked to the displayed utterance. All breath groups are processed in this manner during the first "pass" at the data.

The second step is entirely automatic. First, a short term energy waveform and a spectral stability waveform are calculated. Short term energy is evaluated every 5 msec using a 30 msec window. For each such window, a spectral distance is computed between the first 25 msec (frame indexed by f) and the last 25 msec (indexed by l). The square of this distance measure (discussed in GRAY A. H. et all., 1976) from frame f to l can be expressed as :

$$d_{fl}^{2} = \sum_{k=1}^{M} (c_{fk} - c_{lk})^{2}$$

where c_{ik} is the cepstral coefficient of order k for the frame i and M is the order of truncation of the cepstrum. The signal was multiplied by a hamming window to minimize spectral distortion. The short term energy and spectral stability curves were used in coordination with the phonetic transcription to localize syllable nuclei in the signal. This localization algorithm performs a best match between the symbols of the transcription and peaks, and valleys of the energy curve using sources of knowledge such as :

- mean syllabic duration (total duration/number of syllables)
- energy dips due to closure
- coordination of energy peaks and spectral stability.

A cepstrally smoothed spectrum was estimated for each syllable nuclei on the closest maximally stable segment. The spectrum of oral vowels is conveniently reduced and described by the frequency of the first three formant. This data reduction is

FIGURE 1 : F1-F2 PLOT OF ALL VOWELS OF A SINGLE SPEAKER

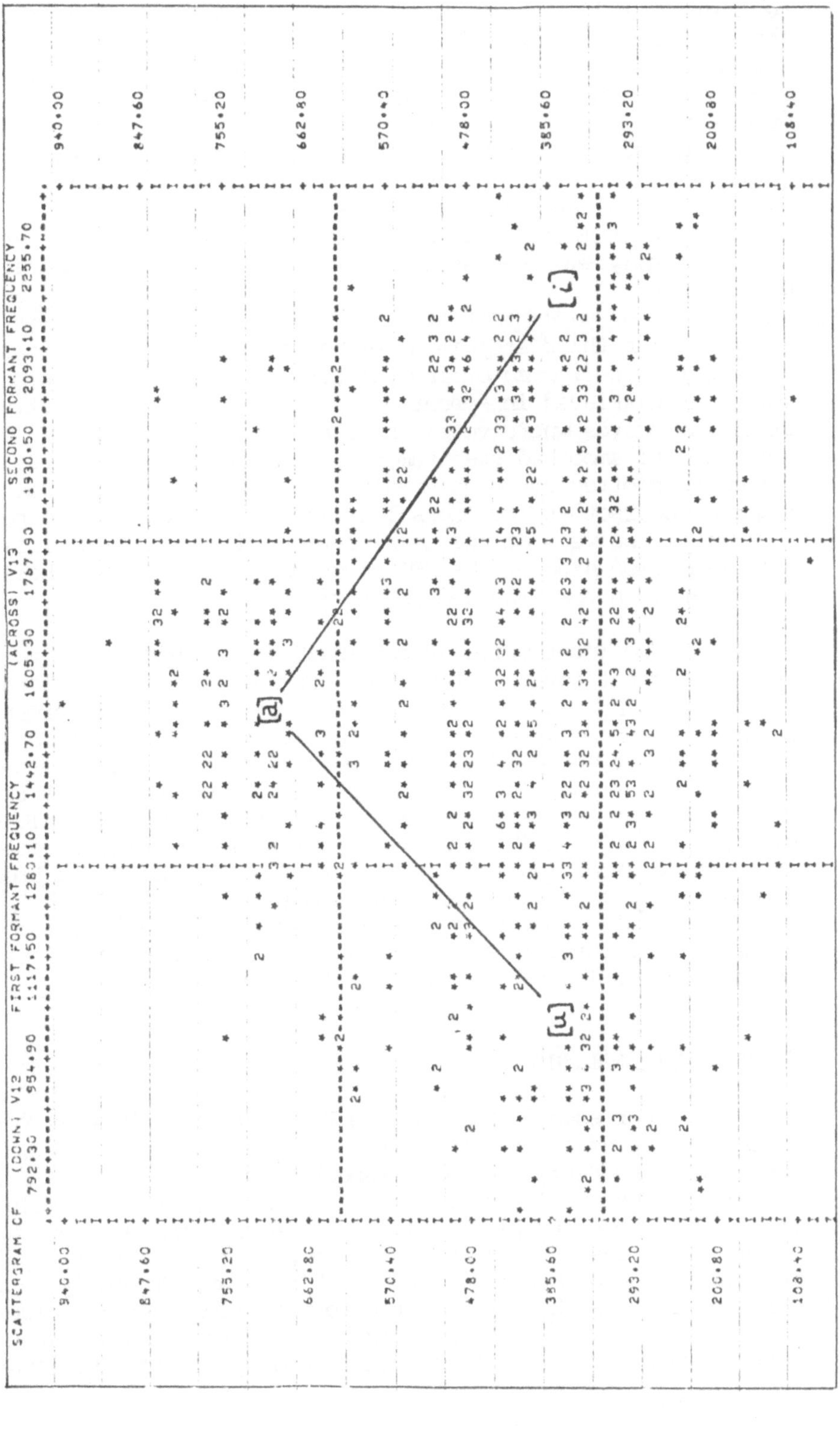

certainly a valid one since these vowels can be adequately synthesized from the knowledge of the three formants and since the perceptual space of oral vowels corresponds directly to the F1-F2-F3 space (POLS L.C.W., 1970). Peaks of the smoothed spectrum were assigned to formants on the basis of peak amplitude and a-priori knowledge of formant frequency range for each vowel type. A temporary computer file keeps for each breath group the energy and spectral stability curves and for each oral vowel the FFT log-spectrum of the frame selected, the cepstrum coefficients, the cepstrally smoothed spectrum, and the formant frequencies measured.

The third step is a verification of the localization and analysis performed during the second step. The speech oscillogram is displayed in synchrony with the energy and spectral stability curves where vowels have been labelled. The FFT log-spectrum and smoothed spectrum, on which formant labels are indicated, are plotted for each vowels. The operator can playback any segment, correct an erroneous localization of vowels, and perform a spectral analysis at the new location. He may also perform a different smoothing of the FFT spectrum or modify the labelling of formants on the smoothed spectrum.

The interactive system ISASS can perform other tasks such as glottal inverse filtering, formant tracking, or parameter verification by synthesis but for this study of vowel formant frequency, the three steps mentionned above seems to be an adequate compromise between accuracy and speed of analysis. Of course, unsupervised analysis (step 2) is only performed because of speed limitations of our minicomputer equipment (we used a SEL 810-B minicomputer with a 790 nsec cycle time). If it could be done at an adequate speed (during the time it takes to display a breath group), steps 1, 2 and 3 could be merged in 1.
The result files contain for each oral vowel : a label, the frequency of the first three formants, the immediate context of the vowel, and two indicators on syllable type.

STATISTICAL ANALYSIS

For each speaker of the Paris-Corpus, the oral vowel files consist of about 1500 entries. On each such data set, we attempted a description of intra-speaker variability of oral vowel formant frequency. A scatter plot of all the vowels of a single speaker in the F1-F2 plane (figure 1) reveals the non separability of vowel classes (a number on this plot represents the population of a bin on this 2 dimensional histogram). The only satisfactory result of such a plot is the verification that F2 values of closed vowels (low F1) have a wider spread than that of open vowels (high F1). This fact was mentionned by DELATTRE P. (1948)

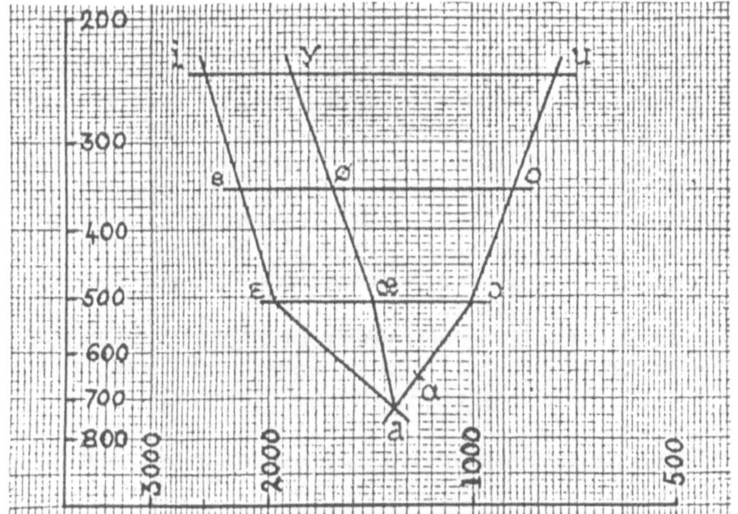

FIGURE 2 : THE FRENCH ACOUSTIC TRIANGLE (DELATTRE, 1948)

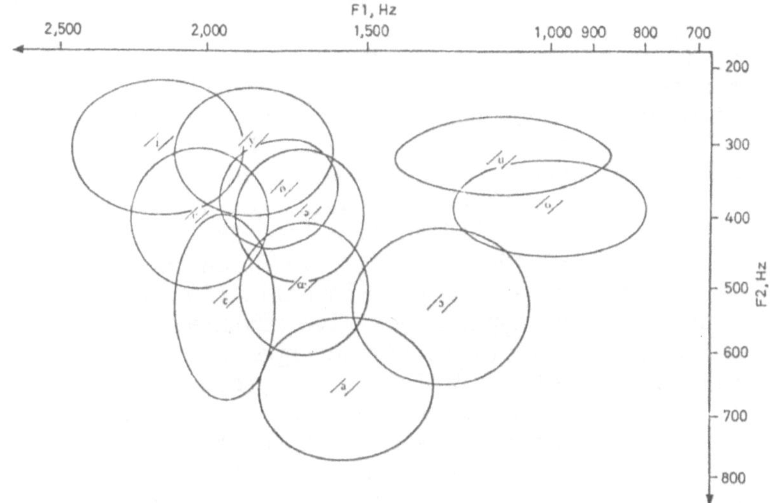

FIGURE 3 : ACOUSTIC DISTRIBUTION OF A SPEAKER'S VOWEL SAMPLES

more than 30 years ago in his discription of the French acoustic triangle (figure 2).

Each vowel group can be modelled in a first approximation by a 2 dimensional gaussian distribution. A visualization of such a model is obtained on figures 3, 4 and 5 by plotting for each vowel type its centroïd and an elipse at one standard deviation from the centroïd. These 3 plots have been chosen to illustrate the range of variability of vowel formant frequency in different speech styles (single but different locutor) :

- figure 3 : from a surreptitiously recorded conversation
- figure 4 : from 700 words of a basic dictionary
- figure 5 : from isolated vowels imitating those of the words of the dictionary.

These plots deserve further comments but let us first discuss the validity of the gaussian model. The series of histograms shown on figure 6 represent the frequencies of the first two formants of the vowels [i, e, ε, ə, ɔ, o, u] uttered by a single speaker in natural conversation. Some of these distributions are sufficiently unimodal to be described reasonably well by their means and standard deviations. Some others, such as F1 of [e] and F2 of [ɔ] are clearly bimodal. Some of these non normal distributions can be described tentatively as the result of contextual effects (CHOLLET G., I976, I977) but nevertheless the gaussian model is quite questionable. In particular, the visual representation of figure 3 is clearly unsufficient to quantify the overlap between vowel types (not only it is based on a gaussian model but higher formants are discarded). A measure of overlap of distributions in an n- dimensional space is given by the confusion matrix of a classification algorithm. Most discriminant analysis programs from statistical packages such as BMDP (DIXON W.J., I975) or SPSS (NIE N.H. et al., I975) which are based on multinormal distributions with equal covariance matrices are clearly inadequate. Discriminant functions that do not assume normality can be estimated usign the so-called kernel method (MEISEL W.S., I972). The ALLOC program (HERMANS J. et al., I976) can be used to compute such functions and perform the classification of all vowels of a speaker according to type. A decisive advantage of this technique is that confusion matrices of many speakers of the same dialect can be collapsed together. Table I is the result of the classification of some I4 000 vowels uttered by 20 parisian speakers. The classification scores give a measure of the discreteness of each vowel. The percentage of vowels i misclassified as j gives a raw measure of closeness (not a distance) of i to j. A better estimate of the overlap between two vowel groups is obtained by performing a discriminant analysis between these two groups only. The recognition scores range from 60 % for [i] to 15 % for the mute-e in our sample of surreptitiously recorded conversational speech (table I). For vowels from our sample of words and sentences read,

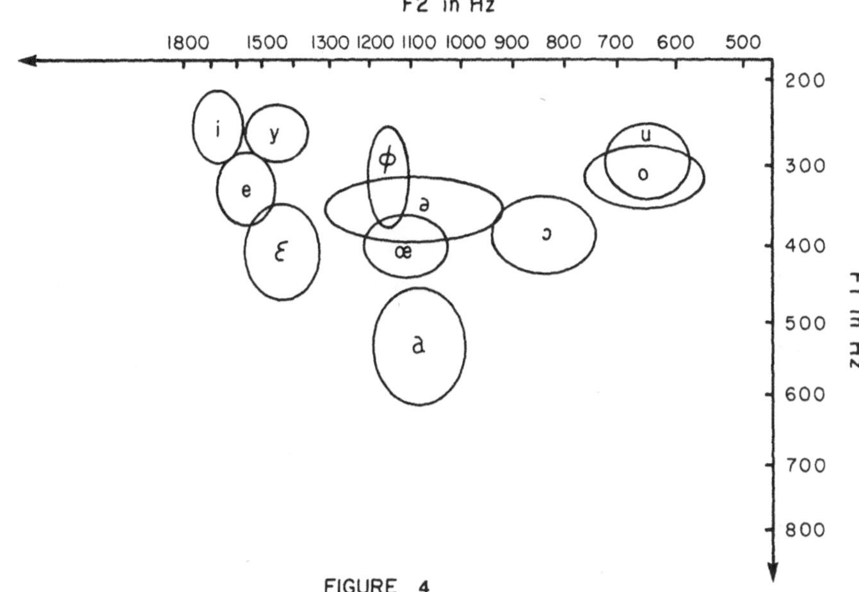

FIGURE 4
F1-F2 PLOT OF THE VOWELS OF A SPEAKER READING A LIST OF 700 WORDS

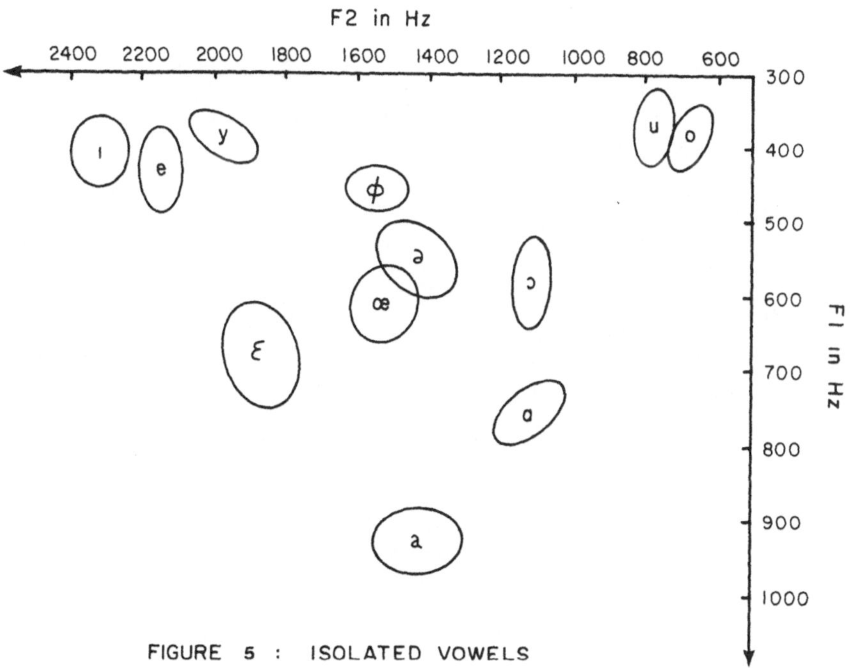

FIGURE 5 : ISOLATED VOWELS

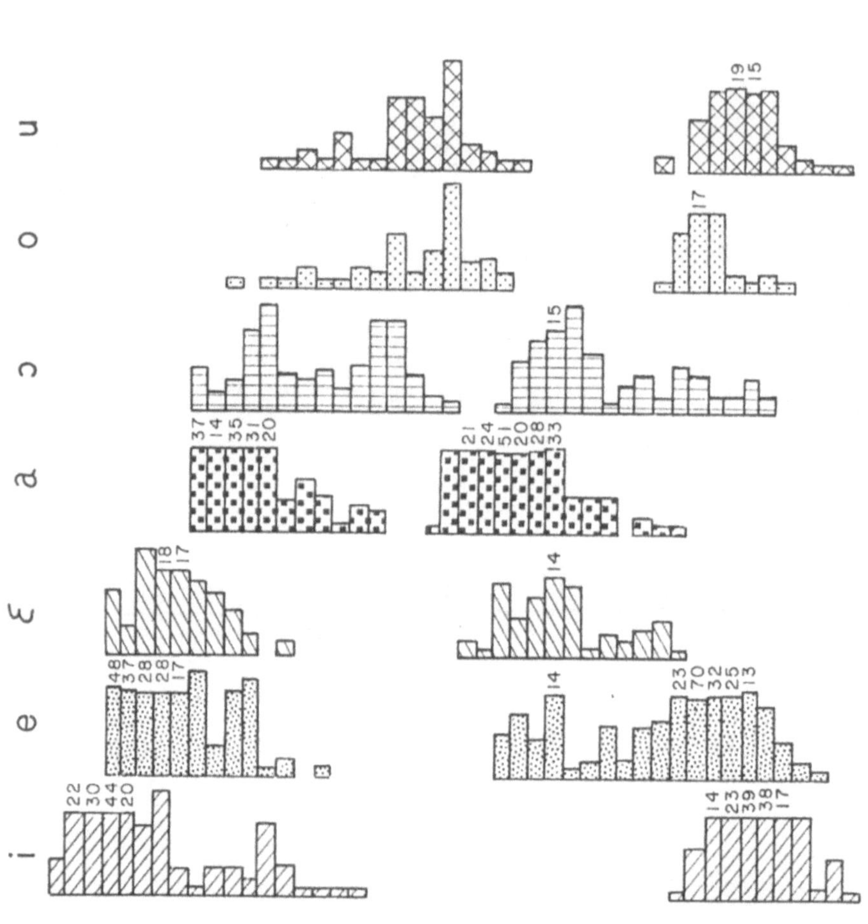

FIGURE 6 : HISTOGRAM OF F1 AND F2 VALUES FOR A MALE SPEAKER

Table 1 Classification Matrices Pooled Together for 20 Speakers (Each cell contains the corresponding classification score)

Vowel	Classified As											Tot. No. of Vowels
	i	e	ɛ	a	ɔ	o	u	y	ø	œ	ə	
i	60	9	8	2	1	2	5	8	2	1	2	1780
e	21	23	19	4	3	2	3	9	6	4	6	3411
ɛ	12	12	45	8	1	1	2	6	3	7	3	1198
a	1	2	5	50	6	6	7	3	4	13	3	2834
ɔ	1	3	2	17	26	14	9	4	7	10	7	999
o	1	2	2	4	8	46	17	4	8	4	4	426
u	3	2	1	2	5	25	37	12	7	2	4	980
y	23	6	3	1	3	5	10	35	8	2	4	831
ø	10	9	6	4	4	5	4	11	30	6	11	203
œ	4	5	13	17	8	5	5	3	10	23	7	182
ə	6	11	8	5	8	7	7	10	14	9	15	1606
												14286

the overlap between vowel types is considerably reduced and consequently the correct classification scores are quite a bit higher (table 2). Isolated vowels show practically no overlap with other vowels, a result which is in good agreement with PETERSON G.E. & BARNEY H.L. (1952) observation of English vowel formant frequency in an [h-d] context.

DISCUSSION AND CONCLUSION

A cooperative user of an oral man-machine communication system would modify his articulation to optimize the machine recognition of his message. Therefore, it would be surprising that any of the speech styles we have studied could apply to this situation. Nevertheless, his performance should fall within the limits we have studied. It is the task of the engineer, with the help of psychologists, to find a compromise between system complexity and naturalness of its use. Man-machine communication systems should be simulated to obtain speech samples of actual usage. These samples could be analysed with our methodology to discover robust acoustic features.

Our results tend to demonstrate that a speaker modifies his articulation according to the intended purpose of his utterance. He may under -or over- articulate depending on whether or not the listener(s) is (are) familiar with his dialect or the subject of the message. In natural conversation, the interlocutor is an active participant and little effort is required to convey the message. If this conversation is recorded, a speaker articulates more distinctly knowing that many listeners could potentially listen to him. If the recording is made in a laboratory, he makes a further effort knowing that his speech samples may be analysed. In fact, in a usual laboratory situation, the locutor articulates utterances without conveying a message. His attention is thus focussed on his articulation. We must henceforth distinguish between performance patterns and realize that any recording made in the laboratory will yield results that more closely approximate preference patterns than natural performance patterns, regardless of how convinced one may be of the naturalness of the utterances.

Recognition scores, as presented here, have a relative value and do not measure only the performance of classification algorithms but also the quality of the speech samples being analyzed. Scores have been reported here to stress the differences between speech styles (a single classification algorithm being used). Error rates are often reported to evaluate automatic speech recognition systems. Such an evaluation could be misleading as test materials could differ widely. These test materials should be standardized and should cover the full range of speech samples that must be processed by such systems.

Table 2 Classification Scores for the Vowels of a Dictionary of 700 Words.

Vowels	Classified As											Number of Vowels
	i	e	ε	a	ɔ	o	u	y	ø	œ	ə	
i	82	8	1					9				135
e	5	68	15					6	4		2	213
ε		13	77	2				1	1	2	4	87
a			3	85	3					7	2	226
ɔ				2	70	7				5	16	74
o					7	64	24				5	33
u					3	29	66				2	70
y	9	7	4				1	71	5		3	66
ø		5	5					5	65	8	14	22
œ			5	5	10				10	55	15	20
ə		3	4		4	3		3	16	19	48	93
												1039

A study of vowel formant frequency is probably sufficient to illustrate some acoustic effects of speech style; but more fundamentally, the variability of all measurable primitive parameters should be investigated to evaluate the robustness of acoustic cues. Of course, the internal (cortical) representation(s) of speech sounds is still a current issue of the neuro-phonetic sciences. Nevertheless, it seems useful to hypothesize articulatory and/or acoustic invariants for descriptive and practical purposes. For example, target values for vowel formant frequency could be chosen as the average values for isolated samples. Deviation from these targets in connected discourse have been primarily observed as a function of segmental duration (CHOLLET G., 1976). The last syllable(said to be stressed)of a prosodic group in French is significantly longer than other syllables. The targets are more often reached for stressed than for unstressed vowels. TIFFANY W.R. (1951) suggested that the dimension of the acoustic space occupied is related to the quality of articulation of the vowels. Our data agrees with this statement as the acoustic triangle for unstressed vowels is enclosed within the one for stressed vowels (certainly more clearly articulated). Unstressed (short) vowels are more sensitive to the immediate phonetic context than stressed ones. They are systematically under-articulated (CHOLLET G.., 1977). The mute-e, the most instable French vowels (MALECOT A. et al., 1977), is very sensitive to these coarticulation effects. Some effects of style have also been described in this paper but theses series of studies have by no means exhausted all the possible causes of variability of acoustic speech parameters.
We believe our Interactive Speech Analysis and Synthesis System (ISASS) and the statistical techniques described are adequate tolls to investigage the variability of any acoustic parameter. A major difficulty remains : the gathering of a statistically valid corpus.

REFERENCES

BLACK J.W. The effect of the consonant on the vowel JASA 10, 203-205 (1939)

BOND Z.S. Identification of vowels excerpted from (l) and (r) contexts - JASA 60 : 4, 906-910 (1975)

BOND Z.S. Identification of vowels excerpted from Neutral and nasal context - JASA 59 : 5, 1229-1232 (May 1975)

CHOLLET G. Computerized acoustic analysis of large corpuses AAPS meeting, San Diego (1976)

CHOLLET G., Effect of syllable type, stress, and phonetic context on acoustic French vowel targets. JASA 60, Suppl n° 1, S44 (Fall I976)

CHOLLET G., La coarticulation et ses effets acoustiques sur les cibles des voyelles orales françaises. 9th ICA, Madrid (July I977)

DELATTRE P., Un triangle acoustique des voyelles orales du français. The French Review 21 :6, 477-485 (I948)

DIXON W.J., (ed.) BMDP : Biomedical computer programs. Univ. of California Press, Berkeley (I975)

GRAY A.H., MARKEL J.D., Distance measures for speech processing IEEE Trans. on ASSP-24 : 5, 380-390 (oct I976)

HERMANS J., HABBEMA J.D.F., Manual for the ALLOC discriminant analysis programs. Dept. of Med. Stats Univ. J Leiden (I976)

LADEFOGED P, KAMENY I., BRACKENRIDGE W., Acoustic effects of style of speech. JASA 59 :1, 228-231 (I976)

LINDBLOM B., Spectrographic study of vowel reduction JASA 35, 1773-1781 (I963)

MALECOT A., New procedures for descriptive phonetics in : Papers in Ling. and Phon. to the memory of P. DELATTRE (Mouton, The Hague, I972)

MALECOT A., CHOLLET G., The acoustic status of the mute-e in French. - Phonetica 34 : 1, 19-30 (I977)

MEISEL W.S., Computer-oriented approaches to pattern recognition. Academic Press, New-York (I972)

NIE N. H. et al., SPSS : Statistical package for the social sciences - 2nd ed., New-York, Mc Graw Hill Book Co (I975)

OHMAN S.E.G., Coarticulation in VCV Utterances : Spectrographic measurements. - JASA 39, 151-168, (I966)

OHMAN S.E.G., Numerical model of coarticulation JASA 41, 310-320 (I967)

PETERSON G.E., BARNEY H.L., Control methods used in a study of the vowels - JASA 24 : 2 (I952)

POLS L.C.W., Perceptual space of vowel like sounds and its correlation with frequency spectrum - in Freq. Anal. & periodicity detect. in Hearing, PLOMP R. & SMOORENBURG G. F. Leiden (I970)

TIFFANY W.R., Non-random sources of variation in Vowel quality JSHR 2, 305-317 (I959)

ZIPF G.B., Human behavior and the principle of Least effort (I949)

§ 3

LEXICON, SYNTAX AND SEMANTIC

THE REPRESENTATION AND USE OF A LEXICON IN AUTOMATIC SPEECH RECOGNITION AND UNDERSTANDING

Jean-Paul HATON

Centre de Recherche en Informatique de Nancy (CRIN)
Université de Nancy I

ABSTRACT

The lexicon is one of the knowledge sources which has to be taken into account in a speech recognition/understanding system. The organization of the two kinds of lexicon -non phonological and phonological- is first discussed. We then present the techniques involved in the use of a lexicon for emitting and verifying word hypotheses : word verification and word spotting. Examples from actual systems illustrate the various aspects of the paper.

1. INTRODUCTION

An automatic speech understanding system has to take into account various knowledge sources (KS) in order to process a sentence spoken in a specified language. These various KS carry some information which contributes to the global understanding. Among these KS the lexicon and the lexical processing level are of particular importance since it is necessary to go through the level of the words at one time or another during the understanding process of a

J. C. Simon (ed.), Spoken Language Generation and Understanding, 311-335.

sentence. This is true as much for written language as for spoken language. Moreover spoken language presents some specific features: first, the pron unciation of a word is highly affected by different factors such as accent or emotional state of the speaker, etc... Each utterance of a language can therefore be produced in a great number of ways. Most of these productions are described by phonology which turns out to be an important component of lexical level for speech.

Secondly, the speech signal is essentially continuous and does not contain any information on word boundaries except some indications given by prosody. This fundamental fact of spoken language introduces some indeterminism (another one is introduced by the automatic acoustic-phonetic decoding of speech) which will affect the techniques used in lexical representation and lexical retrieval for speech compared with similar techniques used in written text processing.

In what follows we shall refer to lexicon as a set of words represented in a specified code (e.g. phonetic transcription of the words) and containing pieces of information of syntactic nature (e.g. place of the word within a sentence) and of semantic nature (e.g. meaning of the word). It is usually efficient to organize a lexicon in such a way that the data-structures involved in this organization make it easier to access any information contained in this lexicon.

In these conditions a vocabulary will be considered as a set of words, i.e. as a sub-set of a lexicon.

The purpose of this paper is to present a unified view of the lexical level in speech understanding with respect to both aspects of lexical representation and lexical retrieval of words.

After having first defined the role of a lexicon in the emission of hypotheses we shall look at phonology : its importance, the definition of phonological rules, the various levels of representation. Different solutions for lexicon representation -both phonological and non phonological- will then be presented and discussed. Finally we shall review the various techniques involved in the use of a lexicon : word recognition or verification, construction of a word-lattice from the phonetic output of a speech recognizer, word spotting.

2. ROLE OF THE LEXICON

Usually -and specially for written language- the lexicon is intended to give for each word its grammatical class and its various meanings. The problem is slightly different in the case of speech. An automatic speech understanding system always incorporates some kind of word hypothesis generator using the various KS and,among them, the lexicon. The lexicon must yield all the terms satisfying one or several given properties of various nature :

phonetics : words of a given length (counted as the number of phonemes)or includind fricatives,...
syntax : verbs, subjects,...
semantics : category of words,...

It will therefore contain the description of the words at the various levels : phonetic, syntactic, semantic. The data structures involved in the representation must moreover be chosen in order to speed up access to these pieces of information.

The lexicon is so used for emitting hypotheses about words. Another complementary role is also to verify or cancel the words which were hypothesized at other levels. Finally, since word boundaries do not clearly appear in connected speech, the lexical level also has to spot either a particular key word within a sentence or all the possible words of the vocabulary. The former case may correspond to an island-driven strategy for sentence parsing ; in the latter one derives a word lattice from the phoneme lattice produced by the acoustic-phonetic level. These techniques will be studied in details later.

Each word of a vocabulary may be realized acoustically in several different ways. Moreover a particular word may be affected by phonological alterations when pronounced in context. These various phenomena are related to phonology. It is then possible to use two kinds of lexicons in speech recognition -i.e. phonological lexicon and non phonological lexicon- the structures of which will of course be quite different. Before looking at the representation of the different kinds of lexicons we will first say a little more about phonology.

3. PHONOLOGY AND PHONOLOGICAL RULES

3.1. General presentation

As stated previously, the phonological alterations which may affect words occur both within a word and in the juncture between two words. Except in limited applications it is obviously impossible to store all pronunciations of words and associations of words. But the various pronunciations of a word are related to one another so that it is possible to define for each word a base form from which all variations (i.e. various pronunciations, tenses of a verb, etc...) can be derived by applying a set a phonological rules. At least a hundred such rules are necessary in order to represent the most common phonological phenomena in languages such as English or French. This way of lexicon representation makes it possible to decrease by a factor of at least 5 the memory space necessary for its storage. It can be noted that phonological rules are generative in nature since they are intended to generate the various permitted utterances of a word. This is not well suited for speech recognition where one would need analytical rules rather than generative ones. This explains the importance of analysis-by-synthesis methods in word verification.

A number of different formalisms have been used for phonological rules. The following are two examples, the first from English, the second from French :

$\star$ unstressed i $\longrightarrow \varepsilon / \left\{ \begin{smallmatrix} \# \\ \mathrm{I} \end{smallmatrix} \right\}$ - consonants

This rule expresses the unstressed vowel reduction which is for instance responsible for the transformation :

/i l ɛ k t r i k/ $\longrightarrow$ /ɛ l ɛ k t r i k/

$\star$ ã $\longrightarrow$ $\tilde{a}^{n}$ / -t or -d

This rule represents the insertion of a nasal consonant after a nasal vowel which is common in southern France, for instance :

/ l æ̃ d i / $\longrightarrow$ /l $\tilde{æ}^{n}$ d i/

In both cases the rules appear as contextual rewriting rules with a left-hand side and a right-hand side which are usually phoneme strings (possibly empty) followed (after the symbol/) by the right and left contexts in which the rules apply.

A great amount of effort has been done in collecting and formalizing such rules for the English language (1) (2). Work is also in progress for French : Perennou and colleagues are building up a very complete lexicon with several thousand entries (3) (4). Haton and colleagues are working on a more restricted lexicon (about 400 words) as a component of a general continuous speech understanding system. In this lexicon phonological rules are implemented in the form of procedures (5) (6).

In fact there is no unified information about pronunciation of words, especially in connected speech which is the case we are concerned with. Linguist's work, examination of spectral analysis of speech and personal experience are the various data sources used in establishing the phonological rules.

The process of applying the phonological rules to the base form of a word accounts for the different pronunciations of this word but it does not give any information on the frequency with which each is produced, i.e. on the conditional probability P(w/p) that p is produced as a particular pronounciation of word w.

It is necessary in a statistical or information theory approach of speech recognition. Cohen and Mercer (1) associate probabilities to both base forms and phonological rules and thus define a kind of speaker model (may be it would also be interesting to define a model of the acoustic recognizer). These probabilities together with combined probability rules make it possible to determine for a base-form string the probabilities of the different surface strings that can be produced.

3.2. From base-form to surface-form of an utterance

The main problem which arises in the use of a lexicon in speech recognition is to derive possible phonetic surface-form strings of an utterance from base-forms of the different words and phonological rules. One can distinguish between three steps in this process (7) :

a) defining a general phonological description of the utterance
b) taking into account the possible personal alterations of the previous description
c) transforming the phonological description into an acoustic production.

a) At this level the goal is to give a phonological description of the utterance which is speaker-independent. In a first time each isolated word representation is built from its base-form and indications on the person, tense (for a verb) or the gender, number (for an adjective) or the number (for a noun). The successive words of the sentence are then associated by applying word juncture rules. The following is an example of a juncture rule in French :

me.z # ami.z # ⟶ mezami.z

This rule means that the liaison is mandatory in the association "Mes amis" (My friends) which must be pronounced as /mezami/

Cohen and Mercer, and several groups (BBN,SDC,SCRL) in the framework of the ARPA SUR Project (8) have carried out important work in the formalization of such rules for English ; as did Perennou et al for French.

b) The second level is related to the various possibilities among which a speaker can choose. Let us give some examples in French :
- substitution of phonemes : for instance the article "les" (the) /lɛ/ can be pronounced /lɛ/ or /le/
- deletion : it consists of the elimination of one or several phonemes. The case of phoneme /ə/ is the most common in French, e.g. "tu me vois" (you see me) : /tyməvwa/ ⟶ /tymvwa/. It does also exist in English, for instance /pœris/ ⟶ /pœis/
- liaison : some are mandatory but most are optional, according to the speaker's social or geographical origin
- insertion : it is usually the case of vowels /ə/ or /œ/ which may be inserted in a sequence of consonants ; for instance "à l'Est de la France" (East of France) :

/alɛstdəlafrɑ̃s/ ⟶ /alɛstœdəlafrɑ̃s/

Most of these choices are usually specific to a speaker. When the speaker is known it therefore becomes possible to simplify the lexicon. On the other hand the learning process of a speech understanding system must determine the usual choices of the different users of the system.
c) The third level makes it possible to take into account the articulatory constraints which influence the transformation of a phonetic transcription into an acoustic production.

Some of these constraints are speaker-dependent and may therefore be learned as stated previously. Some others are more general, for instance the assimilation of an articulatory parameter of a phoneme by the corresponding parameter of its neighbour : "abcès" (abscess) /absɛ/ ⟶ /apsɛ/ or, in English
/abstəkəl/ ⟶ /apstəkəl/

4. LEXICON REPRESENTATION

4.1. Logical organization of a lexicon

We have seen that the lexicon is used in a speech recognition system for emitting word hypotheses, in conjunction with other knowledge sources. Moreover, the lexicon must also contain phonetic and, eventually, phonological information in order to verify words hypothesized by other processing levels.

The lexicon must therefore yield all the terms satisfying some given preconditions of various nature : phonetics, syntax, semantics. Moreover the various phonetic forms of a term have to be given.

Several attributes are attached to the different terms of a lexicon. For instance the attributes of a noun could be
- syntactic attributes : number (singular or plural), etc...
- semantic attributes, e.g. pointers in a semantic net
- syntactic and semantic attributes, e.g. whether a noun is a measure unit or a measure (9).

It turns out to be efficient for speech recognition purposes to give the lexicon a tree structure. Let us examine two examples :

- The lexicon of the SRI system (9) is divided into 12 categories : verbs (3 categories), numbers (2), adjectives, prepositions, determiners, nouns (by far the largest one), classifiers and noun phrases. Each category itself comprises several sub-categories which correspond to the major semantic classes in relation with the universe of the application (for instance relational measures is a subcategory of nouns with the words : length, size, speed).

Each term of the lexicon has two sorts of attributes : some are common to all terms of a sub-category, the other ones are characteristic of the term.

• In the lexicon of the MYRTILLE II system (10) there is a hierarchy of four successive levels as indicated on fig. 1 :

- the first level separates the lexicon into three categories : special words which explicitly appear in the definition of the language, "grammatical" words (adverbs, prepositions,...) and all the remaining words which form the largest category,

- the second level corresponds to the classical partition between nouns, verbs and adjectives,

- the third level corresponds to sub-categories defined on semantic and syntactic criteria,

- the fourth level contains the different terms of the lexicon. With each term is associated a pointer to its phonetic transcription and possible phonological alterations.

The logical organization of the lexicon must then be implemented on a particular machine with memory-size and speed limitations. It is therefore necessary to choose the data-structures which make it possible to access the lexicon as quickly as possible. This pratical organization, though important as far as the performance of the system is concerned, will not be studied here since it is very dependent upon the available machine and programming language.

4.2 Phonological vs non-phonological lexicon

So far we have seen the importance of phonological alterations in the phonetic transcriptions of words. It is however possible to build a lexicon without taking phonology into account. We will then find two kinds of lexicons in speech recognition : phonological lexicons (PL) and non-phonological lexicons (NPL).

In fact NPL are sufficient for limited tasks. For instance in isolated word recognition one is freed of the problems of word junctures and, often, of speaker's variations. For small vocabularies the lexicon is usually made up of the spectra of the different words. This is rather expansive in memory place and in computation time for word matching. When the size of the vocabulary increases it is therefore more convenient to store the basic phonetic transcriptions of the words.

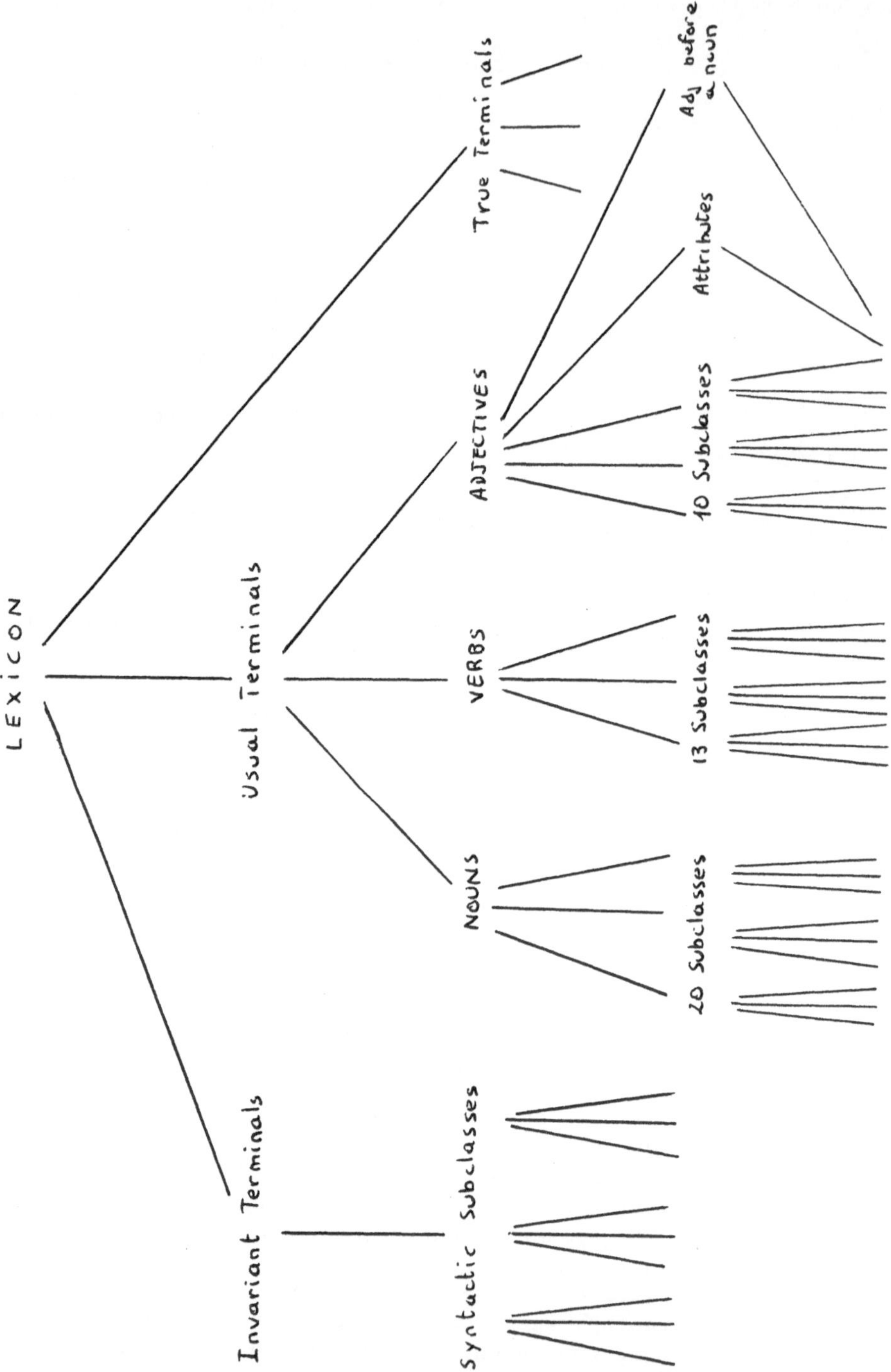

Tree Representation of Syntactic-Semantic Information in the Lexicon of MYRTILLE II (after PIERREL)
Fig. 1

As far as connected speech is concerned NPL can only be used in limited applications characterized by several factors :

- limited vocabulary (some tens up to a hundred words)
- one carefully enunciating speaker for which the system was tu ned during a learning phase
- constrained language allowing only a limited number of grammatical constructions.

An intermediate solution consists of incorporating in the phonetic transcription of each word the most likely insertions, substitutions and elisions of phonemes. A number of systems, as well for isolated words as for connected speech, have used this king of representation. As an example, fig. 2 shows the graph-representation of a word in the MYRTILLE I system (11) together with its Fortran implementation.

Graph-representation:

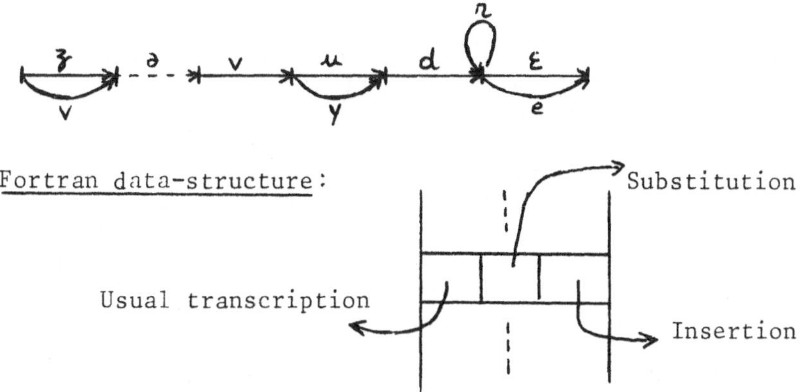

Lexical Representation for the word "je voudrais" (/ʒəvudrɛ/) in MYRTILLE I

Fig. 2

The MYRTILLE I system features another interesting point at the lexical level (12). There are two different representations for the words related to their grammatical roles : one for the ordinary terminals of the language (as indicated on fig. 2) and one for the non-terminals which correspond to different words (e.g. different names of persons in a directory). The latter representation is a tree-structure which makes it possible to merge the common phonetic transcription of different terms.

Phonological lexicons have to be used in more general applications and especially for pseudo-natural language understanding. In this case one has to store base-form phonetic transcriptions together with phonological rules. There are basically two ways in which both types of information can be combined. The first one consists of applying a rule whenever it has to be applied within a word. In this case the application of a rule may be done by a procedure. The second one consists of precompiling all rules and of storing of possible productions in a network. This solution is of course more efficient. Indeed the HARPY system used such a precompiled lexicon and was the most efficient system designed in the ARPA SUR Project (13). Similarly BBN used a diction ary expander which generates all possible pronunciations of words by applying phonological rules to a base-form diction ary (14).

However, this solution leads to very large lexicons and this could be a serious drawback for practical applications.

The design and implementation of a lexicon is a long and tedious job. A number of tools have therefore been developed in order to help in testing and compiling phonological rules, updating or modifying the lexicon. An attempt has also been made to automatically infer a lexicon (15).

An interesting contribution to lexical representation is the BBN's lexical decoding network. In this network common parts of pronunciations of different words are merged, as in MYRTILLE I, and moreover ad hoc procedures are defined for tying words together (16). This makes it possible to search through out the entire diction ary without having to consider each word separately.

5 THE USE OF A LEXICON IN SPEECH RECOGNITION

5.1 Introduction

We have been concerned so far with the representation of a lexicon for spoken language. We will now consider the methods and techniques involved in the use of such a lexicon, essentially in the emission and verification of hypotheses at the level of the words.

The problem of isolated word recognition will be first considered since some techniques developed at this level can be easily generalized to connected speech recognition. As far as continuous speech is concerned we will distinguish between two different operations at the lexical level :

- word verification : i.e. the verification of word hypotheses generated by various KS in the system

- word spotting : i.e. the localization of one or several given words within a sentence.

In fact these two operations can be considered as two different aspects of the same problem. Indeed some techniques apply to both.

A general problem embodied in all cases is the following : given a lattice L of phonemes or phonetic features representing the actual transcription of a sentence by the acoustic-phonetic recognizer and a phonetic transcription of a word W (for instance under the form of a graph or a finite-state automaton) including all possible pronunciations of this word, compute the "distance" between W and L, or a part of L. We will review several methods which were proposed in order to solve this problem.

The problem is highly complicated by the fact that the lattice contains numerous errors which can be formalized under the form of rewriting rules as follows :

deletion of phoneme x : u x v ⟶ u v

insertion of phoneme y : u v ⟶ u y v

substitution of phoneme x by y : u x v ⟶ u y v

5.2 Isolated word recognition

Most systems for isolated word recognition are related to a global approach of the problem which consists of matching patterns extracted from the input utterance for instance by spectral analysis. However recent experiments have shown that analytical recognition based on phonemic transcriptions of the words and lexical retrieval is efficient (if not the only one possible) for large vocabularies of several hundred to some thousand words.

As stated previously, in this case the word boundaries are known which simplifies the problem. The techniques used in word matching can be roughly classified into three categories :

a- heuristic matching
b- stochastic matching
c- analysis-by-syntesis.

a) Heuristic matching

Many different algorithms belong to this class. The basic idea is to define adhoc procedures which can account for the various errors of insertion, deletion and substitution of phonemes. As an example we shall present the algorithm proposed by Haton (17). This algorithm uses the basic principle of dynamic programming which compensates for the variations of length of the strings to be compared. The length variation between a phoneme string representing the transcription of a word given by an acoustic recognizer and the exact phonetic transcription of this word are due to segmentation errors which are quite common in such systems. However, in order to decrease the computation time, this algorithm was modified so that it does not yield the globally optimal solution as usual dynamic programming algorithms do but a local optimum which turns out to be sufficient in this case.

Let

$$R = \left\{ r^j \right\} \qquad j = 1, m$$

be a reference string for a given word W. We want to compute a "distance" between R and an unknown string with multiple answers.

$$S = \left\{ (s_1^1, s_2^1, .., s_p^1), ..., (s_1^n, s_2^n, ..., s_p^n) \right\}$$

The string S is made up of n elements, each with p possible labels and probabilities associated with the different labels.

Let d (r^j, s_k^i) be a metric which reflects the dissimilarity between phonemes r^j and s_k^i. In fact the choice of d is very important for the efficiency of the algorithm. d must take into account the phonetic similarity between two phonemes and also the most common confusions produced by the acoustic recognizer.

Let $D(\ell)$ be the cumulated distance corresponding to the matching of the first j elements of R against the first i elements of S. The next step in the process will be defined by the recurrence relation :

$$D(\ell+1) = D(\ell) + \min_{q=1,k} \left\{ \begin{array}{l} d(r^{j+1}, s_q^i) \\ d(r^{j+1}, s_q^{i+1}) \\ d(r^j, s_q^{i+1}) \end{array} \right\} \qquad (1)$$

The initial condition

$$D(1) = \min_{q=1,k} \left\{ d(r^1, s_q^1) \right\} \qquad (2)$$

and the constraints

$$\begin{array}{l} 0 \leqslant i \leqslant n \\ 0 \leqslant j \leqslant m \end{array} \qquad (3)$$

together with relation (1) define the pseudo-dynamic matching algorithm.

Fig. 3 shows the optimal path obtained by comparing the reference word "Sapin" (/ s a p ɛ̃/) with the following unknown string (with p = 2):

s z ə a b a ã

ʃ s a o p ɛ̃ ɛ̃

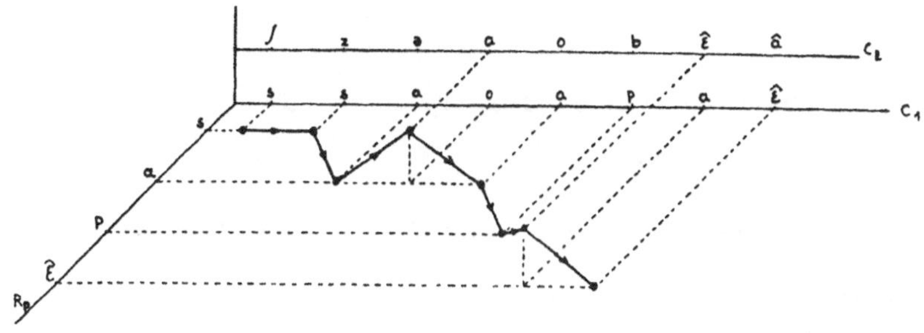

Optimal matching of an unknown word against
a reference word

fig. 3 (after HATON)

This algorithm can be extended to continuous speech. In fact several different algorithms have been proposed for solving similar problems, not only in speech recognition (cf for instance Sankoff (22)) but they cannot all be generalized to the case of continuous speech where the endpoints of the words in the string are not precisely known.

b) Stochastic matching

In this case a word is represented by a graph of alternative pronunciations with transition probabilities associated with all branches of the graph. The matching consists of finding the most probable path in the phoneme lattice corresponding to a reference word. This can be represented by a Markov process, and the Viterbi algorithm (17) yields the optimal solution by dynamic programming, without backtracking. White (19) gives an elementary example of how the algorithm works :

Fig. 4 shows a very simple transition network representing the word "6" (/siks/).

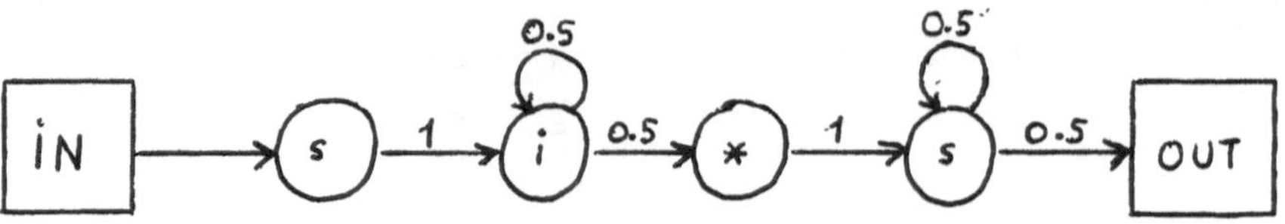

Transition Network for the word "6"

Fig. 4

Let us consider matching the reference network of fig. 4 against the following string given by an acoustic recognizer : /ssi* *ss*/.

Using a measure of phoneme similarity we can establish a matrix C(I, T) which gives the probability of being in state I at time T, as indicated on fig. 5. The computation time can be considerably reduced by eliminating, or pruning away, the improbable paths.

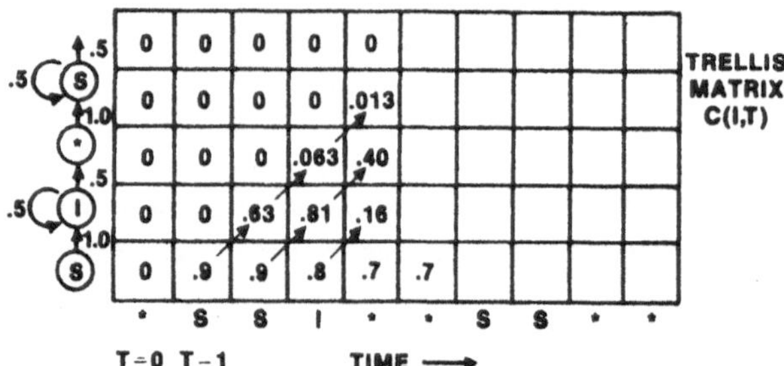

Transition Matrix ("Trellis Matrix") Computed

During the Matching of 2 strings(after WHITE)

Fig. 5

A first technique consists of following the most probable path at each time and of backtracking when necessary (best-first search). Another technique is the one used in the Viterbi algorithm where only very unlikely paths are pruned, all other paths are carried forward in parallel without backtracking.

Stochastic matching is one of the best techniques in word matching. However it is rather time consuming and, moreover, it is highly sensitive to the values of the transition probabilies which have to be carefully determined during a preliminary training phase.

A generalized version is used for continuous speech in the IBM system (20) and in the Dragon system (21).

c) Analysis by synthesis
The application of analysis by synthesis in word matching was proposed by Klatt. The method is very attractive, though time consuming and it will be described in the next section on word verification in connected speech.

5. 3 Word verification

The problem of verification of hypothesized words is encountered in all continuous speech recognition and/or understanding systems. It implies matching the phonetic transcription of a word (including its phonological variations) against a portion of the input phoneme string or lattice, the boundaries of this portion not being known exactly.

As stated previously several methods which were designed for isolated word matching can be generalized to continuous speech. In heuristic matching a common basic idea consists of aligning the reference transcription and the unknown string. This alignment can be based on reliable "anchor-phonemes" such as stressed vowels or unvoiced fricatives.

Once the two strings are aligned, classical methods of matching can be used for computing a distance, or degree of similarity, between them. In this computation the various possible errors of insertion, elision and substitution must be taken into account. This can be done in a number of ways.

Kohda et al. (23) try all possibilities of errors by applying rewriting rules similar to the ones described in section 5.1 to the phoneme strings.

In the Myrtille I system the matching algorithm uses heuristics for computing a score for each possible path in a graph. This score is incremented whenever it is necessary to assume an error of one of the preceeding sorts.

An important problem at the lexical level is to restrict as much as possible the number of words to be matched against the phoneme lattice. In Hearsay II Smith (24) uses the syllables -or more precisely some well defined syllable classes-as minimal units for this purpose. Stressed syllables are used to enter an inverted lexicon and to prune away the words that do not match with adjacent syllables. The remaining words are then rated by taking into account the different syllables of these words. Perennou also uses the syllable as a unit as will be described in the next section on word spotting.

The lexical level developed in the BBN system comprises two parts, corresponding to the tasks of lexical retrieval and of word verification.

The lexical retrieval component determines the most probable words that match against portions of a phonetic segment lattice (25). Each word is given a probability computed by applying the Bayes rule of conditional probabilities. This computation uses the a-priori probability of a given transcription of a word which is determined during the dictionary expansion controlled by the application of phonological rules as seen previously.

The word verification component involves an analysis-by synthesis procedure designed by Klatt (26). The main advantages of this procedure is that it is easier to take into account phonological phenomena since they are mainly described by generative rules. Moreover the fact of matching words at the acoustic level of spectra (and not at the level of phonetic transcription)makes it possible to get rid of the present inaccuracies of acoustic-phonetic recognizers. The algorithm can be summarized as follows:

. the basic transcription of a word to be verified is expanded into all possible detailed productions using phonological rules and syntactic and semantic information,
. the broad phonetic transcriptions of the word are all transformed into sets of parameters necessary for controlling a speech synthesizer by rules,
. the different acoustic productions are then compared with a portion of the input sentence by a global method of pattern matching. The non-linear time normalization is done using the dynamic programming algorithm proposed by Itakura.

This method gives good results but of course it needs a great amount of computation. In the HWIM system the word verification using this method required 15 % of the total CPU time.

The lexical level of HWIM apparently constitutes the first operational realization implementing the idea of analysis by synthesis emitted by Halle and Stevens in 1962 (27) in speech recognition.

The use of fuzzy set theory in word verification was investigated by De Mori and colleagues (28). In this approach the different word hypotheses are expressed as fuzzy relations involving the phonetic elements which are characteristic of each word. It is then possible to compute for each word a degree of worthiness by taking into account the plausibility of the different phonetic elements.

V.4 Word spotting

Although word spotting can be viewed as a particular case of word verification we have chosen to treat it sep rately since it usually corresponds to a particular approach to speech understanding, i.e. a bottom-up approach. In such an approach one tries to build a "theory" for interpreting a sentence by starting from the acoustic-phonetic data. Basically word spotting consists of locating one or several given words in a continuous sentence. This can be done on the incoming acoustic stream of data, e.g. on spectra, but the operation is more usually carried out on phoneme strings or lattices produced by an acoustic recognizer.

The spotting of words in a sentence has three basic uses in a speech understanding system :

1) deriving from the phoneme lattice a word lattice which contains all possible occurrences of words defined in the vocabulary within the sentence to be parsed. This word lattice is then used by higher processing levels in the system. A typical word lattice is shown on fig. 6. This example comes from a bottom-up version of the Myrtille I system (29).

2) locating a specified word anywhere in a sentence. This word is then used as an anchor-point in the case of an island-driven strategy for processing a sentence.

3) verifying word hypotheses emitted at other processing levels without precise indications on the position of these words in the sentence.

Word spotting can be viewed as the process of extracting substrings from the phoneme string (or lattice) which represents a sentence. This is the case in the Keal system designed by CNET in Lannion, France (30). The algorithm operates in three steps :

a) filling a binary coincidence matrix between the phoneme string and the phonetic transcription of a given word. Fig. 7 shows an example of this matrix.

b) detecting all possible beginnings of the given word. A beginning is detected whenever one of the k first phonemes of the word is present. On fig. 7 all possible beginnings are indicated by a circle.

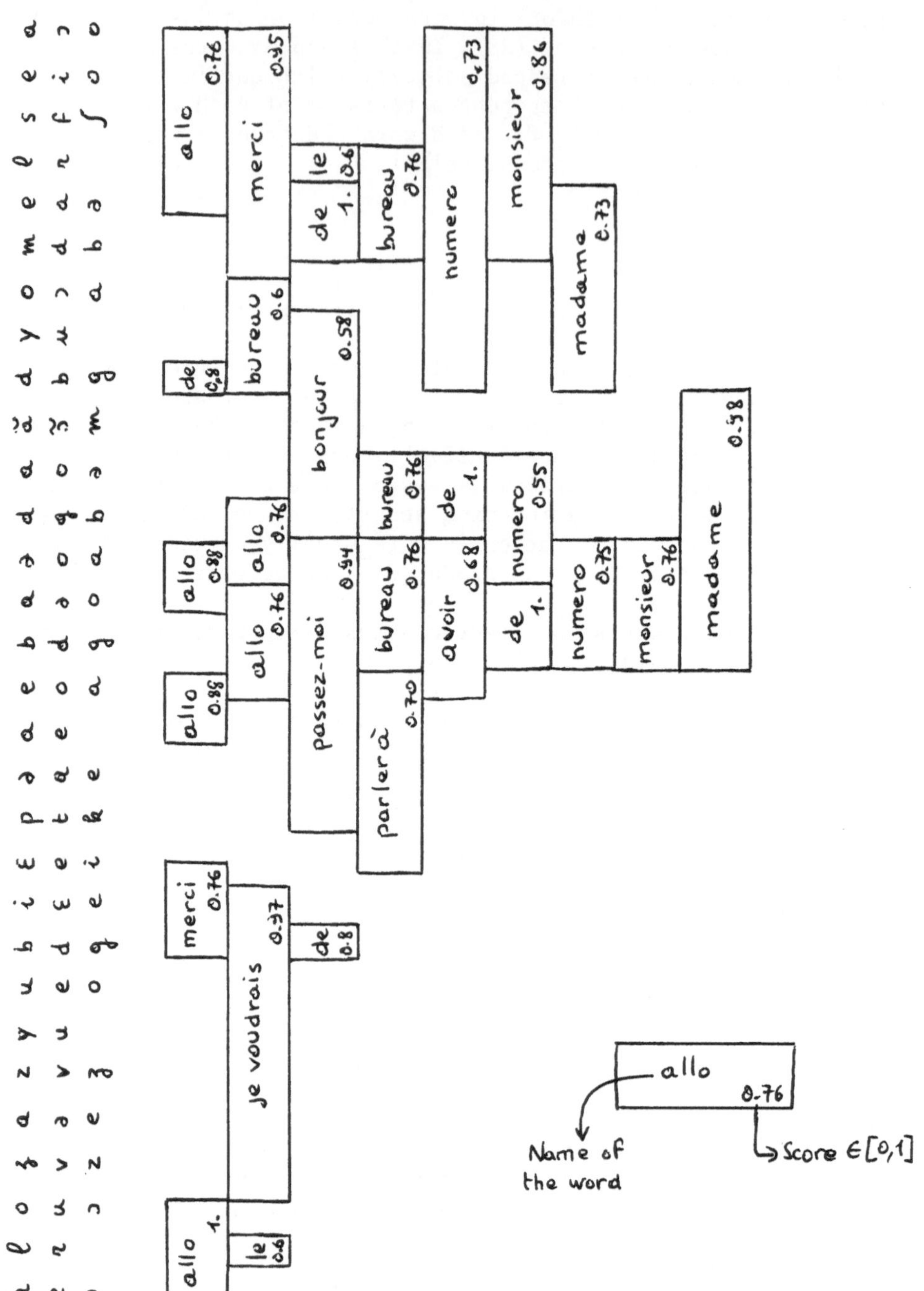

Word Lattice in MYRTILLE I

Fig. 6 (after MARI)

c) computing a similarity index between the word and all possible matchings of this word in the string. This index takes into account the probability of the different phonemes in the string and an experimental similarity between phonemes.

Binary Coincidence Matrix Between

Two Phoneme Strings

(For the Word "Repetition"/repetisjɔ̃/)(after VIVES)

Fig. 7

Efficient methods of word spotting on spectral data have also been proposed. These methods feature a non-linear time warping function using dynamic programming principles. Bridle (31) applies his algorithm to spectral data given by a Vocoder. Christiansen and Rushforth (32) use a similar idea but they work on LPC coefficients. Chiba and colleagues (33) recently reported very good results in word spotting for 40 speakers and a 100 word vocabulary using dynamic programming and a multimicroprocessor.

Related ideas have been applied in the Myrtille systems by Mari (34) but for phoneme strings with possible phonological variations. The problem is still of finding an optimal path for matching a word W against a string S. The various errors which may occur in string S can be represented by the basic patterns of fig. 8.

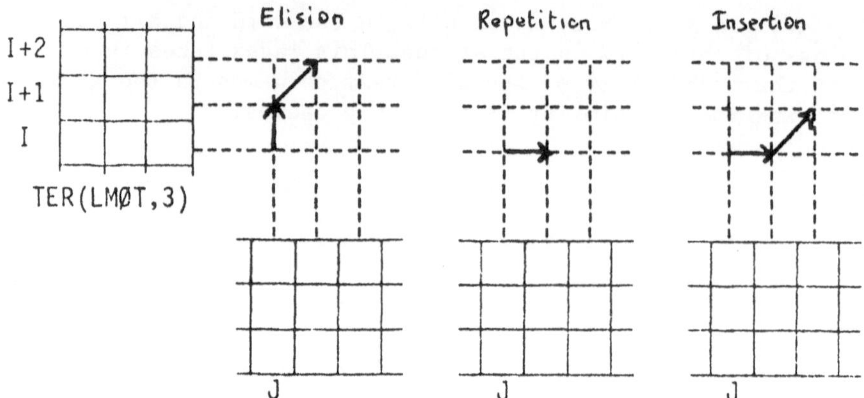

Different Types of Errors in Phoneme Strings
Fig. 8

Let V (i,j) be a local similarity measure between phonemes i and j based on experimental results.

To each matching path ending in point (i, j) is associated a score S(i, j) in which the recent past is preponderant. The following recurrence relation holds:

$$S(i, j) = \alpha_1 \; S(i-a, j-b) + \alpha_2 \; V(i, j) \; K(a,b)$$

Where (a,b) represents the three possible elementary transitions :

$$(a, b) = (0, 1) \; , \; (1, 0) \text{ or } (1, 1)$$

K(a, b) is a weighting factor which makes it possible to emphasize diagonal transitions :

$$K(1, 1) = 1$$

whereas $\quad 0 < K(0, 1), K(1, 0) < 1$

The score S_n at step n is finally defined as a function of the local similarity V_n and of previous decisions at steps n-2 and n-1 as follows :

$$S_n = \beta_1 \; S_{n-2} + \beta_2 \; V_{n-2} \; K_{n-2} + \beta_3 \; V_{n-1} \; K_{n-1} + \beta_4 \; V_n \; K_n$$

Coefficients α_i and β_i are experimentally determined.

A constraint on the slope of any path related to the speed of procunciation of speech was also added in order to increase the efficiency of the algorithm. A typical example of word spotting using this method is shown on fig. 9.

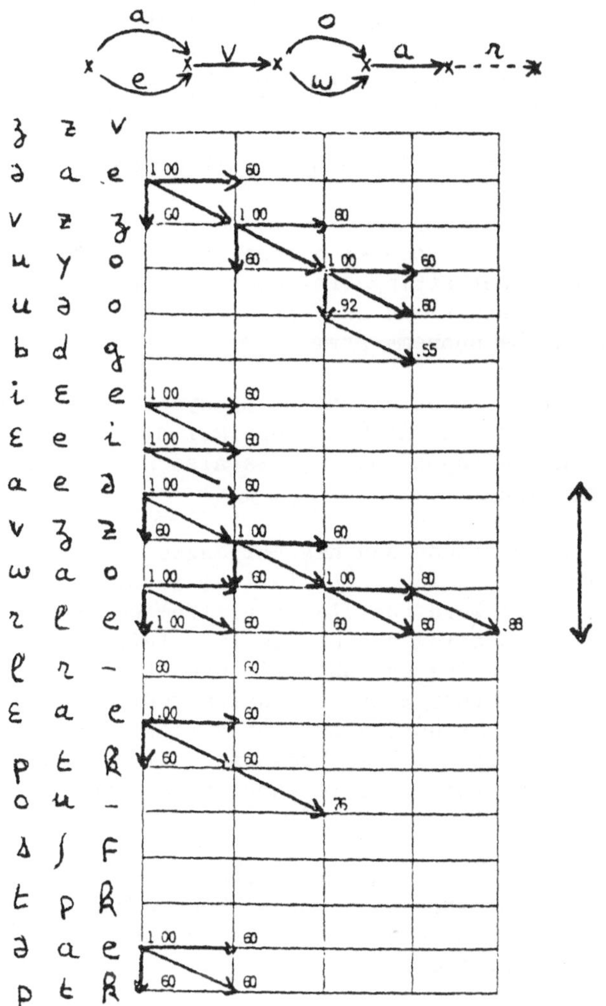

Spotting of word "avoir" (to have) /avwar/ in sentence "je voudrais avoir le poste 340" (I would like to have extension 340) (after MARI)

Fig. 9

The Lithan system designed by Nakagawa and Sakaï in Kyoto, Japan also features a lexical matching procedure using similar ideas (34). A likelihood ratio is dynamically computed for a word in a phoneme string and the method can be modified for word spotting.

Word spotting in the system developed by Univac embodies two successive steps (35) :

- word hypothesization which produces a list of words which match somewhere in the input string. For each word a vowel (usually the one with primary stress) is used as anchor-point for aligning the word with the phoneme string.

- word verification which evaluates the words hypothesized by a spectrum matching procedure in which the optimal alignement of a reference pattern with the input spectrum is also performed through dynamic programming.

The system developed by Perennou and his colleagues in Toulouse, France uses the syllable as unit at the lexical level (36). They have developed an original word spotting algorithm which processes the syllable string representing a sentence from left-to-right. The concatenation of successive syllables is used to enter the lexicon which indicates whether this succession represents a word or the beginning of a word or not. A list of different actions is precompiled and stored in a table. These actions are applied according to the nature of the syllable succession actually examined.

In fact this lexical search procedure is intended as a part of a unified system in the sense that some words which were spotted by this algorithm will then be pruned away by syntactic filters when these filters are designed.

VI- CONCLUSION

We have reviewed in this paper the two complementary aspects of the lexical processing level of a speech recognition-understanding system, i.e. the structure and organization of a lexicon and the use of this lexicon in word verification and word spotting.

Limited tasks can use simple, non phonological lexicons in which the words are represented by their phonetic transcriptions. But for speech understanding systems it is necessary to take into account phonological phenomena. Phonological rules can be stored in the lexicon or precompiled in order to yield all possible procunciations of words before use.

The lexical level is of primary importance in automatic speech recognition/understanding. It is used both for emitting word

hypotheses and for verifying or cancelling word hypotheses emitted by other processing levels.
Dynamic programming and related techniques are often used in this domain. Some uncertainties remain about whether word verification or spotting is most accurate at the phonetic or acoustic level, and best done only when the word is highly probable or for every word at every possible point in a sentence.

Finally much work is still to be done in lexicon representation and lexical retrieval techniques in order to make it possible to handle very large or unlimited vocabularies which now have to be considered.

REFERENCES

(1) P. COHEN and R. MERCER "The Phonological Component of an Automatic Speech Recognition System" pp. 275-320 in "Speech Recognition" D.R. REDDY ed., Academic Press 1975.

(2) B.T. OSHIKA et al. "The role of phonological rules in speech understanding research" IEEE Trans. ASSP, 23, n° 1, pp. 104-112, 1975.

(3) G. CAUSSE, G. GOUARDERES, G. TEP "L'utilisation du lexique pour la génération de phrases parlées" Rapport CERFIA, Oct. 1976.

(4) G. GOUARDERES "Organisation d'un lexique en vue de l'analyse de la parole en continu" Thèse CNAM,Toulouse,janvier 1977.

(5) J.M. PIERREL "MYRTILLE II : un système de compréhension du discours continu" Internal Report, CRIN, july 1978.

(6) J.P. HATON and J.M. PIERREL "Data-Structures and Architecture of MYRTILLE II Speech Understanding System" 4th IJCPR, Kyoto, Japan, Nov. 1978.

(7) J.P. HATON and G. PERENNOU "Reconnaissance automatique de la parole" AFCET Summer Institute, Namur, july 1978.

(8) J. SHOUP "Phonological Aspects of Speech Recognition" In "Trends In Speech Recognition" W.A.LEA ed.,Prentice-Hall,1979.

(9) D.E. WALKER, ed. "Understanding Spoken Language", North-Holland, 1978.

(10) J.P. PIERREL, J.F. MARI, J.P. HATON "Le niveau Lexical dans le système MYRTILLE II : représentation du lexique et traitements associés" 10è Journées d'Etude sur la Parole, GALF, Grenoble, Mai 1979.

(11) J.P. HATON, J.M. PIERREL "Organization and Operation of a Connected Speech Understanding System at Lexical, Syntactic and Semantic Levels" Proc.IEEE ICASSP, Philadelphia, April 1976.

(12) J.M. PIERREL "Contribution à la compréhension automatique du discours continu" Thèse de 3ème cycle, Université de Nancy I, Nov. 1975.

(13) B.T.LOWERRE "The HARPY Speech Recognition System" Ph D Dissertation, Carnegie-Mellon University,Pittsburgh,1976.
(14) W.A. WOODS and V.W. ZUE "Dictionary Expansion via Phonological Rules for a Speech Understanding System" Proc. IEEE ICASSP, Philadelphia, April 1976.
(15) M.S. FOX, F. HAYES-ROTH "Approximation Techniques for the Learning of Sequential Symbolic Patterns" Proc. 3d IJCPR, Coronado, Nov. 1976.
(16) D.H. KLATT "Review of the ARPA Speech Understanding Project" JASA, 62, p. 1345-1366, 1977.
(17) J.P. HATON "Une méthode dynamique de comparaison de chaînes de symboles de longueurs différentes : application à la recherche lexicale" C.R.Ac. Sci., A , 278, p. 1527-1530, 1974.
(18) D. FORNEY "The Viterbi Algorithm" Proc.IEEE,61, n°3, p. 268-278, 1973.
(19) G. M. WHITE "Dynamic Programming, the Viterbi Algorithm and Low Cost Speech Recognition" Proc. IEEE ICASSP, Tulsa, April 1978.
(20) F. JELINEK "Continuous Speech Recognition by Statistical Methods" Proc. IEEE,64,n°4, p. 532-556, 1976.
(21) J.K. BAKER "The DRAGON System-An overview" IEEE Trans. ASSP, 23, n° 1, pp. 24-29, 1975.
(22) D. SANKOFF "Shortcuts, Diversions, and Maximal chains in Partially Ordered Sets" Discrete Math., 4, n° 3, pp. 287-293, 1973.
(23) M. KOHDA, R. NAKATSU, K. SHIKANO "Speech Recognition in the Question-Answering System Operated by Conversational Speech" Proc.IEEE ICASSP, Philadelphia, April 1976.
(24) A.R. SMITH "Word Hypothesization in the Hearsay II Speech Understanding System" Proc. IEEE ICASSP, Philadephia, April 1976.
(25) W.A. WOODS et al. "Speech Understanding Systems, Final Technical Progress Report"Report n°3438 BBN, 1976.
(26) D.H. KLATT "Word Verification in a Speech Understanding System" in Speech Recognition, D.R. REDDY ed., Academic Press, 1975.
(27) M. HALLE, K.N. STEVENS "Speech Recognition : A Model and a Program for Research" IRE Trans. Inform. Theory, 8, p. 155-159, 1962.
(28) R. DEMORI, P. TORASSO "Lexical Classification in a Speech Understanding System Using Fuzzy Relations" Proc. IEEE ICASSP, Philadephia, April 1976.
(29) J.F. MARI "Contribution à l'analyse syntaxique et à la recherche lexicale en reconnaissance du discours parlé continu" Thèse de 3ème cycle,Université de Nancy 1, 1979.
(30) R. VIVES "L'analyse lexicale dans le système Kéal pour la reconnaissance de la parole continue" 7èmes Journées d'Etude sur la Parole, GALF, Nancy, Mai 1976.

(31) J.S. BRIDLE "An Efficient Elastic-Template Method for Detecting Given Words in Running Speech" Proc.Brit. Acoust. Soc. Meet. Tome 2, 1973.

(32) R.W. CHRISTIANSEN, C.K. RUSHFORTH "Detecting and Locating Key Words in Continuous Speech Using Linear Predic tive Coding" IEEE Trans. ASSP, 25, n° 5, 1977.

(33) S. CHIBA, M. WATARI, T. WATANABE "A Speaker-Independent Word Recognition System" 4th Int. Joint.Conf. Pattern Recognition, Kyoto, Nov. 1978.

(34) S. NAKAGAWA, T. SAKAI "A Word Recognition Method From a Classified Phoneme String in the Lithan Speech Understanding System" Proc. IEEE ICASSP, Tulsa, April 1978.

(35) M.F. MEDRESS et al. "An automatic Word Spotting System for Conversational Speech" IEEE ICASSP,Tulsa,April 1978.

(36) B. CAUSSE et al. "Evaluation d'une méthode ascendante d'analyse lexicale dans le discours continu" 7èmes journées d'Etude sur la Parole, GALF, Nancy, Mai 1976.

CONTROL OF SYNTAX AND SEMANTICS IN CONTINUOUS SPEECH UNDERSTANDING

W. A. Woods
Bolt Beranek and Newman Inc.
Cambridge, Massachusetts USA

1. Characteristics of the Speech Understanding Process

A naive view of speech understanding might consider it as a process of successively recognizing speech sounds (called phonemes), grouping phonemes into words, parsing word sequences into sentences, and finally interpreting the meanings of those sentences. However, considerable experience now indicates that the acoustic evidence present in the original speech signal is not sufficient to support such a process [Woods and Makhoul, 1974]. For sentences recorded from continuous speech, it is not generally possible to reliably determine the phonetic identity of the individual phonemes (or even to be sure how many phonemes are present) using the acoustic evidence alone. Experiments in spectrogram reading [Klatt and Stevens, 1971] indicate that the reliability of such determinations can be increased by use of the redundancy provided by knowledge of the vocabulary, the syntax of the language, and semantic and pragmatic considerations.

Tape splicing experiments [Wanner, 1973] seem to indicate that this low-level acoustic ambiguity is an inherent characteristic of continuous speech and not just a limitation of human spectrogram-reading. Specifically, intelligibility of individual words excised from continuous speech is very low, but the intelligibility increases when sequences of two or three words are used. It appears that the additional constraint of having to make sense in a larger context begins to resolve the ambiguities that were present when only the acoustic evidence was considered. This processing, however, happens below the level of introspection and has all of the subjective characteristics of a wholistic or Gestalt phenomenon. That is, if a sufficiently long

J. C. Simon (ed.), Spoken Language Generation and Understanding, 337-364.

sequence of continuous speech is heard, its correct interpretation usually appears immediately and effortlessly, without conscious awareness of the details of the process. The vast majority of our spoken communications are understood in this manner, and it is markedly contrasted with those cases where an utterance is garbled sufficiently to invoke conscious effort to decide what was said.

Recently, speech understanding research has taken a direction that recognizes the importance of syntactic and semantic constraints as an essential part of the process which deciphers speech signals into sequences of sounds [see Newell, et al., 1973]. Consequently, it has become important for speech researchers to be acquainted with the work that has been done in the area of computational linguistics, attempting to construct computer programs to model the process of natural language understanding. In this paper, I will attempt to provide an introduction to some of the ways that these "higher level" sources of knowledge can be used in speech understanding.

2. Syntactic and Semantic Analysis

There are two parts to the problem of syntactic and semantic analysis - one is a component of judgment or decision (whether a given string of words is a possible sentence or not), and the other is a component of representation or interpretation (deciding how the pieces of the sentence relate to each other and what they mean). In speech understanding, the former function is especially important. This judgmental function is critical in distinguishing possible word sequences that a speaker might have uttered from mere random sequences of words that happen to match the acoustic input. Without this ability to discriminate well-formed, meaningful sentences from "word salad", a speech understanding system would frequently (perhaps even usually) produce interpretations of the input that are incomprehensible.

Rules for expressing syntactic constraints on possible sentences can be expressed in several formal grammar models, such as context-free phrase structure grammars, transformational grammars [Chomsky, 1965], and augmented transition network (ATN) grammars [Woods, 1970]. A discussion of various grammar models and parsing methods is given in Woods [1975]. The ATN model is well suited both to expressing sophisticated grammars of natural language and to efficient computational use. In BBN's HWIM speech understanding system, [Woods et al., 1976], a parser was developed that can use an ATN to parse (from the middle out) an isolated fragment of an utterance and determine whether such a fragment is a possible fragment (not necessarily a well-formed constituent) of some complete sentence. Moreover, this algorithm could predict possible words and syntactic classes that could be used to extend such a fragment.

Semantic constraints on possible utterances (i.e., constraints that they be meaningful as well as grammatical) can be expressed by formal semantic interpretation rules [Woods, 1978a]. They can also be expressed as an ATN - either in a combined syntactic/semantic/pragmatic ATN (as in HWIM) or as a separate ATN in a cascade of ATN transducers [Woods, 1978b]. Since space here does not permit a full treatment of semantic rules and semantic interpretation, the reader is referred to the above two references for further details.

3. Theories, Monitors, Notices, and Events - A Computational Framework for Perception

The BBN speech understanding system [Woods et al., 1976; Wolf and Woods, 1977] has evolved within a general framework for viewing perceptual processes. Central to this framework is an entity called a **theory**. A theory represents a particular hypothesis about some or all of the sensory stimuli that are present. Perception is viewed as the process of forming a believable coherent theory which can account for all the stimuli. This is arrived at by successive refinement and extension of partial theories until a best complete theory is found.

In general, a high-level perception process requires the ability to recognize any member of a potentially infinite class of perceptible objects that are constructed out of elementary constituents according to known rules. That is, the object perceived is generally a compound object, constructed from members of a finite set of elementary constituents according to some kind of well-formedness rules. These elementary constituents, as well as the relationships among them that are invoked in the well-formedness rules, must be directly perceptible. Thus, a perceptual system must incorporate some basic epistemological assumptions about the kinds of things that it can perceive and the rules governing their assembly. The well-formedness rules can be used to reject impossible interpretations of the input stimuli, and may also be useable to predict other constituents that could be present if a given partial theory is correct.

This perception framework assumes mechanisms for using subsets of the input stimuli to form initial "seed" hypotheses for certain elementary constituents (stimulus-driven hypothesization) and mechanisms for deriving hypotheses for additional compatible constituents from a partial theory (theory-driven, or predicted, hypothesization*). It also assumes mechanisms for verifying a hypothesis against the input stimuli and evaluating the well-formedness of a compound hypothesis to assign it some measure of quality and/or likelihood. A theory may therefore be thought of as a hypothesis that has been evaluated in this way and assigned a measure of confidence.*

In the case of speech understanding, a theory can range from an elementary hypothesis that a particular word is present at a particular point in the input (a **word match**) to a complete hypothesis of a covering sequence of words with a complete syntactic and semantic interpretation. (In general, a theory can be a set of compatible word hypotheses with gaps between them and with partial syntactic and semantic interpretations.) A partial theory may be able to generate predictions for appropriate words or classes of words either adjacent to the words already hypothesized, or possibly elsewhere in the utterance.

Predictions are dealt with in our computational framework by two kinds of devices: **monitors**, which are passively waiting for expected constituents, and **proposals**, which are elementary hypotheses that are to be evaluated against the input. Proposals result in actively seeking stimuli that would verify them, while monitors passively wait for such hypotheses to be formed. The functioning of monitors assumes that there is an organizing structure into which all derived partial hypotheses are placed as they are discovered and that the monitors can essentially set "traps" in this structure for the kinds of events that they are watching for. This is to be contrasted with continuous parallel evaluation of special processes (frequently called "demons") to watch for expected patterns in the input stream. Monitors perform no computation until and unless some other process makes an entry of the kind they are waiting for in some data structure.

The functioning of monitors is illustrated by an early speech understanding system at BBN dealing with concentrations of chemical elements in lunar rocks. There, for example, a word match for "concentration" would set monitors on the concept nodes for SAMPLE and CHEMICAL ELEMENT in a semantic network. If a word such as "Helium" was subsequently found anywhere else in the utterance, a check in the semantic network starting with Helium would lead to the superset category CHEMICAL ELEMENT where it

---

* Our notion of stimulus-driven hypothesization is essentially the same as that of "bottom-up" processing referred to in many discussions of such processes. However, our notion of theory-driven hypothesization is slightly different from the sense usually given to "top-down" processing in that it does not necessarily imply any global ("topmost") hypothesis, but only predictability by some other hypothesis, which may itself have been derived "bottom-up". The terms "top-down" and "bottom-up" in this sense come from the literature on formal parsing algorithms.

would wake up the monitor from "concentration", thus detecting the coincidence of a detected hypothesis and a predicted hypothesis [Nash-Webber, 1975].

When a monitor is triggered, an **event** is created calling for the evaluation of a new hypothesis and the creation of a new theory if the hypothesis is not rejected. In general, a number of events are competing for service by the processor at any moment. In human perception, there may be full parallel processing of such events, but in a serial machine, these events must be queued and given processing resources on the basis of some priority ordering. (Even in human perception, there is probably some sort of priority allocation of resources, since various kinds of interference can occur.) In our computational framework, events are maintained on a queue in order of priority, the top event being processed at each step.

The processing of an event can result in new proposals being made, new monitors being set, and existing monitors being triggered to produce new events. Since so much hinges on the event chosen for processing, a major issue is that of assigning priorities to events in order to find the most likely interpretation of the input. In the BBN system, priority scores are assigned on the basis of Bayesian estimates of the probabilities of the competing theories, and certain control strategies and priority scoring metrics can be guaranteed to discover the most probable interpretation of the input.

4. Control Strategies

The above discussion leaves open issues such as when should initial one-word "seed" theories be formed, how many should be considered, should all seeds be worked on in parallel, etc. These issues we refer to as control issues. They have been critically important in computerized speech understanding systems. In the BBN system, for example, there are a variety of different control strategies that all fit within the above paradigm. Figure 1 illustrates one class of strategies in which seeds are formed anywhere in the utterance that sufficiently salient word matches are found. The figure shows the seed events formed as a result of an initial scan of an utterance for high likelihood word matches anywhere in the utterance. Each theory is assigned a score expressing its likelihood of being correct (actually a logarithm of the ratio of the likelihood of the acoustic evidence given the theory over the a priori likelihood of that evidence occurring independently). The region of the utterance covered by the theory is indicated by specifying its left and right boundary positions in a list of potential boundary positions (the left end of the utterance is numbered 0 and in this case the right end is numbered 18). The exclamation marks

#	SCORE	REGION	THEORY
1	0	11-14	ADD
2	0	4-7	NEED
3	-.455	0-3	SHOW !
4	-.605	12-17	TRIP
5	-.727	1-5	ROME
6	-.769	8-11	THERE
7	-1.25	12-18	TRIP-S !
8	-1.47	0-5	SHELLY
9	-1.65	15-17	END
10	-1.72	11-14	AND
11	-1.73	1-5	ANN
12	-1.74	0-5	CHEYENNE
13	-2.19	8-14	BERT
14	-2.26	2-6	ANY
15	-2.82	0-5	SOME
		+15 ADDITIONAL EVENTS	

Fig. 1. Seed events for middle-out strategy.

indicate the theories that are actually part of the correct interpretation.

For this general class of control strategies, referred to as "middle-out" or "island-driven" strategies, theories are grown by starting with a seed word, asking a higher-level linguistic component to predict categories of words that can occur on either side of it, asking a lexical retrieval component to find the best matching words in those categories on the appropriate sides, and generating events for each such word found to extend the theory by adding that word. Thus, events will be placed on the event queue to add words both on the left and on the right ends of given theories. These "new word" events will compete with each other and with the remaining seed events on the basis of score to determine which event will be processed next, causing the processor to sometimes continue adding words to a given theory and at other times to shift its processing to a different competing theory.

Figure 2 shows the sequence of theories that are formed as a result of this process, starting with the event queue of Figure 1. (Brackets in the figure indicate theories that include the hypothesis that the left or right ends of the utterance have been reached. A number in parentheses after a theory number is the number of a preceding theory from which the indicated theory was formed by the addition of a new word.) Notice that the final theory is developed in this case by working independently on two different portions of the utterance starting from the seeds "show" and "trips". The final theory in Figure 2 is in fact derived from a kind of event called a **collision** event which combines the theories "show me" and "trips" when they both notice the word "her" filling the gap between them. This event is formed during the processing of theory 13, although its score is such that it does not reach the top of the queue until theory 23.

Figure 3 shows the seed theories for a hybrid strategy in which seeds are started within a bounded distance from the left end of the utterance, and are grown right-to-left until they reach the left end, after which the remainder of the processing is left-to-right. Figure 4 shows the sequence of theories developed in the course of understanding this utterance using the hybrid strategy. The basically left-to-right nature of the hybrid strategy, except for a bounded initial delay in getting started, seems to be a reasonable possibility for a model of human speech understanding, since it is clear that human processing of speech does not involve the buffering of a complete sentence before understanding begins.

5. Priority Scoring

The scoring assigned to a theory by the summation of individual word scores (essentially the log probability of its words being correct), we refer to as the **quality score** of the theory. We distinguish from this a possibly separate score called the **priority score**, which is used to rank order events on the event queue to determine the order in which they are to be done. In early versions of HWIM, we used the quality score itself as the priority score. However, we have developed several algorithms with interesting theoretical properties using priority scores that are derived from, but not identical with, the quality score. The first measures the difference between the particular quality score for a theory and an upper bound on possible quality score for any theory covering the same portion of the utterance. We call this the **shortfall score**, and it can be shown that using the shortfall score as a priority score under appropriate conditions guarantees finding the best scoring interpretation of the input utterance [Woods et al., 1976, Woods, 1977]. Using the quality score itself as a priority score does not guarantee this. Other priority scores are obtained by dividing either the quality score or the shortfall score by the time duration of the island to give **quality density** and **shortfall density scoring**, respectively. Since a fairly complete derivation of the shortfall and shortfall density scoring strategies together with

```
THEORY#1                            ADD
THEORY#2                    NEED
THEORY#3                SHOW !
THEORY#4(3)             [SHOW !
THEORY#5(4)             [SHOW ALL
THEORY#6(3)             SHOW ALL
THEORY#7                            TRIP
THEORY#8                 ROME
THEORY#9(4)             [SHOW ME !
THEORY#10(3)            SHOW ME !
THEORY#11(7)                    HER TRIP
THEORY#12                           TRIP-S !
THEORY#13(12)                       TRIP-S] !
THEORY#14(13)                   HER TRIP-S] !
THEORY#15(12)                   HER TRIP-S !
THEORY#16               SHELLY
THEORY#17(16)           [SHELLY
THEORY#18               [SHOW ME HER TRIP
THEORY#19               SHOW ME HER TRIP
THEORY#20(11)                   HER TRIP IS
THEORY#21(20)                   HER TRIP IS]
THEORY#22(20)                OF HER TRIP IS
THEORY#23(9,13)         [SHOW ME HER TRIP-S] !
```

Fig. 2. Theories formed for middle-out strategy.

proofs of their theoretical properties is given in [Woods, 1977], we will present here only a brief recapitulation of the strategies, and discuss differences we have observed between them.

5.1 Shortfall Scoring

The shortfall score measures the amount by which the quality score of a theory falls below an upper bound on the possible score that could be achieved on the same region. When shortfall scoring is being used, a MAXSEG profile is constructed having the property that the score of a word match between boundaries i and j will be less than or equal to the area under the MAXSEG profile from i to j (call this latter the MAXSCORE for the region from i to j). The shortfall score for a theory is then computed as the sum over all the word matches in the theory of the difference between the score of the word match and the MAXSCORE for the same region. The preferred theory is the one with the smallest magnitude of shortfall.

The MAXSEG profile can be constructed incrementally by adding to the profile whenever a word match is found whose score

#	SCORE	REGION	THEORY
1	3.53	1-5	WHO
2	1.92	3-6	WE !
3	0.0	0-1	-PAUSE- !
4	-2.43	2-3	A
5	-3.24	5-10	ELEVEN
6	-4.32	5-9	IRAQ
7	-5.36	1-3	HER
8	-6.00	1-4	WHOLE
9	-6.18	1-5	DO !
10	-6.21	3-6	WERE
11	-6.53	3-7	WORK
12	-6.85	1-4	HIS
13	-7.00	1-5	HOW
14	-7.12	1-6	HAWAII
15	-7.21	3-6	WHERE
	+39 ADDITIONAL EVENTS		

Fig. 3. Seed events for hybrid strategy.

is not bounded by it. Whenever the score of a word match exceeds the MAXSCORE for its region, the excess score is distributed over the region to raise its MAXSCORE to that of the word match. In HWIM, an initial profile is constructed during the initial scan for seed words and this profile is substantially correct. Occasionally, a word match is found later which raises the profile, and in this case all events overlapping the changed region are rescored.

In order to satisfy the theoretical claims of the algorithm, the way in which the excess score of a word match is distributed to raise the MAXSEG profile does not matter. However, it is desirable to do it in such a way as to minimize the amount by which the shortfall of other words that overlap the region is raised. Our current algorithm is to distribute the excess score over the segments covered by the word match that are not already bounded by the profile and to divide it proportional to the durations of the segments. Other distribution algorithms are possible, some of which have been tried. This one is better than some, but there are probably better strategies to be found. Keeping the MAXSEG profile as low as possible while still satisfying the upper bound condition is important since excessively conservative upper bounds translate directly into an unnecessary increase in the breadth-first nature of the search, requiring more events to be processed before finding the chosen interpretation.

```
THEORY#1          WHO
THEORY#2(1)       [WHO
THEORY#3            WE !
THEORY#4(3)       DO WE !
THEORY#5(4)       [DO WE !
THEORY#6(3)         [WE !
THEORY#7(5)       [DO WE HAVE !
THEORY#8(7)       [DO WE HAVE A !
THEORY#9         -PAUSE- !
THEORY#10(9)     [-PAUSE- !
THEORY#11(3)       ARE WE
THEORY#12(8)      [DO WE HAVE A SURPLUS !
THEORY#13(12)     [DO WE HAVE A SURPLUS] !
```

Fig. 4. Theories formed for hybrid strategy.

The theoretical characteristics of the shortfall scoring algorithm are that if the words are returned by the Lexical Retrieval component in decreasing order of quality and events are processed in order of increasing magnitude of shortfall (plus a few other assumptions, documented in Woods [1977]), then the first complete spanning interpretation found will be the best scoring interpretation that can be found by any strategy. We refer to this condition as "completeness" (a more traditional term is "admissibility"). For speech understanding applications, completeness is a desirable property, but not necessarily essential if the cost of its attainment is too great. Shortfall scoring has the property of being complete without searching the entire space. Its completeness proof depends only on the fact that when the first complete spanning theory is found, all other events on the queue will already have fallen below the ideal maximum score by a greater amount. Thus, the result does not depend on the scores being likelihood ratios, nor does it make any assumption about the nature of the grammar (e.g., that it be a finite state Markov process) provided a parser exists that can make the necessary judgments. The completeness also does not depend on the order of scanning the utterance - it is satisfied both for middle-out and for left-to-right strategies.

5.2 Density Scoring

Another type of priority scoring is density scoring. Here the score used to order the event queue is some basic score divided by the duration of the event. Conceptually, we can think of this priority scoring metric as predicting the potential score for the region not covered by a theory to be an extrapolation of the same score density already achieved. (In these terms, the shortfall strategy can be thought of as predicting that the upper bound for the uncovered region will be achieved.) Unlike the shortfall scores, density scores can get bad and then get better again as new words are added to a theory. Hence, the density score is certainly not guaranteed to be an upper bound of the expected eventual score. However, it has another interesting property: in exactly those cases where it does not bound the eventual score, there is a word to be added somewhere else that has a better score density and whose score density does bound the eventual score. This arises from the property of densities that the density of two regions combined will lie between the densities that they each have. It turns out that this alone is not sufficient to guarantee completeness for a density scoring strategy since it is still possible for the density score starting from the best correct seed to fall below that of some other less-than-optimal spanning theory before it can be extended to a complete theory itself. However, with the addition of a facility for island collision that start from separate seeds when they collide with each other, the density scoring strategy, working middle-out from multiple seeds can be shown to be complete. Again, density scoring does not depend on any assumptions about the basic scores to which it is being applied other than that they be additive (and capable of division). Hence, the density method can be applied to either the original quality score or to a shortfall score. The combination of the two methods in a shortfall density strategy seems to be more effective then either shortfall or density scoring alone.

6. Other Heuristics

In addition to the basic choice of priority scoring metric used for ranking the event queue, there are several additional heuristics that can be used to improve the performance of the island-driven strategies (although sometimes with a loss of admissibility guarantees). Two of these are the use of "ghost" words, and the selection of a preferred direction for events from a given theory.

6.1 **Ghost Words**

The ghost words option is a feature that can be added to any island-driven strategy, although it may affect some admissibility of the strategies to which it is added. Every time a theory is given to the linguistic consultant for evaluation, proposals are made on both sides of the resulting island (unless the island is already against one end of the utterance). Although events can only add one word at a time to the island, and this must be at one end or the other, eventually a word will have to be added to the other end, and that word cannot score better than the best word that was found at that end the first time. The ghost words feature consists of remembering with each event the list of words found by the Lexical Retrieval component at the other end and scoring the event using the best of the ghost words as well as the words in the event proper. The result is that bad partial interpretations tend to get bad twice as fast, since they have essentially a one-word look ahead at the other end that comes free from the linguistic consultant each time an event is processed. On the other hand, an event that has a good word match at the other end gets credit for it early so that it gets processed sooner. The ghost words feature, thus, is an accelerator that causes extraneous events to fall faster down the event queue and allows the desired events to rise to the top faster. Experimental use of this feature has shown it to be very effective in reducing the number of events that must be processed to find the best spanning event.

6.2 Choosing a Direction Preference

When a theory is evaluated by the linguistic consultant, predictions are made at both ends of the island. When one of the events resulting from these predictions is later processed, adding a new word to one end of the island, the predictions is later processed, adding a new word to one end of the island, the predictions at the other end of the new island will be a subset of the predictions previously made at that end of the old island. In general, words noticed by this new island at that end will also have been noticed by the old island, and if the score of the new island is slightly worse than that of the old island (the normal situation), then the strategy will tend to revert to the old island to try picking up a word at the other end. This leads to a rather frustrating derivation of a given theory by first enumerating a large number of different subsequences of its final word sequence. For example, to derive a theory (abcd), one might first start with the seed (b), then add a to get (ab), then go back to (b) to get (bc) then go back to (ab) and add c to get (abc), then go back to (bc) and notice a, but also notice that (abc) has already been made and then go back to (abc) to pick up the d.

Since any eventual spanning theory derived from an island must eventually pick some word at each end of the island, one could arbitrarily pick either direction and decide to work only in that direction until the end of the utterance is encountered, and only then begin to consider events in the other direction. This would essentially eliminate the duplication described above, but could cause the algorithm to work into a region of the utterance where the correct word did not score very well without the benefit of additional syntactic support that could have been obtained by extending the island further in the other direction for a while.

Without sufficient syntactic constraint at the chosen end, there may be too many acceptable words that score fairly well for the correct poorly scoring word to occur within a reasonable distance from the top of the queue. By working on the other end, one may tighten that constraint and enable the desired word to appear, although this can never cause a better scoring word to appear than those that appeared for the sorter island.

The CHOOSEDIR flag in the HWIM system causes the algorithm to pick a preferred or "chosen" direction for a given theory as the direction of the best scoring event that extends that theory, and to mark the events going in the other direction from that theory so that they can only be used for making tighter predictions for words at the chosen end. This is accomplished by blocking from consideration any notices for one of the ghost words at the inactive end of an event if that event is going counter to the chosen direction. This blocking, alone, eliminates a significant number of redundant generations of different ways to get to the same theory. An even greater improvement is obtained by rescoring the events that are going counter to the chosen direction by using the worst ghost at the other end rather than the best ghost. Since only word matches that score worse than any of the ghosts at that end will be permitted by these events, this is a much better estimate of the potential score of any spanning theories that might result from these events

The effect of rescoring the events in the non-chosen direction using the worst ghost is that, in most cases, these events fall so low in the event queue as to be totally out of consideration. Only in those cases where there was little syntactic constraint in the chosen direction and the worst matching word at that point was still quite good, do these events stay in contention, and in those cases, the use of the worst ghost score provides the appropriate ranking of these events in the event queue.

7. Empirical Comparison of the Different Strategies

In the HWIM Speech understanding system, approximations to the above algorithms have been implemented. Details of the HWIM system and its general performance are found in [Woods et al., 1976]. Comparative performance results on a set of 10 utterances for the shortfall (S), shortfall density (SD), and quality density (QD) scoring strategies are shown in Table 1 below. The option of using the quality score (Q) alone as a priority score is given for comparison.

Table 1. Comparison of different priority scoring functions.

	Q	QD	S	SD
Correct first interpretation	4	3	0	5
Incorrect first interpretation	2	0	0	0
No response	4	7	10	5
Average number of theories processed*	49	82	100	73

These experiments were run using the ghosts, island-collision, and preferred direction heuristics with a resource limit of 100 theories to process before the system would give up with no response. The ten sentences used for the test were chosen at random from a test set of 124 recorded sentences.

Although a test set of only ten utterances is admittedly too small, I believe that the trends indicated in the figure are generally correct. Specifically, while the quality score as a priority alone leads to a spanning interpretation in relatively fewer theories, it does so without much assurance of getting the best interpretation. In this case, only two out of three of its answers are correct. All of the other methods spend additional effort in making sure that the best interpretation is found, and consequently found fewer spanning interpretations within the resource limitation. We did not try running the quality scoring strategy beyond the first interpretation to see if a better interpretation could be found since, among other things, it is not clear when to terminate such a process. Running in this mode, one could easily enumerate more theories than the other methods and still not have any guarantee that the best interpretation had been discovered.

---

* Average computed over 9 sentences, omitting one for which the system broke due to a bug.

None of the approximations of admissible algorithms found incorrect interpretations, so the reliability of their interpretations when they get them is 100% (providing the acoustic phonetic analysis of the input utterance does not cause some incorrect interpretation to score higher than the correct one, a situation that occurs sometimes in the HWIM system, but is not a factor in this experiment). Unfortunately, the shortfall strategy alone is so conservative in doing this that it failed to find any interpretations within the resource limit. Both of the density methods are clearly superior to the straight shortfall method.

The shortfall density strategy ranked superior to the quality density strategy in terms of the number of events that needed to be processed to find the first spanning interpretation and consequently found more correct interpretations within the resource limitations.

The effects of the island collision (C), ghosts (G), and preferred direction (D) heuristics are shown in Table 2 (where SD+0 means shortfall density without collisions, ghosts, or chosen direction, SD+C means shortfall density with island collisions, etc.). The inclusion of a heuristic does not always guarantee that the system will understand an utterance in fewer theories, but the pooled results shown (note especially the series SD+0, SD+G, SD+GD, SD+GDC) suggest that the successively added heuristics produce improvements in both accuracy and number of theories required.

Table 2. The effects of island collisions, ghosts, and direction preference.

	SD+0	SD+C	SD+G	SD+GD	SD+GDC
Correct	3	3	3	4	5
Incorrect	0	0	0	0	0
No response	7	7	7	6	5
Average number of theories processed*	83	78	81	76	69

* Average computed over the 8 of the ten sentences - omitting two for which the system broke due to bugs.

8. Comparison with other Speech Understanding Strategies

8.1 BBN HWIM

The admissible strategies presented in this paper are only some of the control strategy options implemented in the BBN HWIM speech understanding system. In addition there are a large number of strategy variations that result in approximate strategies, including strictly left-to-right strategies and "hybrid" strategies that start near the left end of an utterance and work left and then right. For reasons of time and resource limitations, our final test run of the HWIM system was made using one of the approximate strategies. Subsequently, a much smaller experiment was run to compare various control strategies on a set of ten utterances chosen at random from the larger set. Although this sample is much too small to be relied on, the results are nevertheless suggestive. For two comparable experiments using our best left-to-right method (left-hybrid shortfall density) and our best island-driven method (shortfall density with ghosts, island collisions, and direction preference), both with a resource limitation of 100 theories and without using a facility for analysis-by-synthesis word verification, the results were as follows:

	LHSDNV	SD+GCD
Correct interpretation	6	5
Incorrect interpretation	2	0
No interpretation	2	5
Average number of theories evaluated	50.7	75.5

That is, the left-to-right strategy found the best (and in these cases the correct) interpretation within the resource limitation in 6 of the 10 cases, while the island-driven strategy found only 5 (not necessarily a significant difference for this size sample). On the other hand, the left-to-right method misinterpreted two additional utterances with no indication to distinguish them from the other 6. If this strategy were used in an actual application with comparable degrees of acoustic degradation (e.g., due to a noisy environment), the system would claim to understand 80% of its utterances, but would actually misunderstand 25% of those. The island driven strategy, on the other hand, would only claim to understand 50% of the utterances, but would misunderstand a negligible fraction.

The island driven strategy in the above experiments expanded only 50% more theories (and incidentally used only 30% more cpu time) than did the left-to-right strategy. Although as we said before, this test set is much too small to draw firm conclusions, the success rate of the two methods are not much different, except that the island-driven shortfall density method is clearly less likely to make an incorrect interpretation. Moreover, the numbers of theories considered and the computation times are not vastly different. If one considers proposals to improve the performance of left-to-right strategies by having them continue to search for additional interpretations after the first one is found (and thus take the best of several), then the time difference shown above could easily be reversed and there would still be no guarantee that the best interpretation found would be the best possible.

8.2 DRAGON

The DRAGON system [Baker, 1975] is the only other speech understanding system in the ARPA project that provides a guaranteed optimum solution. It does this by using a dynamic programming algorithm that depends on the grammar being a Markov process (i.e. a finite-state grammar). It operates by incrementally constructing, for each position in the input and each state in the grammar, the best path from the beginning of the utterance ending in that state at that position. The computation of the best paths at position i+1 from those at position i is a relatively straightforward local computation, although the number of operations for each such step, for a grammar with n states, is n times the branching ratio (i.e., the average number of transitions with non-zero probability leaving a state). DRAGON performs such a step for each 10 millisecond portion of the utterance using a state transition that "consumes" an individual allophonic segment of a phoneme.

The optimality of the solution found by this algorithm depends on the property of finite state grammars that one sequence of words (or phonemic segments) leading to a given state * is equivalent to any other such sequence as far as compatibility with future predictions is concerned (regardless of the particular words used). It is this property that permits the algorithm to ignore all but the best path leading to each state (even if competing paths score quite well!), and therefore permits it to find the best solution by progressively extending a bounded number of paths across the utterance from left to right. (This is a very attractive property, although in this case it requires one such path for each state in the grammar.) For more

---

* I am using the term "state" a little casually here in roughly the sense that it is used in an ATN grammar [Woods, 1970]. If one takes the condition of having equivalent future predictions as the definition of a "state" of a grammar, then what the finite-state grammar does is guarantee that there are only a finite number of such states, which can therefore be enumerated and named ahead of time. For a more general grammar, the number of such states is open-ended.

general grammars, where there may be context-sensitive checking between two different parts of the utterance (e.g., number agreement between a subject and a verb), the best path leading to a given state at a given position may not be compatible with the best path following it. In this case, second best (and worse) paths must be considered in order to guarantee finding any complete paths at all (much less an optimum).

Although only applicable to finite-state languages, DRAGON's dynamic programming method has the advantage of taking a relatively constant amount of time from utterance to utterance, being simple to compute, and guaranteeing to obtain the optimal solution. The only difficulty is that for a large number of states in the grammar (e.g. thousands for a reasonable size grammar) the amount of computation required is expensive. Except to the extent that the finite-state grammar permits one to eliminate from consideration any path that is not the best one leading to its state, the algorithm exhaustively enumerates all other possibilities.

Although DRAGON's scores are estimates of probabilities of interpretations, its guarantee of optimality does not depend on that, but only on the fact that its grammar is finite-state and that therefore it suffices to carry a record of the best path leading to each state. The same dynamic programming algorithm can be applied at the level of phonemes or words, and can be generalized to apply to an input lattice such as the BBN segment lattice [Woods et al., 1976].

8.3 HARPY

The CMU HARPY system [Lowerre, 1976] is a development on the DRAGON theme which gives up an absolute guarantee of optimality in exchange for computation speed. Like DRAGON, it takes advantage of the unique characteristic of finite-state grammars cited above, so that only the best path leading to a given state need be considered. However, it uses an adaptation of the dynamic programming algorithm in which not all of the paths ending at a given position are constructed. Specifically, at each step of the computation, those paths scoring less than a variable threshold are pruned from further consideration. (In the Itakura-metric version of the system, setting this threshold at 1/100000 of the score of the best path at that point was reported to give the best performance.) This gives an algorithm that carries a number of paths in parallel (the number varying depending on the number of competitors above the threshold at any given point) but is not exhaustive. If the threshold is chosen appropriately, the performance closely approximates that of the optimal algorithm, although there is a tradeoff between the speed efficiency gained and the chances of finding a less than optimal path.

In 1976, the HARPY system had the best demonstrated performance statistics of any continuous speech understanding system. However, it derived this performance in large part from the use of a highly constraining (and advantageously structured) finite-state grammar. This grammar has an average branching ratio of approximately 10, and characterizes a non-habitable, finite set of sentences, with virtually no "near miss" sentence pairs included. For example, "What are their affiliations" is in the grammar, but no other sentences starting with "What are their" are possible. The only two sentences starting with "What are the" are "What are the titles of the recent ARPA surnotes," and "What are the key phrases." These three sentences will almost certainly find some robust difference beyond the initial three words that will reliably tell them apart. Similarly, the grammar permits sentences of the form "We wish to get the latest forty articles on <topic>," but one cannot say a similar sentence with "I" for "we", "want" for "wish", "see" for "get", "a" for "the", "ten" for "forty", or any similar deviation from exactly the word sequence given above.) Most of HARPY's grammar patterns (such as the last one) consist of a particular sentence with one single open category for either an author's name or a topic. A large number of them are particular sentences with no open categories (like the first three above).

The HARPY algorithm makes no guarantee that the correct path will not be pruned from consideration if it starts out poorly, but at least for the structure of HARPY's current grammar (most of whose sentences start with stressed imperative verbs or interrogative pronouns), the correct interpretation is usually found.

The HARPY technique (or variations of it) seems to be the algorithm of preference at present for applications involving carefully structured artificial languages with finite-state grammars and small branching ratios (on the order of 10 possible word choices at each position in an utterance). However, it does not conveniently extend to larger and more habitable grammars. This is due to a number of factors, including: the combinatorics of expanding them into a finite-state network (the branching ratio 10 grammar on which its best performance is reported is about the largest HARPY could hold in its memory), the approximations necessary to represent such a grammar as a finite-state network (most such grammars are at least context-free and usually context sensitive - so that finite-state approximations necessarily accept sentences that the original grammar doesn't or fail to accept some that it does), and the difficulty of dealing with dynamically changing situations such as constraints on utterances that depend on previous sentences.

Neither the DRAGON nor the HARPY system use density normalization or any method to attempt to estimate the potential score that is achievable on the as yet unanalyzed portion of the utterance. Such normalization is not necessary, since they both follow paths in parallel, all of which start and end at the same point in the utterance, and therefore never have to compare paths of different lengths or in different parts of the utterance. Again, it is worth emphasizing that the ability of these algorithms to keep the number of paths that need to be considered manageable depends on the unique characteristic of finite-state languages that requires only the best path to each state be considered.

8.4 IBM

A group at IBM [Bahl et al., 1976] has a speech understanding system based on Markov models of language, which has implemented two control strategies: a Viterbi algorithm (essentially the same dynamic programming algorithm used by DRAGON) and a "stack decoder", a left-to-right algorithm with a priority scoring function that attempts to estimate the probability that a given partial hypothesis will lead to the correct overall hypothesis. The latter apparently does not guarantee the optimal interpretation, but somehow is reported as getting more sentences correct than the other (a circumstance I don't fully understand, but which can happen if there are acoustic-phonetic scoring errors such that the best scoring interpretation is not correct).

Recent experiments with an improved version of one of the IBM systems (incorporating the CMU technique of bypassing a phonetic segmentation to do recognition on fixed length acoustic segments [Bahl et al., 1978] reported performance on the same grammar used in the HARPY system (the "CMU-AIX05 Language") of 99% correct sentence understanding. (This performance is based on recordings in a noise-free environment, however, compared to a rather casual environment for the CMU results). They also report performance of 81% correct sentence understanding on a more difficult, but still small branching ratio, finite-state grammar (their "New Raleigh Language"). Both of these results were obtained in experiments with the system trained for a single speaker and tested on that same speaker. Performance of the system when tested with a different speaker is significantly less.

8.5 Hearsay II

The Hearsay II system [Lesser et al., 1975] permits the kind of generalized middle-out parsing described in this paper, and does so for context free grammars (although apparently not for context-sensitive or more powerful grammars). Moreover, it has a capability for a kind of island collisions [Erman--personal communication]. However, its design philosophy specifically rejects the use of an "explicit control strategy" as "inappropriate" (because it "destroys the data-directed nature and modularity of knowledge source activity") [Hayes-Roth & Lesser, 1976]. It's scoring function for hypotheses, which its authors refer to as the "desirability" of a KS (knowledge source), is an ad hoc combination of functions reflecting intuitive notions of "value", "reliability", "validity", "credibility", "significance", "utility", etc. Specifically, they state: "the desirability of a KS invocation is defined to be an increasing function of the following variables: the estimated value of its RF (an increasing function of the reliability of the KS and the estimated level, duration, and validity credibility of the hypothesis to be created or supported); the ratio of the estimated RF value to the minimum current state in the time region of the RF; and the probability that the KS invocation will directly satisfy or indirectly contribute to the satisfaction of a goal as well as the utility of the potentially satisfied goal." [Hayes-Roth & Lesser, 1976].

They go on to say that the above is not "complex enough" to "provide precise control in all of the situations that arise," and proceed to describe various further elaborations, all of which are vague as to exactly what the system does.

Although it is extremely difficult to tell from the available published descriptions exactly what Hearsay II does, the fact that the "desirability" of a KS invocation is an increasing function of its duration definitely rules out any interpretation of it as implementing the density method. The above allusion to the "current state in the time region of the RF" refers to a parameter that for each point t in the utterance specifies the maximum of the "values" of all hypotheses "which represent interpretations containing the point t." This "state" function at first glance seems similar to the maxseg profile used in the shortfall algorithm (and indeed was what caused me to start thinking along those lines), but in actuality it is quite different. Instead of being an estimate of the maximum possible portion of a score that can be attributed to a segment, Hearsay's state is the maximum total score of any hypothesis found so far that covers it (recall that such scores increase with length of the theory). Its contribution to the desirability of a hypothesis is the ratio of the "value" of that hypothesis to the smallest value of the state parameter in its region.

Since the smallest state value in the region of a hypothesis will always be at least as great as that of the hypothesis being valued (each state is the max value of all covering hypotheses), this ratio is always less than or equal to one, and is strictly less only when every portion of the region covered by the hypothesis has some better covering hypothesis (although not necessarily a single hypothesis that covers the whole region). Consequently, the "state" component of the score has the effect of inhibiting a hypothesis that at every point has a better competitor. Since the values of hypotheses grow with the length of the region covered, the effect will be that hypotheses that get big early will inhibit alternative hypotheses on the regions they cover. With shortfall scoring, on the other hand, the tendency is for big hypotheses to pick up additional shortfall and increase the likelihood of a shift to a competing hypothesis that might ultimately get a better score (this is what makes it an admissible algorithm). Hearsay II's use of the "state" parameter, is more reminiscent of SRI's "focus by inhibition" technique discussed below, which was found to have generally undesirable effects, although it did offset some of the costs of their island driving strategy [Paxton, 1977].

In summary, it is difficult to say exactly what the Hearsay focus of attention strategy does, or how it relates to the methods presented here, except to say that it is certainly not the same as any of these methods.

A superficial comparison of the Hearsay II system performance with that of the BBN HWIM system might lead one to believe that the Hearsay II control strategy is somehow more effective. However, it is more likely that the difference in performance is due to the differences in difficulty of the two grammars or to differences in their acoustic "front end." The reported performance results of the Hearsay II system are based on the same highly constrained, branching ratio 10 grammar used by HARPY. The BBN grammar, on the other hand, is a general ATN grammar with average branching ratio (measured from hypothesis predictions in a running system) of 196, permitting a relatively habitable subset of English which includes such minimal pairs as "What is the registration fee" and "What is their registration fee." Informal conversation with members of the Hearsay II project convinces me that Hearsay II can in principle explore all the alternatives that the SD+C strategy would and would in fact explore at least these if functioning according to its design philosophy of finding a first interpretation and then exploring further any hypotheses that could produce something better.

8.6 The SRI Experiments

At SRI, Paxton [1977] has performed a number of experiments on control strategy options, using a simulated word matching component based on performance statistics of the SDC word matching component to which a speech understanding system at SRI was originally intended to be coupled. Paxton's system is well-documented, and contains a number of interesting and well-done capabilities. He has worked out a very clean representation of the SRI grammar as a collection of small ATN networks (although he doesn't call them that) which do not have the directional left-to-right orientation that conventional ATN's do and in which the association of augments with transitions is more systematized and less procedural. The capabilities of this system for syntactic/semantic/pragmatic constraint are comparable in power to that of HWIM's general ATN grammar, and in several respects the notations used are cleaner and more perspicuous than one usually finds in a conventional ATN. Moreover, the implementation of these grammars contains some very elegant efficiency techniques. The system has a capability for middle out parsing making use of the semantic/pragmatic augments in the grammar, although it doesn't seem to have a capability for island collisions and doesn't construct islands for arbitrary sentence fragments.

In terms of the control strategy framework set up in this paper (as opposed to the terms that he himself uses), Paxton's system makes a distinction between a quality score for a hypothesis and a priority score for an event, although the kinds of hypotheses and events that his system creates are somewhat different than those above. One way of viewing his system in the terms presented here is that his hypotheses are always partially completed constituents (what he calls "phrases), which can make predictions for the kinds of words or constituent phrases that they can use. These phases are incorporated into a structure called a "parse net" in which explicit "producer" and "consumer" links associate such hypotheses to each other, but partially completed phrases are not combined into larger sentence fragments corresponding to our notion of islands (which can be partial at several levels of phrase structure). His events are of two types: operations to look for a word or words at a point (what he calls a "word task", comparable to our proposals to the lexical retrieval component), and events to create such predictions from a phrase (what he calls a "predict task"). Every phrase is implicitly an event for a predict task, and he has a special data type called a "prediction" to represent events for word tasks.

Whereas HWIM, when it processes a hypothesis, will always make all predictions, call the lexical retrieval component to find all matching words, and create word events for each such found word, Paxton's system breaks this cycle up differently. His system schedules separate events for each of the individual word predictions generated by a hypothesis, and whenever a word or completed phrase is found he distributes it immediately to all its "consumers" without waiting. (This difference is probably motivated by his lack of a word matcher that could efficiently find the best matching words at a given position without exhaustively considering each word in the dictionary.)

Paxton's system makes no attempt to guarantee the best interpretation, nor does it stop with the first complete interpretation it finds. Rather it runs until one of several stopping conditions is satisfied (such as running out of storage), after which it takes the best interpretation that it has found so far.

Paxton performed a systematic set of experiments varying four control strategy choices, which he called "focus by inhibition," "map all at once," "context checking," and "island driving." The first was a strategy for focusing on a set of words that occur in high scoring hypotheses and decreasing the scores of all tasks for hypotheses incompatible with those words.

The "map all at once" strategy referred a "bottom up" lexical retrieval strategy that found all possible words at a given point and ranked them taking their word mapper scores into account, rather than proposing such words one at a time in the order in which their proposing hypothesis ranked them (essentially ranking such words according to a priori preferences assigned by the grammar).

"Context checking" referred to a technique of assigning a priority score to predictions of a partial phrase on the basis of a heuristic search for the best possible combinations of higher level constituents that can use that phrase, rather than by basing such priority scores solely on the local quality of the partial phrase alone. (This mechanism gives part of the effect of our use of theories that include arbitrary fragments of a sentence that may cross several levels of phase boundary, but does not apparently permit a fragment that has incomplete phrases at both ends to be prioritized as a whole. It assigns the resulting priority score just to the phrase doing the prediction without apparently remembering the context that justified this score.)

"Island driving," in Paxton's system, referred to the use of a middle-out strategy that looked for a best word somewhere in the utterance to start a seed, and if all hypotheses from that seed scored badly enough would look for another such seed, and so on. However, his system contained none of the features such as island collisions, ghosts, preferred directions, shortfall, or density scoring techniques discussed in this paper, although it may have had something amounting to an absolute direction preference (the documentation is not totally clear on whether both ends of an island can be worked on independently). Hence its version of island driving seems to have all of the disadvantages of a middle-out strategy with almost none of the compensating advantages.

The experiments indicated that the "main effects" of focus by inhibition (i.e., the net effects averaged over all combinations of other strategy options) were negative both in accuracy of the recognition and in number of events processed, and that the main effects of mapping all at once and context checking were positive (the former was more expensive in run time in their system, but might not have been with a suitable lexical retrieval component such as that of HWIM). All three of these experiments showed a statistically significant effect. In addition, the main effect of their island driving feature was found to be negative in time and accuracy, although the result was not statistically significant "because of a large interaction with sentence length." Specifically, Paxton found that island driving improved performance for short utterances, but decreased performance for longer ones, largely due to exceeding the storage limitations before finding the best interpretation. Consequently, it is possible that the implementation of some of the features described in this paper might have improved the performance of the island driving strategy sufficiently to gain a net improvement.

Paxton's results with the focus of inhibition strategy reflect what seems to have been a common experience of the various speech understanding groups in the ARPA project. Although it seemed natural to expect that some word match scores should be good enough that they could be considered correct, thereby eliminating attempts to find alternatives to them, in fact all attempts to implement such an intuition seem to have lead to at best indifferent results and usually to positive degradation. In retrospect, the fact that perfect matches of other words or short word sequences can occur by accident in completely accurate transcriptions of sentences should suggest that there is no magic threshold above which one can consider a given hypothesis correct without verifying its consistent extension to a complete spanning theory. It seems, therefore, that the absolute value of the local quality score is not what matters in deciding the most likely interpretation. The relative scores of competing hypotheses are more relevant, but what really counts is the eventual quality of the complete spanning theory.

9. Conclusions

I have attempted here to provide a perspective on some of the work that has been done in the areas of syntax, semantics and control for understanding spoken natural language by machines and to call special attention to those techniques that have theoretically understood characteristics.

Unfortunately, many of the speech understanding techniques that have been discovered to date have not been sufficiently tested to understand their true value and appropriate ranges of applicability. For example, the pros and cons of left-to-right versus middle-out strategies are not yet clear.

In order to cover the scope of material that I have it has been necessary to treat many issues rather shallowly and others not at all. Hopefully, the references will provide additional detail for the interested reader to follow up. There is now a lot of interest and a number of inviting directions for future research in continuous speech understanding. The HARPY and IBM systems clearly demonstrate feasibility for controlled branching grammars. Moreover, the techniques of the HWIM system hold promise for less restricted input. An important need for the future, however, is much better models of acoustic-phonetics and phonology.

Finally, given the different perspective that the speech understanding task places on the roles of syntax and semantics in language understanding, I believe that the speech understanding problem can eventually have almost as great an impact on research in syntax and semantics as these areas have on the problem of automatic speech recognition.

Acknowledgment

Much of the work described here was supported by the Advanced Research Projects Agency of the Department of Defense and was monitored by ONR under Contract No. N00014-75-C-0533.

References

Bahl, L.R., Baker, J.K., Cohen, P.S., Cole, A.G., Jelinek, F., Lewis, B.L., and Mercer, R.L. (1978)
"Automatic Recognition of Continuously Spoken Sentences from a Finite State Grammar." Conference Record, 1978 IEEE Int'l Conf. on Acoustics, Speech and Signal Processing, Tulsa, OK, April, 1978, IEEE 78CH1285-6 ASSP.

Bahl, L.R., Baker, J.K., Cohen, P.S., Dixon, N.R., Jelinek, F., Mercer, R.L., and Silverman, H.F. (1976)
"Preliminary Results on the Performance of a System for the Automatic Recognition of Continuous Speech," Conference Record, IEEE Int. Conf. on Acoustics, Speech, and Signal Processing, ICASSP-76, Philadelphia, Pa., April, 1976.

Baker, J.K. (1975)
"The DRAGON System -- An Overview," IEEE Trans. Acoustics, Speech and Signal Processing, February, Vol. ASSP-23, No. 1, pp. 24-29.

Chomsky, N. (1965)
Aspects of the Theory of Syntax. Cambridge, MA: MIT Press.

Hayes-Roth, F, and Lesser, V.R. (1976)
"Focus of Attention in a Distributed-Logic Speech Understanding System," Conference Record, IEEE International Conference on Acoustics, Speech and Signal Processing, ICASSP-76, Philadelphia, Pa., April, 1976.

Klatt, D.H. and Stevens, K.N. (1971)
"Strategies for Recognition of Spoken Sentences from Visual Examination of Spectrograms," BBN Report No. 2154, Bolt Beranek and Newman Inc., Cambridge, MA.

Lesser, V.R., Fennell, R.D., Erman, L.D., and Reddy, D.R. (1975)
"Organization of the Hearsay II Speech Understanding System," IEEE Trans. Acoustics, Speech, and Signal Processing, February, Vol. ASSP-23, No. 1, pp. 11-24.

Lowerre, Bruce T. (1976)
"The HARPY Speech Recognition System," Technical Report, Department of Computer Science, Carnegie-Mellon Univ., April, 1976.

Nash-Webber, B.L. (1975)
"The Role of Semantics in Automatic Speech Understanding." In **Representation and Understanding: Studies in Cognitive Science**, D. Bobrow and A. Collins (eds.), New York: Academic Press.

Newell, A. et al. (1973)
Speech Understanding Systems: Final Report of a Study Group, Amsterdam: North Holland/American Elsevier.

Paxton, W.H.. (1977)
"A Framework for Speech Understanding," Stanford Research Institute Artificial Intelligence Center, Technical Note 142, June 1977.

Wanner, E. (1973)
"Do We Understand Sentences from the Outside-in or from the Inside-out?," **Daedalus**, pp. 163-183, Summer.

Wolf, J.J. and W.A. Woods (1977)
"The HWIM Speech Understanding System," Conference Record, **IEEE International Conference on Acoustics, Speech and Signal Processing**, Hartford, Conn., May, 1977.

Woods, W.A. and Makhoul, J.I. (1974)

"Mechanical Inference Problems in Continuous Speech Understanding." In **Artificial Intelligence**, Vol. 5, No. 1, pp. 73-91.

Woods, W., M. Bates, G. Brown, B. Bruce, C. Cook, J. Klovstad, J. Makhoul, B. Nash-Webber, R. Schwartz, J. Wolf, V. Zue (1976) "Speech Understanding Systems - Final Technical Progress Report," BBN Report No. 3438 Vols. I-V, Bolt Beranek and Newman Inc., Cambridge, Ma.

Woods, W.A. (1970) "Transition Network Grammars for Natural Language Analysis," Communications of the ACM, Vol. 13, No. 10, October, 1970, pp. 591-606.

Woods, W.A. (1975) "Syntax, Semantics, and Speech," in D.R. Reddy (ed.), **Speech Recognition: Invited Papers at the IEEE Symposium**, New York: Academic Press. (Also as BBN Report No. 3067).

Woods, W.A. (1977) "Theory Formation and Control in a Speech Understanding System with Extrapolations towards Vision," Proc. of Workshop on Computer Vision Systems, University of Massachusetts, Amherst, June.

Woods, W.A. (1978a) "Semantics and Quantification in Natural Language Question Answering." In **Advances in Computers**, Vol. 17. New York: Academic Press.

Woods, W.A. (1978b) "Generalizations of ATN Grammars," in Woods & Brachman, Research in Natural Language Understanding, BBN Report No. 3963, August.

A CLASSIFICATION METHOD BASED ON FUZZY NAMING RELATIONS OVER FINITE LANGUAGES

R. De Mori L. Saitta

Istituto di Scienza dell'Informazione
Università di Torino (Italy)
C.so Massimo D'Azeglio 42, 10125 TORINO

ABSTRACT

A method for learning memberships of strings belonging to finite languages is proposed. Each language describes a class of phenomena. The learning algorithm attempts to minimize the misclassification error in a sample. Experimental results from the application to automatic speech recognition are presented.

1. INTRODUCTION

In complex situations, such those emerging in speech recognition and image understanding, pattern interpretation is carried out at several levels of abstraction and is based on emission and verification of hypotheses.

The problem considered in this paper refers to those cases for which pattern descriptions cannot be always obtained in a precise way and, furthermore, in which the interpretation of a pattern is controlled by knowledge sources whose content may be also imprecise. Nevertheless, there is a great degree of redundancy in most of the patterns to be interpreted. Such redundancy can compensate the impreciseness of the knowledge models and the pattern description system allowing to

J. C. Simon (ed.), Spoken Language Generation and Understanding, 365-379.

interpret acoustic or visual patterns correctly. Very important for the achievement of such a task is the capability of correctly representing and processing the various degrees of vagueness implicit in the knowledge sources and used during the interpretation of a pattern.

This paper proposes a method for automatically learning the degrees of vagueness of a knowledge source that has to be used for interpreting an imprecise pattern description.

Let $f \in \mathcal{F}$ be the pattern to be interpreted. Preliminary operations, like digitalization, etc., lead to a pattern representation in the form of a matrix A(f) of numbers. Then a process of description takes place giving a description U(f) of the pattern. Such a description can be obtained by asking questions to the numerical representation A(f) of the pattern and by combining the answers, that are linguistic variables, to form a string.

Following Zadeh [1], the pattern descriptions $U_i(f)$, $(1 \leq i \leq N_f)$, are phrases made of fuzzy linguistic variables. Such variables belong to an alphabet Σ and $U_i(f)$ is an element of the power set of Σ. The index i will be omitted in the following for the sake of simplicity.

Each description U is assigned a possibility value $\Pi_{A(f)}(U)$ obtained by composition of the compatibility degrees of the fuzzy variables in U.

Pattern interpretation is seen as a naming relation by which an interpretation $\underline{x} \in S_N$ is assigned to the pattern to be interpreted; S_N is the set of all possible names.

In order to take profits from the information redundancies inside the pattern and to properly infer correct global interpretations from the interpretation of some pattern components, it is necessary to evaluate the Poss $\{f \text{ is } \underline{x}\}$, that is the possibility that the interpretation $\underline{x}$ applies to the pattern f.

Let $R_f(U)$ be the fuzzy restriction associated with the description U of f; let $R(\underline{x},U)$ be the fuzzy relation between a description U of f and an interpretation $\underline{x}$; the meaning of an interpretation $\underline{x}$ of f is represented by the restriction $R_f(\underline{x})$ which is generally obtained by the following general rule of inference:

$$R_f(\underline{x}) = R_f(U) \circ R(\underline{x},U) \tag{1}$$

Following Zadeh's theory of possibility [2], the following rule for computing Poss $\{f \text{ is } \underline{x}\}$ is obtained from (1):

$$\text{Poss}\{f \text{ is } \underline{x}\} = \sup_{U \in \Sigma^*} \Pi_{A(f)}(U) \wedge \mu_{R(\underline{x},U)} \tag{2}$$

where $\wedge$ is the min operator and $\mu_{R(\underline{x},U)}$ is the compatibility of the name $\underline{x}$ with the description U. The main motivation why fuzzy relations are used for assigning names to patterns or subpatterns is because names can be seen as partially overlapped languages, defined over the set $\mathcal{U}$ of possible pattern descriptions. Assume that there are only two names, $\underline{x}$ and $\underline{y}$ defining two languages L_x and L_y over $\mathcal{U}$. For those descriptions U for which:

$$U \in L_x \quad \text{and} \quad U \notin L_x \cap L_y$$

there should be $\mu_{R(\underline{x},U)} = 1$ and $\mu_{R(\underline{y},U)} = 0$; analogously, if:

$$U \in L_y \quad \text{and} \quad U \notin L_x \cap L_y$$

there should be $\mu_{R(\underline{y},U)} = 1$ and $\mu_{R(\underline{x},U)} = 0$;

but, for those cases for which:

$$U \in L_x \cap L_y$$

it is desirable to have $\mu_{R(\underline{x},U)}$ and $\mu_{R(\underline{y},U)}$ different from zero and from one.

This paper deals with the problem of learning the memberships of fuzzy naming relations from imprecise descriptions of the input patterns in order to maximize the number of cases the right (known) name is assigned to a pattern with a degree of possibility higher than the other possible names.

2. FORMAL DEFINITION OF THE PROBLEM

Assume, for the sake of simplicity, that A(f) is a vector of attributes taken on a real-world phenomenon $f \in \mathcal{H}$. Let $\mathcal{A}$ be the universe on which the vector A(f) may vary.
Let $\Sigma = \{ u_1, u_2, \ldots, u_m \}$ be an alphabet of linguistic variables describing A(f).
Let $\mathcal{U}$ be the power set of Σ and let U be a generic string belonging to $\mathcal{U}$.
The possibility $\Pi_{A(f)}(U)$ is a possibility distribution function induced by the proposition:

$$f \text{ is (described by) } U \tag{3}$$

For each symbol u_j of the string U, a fuzzy restriction:

$$R_j(A(f)) = u_j \tag{4}$$

can be considered. The restriction (4) represents the fact that u_j is a fuzzy linguistic variable whose meaning is ill-defined as in the following statements: "a_i is a variable which assume high values", etc..
Due to the (4), the possibility distribution:

$$\Pi_{A(f)}(u_j) = \mu_{u_j}(A(f)) \tag{5}$$

expresses the degree of worthiness of u_j given the vector A(f) of attributes. The string U is a concatenation of the variables $u_1, u_2, \ldots, u_m$; thus its possibility distribution over $\mathcal{A}$ is given by:

$$\Pi_{A(f)}(u_1 \ldots . u_m) = \Pi_{A(f)}(u_1) \ldots \Pi_{A(f)}(u_m) \tag{6}$$

In order to apply (6) it is necessary to know the following functions:

$$\Pi_{A(f)}(u_j) \qquad (1 \leq j \leq m)$$

for every u_J that may describe A(f) and for all the possible values of A(f) and $\mu_{R(\underline{x},U)}$, $\forall \underline{x} \in S_N$ and $\forall U \in \mathcal{U}$. The possibilities $\Pi_{A(f)}(u_J)$ can be established from histograms on the basis of the experience or in a purely subjective way.

The function $\mu_{R(\underline{x},U)}$ represents the vagueness of the knowledge source that is invoked for emitting the hypotheses about the interpretation of the pattern f. Such a function cannot be established in a straightforward way, because f may be described by many strings that are imprecisely related to the pattern to be interpreted. An algorithm for inferring $\mu_{R(\underline{x},U)}$ from experiments will be presented here. It deals with the case of two names, say $\underline{x}$ and $\underline{y}$ and a set $\mathfrak{F}$ of phenomena such that:

$$\mathfrak{F} = \mathfrak{F}_x \cup \mathfrak{F}_y \qquad \mathfrak{F}_x \cap \mathfrak{F}_y = \phi$$

where $\mathfrak{F}_x$ is the set of phenomena having name $\underline{x}$ and $\mathfrak{F}_y$ is the set of phenomena having name $\underline{y}$; ϕ is the empty set.

Generalization to more than two names is straightforward and will be omitted for the sake of brevity.

Let E_x be the number of phenomena having name $\underline{x}$ and for which the (2) gives:

$$\text{Poss}\ \{ f \text{ is } \underline{y} \} \geqslant \text{Poss}\ \{ f \text{ is } \underline{x} \}$$

Conversely, let E_y be the number of phenomena having name $\underline{y}$ and for which the (2) gives:

$$\text{Poss}\ \{ f \text{ is } \underline{x} \} \geqslant \text{Poss}\ \{ f \text{ is } \underline{y} \}$$

Let

$$E = E_x + E_y \tag{7}$$

be the total number of the cases for which the wrong name is assigned to a phenomenon with higher possibility than the true name. The algorithm developped in the following has the purpose of finding a set of functions

$\mu_{R(\underline{x},U)}$ and $\mu_{R(\underline{y},U)}$ for which the value of E, defined by the (7), is minimum. Such a criterion allows one to infer naming relations allowing to assign by (1) the right name to a phenomenon f with the highest plausibility value in the highest number of times.

3. A SIMPLE CASE

A simple preliminary case will be considered in this section to show how the algorithm may work.
Let $\mathcal{F}$ be a set of phenomena, partitionned into two subsets, namely $\mathcal{F}_x$ and $\mathcal{F}_y$.
Let $\Sigma = \{u_1, u_2\}$ be the alphabet of the symbols with which the descriptions of f are made, where f is a phenomemon belonging to $\mathcal{F}$. Let $\Sigma_x = \{u_1\}$ and $\Sigma_y = \{u_2\}$
We are interested in the possibilities of the naming relations, assigning names $\underline{x}$ and $\underline{y}$ to a phenomenon f, that is:

$$\begin{aligned} \text{Poss}\ \{\underline{x} \text{ is } f\} &= \mu_{R(\underline{x},u_1)} \wedge \Pi_{A(f)}(u_1) \\ \text{Poss}\ \{\underline{y} \text{ is } f\} &= \mu_{R(\underline{y},u_2)} \wedge \Pi_{A(f)}(u_2) \end{aligned} \qquad (8)$$

The learning problem consists in inferring $\mu_{R(\underline{x},u_1)}$ and $\mu_{R(\underline{y},u_2)}$ in order to minimize the total error defined by (7). For the sake of simplicity the right-hand possibilities are renamed as follows:

$$\begin{aligned} \mu_{R(\underline{x},u_1)} &= x \\ \mu_{R(\underline{y},u_2)} &= y \\ \Pi_{A(f)}(u_1) &= h \\ \Pi_{A(f)}(u_2) &= k \end{aligned} \qquad (9)$$

Let

$$\varphi_N = \{f_1, f_2, \ldots, f_j, \ldots, f_N\}$$

be the training set of experiments. φ_N may be partitionned into four sets φ'_x, φ''_x, φ'_y, φ''_y according with the following definitions:

$$\begin{cases} f_j \in \varphi'_x & \text{iff} \quad (f_j \in \mathcal{H}_x) \wedge (h \geq k) \\ f_j \in \varphi''_x & \text{iff} \quad (f_j \in \mathcal{H}_x) \wedge (h < k) \\ f_j \in \varphi'_y & \text{iff} \quad (f_j \in \mathcal{H}_y) \wedge (k \geq h) \\ f_j \in \varphi''_y & \text{iff} \quad (f_j \in \mathcal{H}_y) \wedge (k < h) \end{cases} \tag{10}$$

$$(1 \leq j \leq N)$$

If N'_x, N''_x, N'_y, N''_y are respectively the cardinalities of the four defined subsets, we get obviously:

$$N = N'_x + N''_x + N'_y + N''_y$$

Let h_j and k_j be the values assumed by h and k for the phenomenon f_j; for the sake of simplicity the right-hand possibilities of (8) are renamed as follows:

$$\text{Poss}\{f_j \text{ is } \underline{x}\} = \Pi_j^{(x)}$$

$$\text{Poss}\{f \text{ is } \underline{y}\} = \Pi_j^{(y)} \tag{11}$$

The value of $\Pi_j^{(x)}$ depends on x and h_j, the value of $\Pi_j^{(y)}$ depends on y and k_j. Assuming that x and y can vary in the $[0,1]$ interval, an assignment of memberships (x,y) may vary in a square Q of the x-y plane. Given a phenomenon f , Q may be subdivided into two domains Q_j^t, and Q_j^f. Q_j^t is the domain representing the assignments of memberships for which the possibility of the right hyphotesis is higher than the possibility of the wrong hyphothesis. The contrary holds for Q_j^f.

Fig. 1 shows the domain Q^t_j and Q^f_j for the four cases defined by the (10).
By considering the whole set φ_N we will subdivide Q into error domains labelled $Q_k(\varphi_N)$, such that the $Q_k(\varphi_N)$ domain contains the set of assignments (x,y) for which k phenomena of φ_N receive a wrong name with higher possibility.
Subdividing the set φ_N into the fours subsets (10), the phenomena f_j belonging to each subset, can be ordered according to decreasing values of k_j for φ'_x, increasing values of h_j for φ''_x, decreasing values of h_j for φ'_y and increasing values of k_j for φ''_y.
The four ordered sequences will be renamed as follows:

$$f_j \in \varphi_x \begin{cases} k_j \rightarrow k_p \geq k_{p+1} & (0 \leq p \leq N'_x,\ k_o = 1,\ k_{N'_x+1} = 0) \\ h_j \rightarrow h_q \leq h_{q+1} & (0 \leq q \leq N''_x,\ h_o = 0,\ h_{N''_x+1} = 1) \end{cases}$$

$$f_j \in \varphi_y \begin{cases} k_j \rightarrow k_r \leq k_{r+1} & (0 \leq r \leq N''_y,\ k_o = 0,\ k_{N''_y+1} = 1) \\ h_j \rightarrow h_t \geq h_{t+1} & (0 \leq t \leq N'_y,\ h_o = 1,\ h_{N'_y+1} = 0) \end{cases}$$

An example of the domains $Q_k(\varphi_N)$ is reported in Fig.2.
Combining toghether the four error domains of Fig. 2 in a single picture, Fig. 3 is obtained. Here two types of error domains can be considered; their definition is the following one:

$$\Lambda_{p,r} \begin{cases} y > x & \\ k_{p+1} < x \leq k_p & 0 \leq p \leq N'_x \\ k_r \leq x < k_{r+1} & 0 \leq r \leq N''_y \end{cases}$$

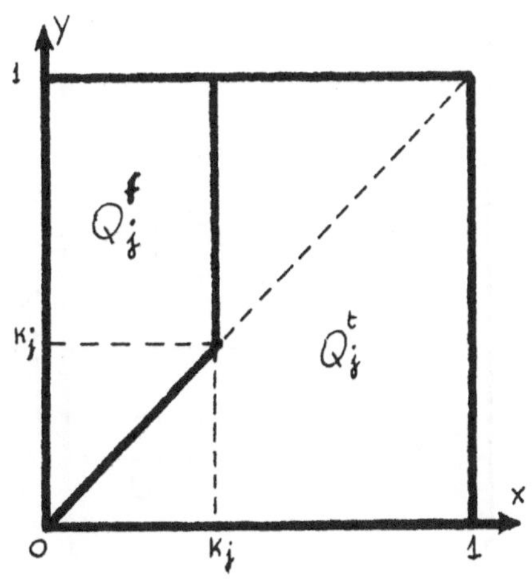

Case 1 : $f_J \in \varphi'_x$

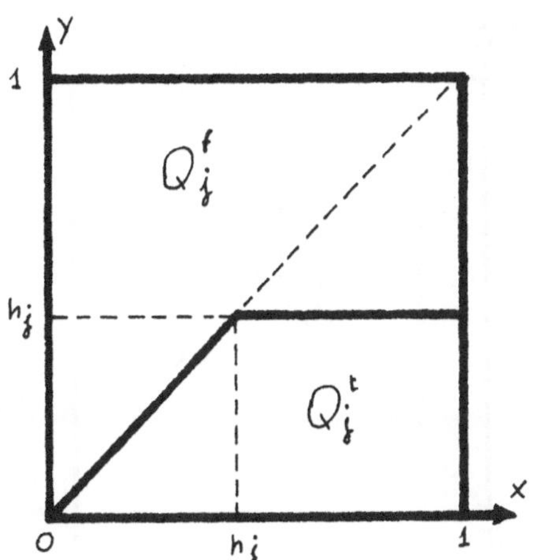

Case 2 : $f_J \in \varphi''_x$

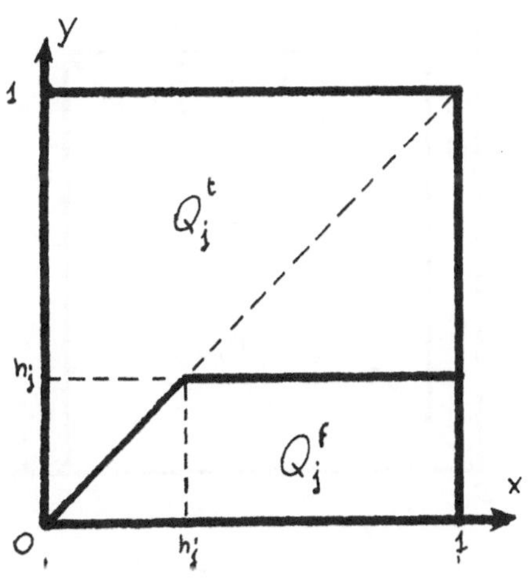

Case 3 : $f_J \in \varphi'_y$

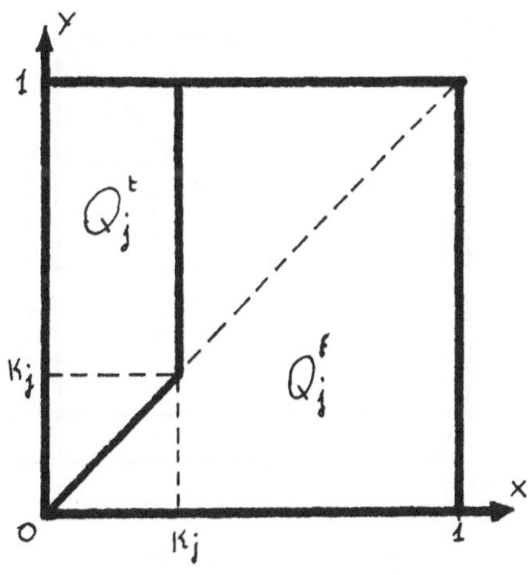

Case 4 : $f_J \in \varphi''_y$

Fig. 1 - Example of domains Q_j^f and Q_j^t

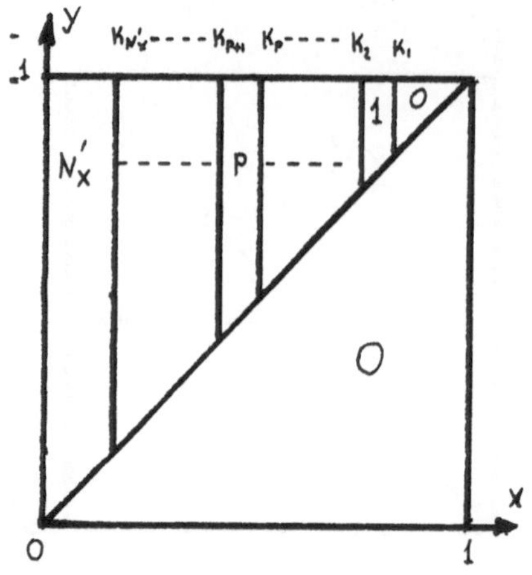

Case 1 : $f_j \in \varphi'_x$

$(1 \leq j \leq N'_x)$

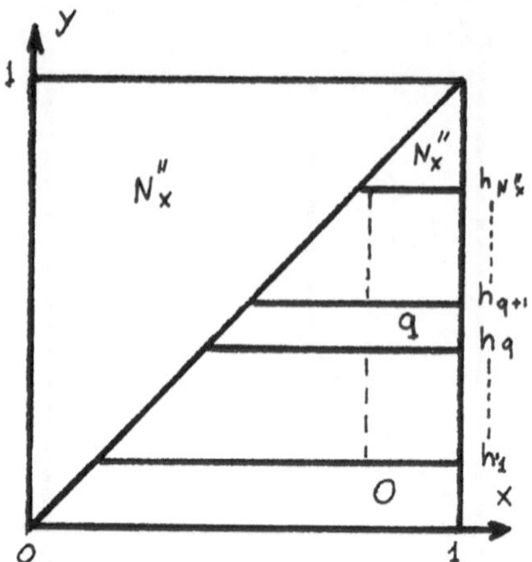

Case 2 : $f_j \in \varphi''_x$

$(1 \leq j \leq N''_x)$

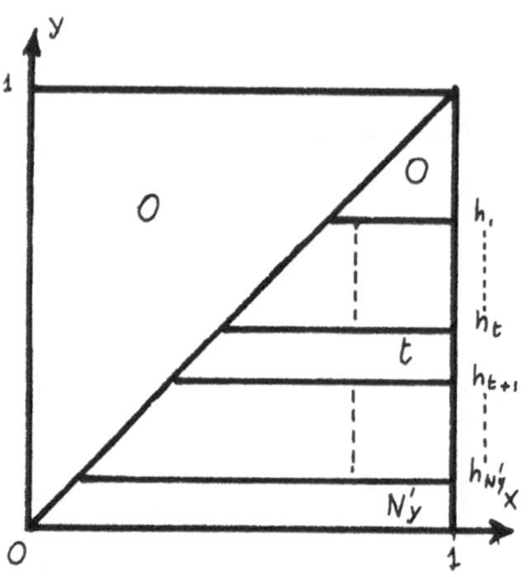

Case 3 : $f_J \in \varphi'_y$

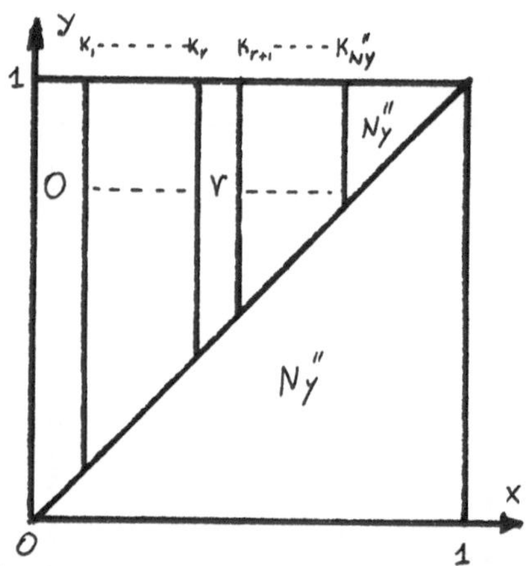

Case 4 : $f_j \in \varphi''_y$

Fig. 2 - Example of the domains $Q_k(\varphi_N)$ for the four cases defined by (10)

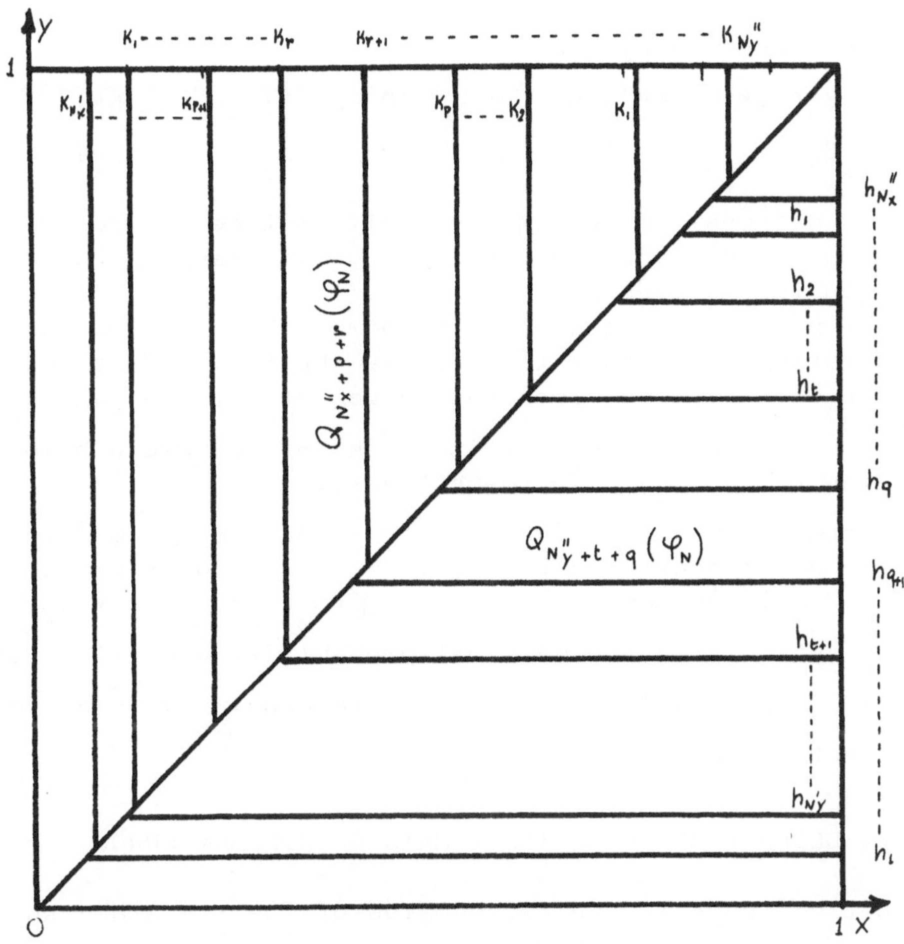

Fig. 3 - Result of the superposition of the four cases in Fig. 2.

$$\Theta_{t,q} \begin{cases} y < x & \\ h_{t+1} < y \leq h_t & 0 \leq t \leq N'_y \\ h_q \leq y < h_{q+1} & 0 \leq q \leq N''_x \end{cases}$$

The number of errors $E_\Lambda(p,r)$ and $E_\Theta(t,q)$ corresponding respectively to the error domains $\Lambda_{p,r}$ and $\Theta_{t,q}$ are expressed by:

$$E_\Lambda(p,r) = N''_x + p + r \qquad (0 \leq p \leq N'_x,\ 0 \leq r \leq N''_y)$$
$$E_\Theta(t,q) = N''_y + t + q \qquad (0 \leq t \leq N'_y,\ 0 \leq q \leq N''_x)$$

The diagonal of the square Q will not be considered, because the error in the case x=y is always greater than in the case $x \neq y$.
The problem of assigning the memberships of the naming relations, in order to minimize the total error E defined by (7), requires the values of k_p, h_q, h_t and k_r to be known for some training set of phenomena belonging to $\mathcal{F}_x$ and $\mathcal{F}_y$ respectively.
Without any loss in generality it may be assumed that the above mentioned possibilities have joint probability densities represented by $\gamma(h_q,k_p)$ and $\delta(h_t,k_r)$.
We define also the marginal probability densities $f(k_p)$, $g(k_r)$, $p(h_t)$, $d(h_q)$ and the corresponding cumulative functions $F(k_p)$, $G(k_r)$, $P(h_t)$ and $D(h_q)$.

4. DETERMINATION OF THE DOMAIN OF MINIMUM ERROR

Let us consider the two domains of the x-y plane shown in Fig. 4 and defined by:

$$\alpha \begin{cases} y > x \\ z_1 \leq x \leq z_2 \end{cases} \qquad\qquad \beta \begin{cases} y < x \\ v_1 \leq y \leq v_2 \end{cases}$$

The purpose of this section is that of evaluating the probability distribution of the errors in α and β, given a generic sample φ_N of N elements.

In this way we can determine the domain where the theorical probability of error (misclassification error) is minimum. This allows an evaluation of the error which affects the "optimal" pair (x,y) inferred from a particular sample.

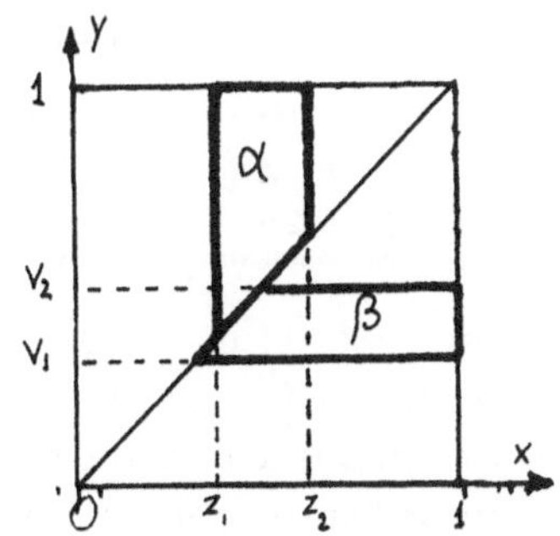

Fig. 4 - Domains α and β defined in section 4.

After some calculations we obtain the expression of the probability of misclassifing a phenomenon f by choosing a pair (x,y) in the domain α ; by taking the limit of the found expression for $z_2 \to z_1$ we get the misclassification error probability when we choose a pair (x,y) with fixed x and $y > x$. The final expression results then:

$$\omega(x,y) = \begin{cases} \mu(x) = p''_x + p'_x\,(1-F(x)) + p''_y\,G(x) & \text{if } y > x \\ \sigma(y) = p''_y + p'_y\,(1-P(y)) + p''_x\,D(y) & \text{if } y < x \end{cases}$$

The pair (or the pairs) which mimimizes the function $\omega(x,y)$ will be the optimal choice of memberships for the defined naming relations.

5. LEARNING ALGORITHM

Let φ_N be a training set of known composition, partitioned according to the (10). Let moreover:

$$M = N'_x + N''_y \qquad\qquad R = N'_y + N''_x$$

Let us order in a non-decreasing way the values of the k values corresponding to the phenomena belonging to φ'_x or φ''_y, i.e.:

$$k_m \leq k_{m+1} \qquad (0 \leq m \leq M,\ k_o=0,\ k_{M+1}=1) \qquad (12)$$

In an analogous way the h values for phenomena belonging to φ'_y or φ''_x are ordered as follows:

$$h_r \leq h_{r+1} \qquad (0 \leq r \leq R,\ h_o=0,\ h_{R+1}=1) \qquad (13)$$

Let us choose x in one of the (M+1) intervals defined by the (12) and y in one of the (R+1) intervals defined by the (13); for each pair of intervals where x and y respectively fall, characterized by the indeces m and r, let us compute the number of misclassifications in φ_N. Let E(m,r) be such a number. Let us now search for a pair $(\hat{m},\hat{r})$ such that:

$$E(\hat{m},\hat{r}) = \min_{0 \leq m \leq M,\ 0 \leq r \leq R} E(m,r) \qquad (14)$$

By applying the Large Number Law it is possible to prove that $E(\hat{m},\hat{r})$ converges to the minimum value of $\omega(x,y)$ which has been found in Section 4. Moreover we can estimate that, with a 95% of probability, the "true" optimal value is within 10% of the estimated one for a sample of about 300 elements; this value has been evaluated for a value of ω minimum of about 10%.

6. CONCLUSION

The previous algorithm, extended to the case of cardinalities of Σ_x and Σ_y different from one and to the case of more than two classes, has been applied in various steps of continuous speech processing. For example, in the decision process SONORANT - NON SONORANT for consonants an experiment has been carried out with 400 consonants extracted from sentences pronounced by four speakers. Using non-fuzzy relations and a definition of error according to the (7), an error rate of about 10% was obtained; this error reduces to 5% using

naming relations which are fuzzy; moreover this error rate did not show significant variations in other experiments consisting of sets of different sentences enounciated by different speakers.
In [3] the details of the computation, the extension to finite languages and to more than two names may be found. Also more examples are given.
In [4] an application of the algorithm to a case of medical diagnosis is described in details.

REFERENCES

[1] - Zadeh L.A. : "Fuzzy Languages and their Relation to Human and Machine Intelligence", Man and Computer Proc. Int. Conf. (Bordeaux 1970), 130-165 (1972)

[2] - Zadeh L.A.: "Fuzzy sets as a basis for a theory of possibility", Fuzzy Sets and Systems 1, 3-28 (1978)

[3] - R. De Mori, L. Saitta: "Automatic Learning of fuzzy naming relations over finite languages", Information Science, in press.

[4] - L. Saitta, P. Torasso: "Fuzzy characterization of coronary disease", Fuzzy Sets and Systems, in press.

§ 4

SPEECH SYNTHESIS

SPEECH SYNTHESIS FROM TEXT

Jonathan Allen

Research Laboratory of Electronics and Department of Electrical Engineering and Computer Science, Massachusetts Institute of Technology, Cambridge, MA 02139 U.S.A.

In this paper, we are concerned with describing a successful approach to the conversion of unrestricted English text to speech. Before taking up the details of this process, however, it is useful to place this task in context. Over the years, there has been an increasing need for speech generated from computers. In part, this has been due to the intrinsic nature of text, speech, and computing. Certainly speech is the fundamental language representation, present in all cultures (whether literate or not), so that if there is to be any communication means between the computer and its human users, then speech provides the most broadly useful modality, except for the needs of the deaf. While text (considered as a string of conventional symbols) is often considered to be more durable than speech and more reliably preserved, this is in many ways a manifestation of relatively early progress in printing technology, as opposed to the technology available for storing and manipulating speech. Furthermore, text-based interaction with computers requires typing (and often reading) skills which many potential users do not possess. So if the increasingly ubiquitous computer is to be useful to the largest possible segment of society, interaction with it via natural language, and in particular via speech, is certainly necessary. That is, there is a clear trend over the past 25 years for the computer to bend increasingly to the needs of the user, and this accommodation must continue if computers are to serve society in the large. The present search for expressive programming languages which are easy to use and not prone to error can be expected to lead in part to natural language interaction as the means best suited to human users, with speech as the most desirable mode of expression.

J. C. Simon (ed.), Spoken Language Generation and Understanding, 383-396.

It is clear, then, that speech communication with computers is both needed and desirable. Within the realm of speech output techniques, we can ask what the nature of these techniques is, and how they are realized. In order to get a view of the spectrum of such procedures, it is useful to consider them as the result of four different constraints, which determine a design space for all possible speech output schemes. Each technique can then be seen as the result of decisions related to the impact of each of the four constraint areas.

1. TASK

The application task determines the nature of the speech capability that must be provided. When only a small number of utterances is required, and these do not have to be varied on line, then recorded speech can be used, but if the task is to simulate the human cognitive process of reading aloud, then an entirely different range of techniques is needed.

2. HUMAN VOCAL APPARATUS

All systems must produce as output a speech waveform, but it is not an arbitrary signal. A great deal of effort has gone into the efficient and insightful representation of the speech signal as the result of a signal source in the vocal tract exciting the vocal tract "system function", which acts as a filter to produce the speech waveform. The human vocal tract also constrains the speed with which signal changes can be made, and is also responsible for much of the coarticulatory smoothing or encoding that makes the relation between the underlying phonetic transcription and the speech waveform so difficult to characterize.

3. LANGUAGE STRUCTURE

Just as the speech waveform is not arbitrarily derived, the myriad possible speech gestures that could be related to a linguistic message are constrained by the nature of the particular language structure involved. It has been consistently found that those units and structures which linguists use to describe and explain language do in fact provide the appropriate base in terms of which the speech waveform can be characterized and constructed. Thus, basic phonological laws, stress rules, morphological and syntactic structures, and phonotactic constraints all find their use in determining the speech output.

4. TECHNOLOGY

Our ability to model and construct speech output devices is strongly conditioned by the current (and past) technology. Speech science has profited greatly from a variety of technologies, including x-rays, motion pictures, the sonograph, modern filter and sampled-data theory, and most importantly the modern digital computer. While early uses of computers were for off-line speech analysis and simulation, the advent of increasingly capable integrated circuit technology has made it possible to build compact, low cost, real-time devices of great capability, and it is this fact, combined with our substantial knowledge of the algorithms needed to generate speech, that has propelled the field of speech output from computers into the "real world" of practical commercial systems, suitable for a wide variety of applications.

With these constraints in mind, we can examine the various approaches to speech output from computers. A great many techniques have been developed, but they can be naturally grouped in an insightful way. Our purpose here is to create a context in which text-to-speech conversion of unrestricted English text using synthesis-by-rule can be considered. This comparison will permit us to highlight the difference between the various approaches, and to compare system cost and performance.

1. WAVEFORM CODING

The simplest strategy would be to merely record (either in digital or analog format) the required speech. Depending on the technology used, this approach may introduce access time delays, and will be limited in capacity by the recording medium available, but the speech will generally be of high quality. No knowledge of the human vocal apparatus or language structure is needed, these systems being a straighforward match of the task requirements to the available storage technology. Since memory size is the major limitation of these schemes, efforts have been made to cut down the number of bits per sample used for digital storage. A variety of techniques has been used, from simple delta modulation, through adaptive delta modulation, adaptive differential PCM, to adaptive predictive coding, which can drop the required bit rate from over 50 Kbit/sec to under 10 Kbit/sec, while still retaining good quality speech. Simple coder/decoder circuits can be used for recording and playback. When the message vocabulary is small and fixed, these systems are attractive, but if messages must be concatenated, then it is extremely difficult to produce good quality speech, because aspects of the speech waveform cannot be "bound" at recording time to the values appropriate for all message situations which use the smaller constituent messages.

2. PARAMETRIC REPRESENTATION

In order to further lower the storage requirements, but also to provide needed flexibility for concatenation of messages, several schemes have been developed which "back up" from the waveform itself to a parametric representation in terms of a model for speech production. These parameters may characterize salient information in either the time or frequency domain. Thus, for example, the speech waveform can be formed by summing up waveforms at several harmonics of the pitch weighted by the spectral prominence at that frequency, a set of resonances can be excited by noise or glottal waveforms, or the vocal tract shape can be simulated along with appropriate acoustic excitation. As compared to waveform coding, more computation is now required at playback time to recreate the speech waveform, but the storage requirements per message are cut down. More importantly, the parametric representation represents an abstraction on the speech waveform to a level of representation where the attributes that contribute to speech quality (e.g. formant frequencies and bandwidths, pitch, excitation amplitudes) can be insightfully manipulated. This allows elementary messages to be concatenated in a way that provides for smooth transitions at the boundaries. It also allows for changes (e.g. in pitch) well within the individual message units, so that substantial changes in prosodic parameters (pitch and timing) can be made. The most popular parametric representations in use today are based on formants or linear predictive coding (LPC), although vocal tract articulatory models are also used. Message units of widely varying sizes are employed, ranging from paragraphs, through sentences, phrases, words, syllables, demisyllables, and diphones. As the size of the message unit goes down, fewer basic messages are needed for a large message set, but more computation is required, and the difficulties of correctly representing the coarticulation across message boundaries go up. Clearly, these schemes aim to preserve as much of the quality of natural speech as possible, but to permit the flexible construction of a large set of messages using elements which require little storage. With the current level of knowledge of digital signal processing techniques, and the accompanying technology, these schemes have become very important for practical applications. It is well to remember, however, that parametric representation systems seek to match the task with the available processing and memory technology, by using a knowledge of models for the human production of speech, but little (if any) use is made of the linguistic structure of the language.

3. SYNTHESIS-BY-RULE

When message units are concatenated using parametric representations, there is a tradeoff between speech quality and the need to vary the parameters to adapt the message to varying environments. Researchers have found that many allophonic variations of a message unit (e.g. diphone) may be needed to achieve good quality speech, and that while the vocabulary of needed units is thus expanding, little basic understanding of the role of structural language constraints in determining aspects of the speech waveform is obtained. For this reason, the synthesis process has been abstracted even further beyond the level of parametric representation to a set of rules which seek to compute the needed parameters for the speech production model from an input phonetic description. This input representation contains, in itself, very little information. Usually the names of the phonetic segments, along with stress marks and pitch and timing are provided, but the latter prosodic correlates are often computed from segmental and syntactic structure and stress marks, plus semantic information if available. In this way, synthesis-by-rule techniques can utilize a very low bit-rate message description (<100 bits/sec) as input, but substantial computation must be used to compute the model parameters and then produce the speech waveform. Clearly there is complete freedom to specify the model parameters, but of course also the need to control these parameters correctly. Since the rules are still imperfect, the resulting speech quality is not as good as recorded human speech, but recent tests have shown that high intelligibility and comprehensibility can be obtained, and when sentence and paragraph-level messages must be synthesized, the rule system provides the necessary degrees of freedom to produce smooth-flowing good quality speech. It is interesting to consider that synthesis-by-rule systems delay the binding of the speech parameter set and waveform to the input message using very deep language abstractions and hence provide a maximum of flexibility, and are thus well suited to the needs of converting unrestricted text to speech. The designers of these systems must, however, discover the relationship between the underlying linguistic specification of the message and the resulting speech signal, a topic which has been central to speech science and linguistics for several decades. Thus synthesis-by-rule both benefits from and contributes to our general knowledge of speech and linguistics, and the steady improvement in speech synthesis-by-rule quality reflects this joint progress. While it is believed that current synthetic speech quality is acceptable for many applications, it can certainly be expected to continue to improve with our increasing knowledge.

4. TEXT-TO-SPEECH CONVERSION

The synthesis-by-rule techniques described above require a detailed phonetic transcription as input. While this input requires very little memory for message storage, a frequent requirement is to convert text to speech. When it is desired to convert unrestricted English text to speech, the flexibility of synthesis-by-rule is needed, so that means must be afforded to convert the input text to the phonetic transcription needed by the synthesis-by-rule techniques. It is clear, then, that first the text must be analyzed to obtain the phonetic transcription, which is then subjected to a synthesis procedure to yield the output speech waveform. The analysis of the text is heavily linguistic in nature, involving a determination of the underlying phonemic, syllabic, morphemic and syntactic form of the message, plus whatever semantic and pragmatic information can be gleaned. Text-to-speech conversion can thus be seen as a collection of techniques requiring the successful integration of the task constraints with other constraints provided by the nature of the human vocal apparatus, the linguistic structure of the language, and the implementation technology. It is thus the most complex form of speech synthesis system, but also the most fundamental in design and useful in application, since it seeks to mirror the human cognitive capability for reading aloud. Other cognitive models attempt to synthesize speech directly from "concept" for those applications where the underlying linguistic structure is already available. These schemes have the advantage of (presumably) more detailed syntactic and semantic structures than can be obtained from text, and are hence of great interest for high quality synthesis, but the pervading presence of text in our culture makes the text-to-speech capability of great practical importance. It is worth emphasizing that both text and speech are surface manifestations of underlying linguistic form, and hence that text-to-speech conversion consists first of discovering that underlying form, and then utilizing it to form the output speech.

In the sections that follow, we will discuss the MITalk-79 text-to-speech system in detail. The aim of this system is to provide high-quality speech from unrestricted English text using the fundamental results of speech science, computing, and linguistics. We aim to do it "right", in the belief that adherence to basic principles will provide more insightful methods, avoid ad hoc "fixes", and produce the best possible quality of speech. We will also discuss the range of possible applications, and the implementation base for both a research system, and a compact, low cost module utilizing state-of-the-art integrated circuit technology.

TEXT ANALYSIS

The overall task of text analysis is to convert the input stream of text characters to a detailed phonetic transcription. The input is not constrained in any way, except that it is assumed to represent English text, and that it can be read aloud by a native speaker of English. There is no restriction on vocabulary or syntax (except in the sense that a native speaker would find it acceptable), and there can even be misspellings (e.g. "recieve"). The linguistic analysis techniques, however, are oriented to procedures which operate on full spellings of English words, so the first task is to insure that all of the input is spelled out in full alphabetic form. Thus, strings such as "$3.17" must be expanded to "three dollars and seventeen cents" and similar conversions must be applied to expressions such as "Mr.", "M.I.T.", "lb", and "1979". A preprocessor program has been written to make these changes. This program may be thought of as a "front-end exceptions" handler, insuring that all following analysis can be performed in a uniform way. Clearly a small lexicon is needed for this purpose, which can be used to "trap" on any input "word" which the designer wants to be pronounced in an exceptional way. In this way, changes to the normal pronunciation procedures, or special pronunciation for technical terms, can be readily accommodated without major changes to the system.

At the next stage, the system must compute the pronunciation of any conceivable alphabetic English word. This is a large task, and is handled by a set of two complementary processes. In essence, an attempt is first made to analyze each word into a set of morphs (prefixes, suffixes, and roots) which "cover" the word. When this fails, a set of letter-to-phoneme rules is applied to derive the correct pronunciation.

It is important to recognize that most words have internal syntactic structure, and that these structural units are called morphemes. The letter representation of a morpheme is called a morph, and these morphs are combined in typical ways to form words. A dictionary of some 12,000 morphs has been formed, where each entry contains the morph spelling, pronunciation, part-of-speech codes, and designation of morph combining roles. At first, an attempt is made to match the input word to some lexical item, and for high-frequency monomorphemic words this is usually possible. In this way, high frequency words of exceptional pronunciation, such as "of" and "the" (in the normal case, the "f" and "th" would be unvoiced) are given a correct analysis, and foreign words (which do not follow English pronunciation rules) can also be accommodated in this way. The morph lexicon thus serves as an "exceptions dictionary", allowing the letter-to-phoneme rules to be clearly structured. Of at least equal importance, however, is the use of the lexicon to represent polymorphemic words as a string of morphs. A recursive decomposition algorithm has been developed which finds all possible "morph coverings" of the given word, and then selects the correct analysis. Thus, for example, the word "formally" is analyzed into "form + ally", "for + mall + y", and "form + al +ly" (the last being correct) by the recursive analyzer, and a running score is kept in order to select the correct analysis. In addition to the ambiguity of analysis shown in the "formally" example, difficulties arise due to the mutating effect on spelling of suffixes which start with a vowel, such as in "scarcity" being analyzed to "scarce + ity". That is, the analyzer must be able to restore these changes correctly. Experience has shown that in excess of 95% of the input words of random texts are correctly analyzed into a string of morphs. The morph dictionary is able to "cover" in this way at least ten times as many words as there are morphs, and can also accommodate new words ("earthrise") which are regularly formed from the morphemic resources of the language by regular processes of affixation and compounding. The morph lexicon is stable with time, new morphs rarely being formed, whereas new words can be freely invented as needs arise. The morph analysis process is thus seen to be an extremely effective means for obtaining the pronunciation of unrestricted English words. The only difficulty with this scheme is the cost in space and access time associated with the lexicon, particularly if a disk is used. Use of modern integrated circuit read-only memory chips, however, completely (and compactly) avoids this problem, so that there is no performance penalty associated with the lexical access.

When words cannot be analyzed into a sequence of morphs, then a set of letter-to-phoneme rules must be used. It must be remembered, however, that the morph analysis is always attempted first, so that high-frequency and other exceptional words do not have to be processed by the letter-to-phoneme rules. The conversion of letter strings to phonemes is only attempted for low-frequency monomorphemic words. In order to help assure that the input to these rules is monomorphemic, affixes are stripped off, so that a word like "changeable" is first analyzed to "change + able". Note that "ea" occurs in many words as a functioning digraph, such as "reach", "tear", "steak", and "leather", but that in "changeable" there is no such functioning digraph, the internal morph boundary having broken up the digraph. It can thus be seen that the letter-to-phoneme rules are not used for either high-frequency exceptional words or polymorphemic words. This greatly simplifies the rule structure, and makes the letter-to-phoneme process much more insightful than would be the case if no morphemic analysis was provided.

Once affix stripping is attempted, the letter-to-phoneme rules convert the letter string to phonemes by first converting the consonant sequences, and then transforming the vowels. It has been found that consonant pronunciations can be reliably formed from spelling alone, but that vowels require both the letter and the phoneme environment for proper performance of the rules. Several sets of letter-to-phoneme rules have been developed, with varying degrees of accuracy, and a version containing about 400 rules has been chosen as sufficient for a broad variety of English texts.

Within the letter-to-phoneme rules, affix stripping must be done with care to avoid incorrect analyses. For example, a straightforward approach might yield "finishing → fin + ish + ing." To avoid such problems, syntactic checks are made. In the present case, "ing" is known to follow only nominals or verbal suffixes, but since "ish" makes adjectives, "ish" cannot be accepted as an affix, and we get "finishing → finish + ing."

The word "subversion" can be used to illustrate all three stages of letter-to-phoneme analysis:

subversion	input
sub + vers + ion	affix stripping
- v-rz -	consonant conversion in root
sʌb vɜz ə n	remaining letters converted

Note that first a prefix and suffix are removed, leaving only four letters. Next, the consonant letters are converted, and there is only one vowel letter ("e") to be converted. Finally, the pronunciation of the affixes is retrieved from a table, and the vowel is combined (by a later rule) with the "r" to yield the final transcription.

After the morphological and letter-to-phoneme analyses have been completed, a number of adjustments to the phonetic transcription must be made. First of all, a number of morphophonemic rules must be applied. These are low-level adjustment rules that modify the phonetic representation in the environment of morph boundaries. Plurals, possessives, and contractions with "is" all involve the "s" letter being attached to a morph, but the pronunciation varies with the last phoneme of the morph on the left of the boundary, as shown by "horses", "dogs", and "cats". Rules of similar form apply to concatenation of the past participle, "ed", as in "mended", "whispered" and "hushed". The prefixes "pre", "re", and "de" are reduced before bound morphs, as in "prefer". Palatalization is a phenomenon that changes stops to fricatives when the suffixes "ion" and "ure" are attached, as in "completion" and "vesture". Finally, the suffix "ic" can change its pronunciation depending on the vocalic environment, as shown in "specificity".

It is, of course, often the case that when morphs are joined together, stress adjustments have to be made. For words analyzed into a complete morph covering, two kinds of change are made. First, compounds (words containing two or more roots) have primary stress assigned to the leftmost root, as in "houseboat". In addition, some suffixes shift the stress pattern of the word, as in "trainee", "mountaineer", and "information". No other stress adjustments are made to morphologically analyzed words, but for those words transcribed by the letter-to-phoneme rules, a complete set of lexical stress rules is applied to derive the correct stress contour. A great deal of study has been applied to these rules, and currently a system based on extensions to rules given by Halle is used. These offer a powerful set of procedures, capable of shifting not only the stress of a given phoneme, but also the phonemic vowel color itself, as is shown in the case of "systematic" and "systematize", where both contain the suffix "ate" (the silent "e" having been dropped before the terminal vocal suffix).

The conversion processes described so far derive a detailed phonetic transcription at the segmental level, but there are many phonetic effects, particularly those which are prosodic in nature, which require a knowledge of the syntactic structure of the sentence in order to specify the appropriate correlates. Ideally, this would mean that a complete clause-level parse of the sentence text should be obtained. There is, however, no known method to accomplish such an analysis for unrestricted English text. In fact, there is no known complete syntax of English available, nor is there likely to be one soon. Nevertheless, the timing and pitch framework of English sentences is closely related to the phrase structure of the sentence, and although it would be better to have additional clause-level structure, the phrase-level information is very helpful. First a part-of-speech set for each word is computed, utilizing suffix information ("ness" makes nouns) and prefix constraints ("em", "en", and "be" form verbs) in addition to local context frame checks. With this information in hand, a left-to-right scan of the sentence is made looking for noun, preposition, and verb phrases. Only one analysis is found, and there is never any backtracking. Augmented transition networks have been formulated for each of these phrase types, and a wide variety of phrases can be detected. It is useful to note that the tightest syntagmatic constraints in English are within the verbal auxiliary and nominal determiner sequences, and these constraints are heavily exploited in this parsing scheme. The following example shows that much of the prosody of a sentence can be related to this phrasal information.

(NG: Most of the exercises) (VG: are) (NG: translations). (VG: There are) (NG: several important changes) (PP: in the way)

(NG: the quantifier rules) (VG: will work) (PP: for the remainder of the course).

There is one last step to take before the analysis phase can be completed. This component deals with a small number of phonological adjustments, plus a formatting of the final phonetic transcription in preparation for the ensuing speech synthesis. At this point, all words are marked with their syllabic and morphological structure, as well as their syntactic role. A final scan of this structure is made to determine the choice of allophones of "l", the presence of flapped alveolar consonants, glottal stop insertion at word boundaries, suppression of stop release when followed by another stop, and selection of the proper pronunciation of "the". These final adjustments constitute the last refinements made to the phonetic transcription.

SPEECH SYNTHESIS

It is now appropriate to start the task of waveform construction. There are several stages in this process including:

a. The determination of the timing structure, including pauses and segmental effects.

b. Computation of the pitch contour.

c. Derivation of phonemic target values for the segmental parameters.

d. Smoothing of target values to obtain parameter contours.

e. Calculation of synthesis model coefficients from the phonetic parameters.

f. Computation of the speech waveforms from the synthesis model.

We start with a discussion of the timing framework. Each segment has its duration computed from several influences. Among these is that of clause final lengthening, non-phrase-final shortening, non-word-final shortening, polysyllabic shortening, non-initial-consonant shortening, unstressed shortening, lengthening for emphasis, influence on vowel of postvocalic consonants, shortening in clusters, and lengthening due to plosive aspiration. In addition to these segmental effects is the insertion of pauses, which are extremely important to the naturalness and ease of comprehension of the speech. Pause insertion is based on the phrase structure as well as the number of syllables and words since the last pause, and these factors are weighted by heuristic considerations to yield appropriate pauses. It is interesting to note that these pauses allow the interpause speech rate to be as high as 180 words per minute without the sensation of hurried speech.

The computed timing framework can now be used to support the determination of an appropriate pitch contour. There are many factors, including semantic effects, needed to completely specify a pitch contour, but a hierarchical approach is used to capture the most important gestures. First, on a clause basis, a declination line is established for declarative sentences. WH questions require a high starting value of the pitch, plus emphasis on the last content word, whereas Yes/No questions require a plateau rise in pitch toward the end of the question. Pitch rises are specified on stressed syllables, and continuation rises are provided where the syntactic and punctuation specifications require it. These rules are quite complicated, and have been derived by carefully adjusting many parameters to obtain the best approximation to natural speech. Contemporary research has posited several abstractions as the basis for a pitch contour model, but as yet, none provides a simple schema while also accounting for the large range of phenomena observed.

We can now turn to the determination of the segmental parameters. A formant synthesizer model is used so that formants, bandwidths, pitch, and excitation amplitudes must be specified. For each vowel, target values for these parameters are obtained from a table. Only the first three formants and bandwidths are specified. For consonants, of course, both steady-state and transition targets are needed, and these are obtained from an examination of a five-phoneme-wide "window" to determine the appropriate contextual effects. In this way, a sequence of target values is obtained, which is then smoothed to yield a full set of parameters every 5 msec.

At this point, a detailed parametric description, using phonetically relevant variables, is at hand, and actual waveform synthesis can proceed using a variety of synthesis models. One could, for example, convert the phonetic parameters to LPC (linear predictive coding) coefficients for use with an LPC synthesizer, but in the present case, a digital resonance model is used so that the parameters are converted to corresponding coefficients for second order difference equations which characterize the resonance. These coefficients are also to be updated at a 5 msec. rate. One has now to use the coefficients with the synthesis model to actually derive a sampled digital waveform. Finally, the samples are passed through a D/A converter and low pass filter to yield the analog speech waveform.

IMPLEMENTATION AND EVALUATION

We strongly believe that it is important to distinguish between a research system and an application oriented system. For the former, perspicuity, ease of modification, relative machine independence, modularization of component procedures to isolate changes, use of high-level computer languages, and transportability are all important, whereas for the latter low cost, small size and power, real-time performance, standard interface, and the use of minimal space are paramount. Our emphasis has been on a research system, where we have used the language BCL for most of the symbolic (e.g. linguistic) calculations, and FORTRAN for the numerical (e.g. parameter and waveform) calculations. The entire collection of modules is coordinated by a well-defined set of file interfaces, and comprehensive trace and debugging facilities are supplied. Needless to say, any subset of the complete system can be used, allowing for ease of experimentation as well as the specification of reduced systems with lower performance. Implemented on a DEC 2040 computer with 256 K words of memory, the entire system runs in about 10 times real time when a real-time hardware synthesizer model is used. This circuit provides up to 32 second order difference equations (for gains or resonances) as well as glottal and noise sources and the capability to arrange the resonances in any desired pattern (e.g. cascade or parallel).

Given that the waveform synthesis is done in special real-time hardware, then most of the remaining delay is due to disk accesses to the lexicon and rule tables. With modern integrated circuit technology, these data bases can easily be accommodated in high density read-only memories. Thus the combination of integrated circuit memory, high performance microprocessors, and custom signal processing circuits can be expected to lead to compact (e.g. 3 by 5 inches) real-time low-power devices which can be produced at low cost. This situation can only improve with time as the degree of circuit integration increases. It is not unrealistic to expect that complete text-to-speech capability will be available within three to five years in the form of an optional circuit board for computer terminals, thereby providing voice output from computers corresponding to any desired text. These circuits will also find use in reading machines for the blind, computer-aided instruction, and interactive data base systems.

There is, of course, no point in pursuing the cost-effective implementation of these procedures unless the output speech is of high quality. Recent intelligibility and comprehensibility has shown that the segmental intelligibility is in the high 90% range, while comprehension is as good as for visual reading of the identical input text. Listeners report that they can readily understand the speech, and also find their place again if they miss a word. No tests for long-term fatigue have been made as yet, but the heavy emphasis on appropriate prosodic correlates helps to make the speech natural-sounding and easy to process in the cognitive sense. Surely there are many improvements that could be made, and undoubtedly will be incorporated with time, but it seems clear that the present speech quality is useful for many applications. Special hardware can be designed in a way to readily incorporate new findings. The coincidence in time of high performance text-to-speech algorithms together with the requisite low-cost implementation technology is indeed the signal to pursue the important applications of speech output from computers with confidence that practical systems, acceptable to naive human users, can now be constructed.

References:

Allen, J. *Synthesis of Speech from Unrestricted Text*. Proc. IEEE, 64, no. 4, April, 1976.

Allen, J., Hunnicutt, M.S., and Klatt, D.C. *MITalk 79*. Notes for MIT Summer Course, 6.69s, given June 25-29, 1979.

Flanagan, J.D., and Rabiner, L.R., eds. *Speech Synthesis*. Dowden, Hutchinson, and Ross, Inc., 1973.

AN OVER-VIEW OF SPEECH SYNTHESIS

Jean-Sylvain LIENARD

Laboratoire d'Informatique pour la Mécanique et les Sciences de l'Ingénieur (L.I.M.S.I.-C.N.R.S.)
B.P. 30 -91406 ORSAY Cedex - FRANCE

ABSTRACT
This paper gives a survey of the main present day approaches to speech synthesis. Several processing levels are envisioned, according to the nature of the input : acoustical synthesis, phonetic and prosodic command, phonetic and prosodic transcription of written text, as well as phonetic message generation from a conceptual formula, and synthesis by words. Intelligibility and quality of synthetic voices cannot be estimated without referring to the specific constraints and problems encountered at each processing level.

1. INTRODUCTION

The general problem of speech synthesis is not yet solved : we know how to transmit speech through low bit-rate channels, and how to transform text into intelligible speech ; but we are far from mastering all the high-level processes which allow the human to express himself with an intelligible, natural, personal and spontaneous voice.
Authentic or spurious, the speaking machines have always fascinated humans. Even the first serious one, KEMPELEN's machine (1780), was surrounded by an atmosphere of mystery ; the success of the speaking machines or devices was - and still is - due to the magic aspect of inanimate objects endowed with the ability to speak. But these marvellous machines never had practical applications : it is so easy to use the natural vocal apparatus ...
However the pioneers had discovered some important features of speech : relevance of the evolution from one speech sound to the next, and influence of a prior knowledge of the message upon its understanding (2).

J. C. Simon (ed.), Spoken Language Generation and Understanding, 397-412.

The great steps forward made during the last forty years in the field of speech synthesis were not motivated by the improvement of speaking machines ; they were the consequence of a strong economical interest in telephone bandwidth reduction and, more recently, in man-machine communication.
In fact the term "speech synthesis" is ambiguous, and covers processes of very different kinds. The present paper will be mainly devoted to a clearer definition of the functional stages of speech synthesis, with regards to the question of intelligibility and quality, as well as to the deep differences in nature between written and spoken communication.

2. WHAT DO WE CALL SPEECH SYNTHESIS ?

2. 1. Functional stages of speech synthesis

Synthesis systems can be classified according to the nature of the input (fig. 1).

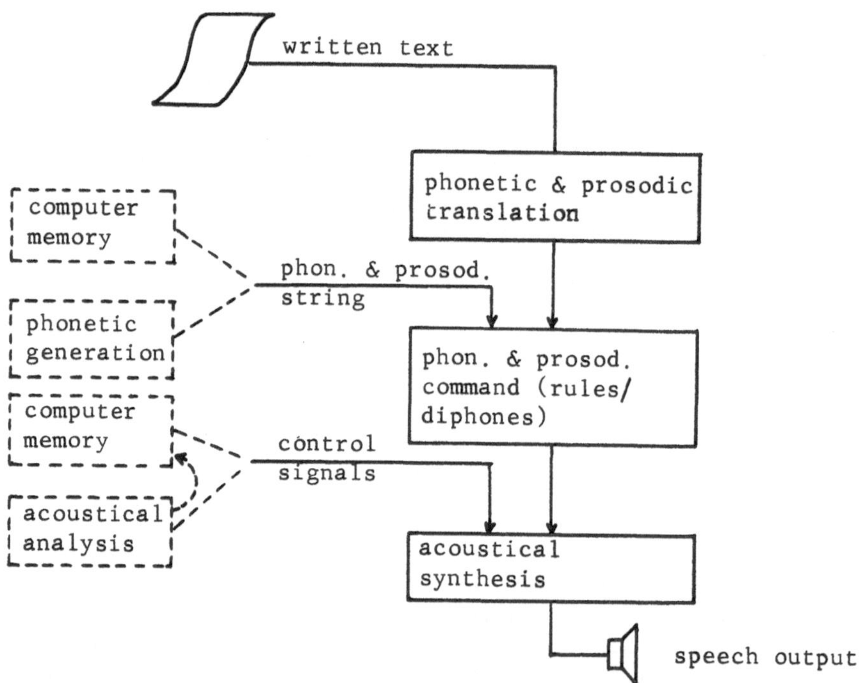

Figure 1. Functional stages of speech synthesis

If the input is a set of acoustical, time-varying control signals, whatever their coding could be, the process will be called acoustical synthesis. It does not depend on the language, but only reflects more or less the acoustical functioning of the human vocal apparatus. We will not include in this category the digital-to-analog conversion, which only achieves a well-known mathematical transformation.

If the input is a string of phonetic and prosodic symbols, the acoustical synthesizer must be preceded by a command system, which transforms this information into control signals appropriate to the acoustical synthesizer. The knowledge embodied in this stage is related to the different sounds of speech,their transitions and mutual relations in the phonetic system of a specific language. We will then speak of "phonetic synthesis".

If the input is a set of graphemic symbols - that is a written text - another different process is required in order to obtain a phonetic and prosodic string. This process is of a higher level than the previous one ; it makes use of pronouncing rules, lexicon, grammar, phonology, semantics ; it implies a correspondence between the written and spoken forms of communication. The usual term is "synthesis from the text".

One can imagine building a phonetic and prosodic string directly from conceptual data, using a phonetic message generation program. No name is still given to this process, but in fact it would be a "speech synthesis from the meaning", in the Artificial Intelligence acception.

Finally a lot of applications use a so-called "synthesis by words", which is nothing but the restitution of a pre-recorded real voice, eventually using the acoustical synthesizer of a VOCODER.

2. 2. Intelligibility and quality

The evaluation of a synthetic voice does not have any meaning if the previous classification is forgotten. A distinction is usually made between intelligibility and quality. Intelligibility is related to the linguistic content of the synthesized message : the measure results from the comparison between the input (phonemes, syllabes, words, sentences) and the output, as estimated by a group of listeners who are competent at the considered language level. In the case of acoustical synthesis, this comparison is meaningless, because the input itself comes from a manual or automatic analysis. Thus, the intelligibility of a VOCODER can only be estimated as a whole, including analysis.

Well-known relations between intelligibility of sentence and intelligibility of syllabes or meaningless words in the presence of noise (1) are not very well adapted to synthetic speech evaluation as defined above ; other tests, such as the Diagnostic Rhyme Test (3), which consider phoneme-pairs at a phonetic or phonological level, could provide a better approach. Anyway the human understanding of a synthetic sentence can be severely reduced by a very few number (10 to 20%) of phonetic errors (phoneme substitutions) (4).

Speech "quality" is a much fuzzier notion. It reflects some degree of acceptability of the synthetic (or transmitted) voice by a group of listeners. The measure is achieved through preference tests, without defining precisely the criteria used by the listeners. Two cases could be distinguished. If speech is unambiguously perceived as emitted by a human, quality is estimated according to all the alterations of the signal, including those related to intelligibility. Thus it is difficult to estimate quality when intelligibility is not perfect. If speech is perceived as artificial - or permits any doubt about its human origin, then quality reflects some degree of naturalness, that is the probability of finding in the linguistic community a human speaking in such a way. In that case the factors of quality are mainly related to prosody.

3. ACOUSTICAL SYNTHESIS

We will distinguish between two categories of acoustical synthesizers : those which are the terminal end of an analysis-synthesis system (VOCODER), and those for which the control signals cannot be obtained automatically. Most of those devices have been extensively described in the literature (1, 2), so we will give only brief functional descriptions of them.

3. 1. The Vocoders

In 1939 DUDLEY, a Bell Telephone Engineer, presented his VOice CODER, designed to reduce by a 10:1 ratio the telephone bandwidth (5). The analysis end of this channel Vocoder is a bank of 10 bandpass filters, equally spaced from 0 to 3000 Hz. The 10 signals obtained after rectification and low-pass filtering at 25 Hz are used at the synthesis end to control the gains of the synthesis filters, which are similar to the analysis ones. A 11th signal, which carries information about the voiced/unvoiced feature and the fundamental F_0, controls the nature (noise/pulses) and the frequency of the pulse source.
Speech transmission by VOCODER is limited in intelligibility and quality in several ways, mainly due to the analysis process :

- The low bit-rate ($\sim$ 2000 bits/s) is achieved through a severe quantization in time and frequency ; in particular the limitation at 25 Hz alters some consonantal features.
- The voiced/unvoiced decision is a binary one, and depends on thresholds, which cannot be adjusted in every case : the decision is often false.
- Pitch detection is also very difficult, depending on the particular speaker and recording conditions, and subject to numerous octave errors.

A lot of improvements have been proposed, the only satisfactory one being the VEV (Voice-Excited Vocoder). In this device the 300-900 Hz band is fully transmitted, so that both voicing and

pitch problems are avoided. But then the gain in bandwidth drops to 3:1, and the expected benefit becomes questionable.
In 1969, ATAL and HANAUER, also from Bell Laboratories, presented a new coding of the speech wave, adapted to the digital form, called Linear Predictive Coding (LPC) (6). The signal is considered to be stationary in successive time intervals. In a given interval, any sample can be estimated as a weighted sum of the preceding p samples (p is usually chosen between 8 and 15). The p weights, or prediction coefficients, are computed from a minimization of the prediction error over the time interval. The pitch and voicing are computed from the evolution of the prediction error, which is maximum at a definite instant of the voicing cycle. Synthesis, or signal reconstruction from the LPC coefficients, is achieved by means of a recursive filter, the input of which is a pulse or noise source. The reconstructed signal is very close to the original ; however pitch and voicing problems still remain, and the computing power required is large, especially to perform the analysis every 10 ms in real time. The LPC synthesizer, as well as the channel synthesizer, can be used in more complex synthesis systems, but the data obtained from the analysis devices or procedures often need expert retouching.

3. 2. The other acoustical synthesizers

In COOPER's PLAYBACK (7) and LEIPP's ICOPHONE (8) the input is a more or less schematized acoustical spectrogram. Synthesis is achieved through the sum of harmonic sinusoidal signals, weighted according to the spectrogram. Synthetic voice has a constant pitch, and a very artificial timber. The main advantage is that the researcher works with control signal which can be easily interpreted, because the information that can be visually extracted from the spectrograms is highly significant, as demonstrated in spectrogram-reading experiments. The formant (or "terminal-analog") synthesizer, due initially to W. LAWRENCE (9),makes use of a small number of bandpass filters (usually 3to5), the center frequency of which can be commanded as well as the gains and also possibly bandwidths. They are connected in a parallel (9,10) or serial form (11, 12, 13) ; in the latter case a particular design of the transfer functions permits an automatic control of the higher filters gains, and a reduction of the number of control signals. The source signal is either a noise, or a periodic signal, the precise shape of which can be very relevant in the synthetic speech quality (10). The formant synthesizers are able to provide a good speech quality. The problem is how to work out the control signals, for which no satisfactory automatic analyzer is available, and which cannot be extracted easily from the spectrogram.
In 1950 DUNN presented an electrical circuit reproducing the propagation of a plane wave in the vocal tract (14).The latter, con-

sidered as an acoustical pipe with a cross-section varying along the abscissa, is divided into cylindrical elements (15 to 30, in different realisations), each of them represented by a resonant circuit. The control parameters are analogous to the cross-section of all the cylindrical elements. The transmission line is excited by a periodic signal. Initially devoted to the study of vowel production, the vocal tract analogs have later received a parallel branch representing the nasal tract, and distributed noise sources. The parameters (area function of the vocal tract) are now controlled by computer, and their relations with experimental measures on the real vocal tract are subject to current work. They become now real articulatory synthesizers, giving good quality speech (15). However the difficulty of obtaining the control signals is still greater than it is in formant synthesis, because the physiological data concerning the vocal tract are very difficult to measure precisely. And the required amount of computing is very large, so that these devices must be considered as good tools for studying phonation, rather than practical synthesizers.

Finally the waveform synthesis is a technique of speech period reconstruction through a sum of elementary waveforms - sinusoids with a time envelop - stored in computer memory. This technique has been used since 1960 in the field of computer music (16), but its application to speech synthesis is rather recent (17, 18, 19, 20). Its advantage is to be entirely software, hence very flexible, and leading to inexpensive realizations, as far as the elementary waveforms of different frequencies can be read in memory by a table lookup. It does not modify the main existing approaches to speech synthesis, using fixed filters or frequencies, like (20), or variable filters - or formants - like (17, 18, 19). A point to be mentioned is that this technique (like the Playbacks) does not attempt to separate in the signal the excitation (glottal source) and the resonance of the vocal tract ; so it avoids some problems associated with the part of each of them in the speech quality. But it does not avoid the problem of obtaining the control signals, as stated above for all the synthesizers which are not parts of Vocoders.

In summary, a number of acoustical synthesizers are presently operating ; some of them can produce a speech signal almost undistinguishable from the original when the analysis deals with a real voice, the control signals being retouched by hand if necessary. The main factor is the information rate : over 10 kbits/s, excellent results can be obtained ; under 2 kbits/s losses of quality and intelligibility are practically unavoidable.

4. PHONETIC AND PROSODIC COMMAND

The input is now a string of phonetic and prosodic symbols, with a bit rate less than 100 bits/s. The output is a set of control signals, with a bit-rate 50 to 100 times higher, adapted to a given acoustical synthesizer.

4. 1. Phonetic command

There is usually an agreement, for a given language, upon a set of phonetic symbols necessary and sufficient to transcribe any oral message without altering its linguistic content - except informations on expression, context, speaker's personality and mood. However these symbols represent abstract linguistic items, called phonemes, whose realization in the physical signal is still not well known or understood. A word cannot be synthesized simply by concatenating speech sounds recorded in isolation. That looks obvious for syllables like [pa] , because it is practically impossible to pronounce a plosive consonant without a subsequent vowel. But it remains true for pairs of sounds like [ʃ] and [a] , or [y] and [i] , which can be pronounced in isolation, in a steady-state way.
The function of the phonetic command is then to transform a string of phonetic symbols into a continuous evolution of the control signals - it must be understood that the transitions cannot be reduced to a simple smoothing of the successive steady-state data, but that they represent a major part of the phonetic information. Two main techniques are currently used : command by rules, and command by diphones.
In the command by rules, transitions are computed between successive steady-state sets of control signals, according to mathematical functions, which, in some cases, take into account the nature of the involved phonemes. Several systems use a more or less sophisticated interpolation between successive steady-state data sets (21, 18, 22). Other approaches provide a greater insight in the speech perception or production processes. In 1965 HOLMES presented a set of rules designed for a parallel formant synthesizer(23).In 1967 COKER presented a model of the vocal tract functioning, described with the articulatory parameters used by the phoneticians. From two or more successive symbols the system computes the evolution of the main articulatory parameters (tongue shape and position, jaws, lips etc) each one with its own time constant ; the acoustical parameters are then computed from the vocal tract model and transmitted to a formant synthesizer.
The phonetic command by rules can provide good intelligibility and quality but at the price of extensive work, done by an expert phonetician. Some synthesis-by-rules softwares, presently on the market, give deceiving results because they do not incorporate a sufficient amount of phonetic research, as well as a corresponding computing power.
Another way to reproduce the necessary transitions is to store in a computer memory the corresponding sets of control signals - except the fundamental F_0 - representing the acoustical evolution

from one steady-state to the next (diphone, or dyad). A word can thus be reconstructed in associating the adequate diphones, like dominoes : [p a r i] = [p a] + [a r] + [r i] . This technique was presented in 1968, by DIXON and MAXEY for English (25), LEIPP and al. for French(26,27). Previous studies (28) foresaw its interest, but met with the difficulty of connecting signal parts extracted from tape recordings : the discontinuities existing at the juncture , especially on the F_0 evolution, altered strongly the intelligibility. This method can only be applied through a synthesizer that separates pitch from spectrum, or maintains a constant pitch. It does not provide any knowledge on speech production, nor takes into account coarticulation effects on more than 2 phonemes. But several experiments and realizations have shown its effectiveness. DIXON and MAXEY used a formant synthesizer and about 1000 hand-made diphones ; LEIPP and al. used a Playback (ICOPHONE) and about 400 diphones, normalized in duration. In both realizations the spectral juncture problem is avoided, because the steady-state spectra are defined before working on transitions. In some respects this approach is very similar to synthesis by rules : the only difference is that the rules are not explicitely defined. It is different in the works by EMERARD for French (30), OLIVE and NAKATANI for English, who use respectively a channel Vocoder and a LPC Vocoder in order to obtain the diphones from a real voice. The steady-state spectra of the same phoneme extracted from different speech sequences never match exactly : the juncture problem is solved by hand in EMERARD's work, by linear interpolation in OLIVE's, who only stores the central part of each diphone.
The command by diphones gives good results ; intelligible, good-quality voice is obtained in some realizations. It could permit to wait for the results of current fundamental researches in the field of articulatory synthesis by rules.

4. 2. Prosodic command

There is no agreement on the prosodic symbols necessary and sufficient to describe a spoken sentence. This is due to the fact that prosody is related to higher linguistic level than the phonetic string : lexicon, syntax, semantics, context etc. No universally accepted prosodic system exists at the present time. However, the notion of prosodic group has obviously some kind of relevance, evoked by a variety of terms : "breathing group", "meaning group" etc. Inside the sentence, the words are regrouped into larger entities. We will suppose, in the present paragraph, that a higher order process works out some prosodic markers, delimiting and identifying some prosodic groups (cf § 5.2 and 6.1). The function of the prosodic command is to associate to those markers the adequate control signals, that is phoneme duration, F_0 evolution, and intensity.

We will restrict our presentation to two realizations in French (for English see J. ALLEN (31)). But some preliminary remarks could be pointed out. Firstly, one aspect of prosody ("microprosody") reflects in fact phonetic (segmental) and articulatory constraints. Secondly, lexical stress is not very important in French. Thirdly, speech intensity (more or less correlated to voice timber) is difficult to define and seems to be of secondary importance in French. So the main prosodic parameters are duration and pitch variations, except for the slight changes due to microprosody.

In EMERARD's system (29), the markers are syntactic, chosen in a repertory of about 10 items, and placed by hand in the phonetic string by an expert operator. Each marker is associated with a specific pitch evolution, and two duration rules, one concerning the duration of the last vowel of the group, the other deciding of a pause and its duration. For example, the marker pointing to the end of a group in a preverbal syntagm is associated with a falling pitch, a slight increase of the last vowel duration and no pause, etc.

The system presented by CHOPPY and al. (32, 33) makes use of 3 different markers only. Prosodic command is achieved in two steps, and concerns mainly F_0 : each prosodic group is first associated with a specific F_0 evolution (rise/fall contour), according to the position of the group in the sentence (final/non final) ; in the second step a F_0 evolution concerning the whole sentence is superimposed on the elementary contours.

In phonetic synthesizers such as VOTRAX (13), the user must enter the string of phonetic symbols and prosodic markers. This requires a linguistic competence, that is a severe limitation of the field of application. In order to overcome this difficulty, it is desirable to transcribe automatically the written text into a phonetic and prosodic string.

5. PHONETIC AND PROSODIC TRANSCRIPTION OF WRITTEN TEXT

This process uses high level linguistic notions, specific to each language. Once more we recommend the paper by J. ALLEN (31) for the work done in English.

5. 1. Phonetic transcription

The subject was first evoked in France, in the frame of speech synthesis, in 1969 (34, 35). TEIL's program, as well as the successors (there are a lot of them, cf JEP GALF, Grenoble 1979 - let us mention, among others, DIVAY and GUYOMARD (36), PROUTS (37)) is based on context-sensitive pronunciation rules such as "P" followed by "H" is pronounced [f] , or "E" followed by "A" and "U" is pronounced [o] . The words having a non-regular pronunciation, such as "MONSIEUR" [m ə s j ø] or "OIGNON" [ɔɲɔ̃]

are stored in a dictionary of exceptions, which is looked up before applying the pronunciation rules. The programs differ in their implementation, in execution time and ease of use. They make very few errors (of the order of 1 ^{o}/oo) on texts from the contemporary French literature, provided that they include a few hundredths to one thousand rules and exceptions.
But two restrictions must be pointed out. Firstly the notion of error can be estimated in two ways : by comparison of the resulting phonetic string to a reference coming from an expert phonetician or from a pronunciation dictionary - and those transcriptions can differ from each other to a large extent (linking, mute "E", stress, etc) -, or by listening to the speech output given by a phonetic synthesizer - and in the latter case several problems are mixed. Secondly there are in French some words that are written using the same string of characters (graphemes), but that are not pronounced the same way, according to their grammatical function (for example the well-known sentence "LES POULES DU COUVENT COUVENT" [l ɛ p u l d y k u v ã k u v] "the hens of the convent are sitting on their eggs"). Such cases are not frequent in the usual language ; clearly, a syntactic analysis of the text is needed to solve them. But it is an expensive, time consuming process, and some researchers consider that its implementation in a speech synthesis system is not worthwhile.

5. 2. Prosodic transcription

As stated above there is no precise definition of the prosodic groups. Consequently there is no universally accepted way to compute the prosodic markers from the text. In EMERARD's system they are placed by the operator, so the problem does not arise. In CHOPPY's system we started from the idea that syntactic knowledge was not essential to work out an acceptable prosody : the prosodic groups are constituted in concatenating the functional words to either the preceding, or the following word. The functional words are chosen in a list of about 50 elements, such as articles, pronouns, conjunctions, etc. This algorithm gives only likely markers, but its evaluation is difficult because it can only be conducted by listening to the resulting synthetic speech; evaluation of prosodic transcription is meaningless if all the lower level processes are not perfect.
There is another, much deeper question, related to the supposed equivalence between written and spoken messages. Is an automatic transcription from written text to fluent speech a legitimate transformation ?
Written communication concerns two persons - writer and reader - who are immersed in separate universes, over long distances or long times, without any feedback or interaction with each other. So the written message must be strongly structured in order to carry the information with as little ambiguity as possible: orthography,grammar, separation of words with spaces, character nor-

malization, etc. On the other hand, spoken communication concerns two persons who are within call ; hence they share the same environment, they are aware of the same situation, they react upon each other by voice and by gesture. It follows that, by nature, speech communication is less structured than written communication because it does not need to ; this is obvious at every processing level, and this is the source of most difficulties in speech recognition. There are in speech some aspects that the mere text cannot give any account of : the context of the situation, the instantaneous interaction between interlocutors, and even the goal of the communication (knowledge transmission for writing, immediate action for speech). Those speech specific features are mainly coded into prosody.

Reading a text aloud is not a usual communication situation ; it supposes a particular attitude of the reader towards the listener,and an interpretation of the text. The reader first understands the meaning of the text, then adapts his elocution to the present situation with respect to the listener, putting stress on some words, pausing here and there, ensuring that the listener still pays attention, etc.

Thus a perfectly "natural" synthesis from the text should include similar considerations, that are far beyond the natural language processing studies of the present day.

But, apart from the reading machine for the blind, is there a large interest in making a machine read a text aloud with intelligence ? We will slightly modify our point of view in the next paragraph, by considering speech synthesis in the frame of man-machine communication.

6. SPEECH SYNTHESIS IN MAN-MACHINE COMMUNICATION

6. 1. Phonetic message generation

Man-machine communication takes place in a limited universe. In fact the operator carries out a task by means of the machine, the function of which is to interpret the operator's instructions in the context of the task, to deduce from them commands towards the task, to interpret the informations coming from the task and to decide a new action, or a question or comment to the operator (38, 39). In the LIMSI project the heart of the system will be a monitor handling conceptual informations. Synthesis and recognition-understanding are symmetrical (fig. 2).

The phonetic generation program transforms a logical-semantic message directly into a phonetic and prosodic string (40) ; for example the following list (where the concepts are labelled in English) :

INTERROGATION(EAT(DEFINED(PLURAL(CHILD)), DEFINED(SOUP)))

is transformed into the French phonetic and prosodic string

[ɛ s k ə l ɛ z ã f ã - m ã ʒ $ l a s u p ?]

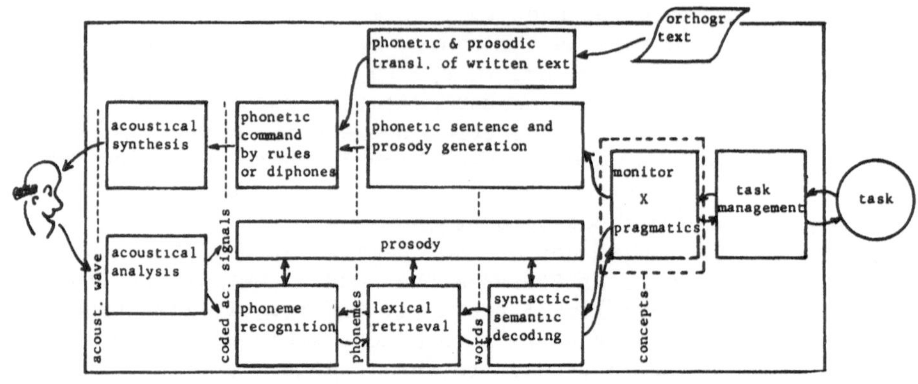

Figure 2. The man-machine speech communication project at LIMSI

that is in written form "Est-ce que les enfants mangent la soupe ?" ("do the children eat the soup ?").
The prosodic markers are "-" and "$" (pre-and postverbal) and "?". This sentence is then transmitted to a phonetic synthesizer, with the adequate prosody.
This work, still in an early stage, indicates a new approach to speech synthesis : the informations related to context, interaction, message structures are present in the system, at least within the limits of the task-operator universe. Such informations, which are so difficult to extract from a written text, or are not contained in it, are precisely those needed to work out prosody.
The relatively simple linguistic structure of the generated sentences is not a drawback in man-machine communication : we have observed that, in the frame of a definite task, short, concise messages are always preferred to wordy, boring sentences.

6. 2. Speech synthesis by words

In this approach, a limited number of messages, recorded by a human speaker, were digitized and stored in a computer. In general this is merely a form of speech recording ; in some cases speech is analysed and restituted by means of a Vocoder, so as to save memory space. Frequently the stored data have been retouched, in order to eliminate the analysis errors. The prosody of the restituted speech is entirely the prosody which was adopted by the reader, except in some systems which use acoustic and prosodic rules in order to insert new words into predefined sentences (40).

In spite of its drawbacks (waste of memory, limited vocabulary), speech "synthesis by words" still is widely preferred by users. There are two main reasons to that. Firstly, the speech output is perceived as "natural", or at least, "acceptable" ; it is, indeed : even unconsciously, the reader understands the meaning of the message, anticipates the type of situation in which the future listener will operate, and expresses this knowledge in his prosody.

Secondly, a limited vocabulary is sufficient in most applications, as it is in speech recognition by words. Why would a potential user choose a more sophisticated, more expensive, less pleasant system ?

However we strongly believe that new techniques and new possibilities such as text processing,man-machine interaction, database interrogation, etc... will very soon emphasize the need of more powerful speech synthesis tools. And the notion of naturalness, according to our definition in § 2.2, will probably evolve in the next years, because the human will become accustomed to synthetic voice, and will no more consider it as a magical operation.

7. CONCLUSION

We attempted some classification of the large set of processes called speech synthesis. We will summarize here the guidelines of our paper.

Acoustical synthesis transforms control signals, resulting or not from an automatic analysis, into a speech wave. Some of the existing techniques are able to reconstitute a quite acceptable speech, provided that the information rate is at least 5000 bits/s.

Phonetic command, which computes the control signals from a phonetic string, uses articulatory or acoustic rules, hand-made or real-voice originated diphones ; the latter approach is more practical, when the former provides a better insight into the fundamental phonation processes.

Automatic phonetic transcription of the written text is practically solved for French, except the problems related to syntax and semantics. On the other hand, computing the prosodic markers from the text, and associating them with appropriate pitch and duration contours, is still in the field of current research.

We tried to show that it is illusory by now to expect a synthetic voice to be as sensitive, expressive, intelligent as a human voice. But the more restricted context of man-machine communica-

tion could provide a new, realistic field of research.
We did not evoke the present-day or future applications, nor the different experiments carried out in all the research laboratories, for which the reader is referred to (42). But we must recognize that artificial speech, which has been for centuries a subject of curiosity, is now becoming a real communication tool, of which the computer is simultaneously the end and the means. Is it the end of a myth and the beginning of the Artificial Intelligence era, or a simple technological illusion ?

REFERENCES

abbreviations :

SS : book of reprints on Speech Synthesis, edited by J.L. FLANAGAN and L. RABINER, Dowden, Hutchinson and Ross, Inc. 1973

JEP : Journées d'Etude sur la Parole, proceedings of the annual meeting of the Speech Communication Group of the Groupement des Acousticiens de Langue Française (GALF)

ICASSP : International Conference on Acoustics and Speech Processing of the Institute of Electrical and Electronics Engineers

JASA : Journal of the Acoustical Society of America

1 - FLANAGAN, J.L. - Speech Analysis, Synthesis and Perception, Springer Verlag, Berlin, 1972
2 - LIENARD, J.S. - Les processus de la Communication parlée, Masson, Paris, 1977
3 - VOIERS, W.D. - Diagnostic approach to the evaluation of speech intelligibility, JEP, Aix-en-Provence, 1971
4 - LIENARD, J.S., MARIANI, J.J., RENARD, G. - Intelligibilité de phrases synthétiques altérées : application à la transmission phonétique de la parole, 9th International Congress on Acoustics, Madrid, 1977
5 - DUDLEY, H. - Remaking speech, JASA 11, 1939, reprinted SS
6 - ATAL, B.S., HANAUER, S. - Speech Analysis and Synthesis by Linear Prediction of the Speech Wave, JASA 50, 1971, reprinted SS
7 - COOPER, F.S., DELATTRE, P., LIBERMAN, A.M., BORST, J.M., GERSTMAN, L.J. - Some Experiments on the Perception of Speech Sounds, JASA 24, 1952, reprinted SS.
8 - LEIPP, E., CASTELLENGO, M., SAPALY, J., LIENARD, J.S. - Structure physique et contenu sémantique de la parole, La Revue d'Acoustique 3-4, 1968
9 - LAWRENCE, W. - The Synthesis of Speech from Signals which have a Low Information Rate, Communication Theory, London, 1953

10 - HOLMES, J.N. - The Influence of Glottal Waveform on the Naturalness of Speech from a Parallel Formant Synthesizer, IEEE Trans. Audio, vol. AU-21, 3, 1973
11 - FANT, G. - Analysis and Synthesis of Speech Processes, Manual of Phonetics, ed. B. MALMBERG, North-Holland American Elsevier, 1968
12 - KLATT, D.H. - Acoustic Theory of Terminal Analog Speech Synthesis, Proc. 1972 Int. Conf. on Speech Processing, Boston, Mass., 1972
13 - GAGNON,R.T. - VOTRAX, Real-Time Hardware for Phoneme Synthesis of Speech, IEEE-ICASSP, Tulsa, 1978
14 - DUNN, H.K. - The Calculation of Vowel Resonances, and an Electrical Vocal Tract, JASA 22, 1950, reprinted SS
15 - MRAYATI, M., GUERIN, B. - Etude des caractéristiques acoustiques des voyelles orales françaises par simulation du conduit vocal avec pertes, Revue d'Acoustique, 36, 1976
16 - MATHEWS, M.V. - The Technology of Computer Music, M.I.T. Press, Boston, 1969
17 - BOURGENOT, J.S., DECHAUX, C. - Codage de la parole à faible débit : le vocoder CIPHON, Revue Technique Thomson-CSF, vol. 7, 4, 1975
18 - RODET, X. - Analyse du signal vocal dans sa représentation amplitude-temps. Synthèse de la parole par règles, Thèse d'Etat, Univ. P. et M. Curie, Paris, 1977
19 - BAUMWOLSPINNER, M. - Speech Generation through Waveform Synthesis, IEEE-ICASSP, Tulsa, 1978
20 - ASTA, V., LIENARD, J.S. - L'Icophone logiciel, un synthétiseur par forme d'onde, JEP, Grenoble, 1979
21 - KELLY, J.L., LOCHBAUM, C. - Speech Synthesis, Speech Communication Seminar, RIT, Stockholm, 1962
22 - RABINER, L.R. - A Model for Synthesizing Speech by Rules, IEEE Trans. Audio, vol. AU-17, 1969
23 - HOLMES, J.N., MATTINGLY, I.G., SHEARME, J.N. - Speech Synthesis by Rules, Language and Speech, 7, 1964, reprinted SS
24 - COKER, C.H. - Synthesis by Rule from Articulatory Parameters, IEEE Conf. on Speech Comm. and Processing, Cambridge, Mass., 1967, reprinted SS
25 - DIXON, N.R., MAXEY, H.D. - Terminal Analog Synthesis of Continuous Speech using the Diphone Method of Segment Assembly, IEEE Trans. Audio, AU-16, 1968
26 - LEIPP, E., CASTELLENGO, M., LIENARD, J.S. - La synthèse de la parole à partir de digrammes phonétiques, 6th Int. Congress on Acoustics, Tokyo, 1968
27 - LIENARD, J.S., TEIL, D., CHOPPY, C., RENARD, G. - Diphone Synthesis of French : Vocal Response Unit and Automatic Prosody from the Text, IEEE-ICASSP, Hartford, 1977
28 - PETERSON, G.E., WANG, W.S.Y., SIVERSTEN, E. - Segmentation Techniques in Speech Synthesis, JASA, 30, 8, 1958
29 - EMERARD, F. - Synthèse par diphones et traitement de la prosodie, Thèse 3e cycle, Univ. de Grenoble, 1977

30 - OLIVE, J.P., NAKATANI, L.H. - Speech Synthesis by Rule, Speech Comm. Seminar, Stockholm, 1974
31 - ALLEN, J. - Synthesis of Speech from Unrestricted Text. Proc. IEEE, vol. 64, 4, 1976
32 - CHOPPY, C., LIENARD, J.S., TEIL, D. - Un algorithme de prosodie automatique sans analyse syntaxique, JEP, Toulouse, 1975
33 - CHOPPY, C. - Introduction de la prosodie dans la synthèse vocale automatique, Thèse de Docteur-Ingénieur, Univ. P. et M. Curie, Paris, 1977
34 - TEIL, D. - Etude de génération synthétique de parole à l'aide d'un ordinateur, Thèse CNAM, Paris, 1969
35 - LIENARD, J.S., TEIL, D. - Les éléments phonétiques et la traduction automatique du message écrit en message parlé, Automatisme, vol. 15, 10, 1970
36 - DIVAY, M., GUYOMARD, M. - Conception et réalisation sur ordinateur d'un programme de transcription graphémo-phonétique du français, Thèse de 3e cycle, Univ. de Rennes,1977
37 - PROUTS, B. - Traduction phonétique de textes écrits en français, JEP, Grenoble, 1979
38 - IRIA-CREST 1978 Course on "Trends in Man-Machine Communication", a collective book to be edited by LIENARD, J.S., Paris
39 - MARIANI, J.J., LIENARD, J.S., RENARD, G. - Speech Recognition in the Context of Two-Way Immediate Person-Machine Interaction, IEEE-ICASSP, Washington, 1979
40 - MEMMI, D., LIENARD, J.S. - Génération automatique de phrases en phonétique à partir de formules sémantiques, JEP, Grenoble, 1979
41 - BOE, J.L., LARREUR, D. - Synthesis by Rule of Enunciative Sentence in French, preliminary Study, Speech Comm. Seminar, Stockholm, 1974
42 - CARRE, R., HATON, J.P., LIENARD, J.S. - Etat de l'art et applications de la synthèse et de la reconnaissance automatique de la parole, Report to be published by IRIA, 1979

SPEECH SYNTHESIS FOR AN UNLIMITED VOCABULARY, A POWERFUL TOOL FOR INQUIRY AND INFORMATION SERVICES

W.K. Endres and H.E. Wolf

Research Institute of the Deutsche Bundespost, Darmstadt

Owing to the progress in digital speech processing, several Telecommunication Administrations are making studies to decide whether or not and to what extent it is possible to improve existing services and offer new ones by using the telephone as terminal. Whereas the problems of automatic speech synthesis have largely been solved, this does not apply to automatic speech recognition. For the near future, however, it seems justifiable to feed in the request with the aid of dial or push buttons.
Among the numerous procedures to be used for the generation of synthetic speech, the synthesis from sound elements has proved to be particularly convenient. A historical survey is given of 30 years of investigations which finally resulted in the development of synthesizers generating very natural speech. The paper introduces such a synthesizer which is known by the name of SAMT (abbreviated for Sprachausgabe in Multiplextechnik). It operates according to the principle of the formant vocoder which is supplied with the texts to be synthesized consisting of 8-bit encoded monophones and prosodic elements. The latter allow the speech rhythm, intonation and stress to be controlled. The implementation of SAMT is based on a very fast microprocessor system (4 x AMD 2901) for sound parameter control supported by a very efficient formant filter hardware. At present, SAMT is able to supply different responses for a maximum of 32 subscribers at the same time.

J. C. Simon (ed.), Spoken Language Generation and Understanding, 413-428

1. INTRODUCTION

Man's growing need for an increasingly faster access to more and more information creates a series of yet unsolved technical, organizational and human problems in the inquiry and information services, which are still largely performed by human beings. This is, above all, due to the following two reasons: The number of telephone subscribers continues to grow very fast; but this process involves, in addition to the new subscriber sets to be installed, frequent changes of the telephone numbers, which can hardly be avoided because of the necessary extensions of the transmission equipment and networks. On the other hand, it is not possible, in general, for an individual subscriber to have the directories of all telephone subscribers, e.g., of his own country. For both reasons, a heavy strain is placed on the directory inquiry service. As a consequence, there are intolerably long queues of inquiring subscribers. Similar phenomena can be observed worlwide in many other telephone inquiry services, e.g., in railway and air traffic. This may be remedied by the additional introduction of automatic announcement and inquiry services which would also satisfactorily solve the problem of round-the-clock availability.

For the reception of the subscriber's request by the computer, there do not yet exist procedures and equipment for the unambiguous recognition of single words or even continually spoken texts. During an intermediate period, however, it is possible to make do with the dial or push buttons of a telephone set, which can be used for the input of a large, although limited number of requests into a computer with the aid of a suitable alphanumeric code. In this way, it is already possible to instal an automatic telephone information service which will be likely to eliminate the bottle-neck in the inquiry service and to allow the introduction of new (announcement and intercept) services.

The most important link between such an automatic device and the user is the speech synthesizer which must be able to generate any speech texts desired in understable and naturally sounding pronunciation and also to provide different texts for several inquiring subscribers at the same time. Efforts have been made for several decades at a number of research institutes all over the world, especially those owned by

Telecommunication Administrations, to develop appropriate procedures and equipment.

In the following, the basic operation of a speech synthesizer developed at the Research Institute of the Deutsche Bundespost will be described. This unit, which performs a synthesis from sound elements, simultaneously supplying up to 32 different texts with an unlimited vocabulary for 32 channels, is known as SAMT (abbreviated for Sprachausgabe in Multiplextechnik). To begin with, we will give a brief survey of the historical development of speech synthesis procedures.

2. SPEECH SYNTHESIS FROM SOUND ELEMENTS HISTORICAL SURVEY

2.1. Sound elements as universally applicable speech components

Among the numerous procedures to be used for a universal speech synthesis, which have been conceived and studied since the publication of W. v. Kempelen's famous book [1], one has recently been developed in such a way that it is helpful for the solution of the above-mentioned problem. This is the procedure of speech synthesis from sound elements, the implementation of which has been attempted by researchers for almost 30 years. The procedure is based on the assumption that it must be possible to concatenate the sound elements of a language, represented by phonetic symbols, according to certain rules in order to obtain naturally sounding speech. This is done in a way similar to the combination of written texts from the letters of the alphabet. As far as we know today, first experiments in this field were made by C.M. Harris [2,3] as early as 1950. In his publications, Harris already pointed out that such sound elements should be allophones in order to ensure a sufficient intelligibility of the synthesized speech. In a series of studies he showed that the time-frequency spectra of the individual sound elements largely depend on the neighbouring sounds and that allophone substitutions create different acoustic impressions.

With the aid of stylized spectrograms, these problems have been thoroughly studied by F.S. Cooper, A.M. Liberman, P.C. Delattre et al. at the Haskins Laboratories. These investigations led to the formulation of the loci theory. The results of this research, which were published in more than 50 papers (e.g. [4-7]) are still of interest for today's scientists.

2.2. First implementations

The concept for the presently used procedures of speech synthesis from sound elements was described by K. Küpfmüller and O. Warns. They discovered that, instead of monophones, transients have to be used as sound elements, each of which begins in the middle of one sound and ends in the middle of the following sound [8]. Such sound elements have later been termed "diphones".

According to Küpfmüller's estimate, the synthesis of monotonously sounding German speech requires approx. 1640 diphones. If prosodic features are included, this figure rises to about 3540. This corresponds to a rate of 200 bit/s. O. Warns designed an analog device which allowed intelligible and rather naturally sounding speech texts to be synthesized, although it was not applicable in practice because of the large memory required and the difficult implementation [9].

For a feasible solution, it would have been necessary to reduce the number of sound elements considerably. Therefore, Küpfmüller suggested that, instead of the real transients, the neutral schwa /ə/ be introduced between the monophones. This speech synthesis procedure would only require 45 sound triads of the type /ə·ə/. Experiments for a speech synthesis from such triads have been reported by B. Cramer. The experiments using analog systems supplied synthetic speech of a fair quality, although insufficient for telephone transmission [10].

2.3. Digitization as a basis for present-day procedures

However, because of the poor speech quality and, even more so, because of the very low processing speed in the composition of synthetic texts, it became soon evident that such analog procedures were not applicable in telecommunication services. Only after cheap digital computer and digital redundancy-reducing signal processing methods had been developed, was it possible to tackle the problem of speech synthesis from sound elements on a more solid technical basis. The experiments using transients obtained from naturally spoken speech such as suggested by Küpfmüller were continued and, on the other hand, procedures were studied in which the transients are only given as spectral parameters.

For the procedure using sound elements stored in the time domain, it is necessary to reduce the memory, which is still very large even for digital processing. Therefore, use was made, in the synthesis, of certain psycho-acoustic effects which allowed the number of sound elements to be reduced from approx. 1300 to about 300 [11-14]. Finally, these experiments led to the development of a device which is relatively easy to implement and supplies monotonously sounding speech of an unlimited vocabulary in a multi-channel audio-response system. The intelligibility is probably sufficient for many applications. However, the introduction and control of prosodic parameters, necessary for increasing the naturalness, seems to be difficult which, at present, still sets a limit to the commercial use of this procedure [15].

For about two decades, attempts have been made to concatenate sound elements in the time-frequency domain, i.e. with the aid of spectral data such as formant frequencies, formant amplitudes, formant bandwidths and control data for the excitation function. The synthesis equipment closely resembles the synthesizer of a formant vocoder provided with controllable pulse and noise excitation sources and with an equally controllable filter system as an analogon to man's speech generation system. Although in this case the individual sound elements are arranged in a time sequence, this procedure is termed speech synthesis in the frequency domain in order to underline the basic difference between both methods. The necessary spectral data of the sound elements are extracted, by means of spectral analyses, from naturally spoken patterns containing all sounds and sound combinations required for the synthesis. These data are stored in a memory to be available for the synthesis.

Apart from the different versions of the OVE devices developed by G. Fant [16,17], first experiments of this kind were carried out, as far as we know, by H.D. Maxey et al. in 1964 [18] and by N.R. Dixon and H.D. Maxey in 1967 [19]. These procedures and especially those described in detail in the following differ essentially from each other in the used filter systems, the rules for the concatenation of sound elements and the number and structure of sound elements, although the basic principle is the same: simulation of the human speech generation system by an appropriately excited filter network and its control with the aid of spectral data extracted from natural speech. In all recently developed procedures, microprocessors are used for digital signal processing and control. Another common aspect is the fact that the naturalness of the synthesized speech can be more easily increased by controlling the prosodic features than in the case of the above-mentioned time-domain procedure. Another approach is being made with a speech synthesizer developed at C.N.E.T., Lannion (France). Here, the synthesizer of a channel vocoder is used for sound generation [20].

The problem of speech synthesis was also thoroughly investigated in recent years at the Deutsche Bundespost Research Institute. The synthesizer designed according to the formant-vocoder principle is known by the name of SAMT. The equipment, whose operation principles will be outlined in the following, is already usable for multi-channel applications in the telephone directory inquiry, announcement and intercept services.

3. THE SAMT AUDIO-RESPONSE SYSTEM

3.1. Integration of SAMT into automatic inquiry systems

The integration of SAMT into an automatic inquiry system is evident from Fig. 1. The heart of this inquiry system is the inquiry computer. The telephone network and appropriate interfaces enable any person calling for information to communicate with the inquiry computer. The input interface may either be designed as a decoding device for keyboard-selection signals or, in the future, also as a speech-recognition system. As a consequence, the subscriber has to

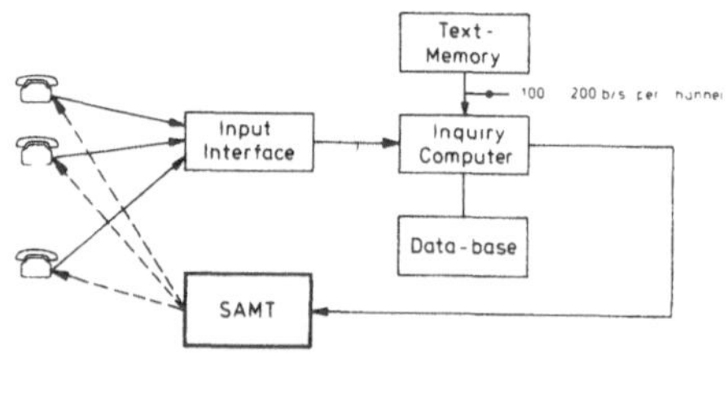

Fig. 1 Integration of SAMT into an automatic inquiry system

feed in his request by using the keyboard for encoding or directly as a spoken question. Once the inquiry computer has recognized the request, it takes the required information out of the data base. To supply this information for the subscriber, the computer uses SAMT operating as output interface together with the text memory. The latter contains the whole system vocabulary consisting of 8-bit encoded monophones and prosodic elements, which require an average storage capacity of approx. 100-200 bits for text durations of 1 sec each. The text to be synthesized is read out of the text memory in units of words or sentences, given a channel address and transmitted to SAMT to be converted into speech signals understandable for the subscribers. At the present stage of development, SAMT is able to supply different responses for a maximum of 32 subscribers at the same time.

3.2. Structure and implementation of SAMT

The main units of SAMT (Fig. 2) are the sound parameter control with the associated parameter memory, the interpolator, the formant synthesis filter and the D/A converter with an analog demultiplexer for 32 channels.

The sound parameter control unit has to perform a very complex task which includes, in addition to the control of the data exchange between the inquiry computer and SAMT, the supply of sound data from the parameter memory, the modification of these data both according to their phonetic environment and prosodic features, a rough pre-interpolation of the parameters and the control of the fundamental frequency f_0. After modification and rough interpolation by the sound parameter control and smoothing in the fast linear interpolator, the control parameters are supplied to the formant filters.

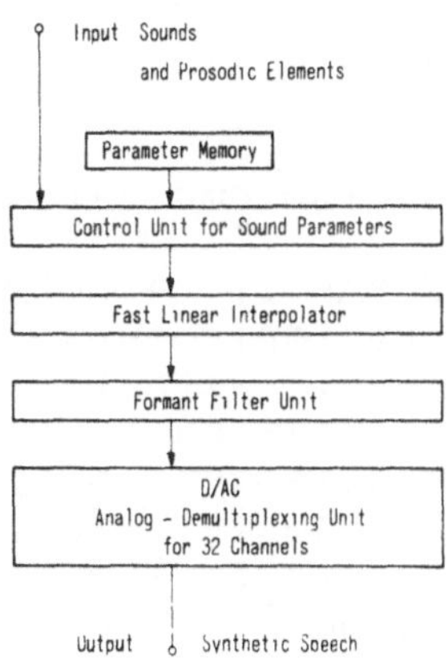

Fig. 2
Main units of SAMT

The structure of the formant filter unit is shown in Fig. 3. It consists of 7 filters arranged in parallel for four vowel, two nasal and one fricative formant, which can be controlled with respect to resonance frequency F (0-4000 Hz), resonance amplitude A (0-100 %) and bandwidth B (50 or 150 Hz for voiced and 150 or 350 Hz for voiceless excitation). They can be excited either by periodic pulses or noise. Since the type of excitation can be adjusted separately for each filter, it is also possible to generate voiced fricatives and plosives by exciting only the first two formants in a voiced and all higher formants in a voiceless manner.

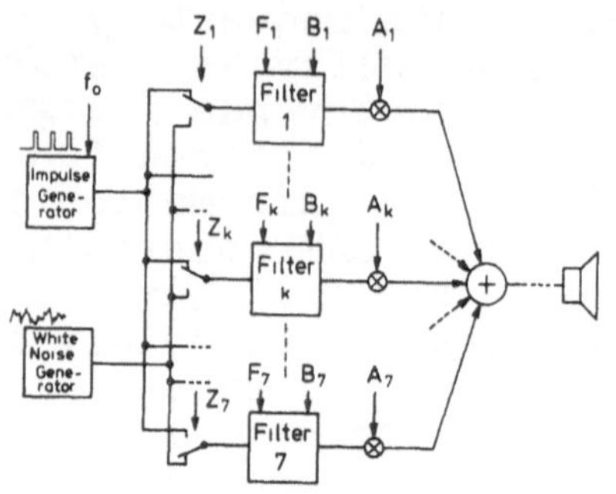

Fig. 3 Structure of the formant filter unit

Since the individual function groups of SAMT have to perform entirely different tasks, their hardware implementations are also different. The large number of complex and heterogeneous functions of sound parameter control are fulfilled by a very fast microprocessor system (4 x ADM 2901). The interpolator and the filter unit, on the other hand, are structured according to the pipeline principle and designed in TTL technique. The whole hardware required has been reduced to a minimum by using a time-multiplex procedure for all channels.

3.3. Implementation of the formant filters

For the filter structure shown in Fig. 3, a very efficient implementation was found, which generates voiced formants with the aid of a transversal filter [21]. The large number of computations required, in general, for this type of filter can be drastically reduced in this special case. To begin with, the number of filter elements is limited according to the duration of one fundamental frequency period. The error caused by this limitation is reduced by a phase correlation in such a way that no losses in speech quality occur. Since a sequence of pulses having a pulse width equal to the duration of one sample is taken for the excitation, only <u>one</u> multiplier of the filter is active at the same time, while <u>all other</u> multipliers perform the trivial multiplication by zero. Consequently, the filter has only one multiplier and even this can be replaced by an adder, if the corresponding data are stored in logarithmic form. A second adder is used for weighting the filtered signals with the logarithmic formant amplitude.

Since this simplification of the transversal filter is only possible in the case of pulse excitation, a modulation procedure instead of a filter procedure is used for the synthesis of voiceless sounds. A low-pass filtered noise signal, whose cut-off frequency is equal to half the formant bandwidth, is modulated by multiplication by one sinusoidal oscillation, whose frequency is equal to the formant frequency. This is done in such a way that the upper and lower sidebands together yield the desired spectrum of the voiceless "formants". Consequently, the generation of this signal also requires only one multiplication, which, in the case of the signals being available in logarithmic form, can again be reduced to an addition. Thus, it is possible to generate voiceless formants with the aid of the same equipment as that used for voiced formants.

This formant filter unit can be designed in Schottky-TTL-technique according to the well-known pipeline principle in such an economical way that 224 filter functions required for 32 speech channels can be implemented on one double Europe board. The number of channels can be easily extended to 128.

3.4. Arrangement and processing of the synthesis parameters

The parameters for the synthesis filters are supplied within the sound parameter control at two different processing stages:
At the first stage, the parameters for transients are supplied. For this purpose, the sound sequence $\nu_1, \nu_2, \ldots, \nu_i \ldots$ (Fig. 4) fed in at the input is converted into a sequence of transient

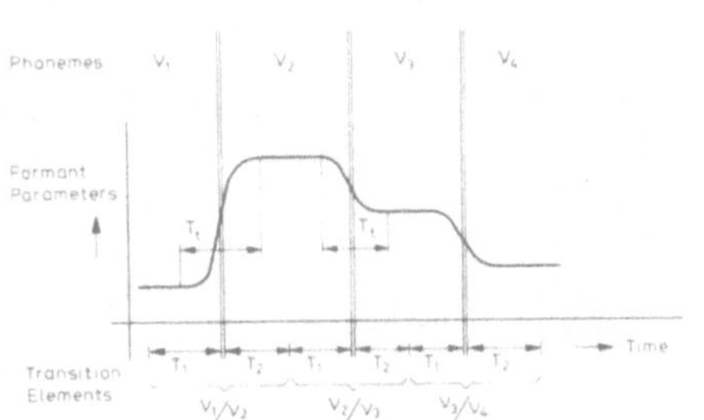

Fig. 4
Sound sequence and definition of time-intervals

elements $\nu_1/\nu_2, \ldots, \nu_i/\nu_{i+1}, \ldots$. This is done according to Küpfmüller's and Warn's abovementioned definition of a transient as the range between the middle of the steady-state of a sound and the middle of the steady-state of the following sound. Very inhomogeneous sounds such as plosives and diphthongs are divided into two parts of sounds (the sound /R/ even into four). Each of them is regarded as homogeneous and can, therefore, be treated in the following as a simple sound.

For each transient element, the following data sets are stored in the parameter memory.

a) for either sound of a transient, one complete filter parameter set with 7 formant frequencies, 7 formant amplitudes, 7 bandwidths, and 7 excitation modes (voiced or voiceless). The parameter values are essentially those of the beginning and end of the steady-state parts of these sounds. The term "steady state" is used here for desired parameter shapes between a starting and an end point, which can be approximated with sufficient accuracy by linear interpolation. Sounds like plosives which do not allow this approximation have to be split up into parts of sounds;

b) the durations T_{t_1} and T_{t_2} of the transition range $T_t = T_{t_1} + T_{t_2}$ (Fig. 4) in which the formant parameters of a transient element have to be interpolated between the values of the first and those of the second sound. This interpolation is performed at the second stage of the parameter supply by functions of the second degree;

c) two intervals T_1 and T_2 as a basis for the calculation of the sound durations. The interval T_1 of transient ν_2/ν_3 plus the interval T_2 of the preceding transient ν_1/ν_2 yields the duration of sound ν_2 and, accordingly, the interval T_2 of the same transient ν_2/ν_3 plus the interval T_1 of the following transient ν_3/ν_4 yields the duration of sound ν_3. The division of the sound durations into these two intervals is necessary, because the durations of certain sounds depend very much on their phonetic environment.

The procedure used for storing and processing the control parameters allows coarticulation effects, whose significance for the intelligibility and naturalness of synthetic speech is well known, to be taken into account in a threefold way: Since,the parameter sets are stored for all transients, they can be optimized independently of each other. Since on the other hand, the interpolation ranges of two succeeding transients are allowed to overlap, it is possible to simulate the smoothing of formant tracks caused by the inertia of the human speech generation system even in cases where the steady state is not reached. This phenomenon is frequently observed in the utterance of short vowels. Finally, a special control parameter allows the formant frequency parameters of a parameter set to be replaced by the corresponding parameters of adjacent parameter sets. In this way, it is possible, in certain cases, to make the parameters influence each other even through several sound boundaries, which normally can only be achieved by a much more expensive synthesis from triphones.

At the second stage of the parameter control, the given sound durations and transition intervals are used for a rough interpolation between the parameter samples supplied at the first stage. This interpolation is carried out by functions of the second degree at intervals of 4 ms. These functions may superimpose each other in part, because of possible overlappings of the interpolation ranges.

At the boundary between voiced and voiceless sounds and viceversa, it is necessary to reduce the formant amplitudes to zero during the short interval required for switching over the excitation in order to avoid disturbing switching pulses.

After the rough interpolation of the parameters at the second stage of the sound parameter control, the parameters are again linearly interpolated, as mentioned above, in the fast interpolator and then directly fed into the formant filter unit.

H.-J. Braun [22] has carried out comparative intellibility measurements of original, telephone bandwidth-limited speech, speech reproductively synthesized according to an LPC procedure and SAMT speech (synthesis by rules). He found the intelligibility rate achieved by SAMT to be as high as 83% which is only 9.7% below that of original speech and only

8.5% below the rates measured for the LPC procedure.

3.5. Prosody control

The quality of synthesized speech, however, cannot only be judged by the intelligibility of meaningless vowel-consonant utterances. In addition, the proper control of prosodic features plays an important role in the generation of fluent speech. Only if rhythm, intonation and stress are reproduced in a natural way, is it possible for untrained, non-adapted listeners to understand longer texts without any difficulty. To meet this requirement, SAMT provides three types of prosody control, i.e.

1. sound duration elements used to shorten or lengthen sounds and transients
2. amplitude elements for emphasizing or de-emphasizing sounds
3. melody elements.

For the computation of sound durations and transition intervals, sound duration elements are used which have the effect of a non-linear time factor. The non-linearity results from differences in the influence of changes in speech velocity on short and long vowels. In the supply of sound parameters, the amplitude elements are accounted for by a formant amplitude factor which is also non-linear. In this case, however, the non-linearity allows for the dependence of the excitation spectrum on the excitation power, which is observable in natural utterances.

The control of the speech melody is based on the following concept: The pitch level falls slowly both within a sentence or a phrase (termed sentence cadence in the following) and within single words (word cadence) [23]. Sentence cadence and word cadence superimpose each other. In addition, main and secondary stresses as well as continuations (rises in pitch level at the ends of phrases not coinciding with ends of sentences) are superimposed on the cadences as Gaussian melody curves.

Accordingly, two types of melody elements have been implemented:

1. the cadence element having the effect of a linear rise or fall of the fundamental frequency. The starting point can be adjusted between 100 and 150 Hz.
2. The stress element with a Gaussian melody curve

which is variable in amplitude, duration and starting point depending on the sound boundaries. The cadence and stress elements are superimposed upon each other for the calculation of the fundamental frequency f_0.

This type of melody control has already yielded very good results in manual text preparation. Nevertheless, this method can only be regarded as a tentative solution. Therefore, measures are being taken to do extensive research, above all, for the purpose of automatic melody control.

3.6. Parameter extraction

The quality of the synthesized speech does not only depend on the structure of the synthesizer and the efficiency of the parameter processing, but also to a very large degree on the stored sound parameters. Therefore, a very large corpus was collected, elaborated and evaluated in the course of several years in order to extract the parameters for SAMT. Tape recordings were made of approx. 3000 VCV and VV utterances spoken by five persons. These recordings were digitized and afterwards processed with the aid of a DEC, PDP 11/34 computer. The following operations were performed:

1. LPC analysis according to the autocorrelation method [24]
2. root solving for computing the formant frequencies and bandwidths for a series filter structure
3. determination of the formant amplitudes for the set of parallel filters [25] for given bandwidths and constant formant powers
4. manual selection of appropriate parameter samples, optimization of the data by synthesis trials and listening tests.

The last operation was mainly based on the data obtained from only one of the five speakers. The material of the other four speakers was used, above all, to eliminate uncertainties in the interpretation of the data and to compensate for undesired speaker peculiarities (dialect, mispronunciations). These experiments, which, so far, have only been carried out for male voices, will be continued with female speakers.

3.7. Text preparation

For the use of SAMT in high-performance inquiry systems, it is necessary, in most cases, to prepare large vocabularies. This means that the available orthographic texts have to be transcribed into strings of encoded monophones and prosodic elements. In order to achieve an optimum speech quality, it is necessary that the texts to be synthesized are spoken by a human speaker. This material is being evaluated, at present, with respect to the prosodic features and the results of this evaluation are used to improve the prosody of the synthesized texts. In order to reduce the manpower and hardware needed for this purpose, research is being done to develop automatic transcription and prosody control procedures for the German language [26-29].

Although these procedures are not able, at least, in the near future, to replace man in the final checking and manual improvement of texts, they allow a very large number of man-hours to be saved. Moreover, for audio-response systems like SAMT, they offer new possibilities of application such as reading aids for the blind, etc.

References:

[1] Kempelen, W. von: Mechanismus der menschlichen Sprache. J.V. Degen-Vlg. Wien

[2] Harris, C.M.: A study of building blocks in speech. J. Acoust.Soc.Am. 25 (1953) No.5, Sept. pp. 962-969

[3] Harris, C.M.: A speech synthesizer. J.Acoust. Soc.Am. 25(1953) No.5, Sept., pp. 970-975

[4] Cooper, F.S. et al: Some experiments on the perception of synthetic speech sounds. J.Acoust. Soc.Am. 24(1952) No.6, Nov., pp. 597-606

[5] Delattre, P.C., Liberman, A.M. u. Cooper, F.S.: Acoustic loci and transitional cues for consonants. J.Acoust.Soc.Am. 27(1955) No.4, July, pp. 769-773

[6] Delattre, P.C., Liberman, A.M. u. Cooper, F.S.: Formant transitions and loci as acoustic correlates of place of articulation in American fricatives. Studia Linguistica (1963) pp. 104-121

[7] Liberman, A.M., Delattre, P.C., Cooper, F.S. u. Gerstman, L.J.: The role of consonant-vowel transitions in the perception of stop and nasal consonants. Psychological Monographs: General and Applied 68(1954) No. 8, Whole No. 379, pp. 1-13

[8] Küpfmüller, K. u. O. Warns: Sprachsynthese aus Lauten. Nachrichtentechn. Fachberichte, 3(1956) pp. 28-31

[9] Warns, O.: Über die Synthese von Sprache aus Lauten und Lautkombinationen. Darmstadt, Techn. Hochschule, Thesis 1957

[10] Cramer, B.: Speech synthesis for transformation with very low channel capacity. Nachrichtentechn. Z. - Communication J. (1965) No.3, pp. 130-141

[11] Endres, W.K.: Die Übergangslaute der deutschen Sprache als verbindende Elemente für eine Sprachsynthese. 7th Intern. Congr. on Acoustics Budapest (1971) Vortrag 25 C 4, pp. 301-304

[12] Endres, W.K. u. E. Großmann: Manipulation of the time functions of vowels for reducing the number of elements needed for speech synthesis. Speech Comm. Seminar Stockholm SCS-74 (1974) pp. 267-275

[13] Endres, W.K.: The transitional sounds of the German language as link elements for a speech synthesis. Acustica 26(1972) No.1, pp. 33-36

[14] Besier, H.A. u. W.K. Endres: Untersuchungen zu einer Sprachsynthese aus Einzellauten, Frequenz 32 (1978) No.5, Mai, pp. 136-140

[15] Großmann, E.: Synthese von Sprache aus Sprachlauten. Acustica 35 (1976) No.4, pp. 258-265

[16] Fant, G.: Speech communication research. IVA 24 (1953) No.8, pp. 331-337

[17] Liljencrants, J.: The OVE III speech synthesizer Speech Transm. Lab., Quarterly Progress & Status Rep., Royal Inst. of Techn. STL-QPSR (1967) H. 2/3, pp. 76-81

[18] Estes, S.E., H.R. Kerby, H.D. Maxey u. R.M. Walker: Speech synthesis from stored data. IBM J. of Res. & Dev. (1964) Jan., pp. 2-12

[19] Dixon, N.R. u. H.D. Maxey: Terminal analog synthesis of continuous speech using the diphone method of sequent assembly. IEEE Trans. on Audio & Electroacoustics, UA-16 (1968) No.1, March, pp. 40-50

[20] Génin, J.: Les études de synthèse de parole au CNET, l'écho des Recherches 85 (1976) Juillet, pp. 40-49

[21] H.E. Wolf: Multiplexbarer Sprachsynthetisator nach dem Formantvocoderprinzip. DAGA'76 Heidelberg, pp. 495-498

[22] H.-J. Braun: Messung der Lautverständlichkeit bei synthetischer Sprache mit einem Diphontest. Technischer Bericht des Forschungsinstitutes der Deutschen Bundespost, 13 TBr 13, März (1979)

[23] Bierwisch, M.: Regeln für die Intonation deutscher Sätze, Studia grammatica VII, Akademie-Verlag, Berlin 1973

[24] Markel, J.D. u. A.H. Gray Jr.: Linear Prediction of Speech, Springer-Verlag Berlin, Heidelberg, New York 1976

[25] H.-J. Braun: Anwendung steuerbarer Filternetzwerke zur Sprachsynthese in automatischen Sprachausgabeeinrichtungen. Der Fernmeldeingenieur 29. Jahrgang, H. 12, Dez. (1975), Vlg. f. Wissenschaft u. Leben, Georg Heideker, Bad Windsheim

[26] Isacenko, A.v. u. H.J. Schädlich: Untersuchungen über die deutsche Satzintonation. Studia grammatica VII, Akademie-Verlag (1973) Berlin

[27] Kiparsky, P.: Über den deutschen Akzent. Studia grammatica VII, Akademie-Verlag (1973) Berlin

[28] Kretschmar, J.: Untersuchungen zum Tonhöhenverlauf deutscher Sätze für die Sprachsynthese. DAGA'78, Bochum, pp. 455-458

[29] Uhle, M.: Die akustische Ausgabe orthographischer Texte. DAGA'78, Bochum, pp. 473-476

TIME - DOMAIN FORMANT - WAVE - FUNCTION SYNTHESIS

Xavier Rodet

IRCAM, 31 rue St. Merri, 75004 Paris, France

1. INTRODUCTION

Formant-wave-Function (FOF) synthesis is a method for directly calculating the amplitude of the waveform of a signal as a function of time. Many signals can be modeled as a pair: Excitation Function-Parallel Filter. In the FOF method this pair is replaced by a unique formula describing in a more or less approximate way the output of the filter. Two types of advantages have motivated the first uses of the FOF technique: on the one hand, in certain cases, the FOF formula can beplified to a point where calculations are fast and easy ([1], [2], [4], [5]). On the other hand, the FOF method allows modeling of signals without the need of separating "a priori" the excitation function and the filter ([3], [5]).

In fact, in Formant-wave-Function synthesis, the term "Formant" is taken in a broad sense. It designates a part of the frequency spectrum of the signal which is considered as a whole for a given application as shown in section 3.

2. THEORY

Production of different kinds of signals (speech or instruments for instance) can be represented in the form of an excitation function e(k) (with z transform e(z)) and a filter, the transfer function of which is H(z). The response of the filter to the excitation e(z) is

$$s(z) = e(z) \cdot H(z)$$

J. C. Simon (ed.), Spoken Language Generation and Understanding, 429-441.

Fig.1. Model of Signal Production

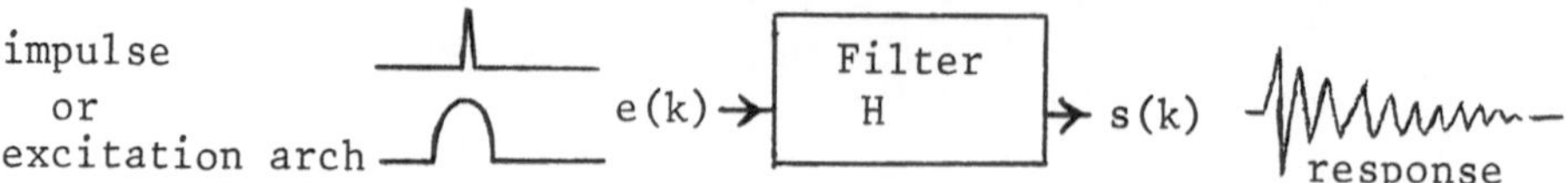

Supposing the filter H to be linear, if the excitation is a succession of impulses or arches $E(k) = \Sigma\, e_n(k)$ (where n indexes successive arches), then the response of the filter can be calculated easily as the sum of the responses $s_n(k)$ offset by one period of the fundamental $\tau_o = \frac{1}{F_o}$ (F_o being the fundamental frequency of the excitation and of the response).

In general, $H(z)$ is considered to be a set of parallel filters $H(z) = \sum_{i=1}^{q} h_i(z)$ and the response $s_n(k)$ is then the sum of q partial responses $s_n(k) = \sum_{i=1}^{q} s_{in}(k)$.

Finally the waveform at period number n is the sum of q partial <u>Formant-wave-Functions</u> $s_{in}(k)$ each for example modelling a formant or a more or less broad portion of a given spectrum so as to obtain the best fit to this spectrum or to its most important characteristics.

It is impossible to consider the responses $s_n(k)$ as infinite. Generally, the quantization of amplitudes limits the non zero duration of $s_n(k)$ to a finite value. Thus we call δt_n the duration during which $s_n(k)$ is non zero.

This model holds true even if the fundamental period is not constant, that is if the delay between the successive excitation arches $e_n(k)$ is not a constant τ_o but a value τ_n which varies with the index n of the excitation arch: in this case, the successive responses to be summed will be delayed respectively by the different τ_n.

The same is true if the shape of the excitation arch $e_n(k)$ varies with time: to each arch corresponds a particular response $s_n(k)$ which has to be calculated. Particularly, if the arches of the excitation signal differ solely by a multiplicative coefficient Q_n, the total response is the sum of the responses: $\Sigma Q_n . s_n(k)$.

Finally, in the case of the filter varying with time, these variations have to be taken into account in the formula defining $s_n(k)$. However, if the variations of the filter during one fundamental period are small, one can consider the characteristics of the filter as constant during the time interval δt_n during which the response $s_n(k)$ is calculated. But each successive response has to be calculated as a function of the characteristics of the filter at the corresponding excitation time.

One of the advantages of the FOF method is that it allows the modeling of a spectrum of complex shape by means of a limited number of functions (as few as one - see paragraph III). Furthermore, in many cases the cost of calculation on a general purpose computer is considerably less with the FOF technique than with the Excitation function-Filter model. It should also be mentioned that the precision of calculations necessary for the FOF method is less than that required for filters: usually it is only the precision required for the final signal (for instance a 12 to 16 bits integer for a High-Fidelity audio signal). Generally, the calculations can be made in fixed point and with tables of limited size ([5]). Finally, as will be seen in section 3 the sum of several simple FOF permits the modeling of spectra of complex shapes with a very great precision (for example those of vocal and instrumental sounds). It is noticeable that the parameters of the model are particularly representive from a perceptual point of view. Generally two types of FOF have been utilised up to now. They will be examined in the following paragraphs.

3. EXAMPLE OF DECOMPOSITION OF A SIGNAL INTO PARTIAL FORMANT WAVE FUNCTIONS ([5], [6])

This method allows one to separate a signal P(k) into a number q of partial FOF's corresponding to q regions Φ_i on the frequency axis. Those regions are distinct but any given region is not necessarily contiguous. The chosen portion of signal P(k) corresponds exactly to one fundamental period. The spectrum of the signal is modeled in the form of an m-pole linear predictive filter ($\frac{G}{A(z)}$). The parallel structure of this filter

$$\left(\sum_{\ell=1}^{m/2} \frac{c_\ell + d_\ell z^{-1}}{1+a_\ell z^{-1}+b_\ell z^{-2}} \right)$$ can be decomposed into q

groups respectively corresponding to the q regions Φ_i: Each group γ_i is the sum of those sections $\frac{c_\ell + d_\ell z^{-1}}{1+a_\ell z^{-1}+b_\ell z^{-2}}$

the poles of whose frequencies are contained in the corresponding region Φ. Filtering of the prediction error E(k) through respectively each of the q groups γ_i gives q partial wave functions $p_j(k)$. Thus they are the responses of the q parallel groups γ_i to one arch of the excitation signal with the property that their sum is equal to the original signal to be modeled

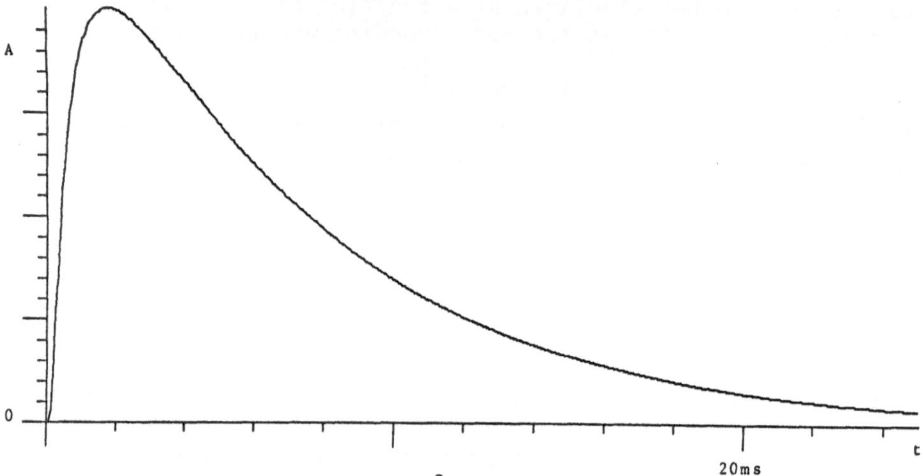

Fig. 2. ENVELOPE $A\, e^{-\alpha \frac{(t-c)^2}{t}}$

$(\frac{\alpha}{\Pi} = 50\text{Hz},\ C = 1.8\text{ms})$

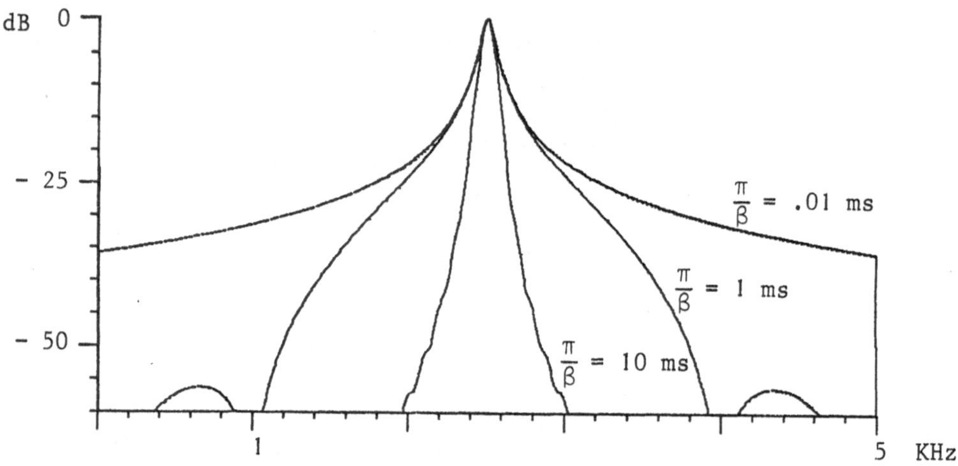

Fig. 4. Power Spectrum of the FOF $A(k)\sin(\omega_c . k + \phi)$

for different values of $\frac{\pi}{\beta}$ $(\omega_c = 2500\text{ Hz},\ \frac{\alpha}{\pi} = 80\text{ Hz})$

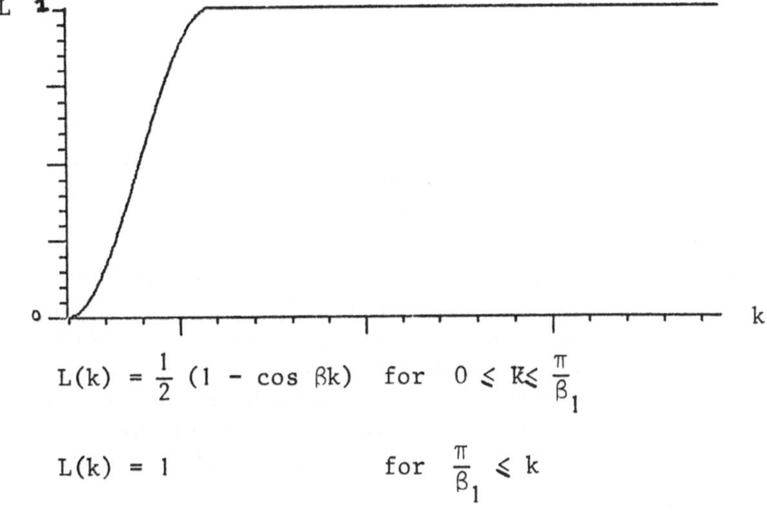

$$L(k) = \frac{1}{2}\,(1 - \cos \beta k) \quad \text{for} \quad 0 \leqslant k \leqslant \frac{\pi}{\beta_1}$$

$$L(k) = 1 \quad \text{for} \quad \frac{\pi}{\beta_1} \leqslant k$$

Fig. 3.a. Attack L(k)

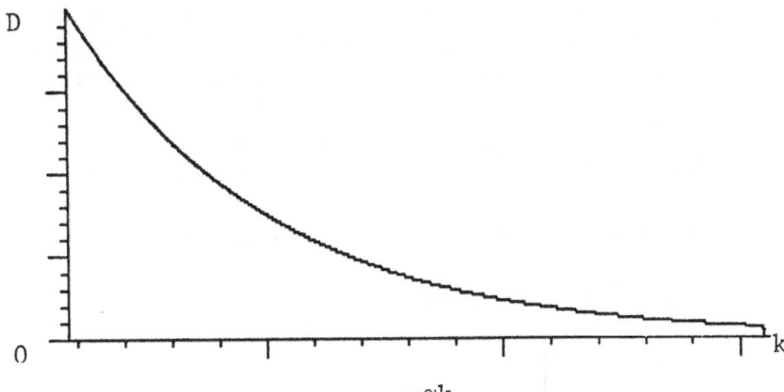

Fig. 3.b. Decay $D(k) = e^{-\alpha k}$

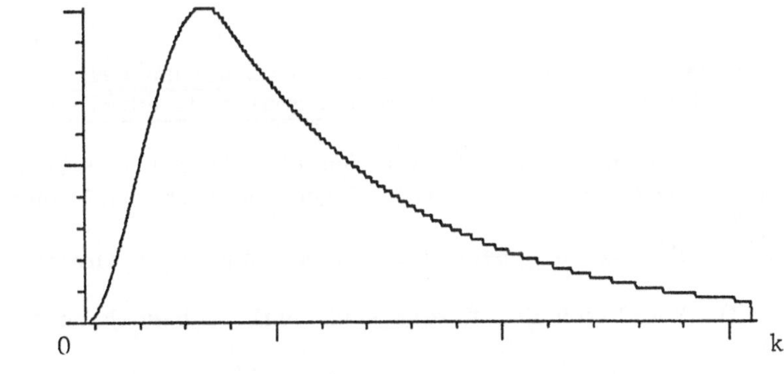

Fig. 3.c. Envelope $A(k) = L(k).D(k)$

$$\sum_{i=1}^{q} p_i(k) = P(k).$$

Thus, these partial wave functions p(k) can be considered as FOF's and, after being tabulated can be used according to the method described in paragraph II. The presence of zeros in the transfer function is naturally taken into account by the model. Each of these FOF functions can independently be modified in amplitude or in time so as to change the spectral shape or so as to displace a "formant" region Φ_i. If it is not necessary to modify these FOF's independently, their number can be reduced to 1(q=1). This allows the synthesis of a sound, even of one with a very complex spectral shape, by means of a single table lookup . The possibility still remains of changing the fundamental frequency or of applying a homothetic transform to the spectrum.

4. AN APPLIED EXAMPLE

Certain models (parallel formant speech synthesizers for instance) utilise a set of second order sections disposed in parallel. The transfer function of such a section is:

$$H(z) = \frac{c+d\ z^{-1}}{1+a\ z^{-1} + b\ z^{-2}} = \frac{c+d\ z^{-1}}{(1-p\ z^{-1})(1-p^{*}z^{-1})}$$

Its impulse response is the Formant wave Function:

$$s(k) = G\ e^{-\alpha k}\ \sin(\omega k+\phi)$$

with
$$\alpha = -\log(P)$$
$$\omega = \mathrm{Arg}(p)$$
$$\phi = \omega - \mathrm{Arctg}\ (\sin \omega \frac{2d\sqrt{b}}{2bc-ad})$$
$$G = \frac{\sin \phi}{c}$$

Thus, some projects have made use of a sinusoïdal wave multiplied by an exponentially damped amplitude envelope.

For natural sounds the excitation function is not a unit impulse. To obtain a more precise control of the spectrum, the function

$$e^{\frac{-\alpha(t-c)^2}{t}}$$ (Fig. 2) has been used as an amplitude envelope

([5]),[6]). Vocal sounds of very high quality have been obtained this way.

For modeling arbitrary spectra one can use other types of envelopes A(t) chosen according to their spectra.

In effect, the spectrum of the FOF so obtained $A(t).\sin(\omega_c t + \phi)$, is that of the envelope translated to the center frequency $\frac{\omega c}{2\pi}$. In particular, the windows used in spectral analysis ([7]) are interesting envelopes. They can be chosen:

- to <u>concentrate the energy</u> in a narrow frequency band and during a temporal window as small as possible (in order to reduce the calculations and obtain a close fit to formant values during transitions)
- <u>to approximate a portion of the spectrum of any shape</u> with a principal peak and lateral lobes of greater or lesser magnitude.

5. ANOTHER EXAMPLE

The following FOF is preferred for its interesting properties:

$$\begin{cases} s(k) = 0 & \text{for} \quad k \leqslant 0 \\ s(k) = \frac{1}{2}(1 - \cos \beta k)e^{-\alpha k}\sin(\omega_c.k + \phi) & \text{for } 0 \lesssim k \lesssim k_1 \\ & k_1 = \frac{\pi}{\beta} \\ s(k) = e^{-\alpha k}\sin(\omega_c.k + \phi) & \text{for} \quad k \geqslant k_1 \end{cases}$$

It is again a sinusoid function shaped by an amplitude envelope A(k). This envelope is a damped exponential, the initial disconti-nuity of which is smoothed by multiplication by $\frac{1}{2}(1-\cos\beta k)$ for a duration of k_1 samples so that $\beta k_1 = \pi$ (Fig.3)

One obtains thus an amplitude envelope A(k)

- which presents an attack for a duration of approximately k_1 samples and a $e^{-\alpha k}$ decay after the attack
- which presents no discontinuity of 1st or 2nd order
- which can be obtained very simply by table-lookup for $\frac{1}{2}(1 - \cos \beta k)$ and sin (ωk), and by successive multiplications by $e^{-\alpha}$ for $e^{-\alpha k}$.

The Fourier transform of this envelope is:

$$\psi(\omega) = \int_{-\infty}^{+\infty} A(t)e^{-i\omega t}dt = \int_0^{\frac{\pi}{\beta}} \frac{1}{2}(1-\cos \beta t)\, e^{-(\alpha+i\omega)t}dt + \int_{\frac{\pi}{\beta}}^{+\infty} e^{-(\alpha+i\omega)t}dt$$

thus $\psi(\omega) = \frac{\beta^2}{2} \frac{e^{-(\alpha+i\omega)\frac{\pi}{\beta}} + 1}{(\alpha+i\omega)((\alpha+i\omega)^2+\beta^2)}$

the modulus of which is:

$$\left|\psi(\omega)\right| = \frac{\beta^2}{2} \frac{\sqrt{M^2 + 1 + 2M\cos(\omega\frac{\pi}{\beta})}}{\sqrt{\alpha^2+\omega^2}\sqrt{(\alpha^2+\omega^2)^2 + \beta^2(\beta^2 + 2\alpha^2 - 2\omega^2)}} \quad (1)$$

with $M = e^{-\frac{\alpha\pi}{\beta}}$

For speech synthesis the orders of magnitude of the values are:

$$\alpha = (\text{Bandwidth}).\pi \simeq 10^2\,\pi \text{ Hz}$$

$$\beta = \frac{\pi}{\text{attack duration}} \simeq 10^3\pi \text{ Hz}$$

Thus one can consider as a first approximation $\beta^2 >> \alpha^2$

Moreover, if $\omega\frac{\pi}{\beta}$ is small with respect to $\frac{\pi}{2}$ that is for frequencies close to the center frequency: $F_c = \frac{\omega_c}{2\pi}$

one can consider $\cos(\omega\frac{\pi}{\beta}) \simeq 1$

and $\beta^2 >> \omega^2$

It follows $\left|\psi(\omega)\right| \simeq \frac{\beta^2}{2} \frac{\sqrt{M^2 + 2M + 1}}{\sqrt{\alpha^2+\omega^2}\sqrt{\beta^4}} = \frac{M+1}{2}\frac{1}{\sqrt{\alpha^2 + \omega^2}}$

Thus, the shape of the power spectrum of our FOF s(k) is of the form $\frac{K}{\alpha^2+(\omega_c-\omega)^2}$ in the neighbourhood of the center frequency $\frac{\omega_c}{2\pi}$. It is independent of β and nearly identical to the transfer function of a second order section, the centre pulsation of which is ω_c and the bandwidth is $\frac{\alpha}{\pi}$. Thus the parameter α controls the -6dB bandwidth of the spectrum of this "Formant". The parameter β controls the width of the "skirts" of the formant peak, and this without modifying the the bandwidth at -6dB, which is a quite remarkable property. (Fig. 4).

Finally, according to (1), given α and β, it is easy to normalize the amplitude of the resulting spectrum.

A software parallel FOF synthesizer has been written. Its structure is as follows:

Fig. 5. Structure of a parallel FOF synthesizer.

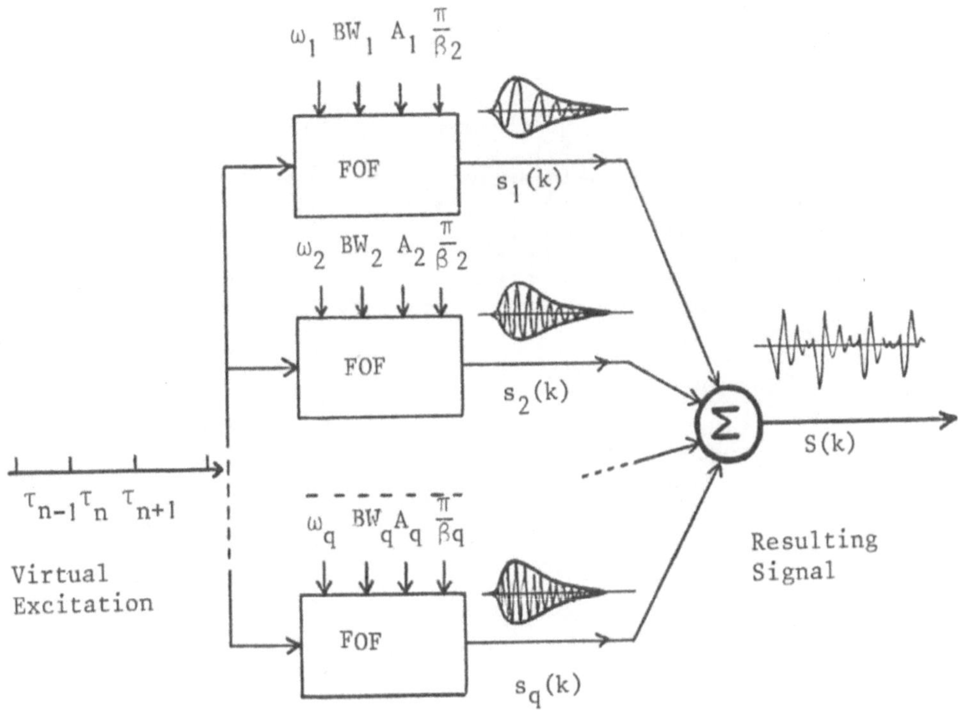

For Formant number i:

ω_i : Centre Pulsation
BW_i : Bandwidth
A_i : Amplitude
$\frac{\pi}{\beta_i}$: Width of the skirts

This program is used for synthesis of singing voices and of instrumental sounds. The different steps of the modeling are as follows:

- Semi-automatic analysis of the spectrum of a given sound and extraction of gross formant characteristics ($\{\omega_i, Bw_i, A_i, \beta_i\}$ for $1 \leqslant i \leqslant q$) and fundamental frequency (Fig. 6).

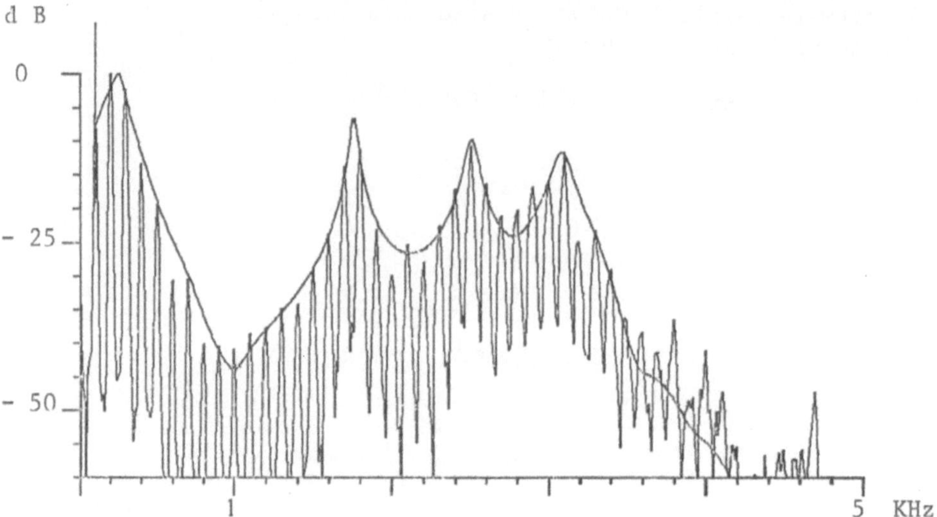

Fig. 7. Adjustment of parameters so as to model the natural vocal spectrum as closely as possible (Parameters in table 1)

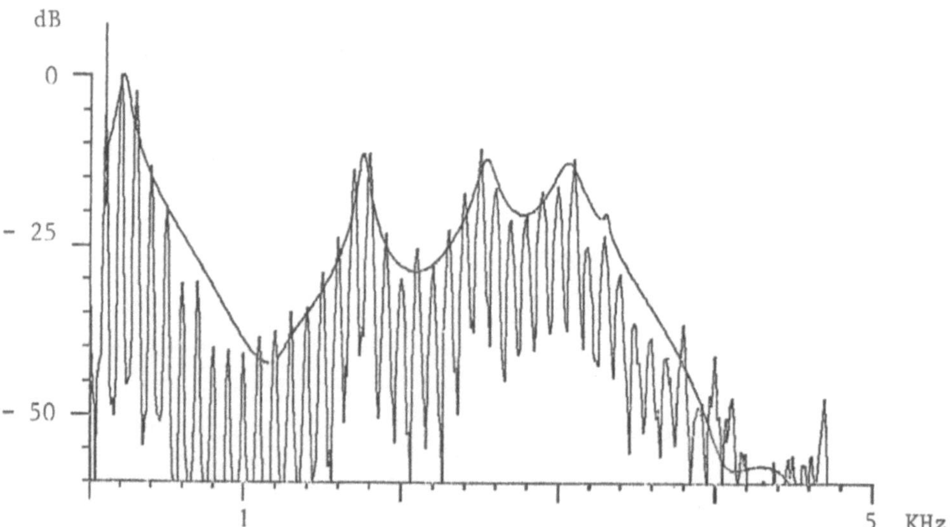

Fig. 6. Comparison of a natural spectrum with the envelope of the synthetic spectrum calculated according (1) from a gross estimation of the parameters.

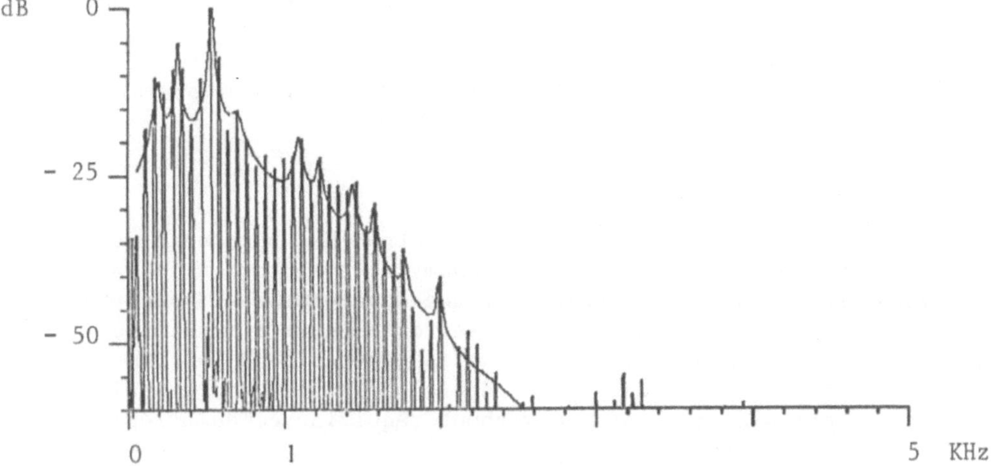

Fig. 9. Same as 7 for Bassoon

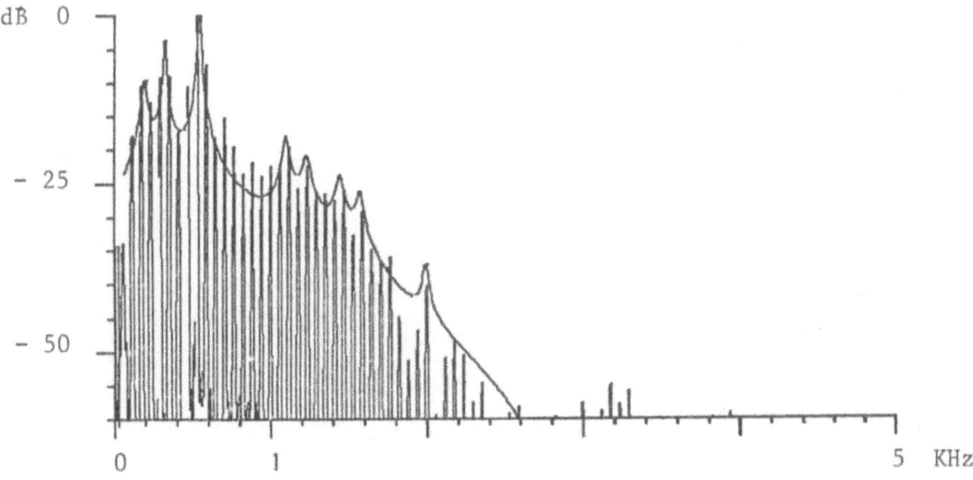

Fig. 8. Same as 6 for Bassoon

- Adjustment of parameters so as to model the natural spectrum as closely as possible (Fig.7) the parameters are shown in table 1.

The figures (8) and (9) show the results of the same operations on the spectrum of an instrument (Bassoon).

6. CONCLUSION

The parallel Formant wave Functions method has been used for a number of years for reasons of economy of computation in speech synthesis. It has been shown that the development of adequate formulas permits a model very close to the excitation-filter model. The FOF model allows synthesis of vocal and instrumental sounds of very good quality on general purpose computers with limited calculation cost and without requiring calculations of great precision.

Table 1.

F1	= 260	Hz
A1	= .029	
BW1	= 70	Hz
$\frac{\pi}{\beta_1}$	= .002	sec
F2	= 1764	Hz
A2	= .021	
BW2	= 45	Hz
$\frac{\pi}{\beta_2}$	= .0015	sec
F3	= 2510	Hz
A3	= .0146	
BW3	= 80	Hz
$\frac{\pi}{\beta_3}$	= .0015	sec
F4	= 3090	Hz
A4	= .011	
BW4	= 130	Hz
$\frac{\pi}{\beta_4}$	= .003	sec
F5	= 3310	Hz
A5	= .00061	
BW5	= 150	Hz
$\frac{\pi}{\beta_5}$	= .001	sec

The set of parameters in our model is

- Fundamental Frequency
- For each "formant"
  - Centre frequency
  - Bandwidth at -6dB
  - Width of the skirts
  - Amplitude.

It should be remarked that theses parameters are particularly easy to determine and to adjust, and that they are especially pertinent from a perceptual point of view.

Provided that the parameters are correctly chosen, a very high sound quality can be obtained in the synthesis. Thus a wide variety of sounds can be synthesized from percussion instruments to voice, from reed instruments to strings.

REFERENCES

(1) . RODET X., SANTAMARINA C. "Synthèse, sur un miniordinateur, du signal vocal dans sa représentation amplitude-temps."

Actes des 6è journées d'étude sur la Parole du GALF - Toulouse, Mai 1975.

(2) . BOURGENOT, J.S., DECHAUX, C. - Codage de la parole à faible débit: le vocoder CIPHON, Revue Technique Thomson-CSF, vol. 7, 4, 1975

(3) . RODET, Xavier, "Analyse du signal vocal dans sa représentation amplitude-temps.
Synthèse de la parole par règles", THESE Université Paris 6, 1977.

(4) . BAUMWOLSPINNER, M. - "Speech Generation through Waveform Synthesis", IEEE-ECASSP, Tulsa, 1978

(5) . RODET X., DELATRE J., "Time-Domain Speech Synthesis-by-rules using a flexible and fast signal management system"
IEEE.ICASSP Washington D.D. 2-4 April 1979

(6) . RODET X., DELATRE J.L., RAZZAM M., "Construction du signal vocal dans le domaine temporel"
10è journées d'Etude sur la Parole, Grenoble, Mai 1979.

(7) . HARRIS, Frederich, "On the Use of Windows for Harmonic Analysis with Discrete Fourier Transform".
Proc. IEEE Vol 66 n° 1 January 1978.

A PROGRAMMABLE, DIGITAL SPEECH SYNTHESISER

R. LINGGARD and F.J. MARLOW

THE QUEEN'S UNIVERSITY OF BELFAST,
DEPARTMENT OF ELECTRICAL ENGINEERING

1. INTRODUCTION

This paper describes the design and operation of a programmable, digital speech synthesiser, intended as a research tool in a study of speech synthesis methods. A major objective of this project was to build a digital synthesiser at the lowest possible cost consistent with real time operation. The output data rate was chosen to be 10K samples per second to give 5 KHz bandwidth speech. The architecture of the machine is shown in Fig. 1.

In order to be able to simulate the different kinds of synthesiser under study, the machine had to be made programmable. Thus the machine itself is simply a high speed processor which is capable of being programmed to synthesise speech in a variety of different ways.

When in operation the synthesiser is driven by speech parameters stored in a host mini computer which supplies them to the synthesiser upon request. The mini computer is also used to load into the synthesiser the programme which specified the particular type of synthesiser being simulated.

An important feature of the programming is the ability to generate excitation (both periodic and random) entirely by software routines. This flexibility in both excitation and synthesiser structure, greatly facilitates the optimisation of synthesiser design. Changes in a proposed synthesiser structure can be quickly and easily evaluated.

J. C. Simon (ed.), Spoken Language Generation and Understanding, 443-454.

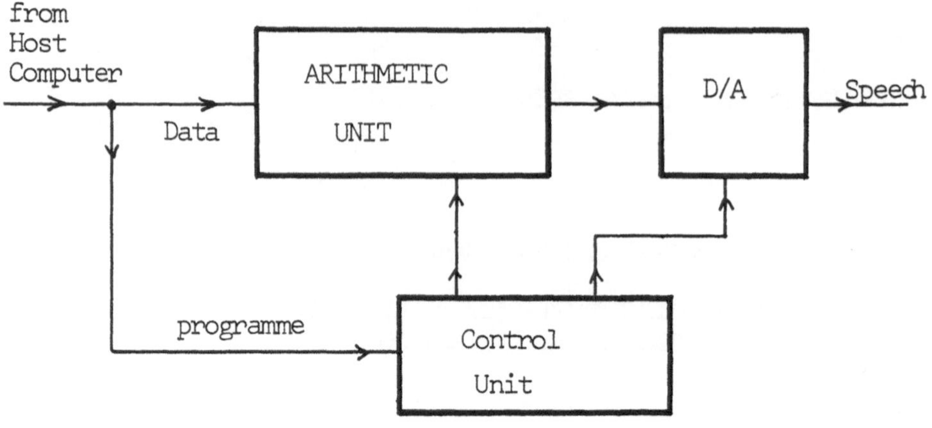

Fig. 1. BASIC ARCHITECTURE

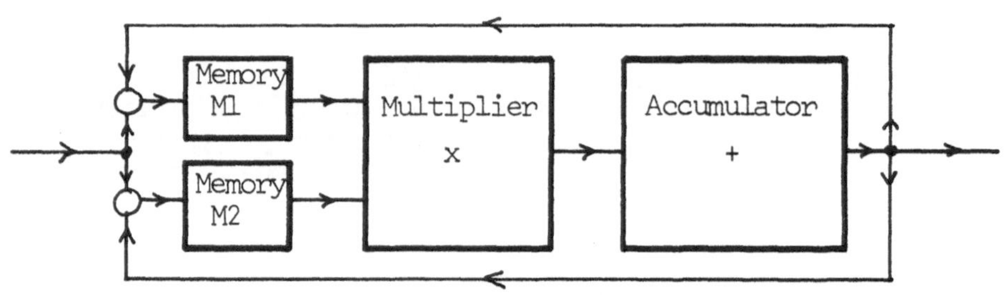

Fig. 2. ARITHMETIC UNIT

2. ARITHMETIC PROCESSOR

A block diagram of the Arithmetic Processor is shown in Fig. 2. Any digital filtering routine can be broken down into a series of delays, multiplications, additions and subtractions. Thus the arithmetic hardware comprises two scratch pad memories, a multiplier and an accumulator. It is convenient to use two memories because almost all the multiply operations are of the kind - coefficient x partial result. Thus by using separate memories for coefficients and partial results, the two inputs to the multiplier can be supplied simultaneously. This not only increases the arithmetic processing speed, but also simplifies the programming.

Gold and Rabiner (2) have shown that, although 12-bits can adequately represent a constant level speech signal, dynamic range considerations within the digital filters, necessitate at least 18-bits to guarantee negligible round-off error. Therefore to ensure sufficient dynamic range at minimum cost, a 16-bit floating point format was adopted, 12-bits representing the mantissa and 4-bits representing the exponent. For ease of computation "twos complement" notation is used in both mantissa and exponent. The multiplier hardware therefore, consists of a 12 x 12 bit twos complement multiplier to process the mantissa and a 4-bit adder to sum the exponents. The resulting 24-bit product is then shifted right a maximum of +7 to give a 28-bit fixed point number that can be added (or subtracted) directly into the accumulator.

The accumulator consists of a 28-bit adder, with one input fed from the multiplier, and the other from its own output via a double latch. The contents of the accumulator can be fed either to the output latch (and hence to the digital-to-analogue converter) or, via a fixed-to-floating point converter, to the scratch pad memories.

Control **signals** to the processor cause data to be fed from the scratch pad memories to the multiplier, then to the accumulator and finally to the output and/or back to the scratch pad memories. The time required for one arithmetic processing cycle (hereafter called a 'microcycle') is 250 n Sec. That is, a scratch pad write, a multiply, or an accumulator operation can be accomplished in 250 n Sec.

The control signals to the arithmetic processor are described in the next section. A complete list of the arithmetic instructions is given in Table I.

3. CONTROL UNIT

The control signals which drive the arithmetic unit are derived from an instruction decoder which consists essentially of a 16 x 16 bit R.O.M. This instruction decoder is addressed by a 4-bit operation code which forms part of the 16-bit instruction word. The 4-bit operation code, plus two 6-bit addresses for the scratch pad memories, make up the full 16-bits of the instruction word. The sequence of 16-bit instructions (the programme) is stored in a 1024 x 16 bit Random Access Memory. This programme memory is addressed by a 10-bit, presettable counter, which is incremented each clock cycle. The whole arrangement of counter, memory and instruction decoder is shown in Fig. 3.

Because of the relatively long access time of the programme memory, a triple-overlap, pipeline structure is employed. Thus the following three operations all take place simultaneously:

a) Instruction 'n' is executed.

b) Programme memory is accessed to produce instruction 'n+1'.

c) The programme counter is incremented to give the address of instruction 'n+2'.

The control unit also provides a means of accessing data from the host computer (in this case, a Computer Automation Alpha 16). Two separate interface modes are used. In the first mode, the host computer loads the desired speech synthesiser programme into the processor programme memory. This is done under manual control. In the second interface mode, the processor itself interrupts the host computer to ask for speech parameters. These are transferred automatically and stored in the processor scratch pad memory.

The instruction set of the decoder is quite small, consisting of only 16 instructions. However, these are quite sufficient to accomplish all the arithmetic processing necessary to synthesise speech. Two instructions in particular, are worthy of note since they greatly increase the flexibility of the processor. These are the unconditional jump and the conditional skip instructions.

The unconditional jump instruction (JMP), is implemented by feeding the address part of the instruction word to the preset inputs of the programme counter, and using the JMP control line to load this address through to the programme memory. Thus the programme 'jumps' to the address given in the JMP instruction word.

The conditional skip instruction (SKP) causes the programme to skip an instruction when the contents of the accumulator are negative. This condition is sensed by gating the M.S.B. (sign bit) of the accumulator with the SKP control pulse and using this to block the clock pulse to the instruction decoder.

By using a SKP instruction followed by a JMP, a conditional jump can be accomplished. The complete instruction set of the processor is given in Table I.

In order to produce speech samples at 10K Hz sample rate, the processor programme must produce an output within 100 μ Sec. This restarts the synthesis programme, which runs and then waits for the next reset.

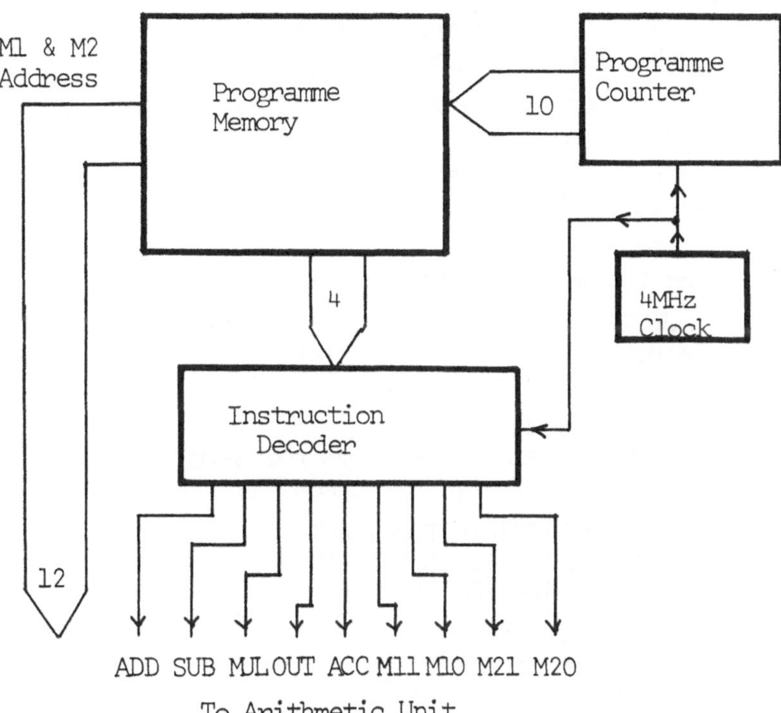

Fig. 3. CONTROL UNIT

TABLE I: PROCESSOR INSTRUCTION SET

MNEMONIC	ACTION	INSTRUCTION WORD OP CODE (HEX)	ADDRESS
NOP	NO OPERATION	0	-
JMP	JUMP - UNCONDITIONAL Programme jumps to address given in instruction word	1	jump address
SPA	SCRATCH PAD-ACCESS Scratch pad memory addresses controlled by instruction word	2	-
SPL	SCRATCH PAD LOAD Scratch pad memories controlled by host computer	3	-
ACC	ACCUMULATOR CLEAR Accumulator output is set to zero	4	-
OUT	OUTPUT LATCH Output latch is set	5	-
SUB	SUBTRACT Multiplier output is subtracted from accumulator	6	-
ADD	ADD Multiplier output is added to accumulator	7	-
INP	INPUT New coefficient from host computer is fed to scratch pad memory	8	
INT	INTERRUPT Host computer is interrupted	9	
MUL	MULTIPLY Two numbers at output of scratch pad memories are latched into the multiplier input	A	M1 & M2
SKP	SKIP - CONDITIONAL If accumulator is negative the next instruction is not executed	B	
M1I	M1 MEMORY LOADED FROM INPUT	C	M1 -
M1O	M1 MEMORY LOADED FROM OUTPUT	D	M1 -
M2I	M2 MEMORY LOADED FROM INPUT	E	- M2
M2O	M2 MEMORY LOADED FROM OUTPUT	F	- M2

4. GENERATION OF SYNTHESISER EXCITATION

The two kinds of excitation required for synthesis of speech are periodic pulses and random noise. It was originally intended to build special hardware to generate these signals, however the flexibility of the processor is such that both types of excitation can be synthesised by software routines.

Periodic pulses are obtained by using a pitch parameter P, equal to the required pitch period in units of 100 μ Sec. P, specified as an integer, is decremented each programme cycle (every 100 μ Sec.) until it becomes negative, whereupon it is reset and a unit impulse is applied to the formant filters. The pitch control parameter P, is obtained from the host computer along with the other speech drive parameters.

The generation of random noise is somewhat more difficult in concept, though in practice quite easy to achieve. Rabiner and Gold (1) have described three techniques for generating pseudo-random numbers which approximate random noise. It is a modification of one of these methods that has been used here. The technique is to use a set of numbers which are added to each other successively using twos complement addition and ignoring overflow. The modification has been to use only two numbers instead of the 50 suggested in reference (1). The routine is as follows.

Two numbers A and B are stored in memory. A is added to B to give C. B is then added to C to give D. D is then added to C to give E, and so on. The output sequence is A, B, C, D, E etc. As the number increases it overflows the accumulator and becomes negative, thus the sequence is kept within bounds and is pseudo-random in amplitude.

5. FORMANT SYNTHESISER SIMULATIONS

5.1. It is well known and documented (1), (2), that digital formant filters can be realised as second order filter sections of the kind

$$\frac{V_o(nT)}{V_1(nT)} = H(z) = \frac{a_o + a_1 z^{-1} + a_2 z^{-2}}{1 + b_1 z^{-1} + b_2 z^{-2}} \quad \ldots (1)$$

where $V_o(nT)$ and $V_1(nT)$ are sampled data output and input respectively, and z is the z-transform variable.

Rewriting this expression in terms of past and present inputs and outputs gives

$$V_o(nT) = a_o\, V_1(nT) + a_1\, V_1((n-1)T) + a_2\, V_1((n-2)T) - b_1\, V_o((n-1)T) - b_2\, V_o((n-2)T) \quad \ldots (2)$$

This equation may be implemented by the programme listed in Table II. For this programme the constant 1 and the filter coefficients a_o, a_1, a_2, b_1 and b_2 are stored in the M1 scratch pad memory in locations 1, 2, 3, 4, 5 and 6 respectively. The filter variables $V_1(nT)$, $V_1((n-1)T)$, $V_1((n-2)T)$, $V_o((n-1)T)$ and $V_o((n-2)T)$ are stored in the M2 scratch pad memory in locations 1, 2, 3, 4 and 5 respectively. The programme consists of 24 instructions which means that a new output sample $V_o(nT)$ is calculated in 6 μ Sec.

Several such second order sections may be realized in parallel or cascade or any other designed configuration. The filter coefficients are derived from the required formant frequencies and bandwidths. They are fed to the processor as speech drive parameters from the host computer.

5.2. As an example of a series formant synthesiser, consider the arrangement of filters shown in Fig. 4. This particular configuration is that used in the well known COMPUTALKER, manufactured by Computalker Consultants of Santa Monica, California. This synthesiser was simulated by connecting three variable resonators (F1, F2 and F3) in series, and adding a fixed, wide bandwidth, low pass filter (FN) in parallel as a nasal resonator for fricative sounds, a variable pole (FP) and zero (FZ) were added, also in parallel. The inputs to these

TABLE II: PROGRAMME FOR SIMULATING SECOND ORDER FILTER

Prog. Addr.	Instr. Mnemonic	M1 Addr.	M2 Addr.	Comments
0	MUL			Multiply $V_1(nT)$ by a_o
1	ACC			Clear acc.
2	ADD			$a_oV_1(nT)$ in acc.
3	MUL			Multiply $V_1((n-1)T)$ by a_1
4	ADD			$a_oV_1(nT)+a_1V_1((n-1)T)$ in acc.
5	MUL			Multiply $V_1((n-2)T)$ by a_2
6	ADD			$a_oV_1(nT)+a_1V_1((n-1)T)+a_2V_1((n-2)T)$ in acc.
7	MUL			Multiply $V_o((n-1)T)$ by b_1
8	SUB			$a_oV_1(nT)+a_1V_1((n-1)T)+a_2V_1((n-2)T)$ $-b_1V_o((n-1)T)$ in acc.
9	MUL			Multiply $V_o((n-2)T)$ by b_2
10	SUB			Calculated output $V_o(nT)$ in acc.
11	MUL			Multiply $V_o((n-1)T)$ by 1
12	M2O			Current $V_o(nT)$ becomes next $V_o((n-1)T)$
13	ACC			Current $V_o((n-1)T$ becomes
14	ADD			next $V_o((n-2)T)$
15	M2O			
16	MUL			
17	ACC			Current $V_1((n-1)T)$ becomes
18	ADD			next $V_1((n-2)T)$
19	M2O			
20	MUL			
21	ACC			Current $V_1(nT)$ becomes
22	ADD			next $V_1((n-1)T$
23	M2O			

three branches were the periodic pulse and pseudo-random number generators described in Section 4.

This simulation has been tested in real time, using speech parameters designed for computalker. These parameters are converted to appropriate digital filter coefficients and stored in the host computer. The processor, is programmed as described above, to conform to the configuration of Fig. 9. It reads in the parameters as required during the speech production process.

The actual programme has four distinct parts:

a) Generation of excitation input
b) Filtering of this input
c) Outputing the resultant speech sample
d) Inputing new coefficients from the host computer.

The programme runs once every sample cycle (100 μ Sec.) giving digital speech samples at a 10K Hz rate.

5.3. An alternative configuration is the parallel structure developed by Holmes (3). This arrangement, shown in Fig. 5 was simulated in a similar manner to that described above. It was tested using speech parameters supplied by the Joint Speech Research Unit, Cheltenham, England.

Informal listening tests suggest that the speech quality of these simulations is not noticeably inferior to that of the original synthesisers.

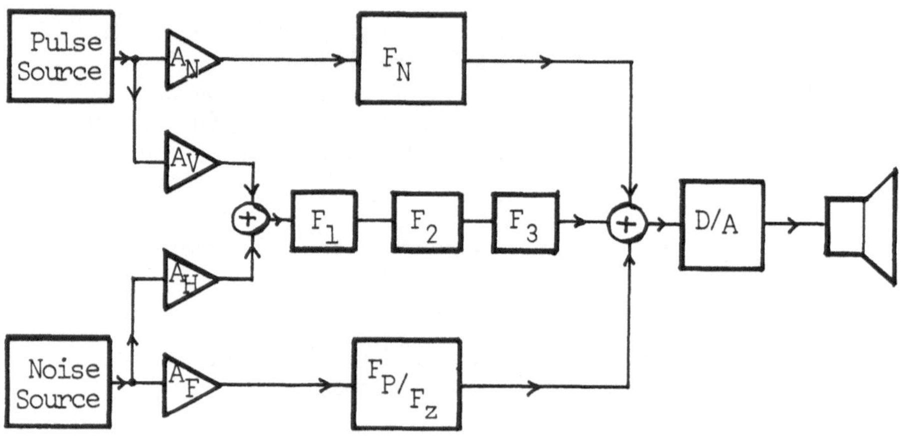

Fig. 4. SERIAL FORMANT SYNTHESISER

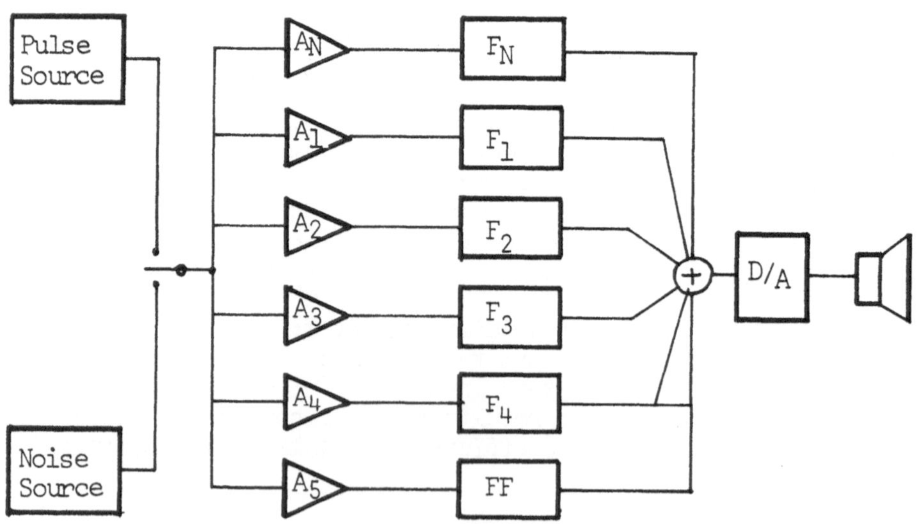

Fig. 5. PARALLEL FORMANT SYNTHESISER

6. VOCAL TRACT SIMULATION

An alternative method of synthesising speech is to model the vocal tract as a series of uniform cylindrical acoustic tubes. The tubes are excited by periodic pulses or random noise, and the area of each section is varied to produce speech output. The discrete time signal flow graph which represents this model given in reference (4), is shown in Fig. 6. The reflection coefficient, r_i, at the ith junction is given by

$$r_i = \frac{A_{i+1} - A_i}{A_{i+1} + A_i}$$

where A_i and A_{i+1} are the areas of the sections i and (i+1) respectively. The forward $U^+_{i+1}(n)$, and backward, $U^-_i(n+1)$, volume velocity are then computed as follows:

$$U^+_{i+1}(n) = (1 + r_i)\, U^+_i(n-1) + r_i\, U^-_{i+1}(n)$$

and $$U^-_i(n+1) = -r_i\, U^+_i(n-1) + (1 - r_i)\, U^-_{i+1}(n).$$

These equations are readily implemented on the processor using the M1 scratch pad memory to store the reflection coefficients, and the M2 memory to store the volume velocity samples.

A ten section tract model was simulated using area data taken from Fant (5). The vocal tract areas were converted to reflection coefficients and these were used to produce continuous vowel sounds. The formant peaks of these vowels were in good agreement with those given in reference (5).

To test the dynamic performance of this model, Wakita's area data for the diphthong /aI/ were taken from reference (6). These were converted to reflection coefficients at 20m Sec intervals. Simple pitch and amplitude contours were used to excite the model and the result was a very natural sounding diphthong.

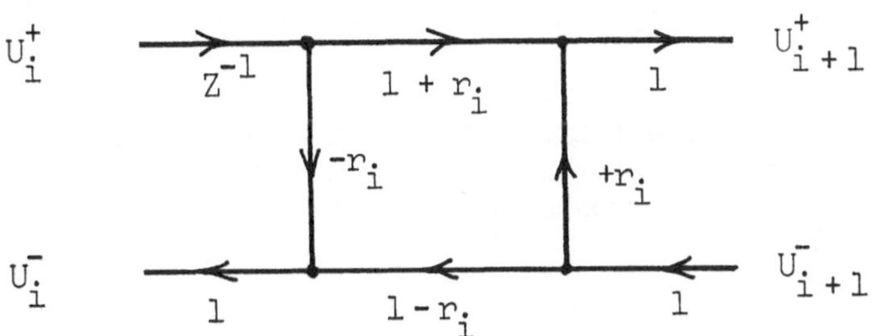

Fig. 6. SIGNAL FLOW GRAPH FOR JUNCTION BETWEEN A_i AND A_{i+1}

7. CONCLUSIONS

A programmable, digital speech synthesiser has been constructed using approximately 200 integrated circuits at a cost of around £350. The machine has been used to simulate various kinds of speech synthesisers, including a vocal tract model, all in real time and with a high degree of success. A principal feature of this machine is that both periodic pulse and random number excitation can be produced entirely by software routines. Programmes to simulate formant type synthesisers used about 180 instruction steps, while the vocal tract model needed 280. The flexibility of a programmable synthesiser permits changes in synthesiser structure to be rapidly evaluated.

So far, the machine has been used to compare different synthesiser types, and to investigate the effect of reducing the word length of the speech drive parameters.

The processor itself is an interesting compromise between complexity and speed. It was found that a very small instruction set suffices for a large class of speech processing.

REFERENCES

(1) Rabiner, L.R. and Gold, B.: "Theory and Application of Digital Signal Processing." Prentice-Hall, 1975.

(2) Gold, B. and Rabiner, L.R.: "Analysis of Digital and Analog Formant Synthesizers", I.E.E.E. Trans. Audio, Electroacoust., vol. AU-16, pp. 81-94, March 1968.

(3) Holmes, J.N.: "The influence of the Glottal waveform on the Naturalness of Speech from a Parallel Formant Synthesizer", I.E.E.E. Trans. Audio, Electroacoust, vol. AU-21, pp 298-305, June 1973.

(4) Rabiner, L.R. and Schafer, R.W.: "Digital Processing of Speech Signals", Prentice-Hall, 1978.

(5) Fant, G.: "Acoustic Theory of Speech Production", Mouton, 1970.

(6) Wakita, H.: "Direct Estimation of the Vocal Tract Shape by Inverse Filtering of the Acoustic Speech Waveform", I.E.E.E. Trans. Audio, Electroacoust., vol. AU-21, pp. 417-427, Oct. 1973.

AN INTELLIGENT SPEECH PROSTHESIS WITH A LEXICAL MEMORY

Daniel Christinaz, Kenneth Mark Colby, Santiago Graham,
Roger Parkison and Linda Chin
Department of Psychiatry, Neuropsychiatric Institute
University of California
760 Weatwood Plaza, Los Angeles, California 90024

INTRODUCTION

We have interfaced a microcomputer with a voice synthesizer to produce a personal, (weakly) intelligent, and quick verbal response speech prosthesis useful for people who are unable to speak (Colby, Christinaz, and Graham, 1978). The intelligent speech prosthesis (ISP) is designed to aid those physically handicapped who are unable to speak due to a great variety of medical, neurological or surgical reasons. In its current state the ISP/1 can be used by users having communication disorders such as aphonia (loss of voice), dysphonia (an impairment of voice), or dysarthria (impairment in articulation) unaccompanied by damage to the semantic syntactic mechanisms of the language system. However, there exists a group of potential users who have some degree of damage in the semantic syntactic mechanisms of the language system. It is for a subgroup of this population, those people having word finding difficulties, that the development of an ISP/2 with a lexical memory and associative network is necessary.

HARDWARE

The ISP system is designed around a microcomputer and a voice synthesizer (Figure 1). Physically the ISP system is composed of four parts: (1) a Z-80 based microcomputer with 24K of EPROM, 4K of RAM, 8K (ISP/1) or 32K (ISP/2) of CMOS RAM; (2) an LCD display; (3) a standard keyboard; and (4) a phonemic synthesizer (Votrax VSH). The size of the unit is 15cm in depth, 25cm in width, and 33cm in length. The weight is approximately 4kg.

J. C. Simon (ed.), Spoken Language Generation and Understanding, 455-466.

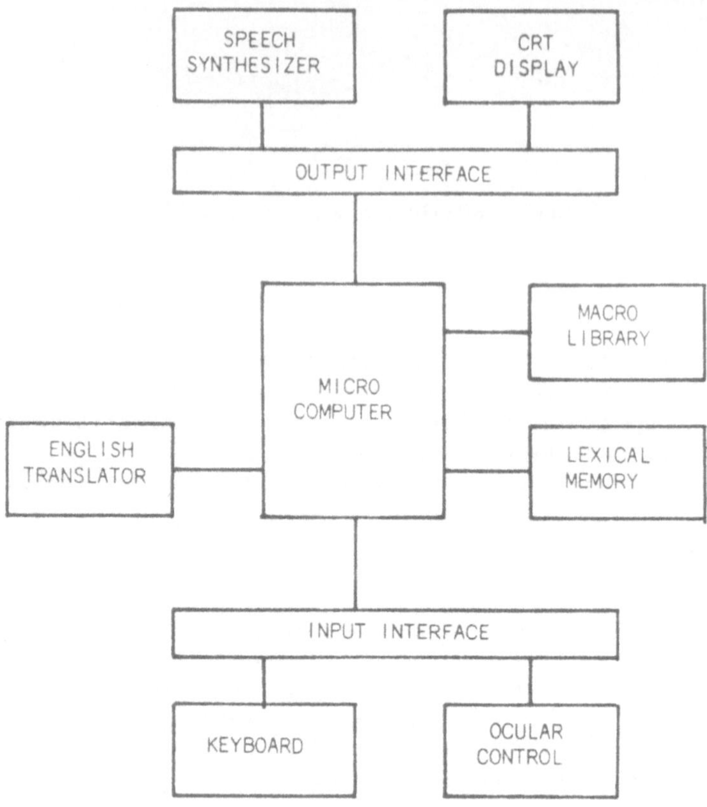

Figure 1. Diagram of the Intelligent Speech Prosthesis Hardware

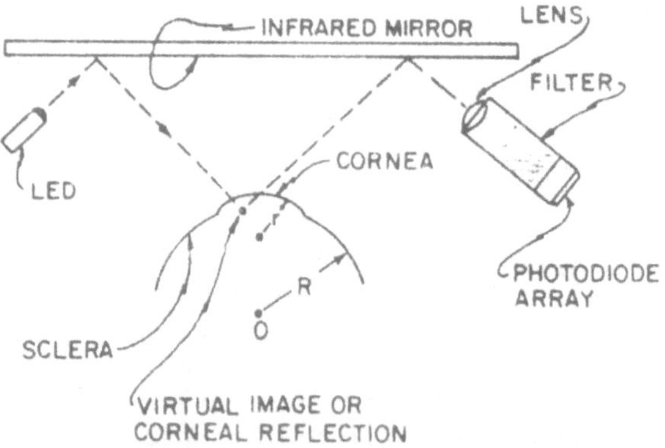

Figure 2. Optics used in Eye Position Monitor

Between the user of an ISP and the speech output lies a mechanical interface, a keyboard. Many potential users have difficulty operating a keyboard because their hand movement is athetoid or spastic. For these people, even the construction of a simple utterance is a long and difficult task. For these users we are constructing a new interface using an ocular control.

The ocular control as proposed by Rinard and Rugg (1977), involves placing an optical sensor on a set of eyeglasses to monitor eye position relative to head position. An infrared light emitting diode is attached to the nosepiece of the eye-glasses. The inside surface of the lens of the eyeglass is coated to form an infrared mirror that still is transparent to visible light. The image sensing unit is mounted on the bow of the eyeglass with an adjustable bracket (Figure 2).

The eye is illuminated by the infrared source reflected from the infrared mirror. The surface of the cornea acts as a spherical mirror and a virtual image of the LED is formed about 4 mm behind the surface. The corneal reflection is detected by the image sensor unit via another reflection from the infrared mirror. The image sensor consists of a lens, a visible light absorbing or interference type infrared bypass filter, and a 3 mm square 32 x 32 element photodector array. The lens produces an image of the corneal reflection on the photodector array. Each array element is sequentially scanned by the eye position monitor and the signal is processed to determine the X and & coordinate of the corneal reflection.

SOFTWARE

The ISP system software is composed of five basic programs: (1) MONITOR a control program; (2) EDIT, updates, builds, and stores previously prepared utterances; (3) TRANS, translates English text to phonetic text; (4) SPEAK, generates the parameters necessary to drive the voice synthesizer from the phonetic text produced by TRANS; (5) DAN, a set of algorithms capable of correcting and compensating for certain types of linguistic input errors.

The MONITOR is a program which calls and activates the appropriate modules of the system. The system is organized so that communication between modules, data entry and validity testing are done by the monitor.

The EDIT program allows the user to build, modify or save utterances which are to be used as macro instructions or mnemonics. The text of a macro instruction can be composed of (1) standard English text; (2) other macro instructions; or (3) standard English text and macro instructions. For example,

(1) #HTG -- HOW ARE THINGS GOING
#FTD -- FEELING TODAY?
#HAU -- HELLO, HOW ARE YOU

(2) #HAU -- #HAU #FTD
(HELLO. HOW ARE YOU FEELING TODAY?)

(3) #IHTG-- I AM FINE. #HTG WITH YOUR FAMILY?
(I AM FINE. HOW ARE THINGS GOING WITH YOUR FAMILY?)

The TRANS program is an English text to phoneme text translator. English is input, translation rules are applied, and phonemic text is returned. Rules in the form of a top down production system similar to those of Ainsworth (1973), Elovitz (1976) and McHugh (1976) are used. Rules have the form of a short character or null string, the target string, and a short character or null string. For example, for the general rule /X(Y)Z=P/, the string Y occurring in the context of X and Z gets the pronunciation P. P is a representation of the International Phonetic Alphabet. Y is a letter or text fragment to be translated. X and Z may be strings of letters and/or other characterist which denote classes of strings. Some of the special characters are:

- vowel cluster
@ - single vowel
+ - E, I or Y
* - consonant cluster
: - zero or more consonants
  - single consonant
, - single voiced consonant
% - A suffex E, ED, ELY
& - possible S ending

The translator first matches the Y target string. It then tries to match in a right to left manner the X string and lastly in a left to right manner the Z string. Once a match is complete, the phoneme or phoneme string is saved and the text target string is fetched. An example of a simple rule is:

/(EY)=AY/

which means that an EY at the beginning of a word is pronounced similar to the letter l in the word hide. An example of a more complex rule is:

/@*(A) 1S &=AX/

which means that an A preceded by a consonant cluster and a

single vowel and followed by a single consonant, the letters l and S, another single consonant and an S or a blank is pronounced similar to the A in about.

After the translation is complete, the phoneme text is scanned and rules for applying stress and the deletion of unnecessary phonemes are applied. These rules are used to make the vocal output sound more natural and easier to understand.

Many computers that speak do so in a cut-off, monotonous, or robot-like manner because the speech is generated by simply joining together a sequence of fixed sounds. This type of speech is intelligible, but is more difficult to understand than natural speech. The SPEAK program is a program which attempts to give the computer a better quality tone using a set of rules which are applied to the phonetic text produced by TRANS.

The speak program has three types of rules. The first type of rule assigns a systematic rhythm pattern to each word. These rules do not allow two high stressed vowels to come in contact with each other; instead, a regularly alternating pattern of medial stress and strong stress occurs. A second type of rules tries to break up the sing song effects when stress alternation is systematically applied. Certain words, such as definite (the) and indefinite articles (a, an) are never marked for high stress or long duration, but marked for flat intonation and low pitch. Other rules recognize juncture words (and, or, but, etc.) and commas within the utterance. A short pause is inserted after the conjunction or comma and the conjunctions are marked for a flat intonation and low pitch. Also, common high frequency non-contentive function words (this, that, etc.) within the utterance are marked for flat intonation and low pitch. A period at the end of an utterance indicates a falling intonation and insertion of a long pause. A question mark indicates a rising intonation and a long pause. If there is no punctuation at the end of an utterance, a falling intonation is assigned. Punctuation within an utterance is literally interpreted. No effort is made to correct errors. For example, if the user inserted a period at the end of the macro #HTG the expression

#HTG WITH YOUR FAMILY?

would expand into

HOW ARE THINGS GOING WITH YOUR FAMILY?

and the speak rules would be applied. A third type of rule modifies the phoneme duration. For example, the final vowel in an utterance is lengthened, or if it is not possible to do so, the preceeding vowel is shortened. The SPEAK program and its rules are tentative and not always reliable because the interpreter does not do a complete content analysis but instead works with a set of simple and intuitively derived rules.

The process of encoding and decoding normal speech can be divided into several stages:

(1) A sound or picture stimulus is received and features that characterize it are abstracted from the total event.

(2) Abstracted information is integrated and mapped into a semantic representation.

(3) The semantic representation is mapped to a lexical item, a descriptive name or word that is its name.

(4) The word that has been accessed is translated into its phonetic representation.

(5) Motor commands are issued that make possible the articulation of the phonological representation.

The DAN program is designed to aid users whose brain lesion affects the speech encoding process in stage 3. It is not designed as an aid to someone who does not have an awareness of the input item or the output item, because of a lesion in the posterior part of the brain that affects stages 1 and 2.

The word finding disorders, in which we are interested, can be divided into the different types, which are weakly correlated with the locus of damage and other symptomatology. The first group has been described by Wepman, Bock, Jones and Van Pelt (1956); Goodglass, Hyde and Blumstein (1969); Marshall and Newcome (1966) and others. It involves damage to the anterior portion of the dominant cerebral hemisphere. These users have many concrete and picturable words available in consciousness and can perform well although slowly in naming tasks. For them, naming disruption is part of a more generalized disturbance, since other symptoms include laborious articulation and a severe disruption of productive syntax. Those whose lesions are in the posterior portion of the dominant hemisphere, can also suffer naming performance that displays a distinct deficiency of concrete or picturable words. Although they often fail to provide the correct name of an object, their responses can be related to the target word. When the words (the target and the word produced) are compared on the basis of semantic features, the two may share features that represent the major semantic categories and they differ on the specific functional distinctions. For example, pen for pencil, or knife for fork

(Rinnert and Whitaker, 1973; Schuell and Jenkins, 1961). Both the clinical and experimental literature on naming disruptions account for the impairment in terms of a failure of the peripheral mechanisms, stages 4 or 5, or retrieval failure, stage 3 (Gardner, 1973; Goodglass, Barton, and Kaplan, 1968; Goodglass, Hyde and Blumstein, 1969). We are primarily concerned with the retrieval failure since the user with a failure only in the peripheral mechanisms can use a speech prosthesis without a DAN network.

How a lexical memory and its corresponding associative network are organized is a problem of great interest to artificial intelligence (Deese, 1965; Schank and Colby, 1973; Fodor, Bever and Garrett, 1974; Fodor, 1975; Miller and Johnson-Laird, 1976). The lexical memory of the DAN network contains (1) contextual information (CXT); (2) generic properties; and (3) semantic properties (SA). This is illustrated in the following table.

Word	Properties				
	CXT	1st letter	Mid letter(s)	Last letter	SA
Steak	Food		T	K	Meat

In step one of attempting to find a target word the system queries the user for whatever clues he can offer about the context, the generic properties, and the semantic properties. The clue pattern is of the form:

CLUE = CXT + L + (L or STRING) + L + SA

Assuming that each slot in the pattern is not null and correct we find a short list of target words (NULL to a maximum of 20 words) meeting the input criteria can be retrieved from a dictionary of 900 words. The candidate list of words is then presented to the user using both an auditory and visual display. If the user sees or hears the intended word, the user enters a number (indicating which item on the list is the target word) and the word is inserted into the output utterance of the ISP/2. We have also found, using a hand simulation, that many words can be successfully retrieved in a 1600 word lexicon using the above described strategy.

This process assumes full and correct entries in the clue-pattern. However, we have been successful in retrieving the correct word list when list of clues is incomplete. However, if too few entries are made the output list becomes unmanageable. For example, given only that the word begins with the letter A, the output list contains all words beginning with the letter A within a particular context.

If the output list is NULL and the clue pattern is complete, the program deletes a clue. The first clue to be deleted is the semantic property. If the output list is still NULL, the context clue is deleted, and the semantic clue is restored. The search then looks at across all contexts for the semantic clue and its associations. If the output is again NULL, the semantic clue is disregarded and only the generic clues are used to search across all contexts. The clues are deleted in this particular order because experimental evidence shows that the semantic clue is the least reliable. This may be due to a breakdown in the users semantic organization (Goodglass and Baker, 1976). Also, the semantic clue is the most constraining, and its deletion quickly leads to an expanded target list. If the above fails, the program returns to the first step and asks the user for some other clues, and if they are input, the search is repeated.

Word associations are highly idiosyncractic to the individual. For example, if one asks 1,000 college students the word they associate to the word table, 691 say CHAIR but the rest of the responses are distributed over 32 other words (Palermo and Jenkins, 1964). Our associative network has been initialized with a set of known high frequency words. To this network we have added many semantic associations being obtained in field testing of the ISP/2. Since we are proposing a dynamic network, a frequency count of word usage is kept by the machine and the pointers used in associating clues to a particular target are being constantly modified. Over time, pointers are added or subtracted as the net is used. For example, suppose the user is seeking the word STEAK. However, he only has a vague notion of it, and is able to give clues that he is in FOOD, it ends with a K and it reminds him of MEAT. He enters the clue string

CLUE = FOOD + ? + ? + K + MEAT

and the network retrieves and displays a list of words semantically associated with MEAT and that ends with a K. If the target is in the list, it is inserted into the output expression. However, if the target is not in the list the more detailed strategy described above is applied.

If the word STEAK is then retrieved the network is modified. A pointer to the STEAK is added to the list of pointers associated with MEAT. In future searches the word STEAK will then be found whenever the user gives the clue MEAT.

CLUE = FOOD + ? + ? + K + MEAT

The repeated association of the word MEAT with the word STEAK will modify the set of pointers coming off MEAT. The least used pointers will be pushed down the list of pointers pointing to the semantic association of MEAT while high frequency associations are pushed to the top. In time, the speech prosthesis "learns".

From a basic general lexicon of a few thousand words a lexical memory can be developed which is individualized to a user's environment from the user's experience. An automatic tracking can be made of how the user operates his lexicon. That is, a frequency count is kept not only on the frequency of a word by a particular user but also on the frequency of his association from one word to another. This allows for the updating or modification of a particular user's lexicon so as to eventually focus upon his particular needs and demands.

Using the STEAK example in a food context an input string could have the form.

#PBM (   ) RARE

where: #PBM = a mnemonic for "please bring me a"
( ) = an indicator of a word to be found
RARE = an orthographic spelling of the English word

In analyzing this input, the <u>MONITOR</u> initially checks for invalid symbols, checks the MACRO dictionary for the existence of the string associated with #PBM, and then calls the <u>DAN</u> program. The <u>DAN</u> program will query the user in the following way.

CONTEXT (   )
FIRST LETTER (   )
LAST LETTER (   )
MIDDLE LETTER OR LETTERS (   )
WORD GOES WITH (   )

In the optimal case the user would reply in the form

CLUE = FOOD + S + E + K + MEAT

The network would then invoke a set of search procedures and would retrieve a list of candidate words among which would be the word STEAK. After the user indicates that STEAK is his target word, control would then be passed back to the control program which would call the TRANS program to translate the word "RARE" to its phonetic equivalence and return control to the <u>MONITOR</u>. Finally, the <u>SPEAK</u> program would be called and the utterance vocalized. The heuristic search for the target list is based on partial and related information. It is much like playing a game of twenty questions to reduce the number of alternatives among a set of candidates. Much depends on the size and content of the lexical memory and the cut-off rules used in the search.

USAGE AND EVALUATION

There is a large and growing number of people in the world who are unable to communicate through speech. In the U.S. alone there are between one and five million people with some degree of a communication disorder. Hence, the population for whom a speech prosthesis would be helpful numbers in the millions.

To develop and improve this prosthesis, we are reducing the size of the hardware, completing the software, and testing it in the field. Empirical experience is needed with a variety of users to explore the determining parameters of the search. This problem lies not only in the algorithms speed and capability to retrieve a target list but also in the area of the capability of the user to interact with the machine. For example, if a word has 50 semantic associations, we want to retrieve and display a subset of these words. However, if a word has only 2 semantic associations we might want to select words two or three stages away in the associative net from the starting node. This problem of breadth versus depth search is familiar to computer scientists. Extensive practical experience is needed to determine (1) which types of patients can and cannot operate the device; (2) what are the real-time communication problems in terms of the amount and rate of output the patient can generate; and (3) what are the situational dependencies that hinder or facilitate the use of the speech prosthesis?

A CLINICAL EXAMPLE

An ISP with an occular control is being used by a 47 year old male who suffered two cerebrovascular infarcts on December 26, 1976. The patient is confined to a wheelchair, has a limited amount of controlled head motion, and has a mild wrist and ankle clonus.

Results of the Minnesota Test for Differential Diagnosis of Aphasia are as follows: good auditory and reading comprehension, severely impaired gestural and graphic skills; and a severe oral apraxia. In test situations using an ISP with occular control and the alphabet displayed on a CRT, the patient, after 5 hours of practice, was able to generate text at a rate of 20 characters per minute. In example, the expression

I WANT A DRINK OF POP<CR>

took approximately 70 seconds for the patient to generate and the machine to say (22 characters). However, when the patient was allowed to create and use macro expressions, the same expression

#DRK POP<CR>

was generated and spoken in less than 30 seconds.

An ISP with the word finding program described above is being used by a 54 year old male who suffered a cerebrovascular accident on February 25, 1979. A neurological examination revealed a left hemisphere infarct. The patient can walk slowly with a cane but cannot use his right hand easily. In social conversation he shows a severe anomia.

Results of the Minnesota are as follows: a minimal auditory comprehension/retention deficit; a mild to moderate verbal apraxia; a mild reading comprehension deficit at the sentence level and a mild to moderate graphic deficit at the sentence level. In test situations the patient is able to retrieve about one half of the words needed to express himself with a moderate response latency.

In a test situation using the ISP with a lexical memory the patient filled in the clue pattern as follows:

FIRST LETTER -- C

LAST LETTER -- none given

MIDDLE LETTER -- none given

WORD GO WITH -- SWEET

The program selected the following words:

1. COOKIE
2. CAKE
3. CHOCOLATE
4. CANDY
5. CARAMEL

with CANDY being the word he was seeking.

ACKNOWLEDGEMENT

This research is supported by Grants MCS 78-09900 and PFR 79-17358 from the National Science Foundation.

References

Ainsworth, W.A. A system for converting English test to speech. IEEE Trans. Audio and Electroacoustics, 1973, AU-21, 288-290

Colby, K.M., Christinaz, D. and Graham, S. A computer driven, personal, portable, and intelligent speech prosthesis. Computers and Biomedical Research, 1978, 11, 337-343

Deese, J. The Structure of Association in Language. John Hopkins Press, Baltimore, 1965

Elovitz, H.S., Johnson, R.W., McHugh, A. and Shore, J.E. Automatic translation of English text to phonetics by means of letter-to-sound rules. NRL Report 7948, 1976, 1-98

Fodor, J.A., Bever, T.G. and Garrett, M.F. The Psychology of Language. McGraw Hill, New York, 1974

Fodor, J.A. The Language of Thought. Thomas Y. Crowell Press, New York, 1975

Gardner, H. The contribution of operativity to naming capacity in aphasic patients. Neuropsychologica, 1973, 11, 213-220

Goodglass, H., Barton, M.E. and Kaplan, E. Sensory modality and object naming in aphasia. J. of Speech and Hearing Resh., 1968, 488-496

Goodglass, H., Hyde, M.R. and Blumstein, S. Frequency, picturability, and the availability of nouns in aphasia. Cortex, 1969, 5, 104-119

Goodglass, H. and Baker, E. Semantic field, naming and auditory comprehension in aphasia. Brain and Language, 1973, 3, 359-374.

Marshall, J.C. and Newcomb, F. Syntactic and semantic errors in paralexia. Neuropsychologica, 1966, 4, 169-176

McHugh, A. Listener preference and comprehension test of stress algorithms for a text to phonetic speech synthesis program. NRL Report 8015, 1976, 1-21

Miller, G.A. and Johnson-Laird, P.N. Language and Perception, Harvard University Press, 1976

Palerno, D.S. and Jenkins, J.J. Word Association Norms, University of Minnesota Press, Minneapolis, 1964

Rinard, G. and Rugg, D. Communication/control applications of the ocular transducer, in Proceedings of the Workshop on Communication Aids for the Handicapped. National Research Council of Canada, Ottawa, Canada, June 1977

Rinnert, C. and Whitaker, H.A. Semantic confusion by aphasic patients. Cortex, 1973, 9, 56-81

Schank, R. and Colby, K.M. Computer Models in Thought and Language. Freeman, San Francisco 1973

Schuell, H. and Jenkins, J.J. Reduction of vocabulary in aphasia. Brain, 1961, 84, 243-261.

Wepman, J.M., Bock, R.D., Jones, L.V., and Van Pelt, C. Psycholinguistic study of aphasia. Journal of Speech and Hearing Disorders 1956, 21, 468-477

§ 5

SYSTEMS AND APPLICATIONS

INDUSTRIAL APPLICATIONS OF SPEECH RECOGNITION
(Panel discussion)

Kasuo NAKATA
Central Research Laboratory,
Hitachi Ltd., Kokubunji, TOKYO, JAPAN.

ABSTRACT

Since Texas Instruments has put its "Speak & Spell" on sale, speech processing machines and devices have become very popular, and managers of industry have begun to think that speech research may become profitable.
The following four questions are discussed in this panel discussion:
1°/- What is a good example of speech recognition application ?
2°/- In which direction should we go in speech recognition research ?
3°/- What technical problems should we attack ?
4°/- What is the best architecture for a speech recognition machine ?
Some speech recognition applications in Japan are described.

1.- INTRODUCTION

Since Texas Instruments has put its "Speak & Spell" on sale, speech processing machines and devices have become very popular. Many non-technical people expect a handy LSI speech recognition machine to be realized and sold in the near future.
Recently, a remote control device which can switch TV channels by voice command was actually demonstrated at a micro-computer show in Japan.
This trend has two contradictory influences on speech research and speech research people.
The first, a desirable influence, is that the managers of industry begin to think that speech research may become profitable in the sense of industrial investment, so that they are apt to invest money and people in speech research. This fact activates speech research and puts new value on the speech research people in industry.
The second, an undesirable influence, is that the research objectives become more application-oriented than in the past, and that the evaluation of results becomes more realistic and severe,

J. C. Simon (ed.), Spoken Language Generation and Understanding, 471-474.

(i.e. the evaluation of results becomes more important than the evaluation of new methods). The duration of the research work is inclined to be determined by external factors such as a schedule for announcing new products or sales.
Thus it becomes more difficult than before to carry out research by the researcher's own will and at his pace. Immediate follow-up and severe assessment of the reported results by many actual data will take place when the results are considered to make a contribution to industrial applications, even when they are carried out at universities or at governmental or non-profit research institutes.
It is commonly said that industrial managers usually take a relatively short-range view, and it is to be feared that after two or three years, investments will be stopped because of critical assessments.
I believe now, is the time to try very seriously to find really excellent applications of speech recognition within the technical level that we can achieve.
I also believe that the only possible way to promote and continue speech research is to increase our knowledge and to find useful applications without overemphasizing the "magic power" of speech recognition and speech understanding.
I asked each panelist to discuss and comment on the following four topics:

1°/- What is a good example of speech recognition application ?
2°/- What directions should we go in speech research with regard to a practical recognition application ?
3°/- Along that direction, what technical problems should we attack in the future ?
4°/- What is the best architecture for real speech recognition machine in the future ?

2.- BACKGROUND

I would like to give some general background on why I did put these four questions at this panel discussion.
For *Question* 1, already I emphasized the importance of this question, and again I repeat my belief that this is the key question to be answered.
For *Question* 2, four directions should be followed for future applications of speech recognition, as shown in Fig. 1. There must be strong correlations between directions and the good examples you conceive in your mind.
Question 3 is a question regarding the technical problems which must be solved before you can proceed to your final goal. For example, if you intend to develop a speech recognition machine for an unspecified speaker, an essential technical problem to be solved is an effective normalization process for speaker-dependent variations of extracted features.
Question 4 is also a very important point in the design of a practical speech recognition machine. One possibility is a

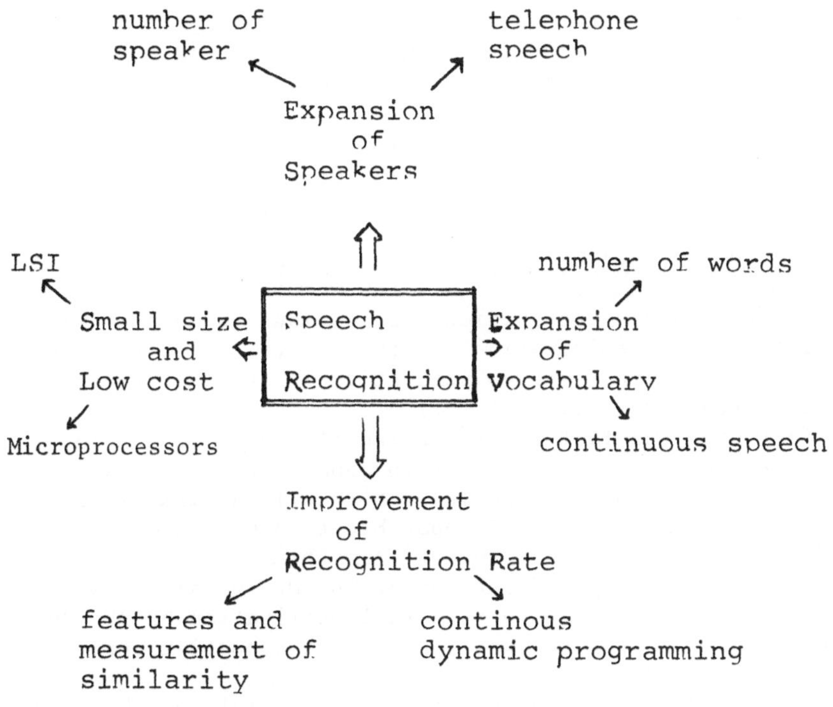

Figure 1: Possible Future Directions of Recognition.

Table 1: Main Specifications of the speech recognition machine DP-100

vocabulary	10 digits and 120 isolated words, and continuous digits
speaker	one specified speaker at a time
registration	1 trial word at beginning
input device	microphone (noise cancelling head set)
analysis	16 channel spectrum analysis (200 Hz---5000 Hz) 20 m sec./frame
time axis alignment	dynamic programming and two-stage dynamic programming
recognition rate (reported by NEC)	10 digits: 99.7% isolated words : 99.3% continuous digits: 99.3%

completely special hardware; another possibility is a combination of a special hardware analyzer and of mini or microcomputers. There may be some other way suitable for special conditions of practical use.
But whatever it may be, consideration should be given to the possibility of an LSI implementation of a machine as a whole or in parts.

3.- SPEECH RECOGNITION APPLICATIONS IN JAPAN.

For your reference, I will say a few words on the present situation of the speech recognition applications in Japan.
You will find an informative reference with regard to a general speech recognition research activities in Japan in the lecture by Sakai (this volume, page 147).
The following two facts characterize the recent situation of speech recognition applications in Japan. Firstly a speech recognition machine has become commercially available for the first time in Japan. NEC (Nippon Electric Company) has put its first commercial speech recognition machine, the DP-100, on the market. The main specifications of the DP-100 are shown in Table 1. Its main feature is a full utilization of DP (Dynamic Programming) and the recognition of continuously uttered digits by a two-stage DP.
Fujitsu and Hitachi also announced their trial speech recognition machines very recently, and soon they will go on the market.
Secondly, intensive sales of speech recognition machine take place in Japan by several USA manufacturers, such as Threshold Technology, Interstate, Mike, and some others. None of them are really used in routine work, but some of them are used under almost real conditions for voice control of baggage sorting and voice data input at inspection lines, and they are reported to be successful.
I am sure to say that now is the beginning of speech recognition application in industry in Japan.

SOME POINTS CONCERNING SPEECH COMMUNICATION WITH COMPUTERS

Joseph MARIANI

L.I.M.S.I. - C.N.R.S.
B.P. 30 - 91406 ORSAY Cedex - FRANCE

ABSTRACT :
We present here several points concerning speech communication with computers, and dealing with different aspects of the problem. We do not intend to write about a definite part of a speech communication system, but rather to give in formation and to report experiments that may be useful for people working in that field.

1. PERSON-PERSON SPEECH COMMUNICATION

I would like first to report an experiment concerning person-person communication. This experiment was conducted by Pr Alphonse CHAPANIS from The Johns Hopkins University (1). Two people were in separate rooms. One had the instructions to execute a task (several tasks were tested) and communicated these instructions to the other one that realized the task. Pr CHAPANIS studied different ways of communication : teleautograph, teletype, video, speech, direct mode (the two persons are in the same room) and the combination of these channels.
The experiments showed that when speech was used in the communication, the task was realized two times faster than when it was not (moreover, it was then almost a fast as direct communication!). The reason is that two times more information was exchanged because the transmission rate was two times faster. More than the speed aspect, the important factor is the naturalness of speech communication (this appears in the fact that the skill of the typist does not interfere in the teletype communication mode). The fact that speech communication may be interrupted is not important. If interruptions are forbidden, the messages are longer and less numerous, and the task lasts the same time.

J. C. Simon (ed.), Spoken Language Generation and Understanding, 475-484.

All these conclusions show that speech is the most natural and comfortable way of communication between two people upon a task. We may induce that it would be the same for person-machine communication upon a task. This would mean the death of Keyboards.

2. APPLICATIONS OF AN ISOLATED WORD RECOGNITION SYSTEM

The isolated word recognition system realized at LIMSI is based on the relevance of the unsteady parts of the signal (2). The frequency analysis, obtained with a 8-channel filterbank, is very rough. Each word of the list is represented by the same number of references (30) and this allows a good recognition score whether the words are long or short.
This system needs a single training pass and gives good results, even on difficult vocabularies :

- 99% on digits
- 87% on very similar words (karo, kado, kaʃo, kavo, kabo, kano, kamo, kalo, kaʒo, pri, tri, kri, bri, dri, gri, fri).

This system runs on a minicomputer IBM S7 (50 words in real time) and, recently, on an INTEL 8080 microprocessor (10 words in real time without optimization (first release)). We should be able to reach 100 words in real time on a 16 bits microprocessor. To get this result, the speech signal is processed during the acquisition. We even plan to do the word recognition during the acquisition, so that some words could be recognized before the end of the pronunciation.
This system has been used in several experiments concerning different tasks :

- SPEECH COMMUNICATION IN ROBOTICS (3)

In this area, the goal was to communicate with a robot, moving in a hostile environment. It was necessary to do the trade-off between security and convenience.
The first idea was that the system would systematically ask the user whether its recognition is right, but it rapidly became very fastidious. In fact, several rejection levels must be introduced : if the recognition score is very good, there is <u>no feedback</u>, if it is less good, the system may give a vocal feedback so that the user may correct a bad recognition.
If the two best words are very close, the system ask the human person to confirm its recognition by saying "yes" or "no". If the score is bad, the system asks the user to repeat the word.
Those different rejection levels are easy to introduce in an isolated word recognizer, it is more difficult in a continuous speech recognition system.

- FORTRAN VOCAL PROGRAMMING (4)

Here, we use isolated words recognition on a 110 word vocabulary, with a syntactic parser that may have two functions :

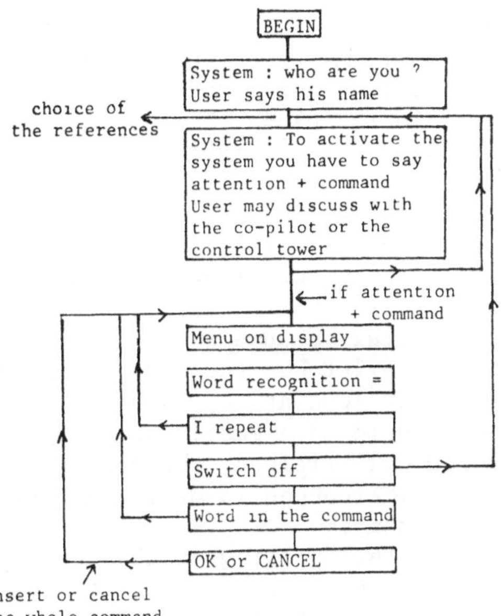

Fig.1- Pilot-plane dialogue system

- if it is used in a predictive way, then it reduces the number of words to be recognized at each time. This gives best scores and fast feedback

- if it is used in a verification way, the system tells the user what kind of programming errors he has made, and tells him the correct words that could be used, in connection with the syntax.

The function of the system is then more pedagogic.

In fact, both should be used, doing first the comparison among the words that are syntactically correct and, in case of rejection, among the other ones, with comments in case of error.

Of course FORTRAN is not the best language for vocal programming. This would be more useful for higher level languages that may appear in the future.

- PILOT PLANE DIALOG (5)

In cockpit functions, safety and convenience are also necessary. The procedure we used tried to take them into account (fig. 1). The pilot pronounces his name, the system recognizes it in the name list and chooses automatically the user's references. The pilot may then speak in a conversational mode. The system will not take his words into account until it recognizes two successive Keywords. Then the system is in the command mode. A menu, presenting the possible words at each step, may he displayed (it would be useful particularly for training the pilots). The pilot pronounces his word. If the word is not in the menu (syntax error), the word is rejected. A command may include several words. At the end of the command, the system will utter it, and the pilot may then insert or cancel it. He may also switch off the system with a command made by voice.

The experiments we made with pilots speaking through an oxygen mask during the air flight, registered on a tape recorder, gave good results, but less good than under the laboratory conditions, as there were no feedback in the plane. The noise environment is surprisingly very low. Over 3 G acceleration, the recognition score becomes bad.
The next step is to experiment it on a flight simulator.
The vocal feedback allows the pilot to have his eyes completely free. The vocal command sets his hands free and multiply his capabilities.
This application is interesting because :

- it is useful : the pilot has too many things to do with his hands, too many things to look at.
- it is possible : there is a small number of users, that are highly specialized and well trained, the command syntax is very much constrained, the microphone is at a constant distance from the mouth in the oxygen mask.

There are still problems in the spelling of the numbers using isolated digits : it is too long. Connected words recognition would be necessary. The pilots that have been contacted were very much interested.

COMPUTER-AIDED SYNTAX GENERATION

In our opinion, the user of an isolated word recongition system must be able to generate easily the syntax he intends to use, without modifying the recognition program. This leads us to realized a computer-aided syntax generation system that will help the user to build his own syntax.
Fig. 2 shows an example of the dialogue between man and machine during this syntax learning session. The system automatically represents the syntax in a list structure, including subtrees, and builds up the lexicon. We are trying now to improve this system by implementing the possibility of modifying the syntax.

3. PHONEME RECOGNITION

The main idea here was to take the unsteady instants into account. From a spectrogram obtained with a filter bank, we calculate a stability function, using the variation between two spectra in a 40 ms window. The extrema of this function give the steady and the unsteady instants. This was reported by HALTSONEN & Al. (6) to give a good segmentation (since they did their own algorithm). From this segmentation, we first recognize two successive steady instants, by comparing, with a Hamming distance, the spectra of these 2 instants and the spectra of the references.
We get n phonemes candidates and this generates n x n possible diphones. We then do the recognition of the transition among the references of these n x n diphones and we keep the k best phoneme strings. If m is the nombre of phonemes at the end of these

possible sentences : selection speed (3 digits) knots
selection altitude (4 digits) feet
acquisition speed
acquisition altitude

session :
System : What do you want to say (WS) first ?
User : Selection OR Acquisition → OR because the words have different syntactic categories
S : WS after Selection ?
U : Speed OR Altitude
S : WS after Selection Speed ?
U : (3 digits) → Introduction of a subtree
S : WS first in (3 digits) ?
U : (digit)
S : WS first in (digit) ?
U : 0=1=2=3=4=5=6=7=8=9 → = because the words have the same syntactic category
S : WS after 0=1=...=9 in (digit)
U : % → Nothing
S : WS after (digit) in (3 digits) ?
U : (2 digits)
S : WS first in (2 digits) ?
U : (digit) → The system already knows what is (digit)
S : WS after (digit) in (2 digits) ?
U : (digit)
S : WS after (digit) (digit) in (2 digits) ?
U : %
S : WS after (digit) (2 digits) in (3 digits) ?
U : %
S : WS after Selection Speed (3 digits) ?
U : Knots
S : WS after Selection speed (3 digits) Knots ?
U : %
S : WS after Selection Altitude
U : ....

Fig. 2 : A syntax generation session .(English translation)

strings, there will be m x n possible diphones for the next transition and so on ...
This allowed us to get about 65% of correct phoneme recognition (7). The problem concerns the time (5 x real time on a big computer), the large amount of data (> 1000 diphones), and the difficulty of doing talker adaptation on these references.
In our speech understanding system ESOPE Ø, we just take the steady instants into account (no consolidation (8)). The score falls to about 55%, on the first phoneme, and 80% among the 4 best phoneme candidates. But it allows us to do it faster than real time, and would simplify a talker ada ptation.

PHONEME VOCODER

The interest of this device is to minimize the information rate in speech communication (fig. 3).

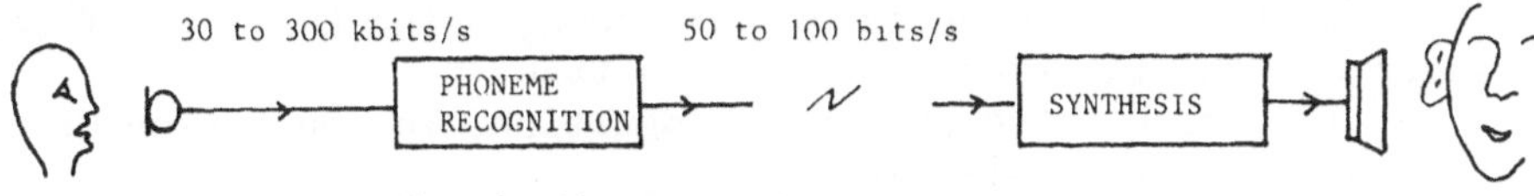

Fig. 3 : Phoneme vocoder .

The quantity is about 300 K bits/s at the source (30 K bits/s through telephone). When the phonemes are recognized and transmitted, then the information rate is 50 bits/s (100 bits/s with some prosodic elements). This phoneme string is transmitted and synthesized for the listener. The economy is about 1/1000. Unfortunatly the experiments we made about transmission of phoneme strings including errors (9) showed us that it was necessary to have less than 15% of errors, and no severe errors (no phoneme class confusion) to get an understandable message. Therefore the results are not sufficient for the moment.

4. ON THE USE OF PRAGMATICS/DIALOGUE IN PERSON-MACHINE SPEECH COMMUNICATION

In our project of Person-Machine speech communication (fig. 4), we want to consider both speech analysis and speech generation to realize a total voice communication. Some parts of it already exist (synthesis from the text, isolated word recognition), other parts are still to be done (monitor, prosodics, semantics).

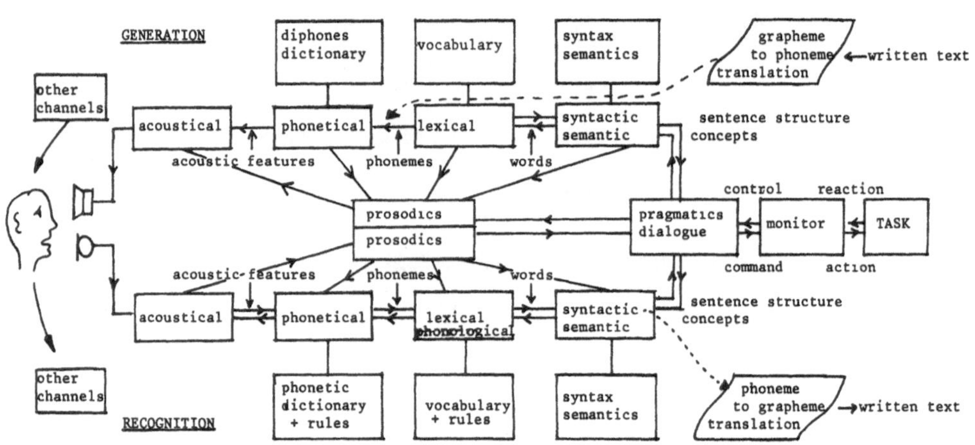

Fig. 4 : The person-machine speech communication project at LIMSI

We include short circuits at different levels (compressed speech, phoneme vocoder ...). Because of the fuzziness of all the levels (phoneme recognition, syntactic rules, words meaning) many ambiguities remain at each level. At pragmatic level, the knowledge about the task to be done clears up the ambiguities.
In fact, the pragmatic level interferes in several ways :
In speech analysis :

- It finds the meaning of sentences, using the past of the dialogue and the information about the task.
- It corrects the phoneme recognition, in connection with the syntactic/semantic level.

In speech generation :

- It asks a question to the user, if ambiguities are still remaining.

- It may ask a question or give information to the user about what he has to do in connection with the task. This involves speech generation from semantic concepts. We did some work on this subject (10).

In task management :

- It interprets the sentence meaning as a command to the monitor, and it interprets the controls from the monitor : this level is an interpreter.

We already use this level, in a very restricted sense, in our system ESOPE ∅.
In word-spotting, on data base interrogation for example, the pragmatic level has to guide the dialogue by asking questions to the user. It also predicts the list of possible keywords at each step of the dialogue. These keywords must be long enough (2 syllables or more), and the task must be very definite. The dialogue is very fast (real time) and the interesting point is that the speaker may use any syntax. It appears that during the dialogue, if the computer does not understand a sentence, the user shortens the sentence to a single keyword (fig. 5 a).
Although this dialogue may be demonstrated in our laboratory, it is mainly an example of word-spotting use in speech dialogue. There is no data base investigation process for the moment and the vocabularies are very small.
In continuous speech understanding, the system predicts at each step of the dialogue not only the vocabulary but also the syntax and may ask the talker to confirm the correctness of its recognition (fig. 5 b). Our system is based on the prediction/verification/modification concept, and actually uses a top-down strategy, with best few search and no backtracking. It works in real time on short vocabularies with constrained syntax.

5. PHONEME-TO-GRAPHEME TRANSLATION

The question we tried to answer is the following : even if we were able to recognize phonemes with a 0% error rate, would it be possible to write the sentence in correct French ? This problem is not simple, as a sentence like "J'ai mal au pied" (ʒemalopye, "my foot hurts me") may have more than 1500 possible word combinations at lexical level. If we use syntax, semantics, phonology, we still have two possibilities : "J'ai mal au pied" and "J'ai mal aux pieds" ("My foot hurts me", "My feet hurt me"). It is necessary to come to reality, to pragmatics, to get the right orthographication.
To investigate this problem, we used an error-free phoneme string and we tried to do first the word segmentation, then the grapheme translation.

L1 : vocabulary of Data Base names
L2 : " of large items
L3 : vocabulary of processes in chemics
L4 : homogeneous - heterogeneous - enzymatic

Recognized word	synthesized sentence	prediction	pronounced sentence
	Which Data Base do you want to consult ?	L1	The "Bulletin Signalétique" from the C.N.R.S., please .
C.N.R.S.	In what domain ?	L2	Hum , concerning chemistry .
nothing	I didn't understand , repeat please .	L2	In chemistry
CHEMISTRY	what kind of process in chemistry ?	L3	catalysis .
CATALYSIS	Homogeneous , heterogeneous or enzymatic ?	L4	heterogeneous catalysis .
HETEROGENEOUS	.....		

Fig. 5.a : Person-machine dialogue using word-spotting (English translation)

SL1 : sublanguage n°1 : question from the user
SL2 : sublanguage n°2 : confirmation from the user

Recognized sentence	synthesized sentence	prediction	pronounced sentence
		SL1	May I speak to Dupont ?
May I speak to Dupont ?	Did you ask for Dupont?	SL2	Yes , please .
yes , please .	Here he is .		
		SL1	Hello !
Hello	Who are you asking for ?	SL1	I'd like to get extension 240 .
I'd like to talk to Durand.	Do you want to speak to Durand ?	SL2	No , extension 240 please !
no,extension 240 , please .	Did you say extension 240 ?	SL2	Yes , sir .
Yes , please .	Here is extension 240		

Fig. 5.b : Person-machine dialogue using continuous speech recognition (English translation)

We did it in connection with linguists (CNRS, ERA 430) who built up a very large full-form dictionnary (170.000 forms now, 300.000 forms soon) with a phoneme translation for each word, and a pseudo-natural language syntax obtained with a learning process, including 160 syntactic categories and formalized by a ternary matrix presenting the possibilities of three successive syntactic categories (12).
The text we studied was about laws, the grapheme-to-phoneme translation was realized by program, and the blanks between words were suppressed.
A morphological criterion was necessary to avoid divergence. The heuristic was that, in the case where the syntax allows several word combination, the one segmenting the smallest number of words was chosen. (ex. "L'arrêté accordant le permis" and "L'arrêté à corps dans le père mis").

The results showed that there were about 5% of errors and ambiguities (fig. 6). This means that, to perform a perfect grapheme translation, it is necessary to take into account semantics and pragmatics and to understand the sentence.

1- LES EXEMPTIONS PREVUES PAR LE PRESENT ARRETES NE SONT PAS APPLICABLES AUX TRAVAUX CONCERNANT LES CONSTRUCTIONS FRAPPEES D'ALIGNEMENT ET SEL/SELLE/SELS SITUE/SITUEE/SITUEES/SITUES DANS LE PERIMETRE DE PROTECTION DES MONUMENTS HISTORIQUES/HISTORIQUE ET DES SITES CLASSES.

2- POUR LES CONSTRUCTIONS EDIFIEES SUR LE TERRITOIRE DE LA VILLE DE PARIS, LA CONSULTATION S'EFFECTUE AU LIEU, JOUR/JOURS ET HEUR/HEURES/HEURS/HEURT/HEURTS FIXE/FIXEES/FIXES PAR ARRETES DU PREFET DE LA CENE/SCENE.

3- LE PRESIDENT DE LA CONFERENCE PEUT ENTENDRE, POUR LES AFFAIRES QUI LES CONCERNENT, TOUTE/TOUTES AUTORITE/AUTORITES OU PERSONNE/PERSONNES COMPETENTE/COMPETENTES POUR EMETTRE UN AVIS SUR CES AFFAIRES.

4- DES L'AFFICHAGE A LA MAIRIE D'UN EXTRAIT DE LA DECISION OCTROYANT LE PERMIS-DE-CONSTRUIRE ET JUSQU'A L'EXPIRATION D'UN DELAI DE UN AN ET UN MOIS APRES CET AFFICHAGE, TOUTE PERSONNE INTERESSEE PEUT CONSULTER LES PIECES SUIVANTES DU DOSSIER, DEMANDE/DEMANDES DE PERMIS-DE-CONSTRUIRE, PLAN/PLANS/PLANT/PLANTS DE SITUATION, PLAN DE MASSE/PLANS DE MASSE, PLAN:PLANS/PLANTS DES FACADES, ARRETES/ARRETE ACCORDANT LE PERMIS-DE-CONSTRUIRE ET EVENTUELLEMENT, ARRETES/ ARRETE DE DEROGATION, CONTRAT/CONTRATS OU DECISIONS/DECISION EN MATIERE D'INSTITUTION DE SERVITUDE DITE DE COURS COMMUNE/COMMUNE - OU DE MINORATION DE DENSITE SURE DES FONDS VOISINS

5- LES TRAVAUX NON EXEMPTES PAR L'ARTICLE PREMIER ET CEUX QUI NE SONT PAS ASUJETTIS A LA PROCEDURE NORMALE DU PERMIS-DE-CONSTRUIRE NE PEUVENT ETRE ENTREPRIS QU'APRES COMMUNICATION DU PLAN DE MASSE AU CHEF DE SERVICE DEPARTEMENTAL DE L'URBANISME ET DE L'HABITATION DANS LE DELAI DE VAIN/VAINS JOUR/JOURS AVANT LEUR/LEURS EXECUTION/EXECUTIONS OU LA PASSATION DES MARCHES.

Fig. 6: Some examples of phoneme to grapheme translation
Accentuation is not represented but is also processed.

BIBLIOGRAPHY

1 - CHAPANIS, A., PARRISH, R.N., OCHSMAN, R.B., WEEKS, G.D. - The effect of four communication modes on the linguistic performance of teams during cooperative problem solving - Human Factors, 1977, 19, pp. 101-126.

2 - LIENARD, J.S. - Speech characterization from a rough spectral analysis - 1979 IEEE-ICASSP, Washington.

3 - OSORIO, A. - Communication parlée en robotique - LIMSI, Internal Report, feb. 1977.

4 - NEEL, F. - Programmation vocale en FORTRAN - LIMSI, Internal Report, feb. 1977.

5 - Final report DRET/ISPENA contract n° 77/455 - Commande vocale à bord d'avions - jan. 1979.

6 - HALTSONEN, S. and Al. - Application of novelty filter to segmentation of speech - IEEE-ICASSP 1978, University of Technology, Helsinky.

7 - MARIANI, J., LIENARD, J.S. - Acoustic-phonetic recognition of connected speech using transient information - IEEE-ICAS SP 1977, Hartford.

8 - SILVERMAN, H.F., DIXON, N.R. - The 1976 Modular Acoustic Processor (MAP) : Diadic segment classification and final phonemic string estimation - 1976 IEEE International Conference.

9 - LIENARD, J.S., MARIANI, J., RENARD, G. - Intelligibilité de phrases synthétiques altérées, application à la transmission phonétique de la parole - 9th ICA, Madrid, 1977.

10 - MEMMI, D., LIENARD, J.S. - Génération automatique de phrases en phonétique à partir de formules sémantiques - 10e JEP, Grenoble, 1979.

11 - MARIANI, J., LIENARD, J.S. - ESOPE Ø : un programme de compréhension automatique de la parole procédant par prédiction vérification aux niveaux phonétique, lexical et syntaxique - AFCET-IRIA, Congress on Pattern Recognition, 1978.

12 - ANDREEWSKY, A., BINQUET, J.P., BOUDEROUX, B. et col., LIENARD, J.S., MARIANI, J. - Les dictionnaires en forme complète et leur utilisation dans la transformation lexicale et syntaxique de chaîne phonétique correcte - 10e JEP, Grenoble 1979.

SECURE SPEECH COMMUNICATION OVER A CCITT-SPEECH CHANNEL

Dr. P. Meier

Swiss Army Signal Corps, Bern

ABSTRACTS

Today's main emphasis in secure speech communication is on vocoders. However there are still a lot of CCITT-speech communication channels in use for which secure speech communication is desired. It has generally been found that to do so a compromise between cost, security and speech quality has to be chosen.

In cases where vocoders cannot be used because digital transmission of the required bit rate (normaly 2.4-4.8 kbits/sec) cannot be safely achieved, the method of speech scrambling has to be looked at.

Starting with an introduction to the "speech iceberg" which shows the elements of speech relevant to communication, the problem of scrambler security is dealt within the context of a communication system. It is shown how scramblers relate to vocoders on the one hand and waveform coding schemes suitable for analog transmission on the other hand. A definition of the security of scràmblers is presented and its application is indicated and compared to the criterions used to evaluate digital ciphering systems.

It is also shown how scramblers have developped and that secure systems are possible with the modern digital technology.

J. C. Simon (ed.), Spoken Language Generation and Understanding, 485-495.

1. INTRODUCTION

The basic problem in speech processing is that speech consists of different but overlapping information elements (eg. linguistic information, speaker characteristics, redundancy) which are to be separatly processed for an optimal solution. In case of speech scrambling an additional restriction, that is the influence of the communication system has to be considered. We speak of speech scrambling if the same communication system can be used for the scrambled speech as well as for the original speech signal. If we represent speech digitally ciphering applies.

2. ELEMENTS OF SPEECH RELEVANT TO COMMUNICATION

In psychology we speak of the "iceberg model" in which the consciousness is compared to the tip above the water and the unconsciousness to the part below the water. In Fig.1 the information contained in speech is divided accordingly into a part which can be perceived and a part which is not perceived. As for speech production we can apply the same idea: there are elements of speech which we can consciously reproduce and others on which we have little influence - a separation which is important for speaker recognition.

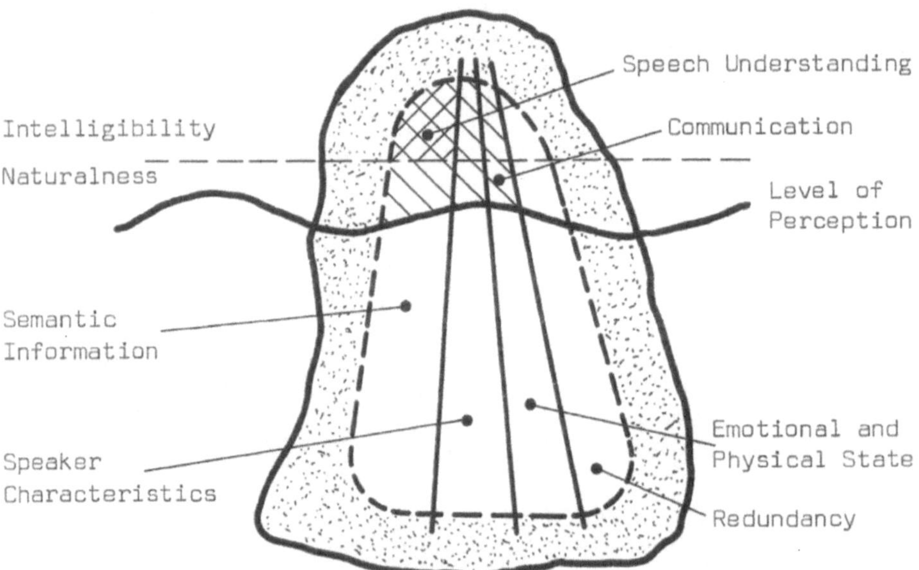

Fig. 1 Elements of Speech

Lost in Communication System

Available for Analysis

The part which we perceive contributes to what we call naturalness and intelligibility. The whole information contained in speech can be separated into semantic information, speaker characteristics, information which reflects the emotional and physical state of the speaker and redundancy which is important for the speech to be understood. If speech is transmitted with a communication system part of it is lost or modified. The parts of interest for communication are marked and compared to the parts relevant for speech understanding systems. Of course those elements are not as clearly separated as in Fig.1; they do overlap.

3. THE PROBLEM OF SECURE SPEECH COMMUNICATION

In Fig.2 the elements of a secure communication system are shown. The information (INPUT) is first transformed with a coding function [C] and then [C] I is adapted with [S] in order to send it over the communication channel.
The channel modifies [S][C] I with its transfer function [H]. In case of a radio link an eavesdropper can easily receive this information available on the channel without the legal sender or receiver noticing it.
The receiver side is designed to invert [S] with the function [R] and to decode [C] with the function [D] giving the output:

$$O_S = [D][R][H][S][C]\ I \qquad (1)$$

In case of speech scrambling it is usually required that the scrambler is a modular addition to the system, which by itself produces the output O_C:

$$O_C = [R][H][S]\ I \qquad (2)$$

The quality Q of a speech scrambler can now be defined as the degradation of O_S relativ to O_C in terms of intelligibility, naturalness and speaker characteristics recognition as a function of different (eg. jamming) channels [H] .

$$Q\ (H) = O_C(H)/O_S(H) \qquad (3)$$

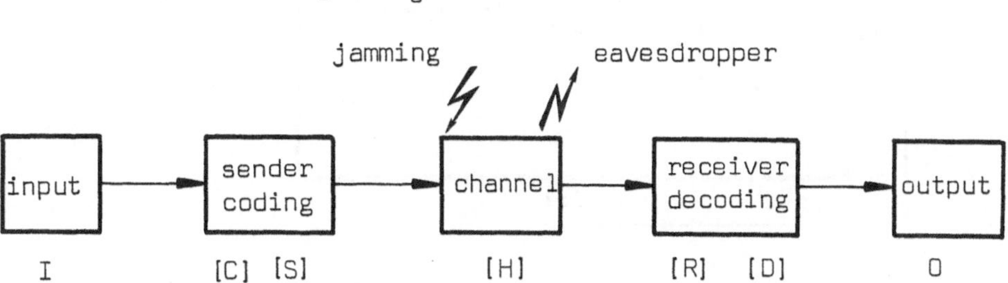

Fig. 2 Elements of a Secure Communication System

The determination of Q is basically subjective and can be done with probes given to listeners as reported by Goodmann, 1976. An objective method of measuring the intelligibility reduction of a communication channel has been proposed by Steeneken, 1975.

To evaluate the security of a communication system the information E available to an eavesdropper has to be considered:

$$E(H) = [H][S][C]\ I \qquad (4)$$

Basically there are three means to make E secure; first by using a secure channel [H] (eg. a courier), second by coding [C] (eg. ciphering or scrambling) and third by a combination of both (eg. spread spectrum technique).

4. THE DEVELOPMENT OF SPEECH SCRAMBLING TECHNIQUES

Most conventional scrambling methods work on speech elements S in the three dimensional representation of speech in terms of time (T), frequency (F) and amplitude (A) (eg. a sonogram),

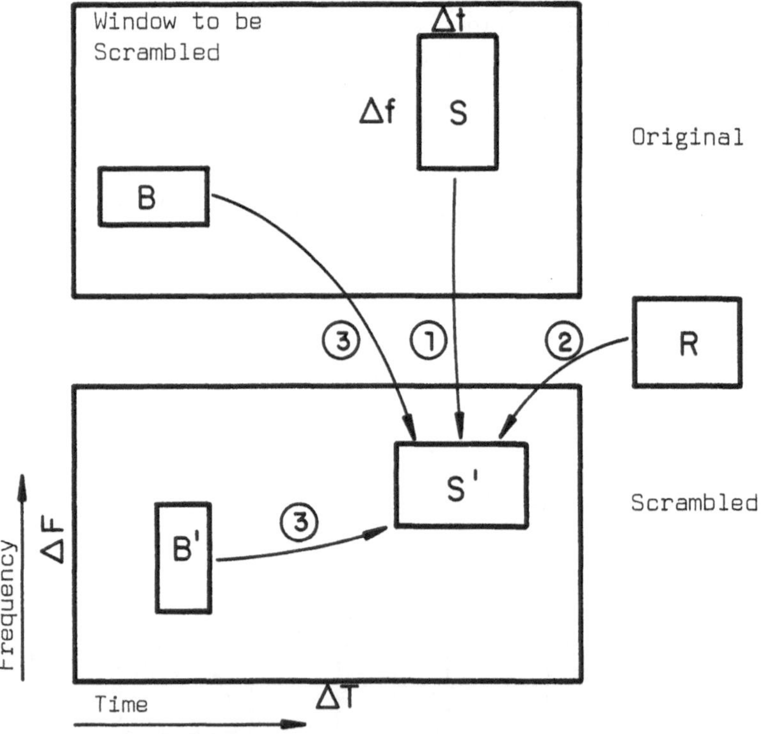

Fig. 3 Basic Method of Speech Scrambling

The three basic scrambling operations possible are shown in Fig.3, they are:

1) Transformation of an element S to S'
2) Adding of pseudo-noise R or a masking signal to S'
3) Adding of past or future elements of the original (B) or the scrambled (B') speech to S' (eg. echo).

If we now look at the existing scramblers, we find a systematic development over 5 generations. From Fig.3 we can deduce the three basic elements to be considered:

Dimension: the dimensions T, F or A in which scrambling is performed.

Control: the way in which the scrambling operation is controlled. It can either be in a fixed manner for each window or the scrambling operations can be varied under the control of a keygenerator (RC = Rolling Code).

Structure: the scrambling operation is performed with a structure in the dimensions used (eg. permutation of time segments of the speech signal). Again this structure can be fixed or rolling code controlled.

The development of scramblers in these terms is shown in Tab.1:

Tab. 1 Scrambler Generations

Generation	Dimension	Control	Structure
1	one	fixed	fixed
2	two	fixed	fixed
3	one	RC	fixed
4	two	RC	fixed
5	two	RC	RC

1. Generation: Workes in one dimension only, with fixed control and structure (eg. frequency inverter).

2. Generation: Works in two dimensions basically to decrease the rest-intelligibility as defined in chap.5 (eg. frequency inverter with masking signal).

3. Generation: Employs a rolling code in order to increase the analytical security (see chap.5).

4. Generation: Works with a rolling code in two dimensions which again decreases the rest-intelligibility, increases the analytical security.

5. Generation: In addition to the 4. generation a varying scrambling structure is used which significantly increases the analytical security.

The security aspects of those five generations are extensively treated by Mc Calmont, 1979. Among others the influence of the product of $\Delta f \cdot \Delta t$ = TWE as compared to $\Delta T \cdot \Delta F$ (see Fig.3) on the analytical security is treated. It is obvious that a small TWE increases the analytical security. Realising a 5. Generation scrambler with a small TWE using conventional methods requires a lot of experience in order to get a cheap good quality scrambler. The minimum TWE achieved is of the order of 20. Such an element still contains information to be used by cryptoanalysts.

5. EVALUATION OF THE SECURITY OF SPEECH SCRAMBLERS

The problem of evaluating the security of scramblers is quite different from the cryptoanalytical problem of evaluating a digital ciphering scheme (Diffie and Hellmann, 1979). The former is connected to the human capability of perceiving highly distorted speech, the latter is a pure mathematical problem.

There are two aspects to the security of scramblers:

A. The rest-intelligibility of the signal E(H) in (4) as perceived by an eavesdropper: The suggested criterions to be used are given in order of increasing rest-intelligibility:

1. How accurately is it possible to decide whether speech is transmitted or not?
2. How much of the speaker characteristics can be recognized?
3. To what degree can the language be recognized?
4. What is the isolated single word intelligibility (eg. for numbers) or the word spotting capability?

5. What is the intelligibility of continous speech at different speeds?

B. What is the analytical security against an attempt to increase the rest-intelligibility of E(H) by using the means of cryptoanalysts. Looking at ciphering where one can either decrypt a message or not, an exact descrambling can usually not be achieved for a secure scrambler. The suggested criterions for the analytical security are:

1. What level of knowledge about the scrambler must be assumed to be available to the eavesdropper.
2. What level of means can be assumed to be used to increase the rest-intelligibility.
3. To what degree can an eavesdropper use information or weaknesses of the keygenerator.

In order to illustrate the problems a cryptoanalyst runs into if he attempts to break a scrambler we consider the well known time division technique in Fig.4 (Baeschlin, 1977). There are three problems to be solved in order to increase the rest-intelligibility by rearranging the time segments:

1. The scrambling structure eg. the time segment boundaries have to be found.
2. A metric which gives a measure for the neighbourhood of adjacent speech segments in the original speech signal (see Gray, 1976) has to be found.
3. Using this metric the puzzle of rearranging the time segments for better rest-intelligibility is to be solved. Mathematically this is a Travelling-Salesman problem with special boundary conditions (Liesegang, 1974).

We have now established the base to extrapolate to the scramblers of the 6. Generation. The idea behind the scramblers of the 5. Generation was to make it very hard to detect the scrambling structure. So what is still required is a method to make it hard to find a suitable metric to solve the puzzle and thirdly to increase the order of the puzzle to make it computationally more time consuming to solve it. If we achieve this purpose we have a scrambler on which it is practically impossible to increase the rest-intelligibility by the above puzzle solving, because this requires the accurate determination of the scrambling structure, the existence of a good metric and the computational power to solve the puzzle to the degree that an improved rest-intelligibility results. If the first step already cannot be solved accurately we cannot achieve this aim.

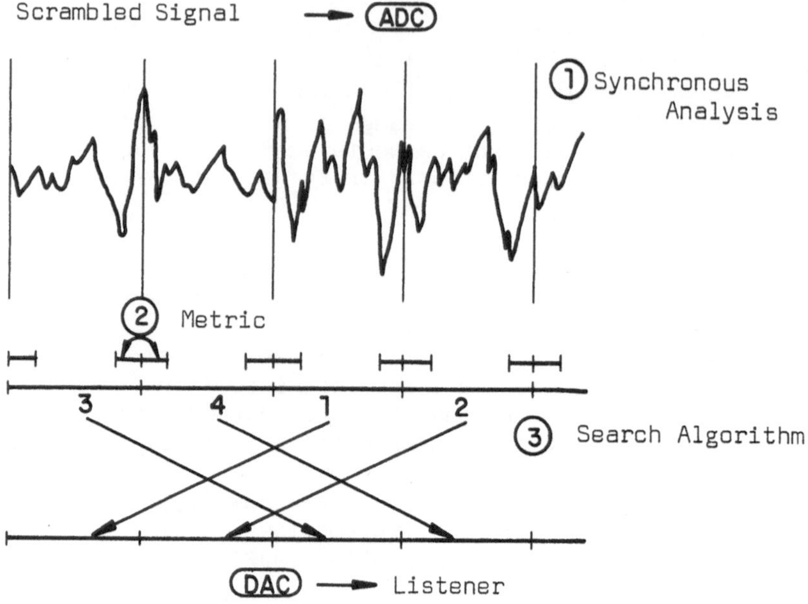

Fig. 4 Problems of Rearranging Time-Division Scrambled Speech:

1. Synchronous Analysis with Scrambling Structure
2. Metric for Neighbourhood
3. Puzzle Solving and Rearranging the Segments

The key problem in finding an appropriate metric is that speech itself contains very abrupt changes which are hard to distinguish from the time segment boundaries. On the other hand the short-time changes (in amplitude) are smoothed over the segment boundaries by the communication system ([R][H][S]).

We are therefore obliged to use some long-time tendency in the speech signal such as the changes in the articulatory parameters. As compared to the problem of speech understanding the following additional difficulties arise:

1. The segmentation cannot be optimally adapted to the nature of speech because of the time division by the scrambler.

2. The metric to be used has to give a measure of neighbourhood of speech segments rather than similarity to reference segments as in the problem of automatic speech understanding.
3. Contrary to the limitations in which speech understanding is considered (good speech quality, restriction of vocabulary, context and speaker) we deal with unlimited natural speech which is usually of poor quality. This means that the puzzle has to be solved on the acoustical level only. However the listener of the hopefully improved scrambled speech signal naturally uses higher level knowledge like semantic, linguistic or contextual information.

As to point 3 it was found that humans have a considerable learning capability to perceive distored speech. This should not be underestimated in evaluating the rest-intelligibility.

6. FUTURE SPEECH SCRAMBLING

As a conclusion of chap.5 a scrambler of the 6. generation has to smear the structure of speech over the elements to be scrambled (see Fig.3). This can be achieved by an invertible transformation that satisfies the quality criterion defined in chap.3. The mathematical problems which arise from this postulate,and possible scrambling schemes are treated by Wyner, 1979. This idea is covered by a patent of Robra 1975.
Computer simulations of such a scrambling scheme using the Fourier transformation (DFT) and an appropriate coding scheme to transmit the Fourier coefficients have proven the feasibility of this idea.
In chap.4 it was stated that a conventional 5. generation scrambler can achieve a TWE of about 20. Using this method TWE = 0.5 is achieved:

- Upper Band limit = F → sampling frequency for a DFT = 2F
- Using a 2^n DFT we obtain a $\Delta t = 2^n/2F$
- The corresponding Δf (see Fig.3) is $F/2^n$

$$\rightarrow \quad TWE = \Delta t \cdot \Delta f = (2^n/2F) \cdot (F/2^n) = 1/2 \qquad (5)$$

It is therefore possible to realise a scrambler with a very low rest-intelligibility (no speaker recognition) as well as a very high analytical security.
The transformation used in such a scrambler influences the spectrum of the transmitted scrambled signal. This fact can be used to adapt the scrambler to the frequency characteristics of the communication system.

7. CONCLUSIONS

Looking at the existing scramblers and the relevant security criterions it was shown how the scrambling technique developped over five generations, out of which a future 6. generation could be deduced. This new scrambling scheme promises a high security at a good quality.
We have defined what the optimal scrambler is in terms of the quality and security criterions. This approach to the problem of secure speech communication can now be compared to the usage of vocoders, where the emphasis is in separating the redundancy from the speech information to be transmitted. To transmit digiticed or parameterized (by a vocoder) speech over a CCITT-speech channel sophisticated equalization and error correction is necessary. An optimal scrambler of the 6. generation represents an optimal combination of those techniques and can be regarded as being based on an "analog vocoder".

ACKNOWLEDGEMENT

The author is deeply grateful to the members of the SWISS ARMY SIGNAL CORPS who cooperated in the research about speech scrambling as well as to Mr. Steinmann through whose effort this research was initiated.

Literatur

- W. Baeschlin, "The Integration of Time Division Speech Scrambling into Police Telecommunication Networks", Proceedings 1977 International Conference on Crime Countermeasures-Science and Engineering, University of Oxford, July 25-29, 1977, p. 141-144,

- A. M. Mc Calmont, "Quantitative Measure of Security for Analog Speech Communications Security Devices", Carnahan Conference on Crime Countermeasures, University of Kentucky, May 16-18, 1979 (to be published).

- W. Diffie, M.E. Hellmann, "Privacy and Authentication; An Introduction to Cryptography", Proceedings of the IEEE, Vol. 67, No. 3, March 1979, p. 397-427.

- D.J. Goodman, B.J. Mc Dermott and L.H. Nakatani, "Subjective Evaluation of PCM Coded Speech", The Bell System Technical Journal, Vol. 55, No. 8, October 1976, p. 1087-1109.

- A.H. Gray, J.D. Markel, "Distance Measure for Speech Processing", IEEE, Vol. ASSP-24, No.5, October 1976, p. 380-391.

- J. Robra, "Verfahren zur Verschleierung von Sprachsignalen mit Hilfe orthogonaler Matrizen", Deutsches Patentamt, Auslegeschrift 25 23 828, Anmeldetag 30.5.1975.

- D.G. Liesegang, "Möglichkeiten zur wirkungsvollen Gestaltung von Branch and Bound Verfahren dargestellt an ausgewählten Problemen der Reihenfolgeplanung", Inaugural-Dissertation der Universität Köln, Promotion 15.7.1974.

- H.J.M. Steeneken, T. Houtgast, "MTF as a Physical Measure of the Quality of Communication Channels", F.A.S.E., Paris 1975, p. 351-359.

- A.D. Wyner, "An Analog Scrambling Scheme which does not expand Bandwidth",
  Part I : Discrete Time, IEEE, Vol. IT-25, No.3, May 1979 p. 261-274
  Part II : Continous Time, IEEE, Vol. IT-25, No. 4, July 1979, p. 415-425.

SPEECH COMPRESSION/RECOGNITION WITH ROBUST FEATURES

Sidhartha Maitra

Systems Control, Inc.,
Palo Alto, California, U.S.A.

Described here are feature extraction procedures that overcome the limitations of linear predictive coding. Problems of speech compression and speech recognition in the feature extraction stage are attacked with similar tools. A uniform evaluation procedure for both disciplines is indicated.

I. INTRODUCTION

Speech recognition technology has seen very little growth in the past two years. The success of the Harpy system [1] has not triggered an avalanche of more sophisticated algorithms with larger vocabularies. One problem might be that most algorithms try to perform the speech recognition on a non-optimal set of features extracted from the speech waveform. Some of these features are currently also being used in techniques for low-bit-rate speech compression. In this paper we will point out the inadequacies of these features in the context of speech recognition/compression and indicate how a more robust set of features can be obtained.

The problem of extracting a compact set of features for speech recognition runs parallel to the problem of designing a low-bit-rate speech compression system [2]. In the past, results from speech compression have been applied immediately as input features for speech recognition (e.g., Itakura reflection coefficients). We believe that recent results in speech compression can also lead to an improvement in the task of speech recognition, i.e., better speech compression immediately reduces the dimensionality of features used for the recognition task. However, there are some

J. C. Simon (ed.), Spoken Language Generation and Understanding, 497-504.

differences in the goals of these twin tasks (speech compression vs. speech recognition). The goal of speech compression is to extract the minimum set of features that preserves all the information in the original waveform. For speech recognition, the features extracted need not preserve information such as speaker's accent, nuances, etc. This difference has prompted researchers in the speech recognition arena to sometimes ignore the advances made and the problems encountered in the speech compression area. This paper is an attempt to bridge precisely that gap.

Section II describes the problems inherent in Linear Predictive Coding (LPC) measures that are currently used in speech recognition; it also describes the properties necessary for a more robust operation. Section III describes four new models/techniques that attempt to extract features for speech compression/recognition that satisfy the properties mentioned in Section II. In conclusion, in Section IV, we make a case for using similar test procedures for speech compression and speech recognition.

II. CURRENT MEASURES - PROBLEMS AND DESIRED PROPERTIES

For best recognition/compression the features extracted should possess the following properties:

1. Noise immunity - The human ear/brain system has an uncanny ability to separate speech from background noise. Even in high noise environments, intelligibility is not reduced for human listeners, whereas the performance of machines deteriorates rapidly. This discrepancy can be reduced by extracting features that are more insensitive to added noise.

2. Robustness under non-linearities - Speech is often distorted (by carbon microphones, clipping, etc.) before it is available for analysis. In such circumstances, measures that depend on the linearity of the input speech are almost unusable.

3. Ability to handle pole-zero models - It is easy to demonstrate that the classical all-pole impulse-excited model is incapable of handling the large variety of speech waveforms that is observed in practice. An efficient pole-zero model is therefore necessary for accurate feature extraction.

4. Efficiency in data compression - Most feature extraction algorithms retain significant redundancy in the parameters; this makes the subsequent task of encoding/recognition more difficult than it ought to be.

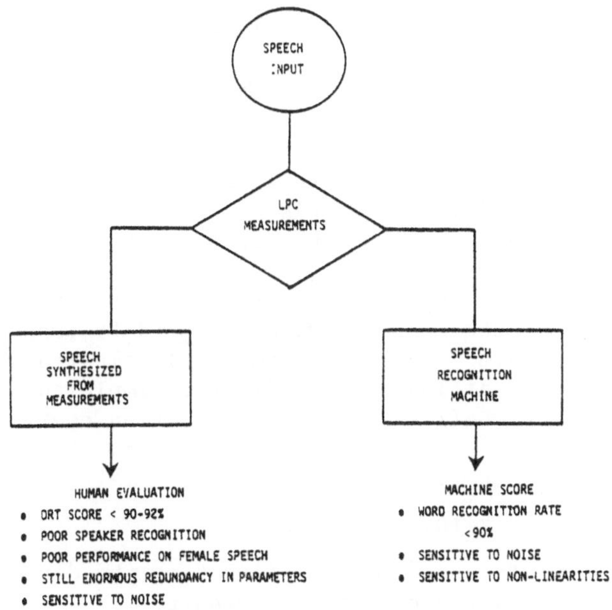

Figure 1. Problems Associated with LPC Measurements

5. Ability to recreate speech from extracted features - With item 4 above (efficiency requirement) being simultaneously satisfied, it would be ideal if the original speech could be recreated from the extracted features with little or no loss of fidelity. This ensures that no information has been lost in the feature extraction process. Moreover, simple checks on the recreated speech, like the Diagnostic Rhyme Test (DRT), can determine the extent of the information loss. The DRT score [2] is based on human evaluators choosing between alternative rhyming words ('catch' vs 'latch,' 'rap' vs 'sap', etc.). Such a test can point out not only the percentage of words recognized correctly but also the weak areas (viz. nasals, glides, etc.).

In the context of the above mentioned properties, Figure 1 points out the deficiencies of the LPC parameters that are currently being used. In particular, it is difficult to expect a better performance of the speech recognition machine than the human evaluator; hence an upper bound of the performance using LPC measurements is surely 90-92% on discrete words from a constrained vocabulary. Performance on connected words is possibly even poorer.

III. NEW MODELS/TECHNIQUES FOR FEATURE EXTRACTION

1. Reducing the Effect of Background Noise

With the addition of noise to speech, the linear prediction parameters of the sum does not agree with either the noise or the speech component. To illustrate this point, consider the simple case of a signal y(i) with a speech component $\mathscr{S}(i)$ and a white noise component $\omega(i)$ $(y(i) = \mathscr{S}(i) + \omega(i))$. The auto-correlation function $R_y(\tau)$ of y(i) and the auto-correlation function $R_{\mathscr{S}}(\tau)$ of $\mathscr{S}(i)$ differ only at the zero-lag value. The difference in the reflection coefficients calculated from $R_y(\tau)$ compared to $R_{\mathscr{S}}(\tau)$ can be eliminated (or at least reduced) by a simple adjustment of $R_y(0)$. For instance, $C_y(1)$, the first reflection coefficient for y(i), is given by

$$C_y(1) = \frac{R_y(1)}{R_y(0)} = \frac{R_{\mathscr{S}}(1)}{R_{\mathscr{S}}(0) + W} \tag{1}$$

where W is the white noise power. Adjusting the denominator by the proper amount achieves the desired result.

For the above adjustment, the noise power term has to be known accurately; speaker-silent sections of the signal have to be determined and W estimated. Even with a proper estimate of W, there may arise a problem of stability due to over-correction: with too large a reduction of $R_y(0)$, we may no longer have a valid auto-correlation function. This can be checked and adjustments made accordingly.

The above discussion points out how the effect of white noise can be reduced/eliminated for LPC speech. When the noise is not white, we have designed a system that reduces the problem to the above white noise case. Figure 2 shows the flow-chart of the algorithm. A linear prediction filter is used to whiten the noise component of the input signal. This linear prediction filter is updated only on speaker-silent segments. When speech is present, the filter also transforms the auto-correlation of the speech component; a simple adjustment at the output stage restores the auto-correlation values for use in speech recognition/compression.

2. Robustness Under Non-Linearities

Speech waveforms have traditionally been deconvolved using a least-mean-square (LMS) criterion. This deconvolution is used in LPC parameter estimation and is susceptible to moderate amounts of non-linear distortion. We have been using a new technique [3,4], which does not use an LMS criterion but rather uses the "peakiness" of the deconvolved waveform to determine the filter parameters.

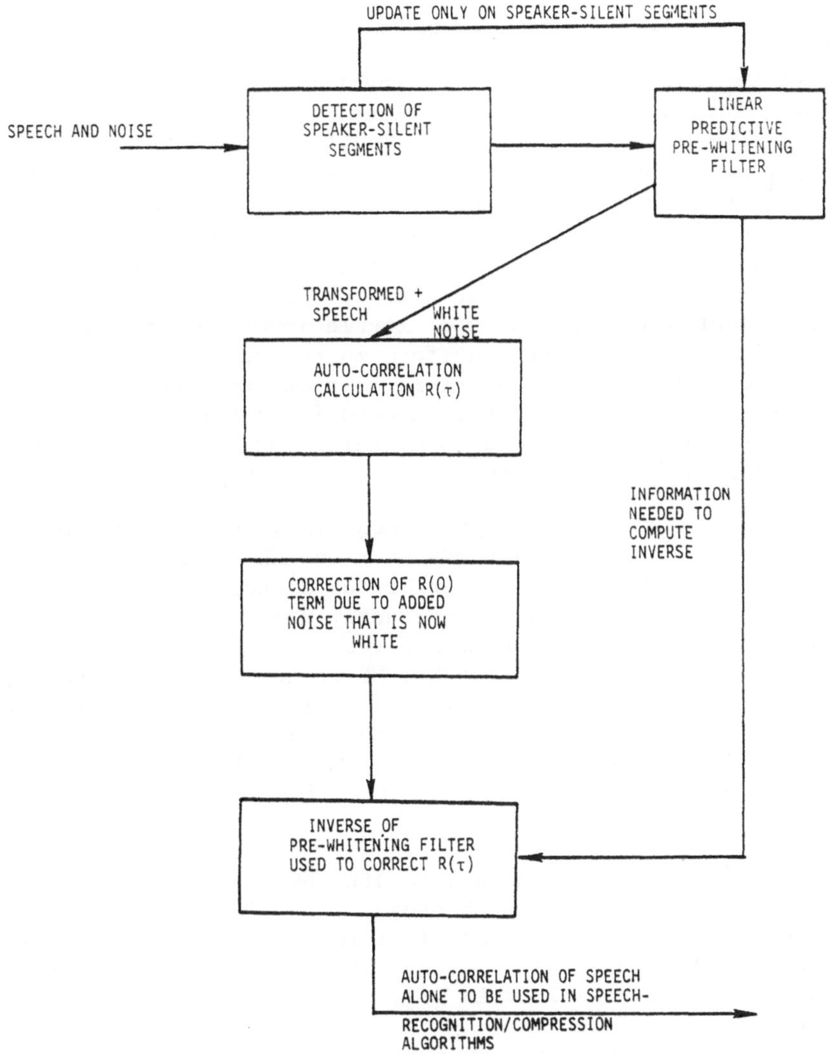

UPDATE ONLY ON SPEAKER-SILENT SEGMENTS
SPEECH AND NOISE
DETECTION OF SPEAKER-SILENT SEGMENTS
LINEAR PREDICTIVE PRE-WHITENING FILTER
TRANSFORMED SPEECH
+ WHITE NOISE
AUTO-CORRELATION CALCULATION R(τ)
INFORMATION NEEDED TO COMPUTE INVERSE
CORRECTION OF R(0) TERM DUE TO ADDED NOISE THAT IS NOW WHITE
INVERSE OF PRE-WHITENING FILTER USED TO CORRECT R(τ)
AUTO-CORRELATION OF SPEECH ALONE TO BE USED IN SPEECH-RECOGNITION/COMPRESSION ALGORITHMS

Figure 2. Reducing the Effect of Background Noise on LPC Parameters

SPEECH INPUT
$f_1(n)$
$f_2(n)$
c_1
c_2
Z
Z
$b_1(n)$

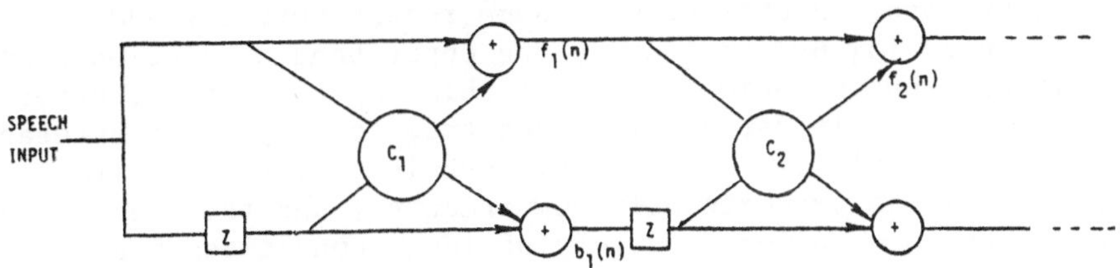

Figure 3. Lattice Filter (c_i are the reflection coefficients, Z an unit delay)

The lattice filter shown in Figure 3 can be used to deconvolve the input speech waveform. The best deconvolution of a voiced speech segment is obtained when the forward prediction error $f_k(n)$ (after k stages of the lattice) looks like an impulse-train. The criterion that is currently used for determining $\{c_i\}_{i=1,k}$, has been some form of least-squares-analysis. The problem of minimizing the energy in $\Sigma f_i^2(n)$ is analytically tractable; this has generated an enormous body of literature in least-squares analysis for minimizing $\Sigma f_i^2(n)$, $\Sigma b_i^2(n)$, $\Sigma f_i^2(n) + \Sigma b_i^2(n)$, etc. (Burg, Itakura, Auto-correlation, etc.). What has been neglected is that it is of little importance whether the energy in the deconvolved waveform is reduced or not. The goal should be to make the deconvolved waveform as impulse-train-like as possible; i.e., the criterion used for determining the reflection coefficients should be to maximize the peak-to-average value in the deconvolved waveform.

This criterion of maximum 'peakiness' (MP) in the deconvolved waveform has several advantages over an LMS criterion. Least-squares analysis is prone to error for distorted speech input whereas the MP criterion is inherently more stable. For speech with narrow bandwidths, least-squares analysis tends to generate unstable realizations whereas the MP criterion steers the filter coefficients in the 'right' direction; i.e., towards values that are more amenable to being driven by an impulse train. Finally, for speech corrupted by added noise, LMS analysis can give erroneous results whereas the MP criterion is still usable under such circumstances.

The MP criterion can be used in a step-by-step maximization of the peak-to-average ratio at each stage of the lattice filter of Figure 3; alternatively, a matrix formulation of the overall maximization process can be used. The latter is computationally efficient and is described in [3].

3. A Two-Source Model for Speech Production

There are segments in speech that cannot be described as either voiced or unvoiced utterances. Some recent work has addressed this problem [5] by considering the upper band as aperiodic and the lower band as periodic. While this may be true for utterances like 'Z', it certainly is not true for aspirated utterances like 'ha', 'gha', etc. Moreover, turbulence resulting from breath-noise is predominantly low frequency and the analysis of speech waveforms under a variety of input conditions has led us to believe that there can be no strict dichotomy in the frequency domain between the periodic and non-periodic excitations. We have developed techniques [6] that decompose the speech input into a periodic component and a non-periodic component; this is followed by an independent analysis of each of the two components.

With this approach, the parameters extracted from each component are more accurate and are more robust under added noise, breath turbulence, etc. The key point worth noting is that compared to technique 1 (Reducing the Effect of Background Noise), the acoustic interference can be time-varying. As long as this interference is not periodic at a rate close to that of the voiced speech component, a proper separation can be obtained.

4. Pole-Zero Models

Pole-zero models are increasingly being used in speech compression [7,8]. This stems from the realization that speech contains information beyond the parameters extracted in an all-pole impulse-excited model; i.e., even after the best all-pole deconvolution, there is information left in the residual waveform. We have developed a new approach for preserving this information without the penalty of using a large number of bits. If the full prediction residual were available (at the expense of a high number of bits), it would maintain all the information in the original and it would be simple to recreate the input with the original quality. However, for a periodic waveform, to recreate it with comparable quality, we need to extract the key information contained in only one pitch period. For a non-periodic waveform we need to extract the average power in the residual waveform as well as any information regarding the location and strength of large clusters of energy. Looked at from a pole-zero point of view, we are first determining the best all-pole deconvolution, and then extracting efficiently the parameters in the excitation system (the zeros).

Even though the above system is highly efficient and robust, it involves more parameters than the simple all-pole model. However, the pole-zero parameters fall very neatly into an optimal vector quantization scheme [9]. This quantization scheme removes most, if not all, of the redundancy in the data to rates below 1000 bits per second. We feel that preserving all the information in the original at such low rates should have a favorable impact on the feature extraction stage of speech recognition and speech compression algorithms.

IV. CONCLUSIONS

In the preceding section we have presented techniques that can be applied to both speech recognition as well as speech compression. In applying these techniques, one is faced with the problem of judging their efficacy. In speech recognition, beyond the feature extraction stage, the subsequent algorithms can be so complex and variable that basing the evaluation of the extracted parameters on a recognition rate can be misleading. In our opinion, tests designed for speech compression systems can be applied to this problem. As mentioned before, a human evaluation of the information content in the extracted features (a DRT score) provides a stable and repeatable guideline. The only restriction imposed by the use of this test procedure is that the extracted features should be capable of recreating the original speech with some degree of intelligibility retained. We feel that it is indeed advisable to have this restriction on the algorithms; it will lead to a more uniform assessment of the capabilities of each aogorithm. The 'black-art' of speech recognition may yet be reduced to a science!

REFERENCES

[1] B. T. Lowerre, "The Harpy Speech Recognition System," Ph.D. Dissertation, Carnegie Mellon University, 1976.

[2] Sidhartha Maitra and C. R. Davis, "A Speech Digitizer at 2400 Bits per Second," IEEE Transactions on Acoustics, Speech and Signal Processing, December, 1979.

[3] Sidhartha Maitra, et al., "A Non-Least Squares Criterion for Speech Coding," 19th IEEE Computer Society International Conference, Washington, D.C., September, 1979.

[4] Sidhartha Maitra, et al., "A Maximum Peakiness Criterion for Deconvolving Speech Waveforms," IEEE International Conference on Acoustics, Speech and Signal Processing, Denver, April, 1980.

[5] J. Makhoul, et al., "A Mixed Source Model for Speech Compression and Synthesis," Journal of the Acoustical Society of America, Vol. 64, No. 6, pp. 1577-1581, December, 1978.

[6] Sidhartha Maitra, et al., "Improvements in the Classical Model for Better Speech Quality," IEEE International Conference on Acoustics, Speech and Signal Processing, Denver, April, 1980.

[7] M. Morf and D. T. Lee, "Fast Algorithms for Speech Modeling," Final Report to Defense Communications Agency, Contract DCA100-77-C-005, Information Systems Laboratory, Stanford, California, November, 1978.

[8] M. Morf and D. T. Lee, "Recursive Least Squares Ladder Forms for Fast Parameter Tracking," Proc. of the 7th IEEE Conference on Decision and Control, pp. 1362-1367, January, 1979.

[9] A. Buzo, et al., "A Two-Step Speech Compression System with Vector Quantization," Proceedings of 1979 IEEE International Conference on Acoustics, Speech, and Signal Processing, pp. 52-55, April, 1979.

A REAL-TIME SYSTEM FOR SPEECH RECOGNITION

Camille BELLISSANT

IMAG - Université of Grenoble, France

ABSTRACT

This paper describes the organization of a real time sentence recognizer. The sentences belong to a corpus of 17 student's answers extracted from a Computer Assisted Instruction dialog in physics. The overall operation includes two main loops : the learning loop and the decision loop. Each of these interactive process calls a segmentation procedure for gathering the minimal intervals representing the extracted features into larger segments which are the prefiguration of phonemes. While the learning loop asks the user to introduce a phonemic label for each segment and enlarges a phonemic lexicon, the decision loop produces after a segmentation of each unknown utterance, a phonemic lattice. Each element of this lattice holds the best six candidates of the phonemic lexicon to match the characteristics of the unknown segment; each candidate exhibits its phonemic label and its score. The phonemic lattice is used as input by the word and sentence recognizer.

FEATURES EXTRACTION

The segmentation process presented here operates on parameters extracted from the speech signal during an electronic preprocessing and transmitted in real time to the Central Processor (IBM 360/67). The parameters consist in measurements done during fixed lenght time intervals at the time of the speech input. This duration of intervals is the time unit for the preprocessing system and its value is adjustable from 0.5 millisecond to 15 milliseconds. The speech signal is filtered in three frequency bands (1 : 150 - 900 Hz; 2 : 900 - 2200 Hz, 3 : 2200 - 5000 Hz). Within each of these bands and for each time-unit interval, the device measures the

J. C. Simon (ed.), Spoken Language Generation and Understanding, 505-515

maximum peak-to-peak amplitude, the number of zero-crossings of the signal and the number of relative extrema of the signal, i.e. the number of zero-crossings of the first derivative of the signal. This last measurement improves the classification but does not interfere in segmentation. The numerical values given in this paper refer to a time-interval of 10 ms.

NOTATIONS

Let us denote by $A1_i$, $A2_i$, $A3_i$ the peak-to-peak amplitudes within the bands 1, 2, 3 for the time interval number i. Likewise, $Z1_i$, $Z2_i$, $Z3_i$ and $D1_i$, $D2_i$, $D3_i$ will denote the zero-crossings rate and the relative extrema rate within the three bands and for the same time interval. All these numbers are positive or null and extend on 8 bits.

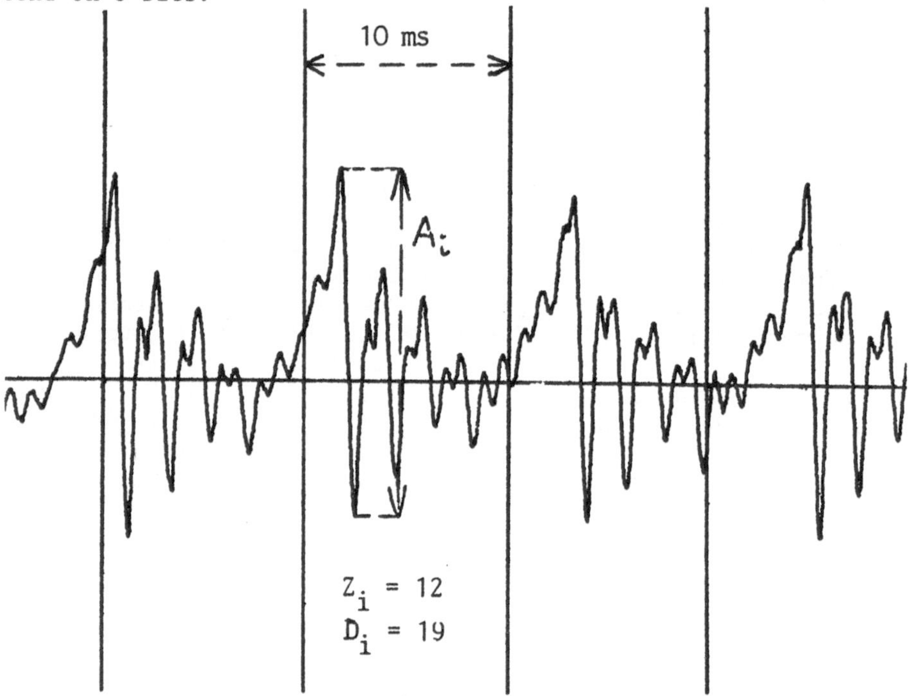

The data to be processed are organized in 9 one-dimension arrays representing N intervals of 10 ms. During the segmentation, only 5 arrays will intervene in the processing : A1, A2, A3, Z2, Z3. The segmentation process consists in building from these 5 applications

$$i \in [1, 2, \ldots, N] \rightarrow A1_i, A2_i, A3_i, Z2_i, Z3_i$$

other applications bringing out larger segments homogeneous in time and separated by boundaries.

NORMALIZING

The first step following the electronic preprocessing consists in normalizing the amplitudes within the three frequency bands. This normalizing occurs only if the maximum value A_{max} of the amplitudes (generally in the first band) is clearly too weak :

$$\text{if } A_{max} < 240 \text{ then}$$

$$AK_i \leftarrow AK_i \cdot \frac{255}{A_{max}}$$

$$\forall K \in [1, 2, 3], \ \forall i \in [1, 2, \ldots, N]$$

INITIAL SEGMENTATION

The initial segmentation consists in scanning the intervals and assigning to each of them a type within the five following values: S for silence, F for fricative, N for nasal consonant, O for voiced occlusive and V for everything else. It's an application ϕ of $[1, 2, \ldots, N]$ in {S, F, N, O, V}.
The setting of the value of Φ at interval i is :

$$\forall i \in [1, N],$$

$$\Phi(i) \leftarrow \text{if } (A1_i \geq 48) \text{ and } (Z1_i \leq 5) \text{ and } (A1_i \geq \max(A2_i, A3_i)) \text{ then N}$$

$$\text{else if } (Z2_i + Z3_i \leq 2) \text{ then}$$

$$\text{if } (A1_i + A2_i + A3_i < 18) \text{ then S else O}$$

$$\text{else if } (A3_i > 2A1_i) \text{ and } (Z3_i > 55) \text{ then F}$$

$$\text{else V.}$$

These threshold values have been fixed after numerous experiments. After this type assignement, it occurs sometimes that an isolated interval is different in type than adjacent ones. As each interval has a duration of 10 ms, such a singularity could not have a phonetic meaning and it's better to delete it by a kind of smoothing:

$$\forall i \in [2, n-1], \text{ if } (\Phi(i) \neq \Phi(i-1)) \text{ and } (\Phi(i-1) = \Phi(i+1)) \text{ then}$$
$$\Phi(i) \leftarrow \Phi(i-1).$$

After that, boundaries are set up according to the different values of Φ. A "segment" will now refer any sequence of contiguous intervals for which Φ holds a constant value. See diagram I.

DIAGRAM I

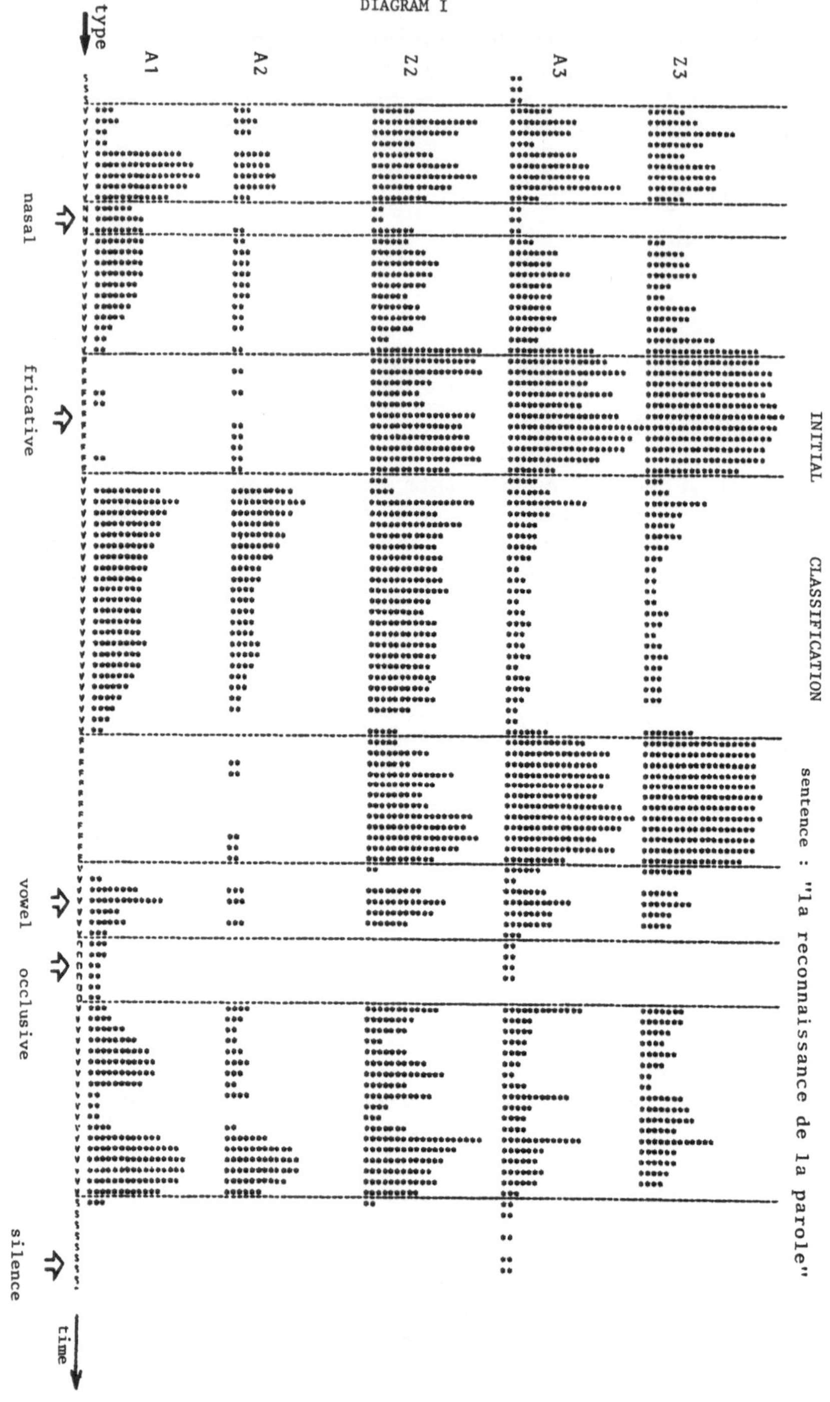
type
A1
A2
Z2
A3
Z3
nasal
fricative
vowel
occlusive
silence
time
INITIAL CLASSIFICATION
sentence : "la reconnaissance de la parole"

DIFFERENCING VOWELS AND CONSONANTS

This step of segmentation is the most complex one. Il operates on the V-type segments as they are produced by the initial phase, i.e. neither S, F, N nor 0. Among these segments, only those with a duration greater than 100 ms will be considered. Only the parameters A1 and A2 representing the peak-to-peak amplitudes in the bands 1 and 2 will be used during the next processing. The leading idea of this processing is to carry out a new segmentation according to the "variability" of A1 and A2. If A1 is "the most variable" in a certain domain, it's this parameter wich will be decisive for fixing a boundary between 2 segments and A2 will be ignored in the vicinity of this boundary, and vice versa.
We must now define numerically what is expressed by "the most variable". Intuitively one can conjecture the priority of large variations upon small ones.
Let us design by $P1_i$ and $P2_i$ the last squares fitting polynomial of degree 2, centered on the point i and built from the 5 abscissae i-4, i-2, i, i+2, i+4 for the corresponding values A1 and A2. The graph in $\mathbb{R} \to \mathbb{R} \times \mathbb{R}$ of these polynomials is a piece of parabola smoothing the experimental values A1 and A2 within the domain $[i-4, i+4] \subset \mathbb{R}$. For each point belonging to a V-type segment of duration > 100 ms, we shall compute the second derivative of these fitting polynomials at point i. The least squares method for fitting the 5 experimental values $A_{i-4}, A_{i-2}, A_i, A_{i+2}, A_{i+4}$ by a polynomial
$P_i = a_{0,i} + a_{1,i}x + a_{2,i}x^2$, provides the value of the second derivative :

$$P''_i = 2a_2 = \frac{1}{7}[2(A_{i+4} + A_{i-4} - A_i) - A_{i+2} - A_{i-2}]$$

Thus for each point i, we compute the values

$$B1_i = 2(A1_{i+4} + A1_{i-4} - A1_i) - A1_{i+2} - A1_{i-2}$$

and

$$B2_i = 2(A2_{i+4} + A2_{i-4} - A2_i) - A2_{i+2} - A2_{i-2}$$

Let us call "concavity" the numerical value of this derivative at point i, expressed by a positive number if the fitting parabola is concave, a negative one if it is convex, the sign variation for two adjacent intervals pointing out the possibility of a boundary between two phonemes.
Diagram II shows that sign variations of B1 and B2 are generally distinctly stated and one could think that this is enough to set up phonemic boundaries at these intervals where sign changes occur. That would come to consider the domains in which the "concavity" of A1 or A2 holds continually the same sign as a prefiguration of the future phonemic segments. In fact, a segmentation performed at this stage would be too biased by small variations of the "concavity" of A1 or A2, and would bring out too many small duration

DIAGRAM II

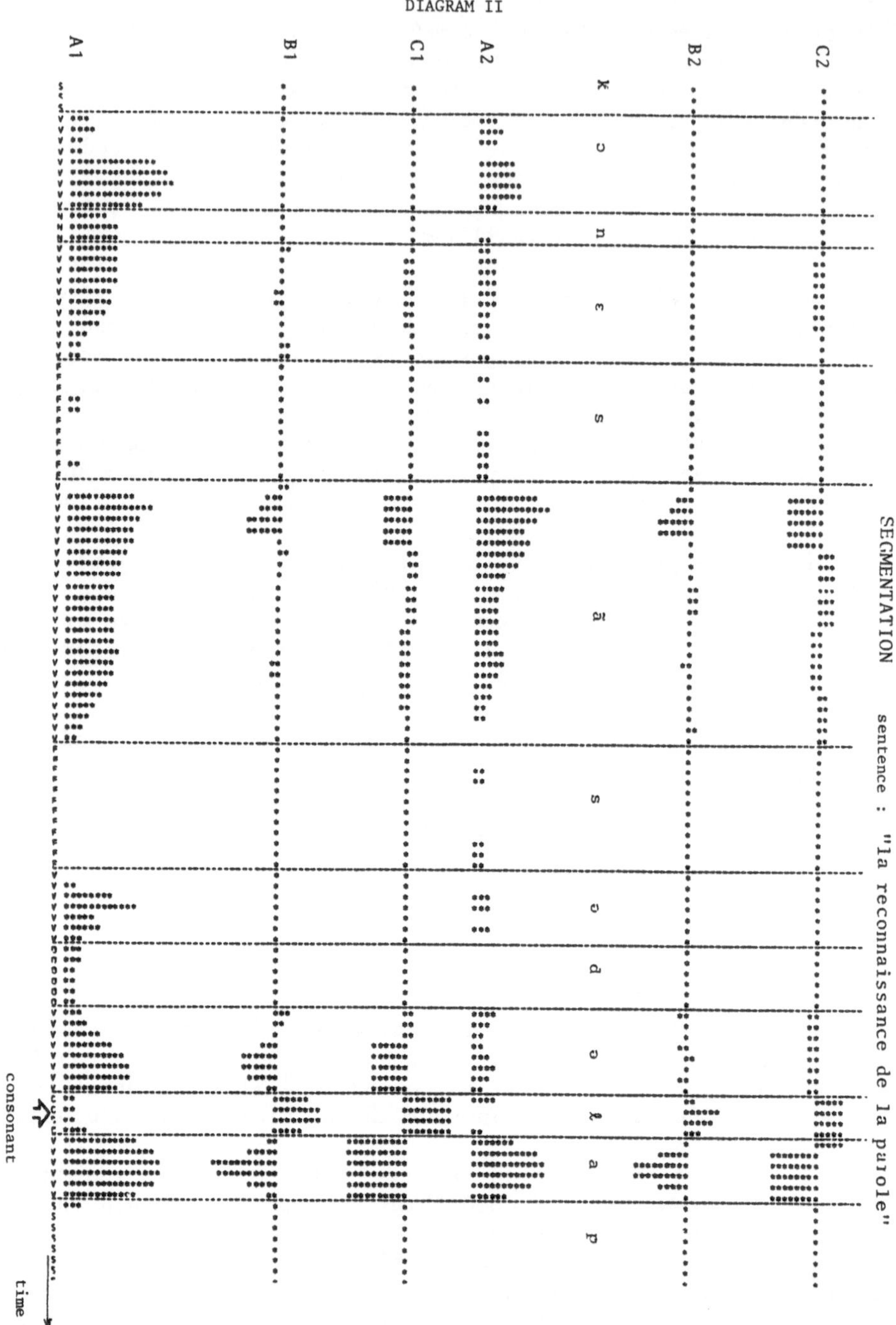
SEGMENTATION sentence : "la reconnaissance de la parole"
A1
B1
C1
A2
B2
C2
consonant
time

domains with characteristics near from one another, due to the rapid oscillations of the values B1 and B2 near zero. In order to avoid this local fragmenting, a particular smoothing on B1 and B2 is carried out in the following way : Within a domain where B1 holds the same sign, one computes the sum X of the values $B1_i$:

$$X = \sum_{i=K}^{i=\ell} B1_i .$$

Then, each value $B1_i$ is replaced by $C1_i = X$. The graph of C1 represents a step function. The same processing is applied to B2. Looking at the phoneme [d] in diagram II, it is easy to compare the greater "variability" of A1 in comparison to A2 at this place. The small oscillations of B2 are responsible for the weaker value of C2, compared to C1, in module. Thus the values C1 and C2 illustrate the more or less large "coherence of concavity". We shall use this criterion of greatest "coherence of concavity" for determing which parameter, A1 or A2, is the "most variable" in a given point.
After all, a "boundary" will be set up at point i according to the "variability" of A1 under the following conditions :

1 - $C1_i \cdot C2_i < 0$

2 - $|C1_i| + |C1_{i+1}| > |C2_i| + |C2_{i+1}|$

or

$$\frac{|C1_i| + |C1_{i+1}|}{A1_i + A1_{i+1}} > \frac{|C2_i| + |C2_{i+1}|}{A2_i + A2_{i+1}}$$

These last two inequalities express a "coherence of concavity" greater for A1 than for A2. The last inequality allows to eliminate by normalizing the preponderance of the values A1 which are generally higher than the values A2. A third condition :

3 - $\dfrac{|A1_{i+2} - A1_{i-1}|}{A1_i + A1_{i+1}} > \dfrac{1}{6}$

avoids a boundary creation where the declivity of A1 is too weak, which would lead to segment within a phoneme. These three conditions on A1 exist for A2, too, by inverting A1 and A2, C1 and C2.

The last step of segmentation process consists in differencing between vowels and consonants according to the "general concavity" of the phonemic segment. Of course a positive concavity will correspond to a consonant while a negative concavity will express the presence of a vowel. This "general concavity" G is computed between the two boundaries of a segment, built up at intervals K and ℓ

$$G \leftarrow \sum_{i=k}^{i=\ell} C1_i + C2_i .$$

The corresponding segment will hold a V-type if $G \leq 0$ (vowel) and a C-type (consonant) if $G > 0$ (See diagram II).

RESULTS OF SEGMENTATION

The main interest of this method is to proportion the performance to the difficulty of segmentation. In fact, in sentences with a good distribution of silences (plosives) and fricatives, the initial classification allows to gain time. In the general case, the discrimination between vowels and consonants is done without errors when there is an alternance in time of vowels and consonants. In diphtongs or sequences of vowels, the method still leads to satisfatory results, due to the "greatest variability" criterion. In sequences of consonants, this algorithm has a tendancy to fragment some phonemes in smaller segments. This unwanted fragmentation can be opposed during the subsequent phase of phonemic labeling. At the end of the segmentation process, the speech input is partitioned into segments belonging to six categories : silence (stop or unvoiced occlusive), ficative, consonant, nasal, voiced occlusive, vowel.

LEARNING

This learning phase consists in creating and augmenting a phonemic lexicon in which each record represents a phoneme. This lexicon will be used during the Recognition phase.

For each utterance prononced within this learning context, the system begins to call the already described segmentation procedure. For each segment, the system provides the speaker with type and asks him to enter the corresponding phonemic label. This phonemic label refers to a record in the lexicon, in which data are then accumulated in a six-dimensional weighted cluster. The six dimensions which are taken in account are the three zero-crossings densities Z1, Z2, Z3, and the three relative extrema densities D1, D2,D3.
These six components describe, for each 10 ms interval, a point X in the $\mathbb{N}^6$ space. The set of these points forms a cluster in $\mathbb{N}^6$. The centre of gravity Ω of this cluster has the following coordinates :

$$\omega^{(K)} = \sum_{i=1}^{N} \frac{x_i^{(K)}}{N} \qquad K = 1,\ldots,6$$

where $x^{(K)}$ are the coordinates of X, and N the number of intervals used in the learning of the considered phoneme.
In parallel with this computation of the centre of gravity for each phonemic cluster, a uniform amplitude profile is calculated by averaging for this phoneme the three amplitudes A1,A2,A3 then normalizing in order to get a sum of value 100 :

$$UAPK = \frac{100}{A1+A2+A3} \cdot \sum_{i=1}^{N} \frac{AK_i}{N} \qquad K = 1,2,3$$

At the end of the learning phase, each phoneme is thus accumulated into a six-dimension point, paired with a three-dimension uniform amplitude profile.

CLASSIFICATION

The purpose of the classification phase is to build up for each new utterance a "phonemic lattice" holding the sequence of the best candidates among the phonemes in the lexicon, to match the unknown segments in the utterance.

First of all, the system calls the segmentation procedure in order to partition the unknown utterance into segments belonging to the categories S,F,C,N,O,V. For each produced segment, the coordinates of its centre of gravity in $\mathbb{R}^6$ and uniform amplitude profile in $\mathbb{R}3$ are calculated and euclidian distances are evaluated with the clusters of the phonemic lexicon. Only those phonemic clusters belonging to the same category than the unknown segment are taken in account. Finally the best six candidates are extracted from the clusters according to a score S_ℓ :

$$S_\ell = \frac{1}{D_\ell + d_\ell + 1}$$

Where D_ℓ and d_ℓ represent the euclidian distances between the

centres of gravity and uniform amplitude profiles between the unknown cluster and the phoneme candidate n° ℓ. Each score is then normalized in order to get a uniform sum of scores having the value 100. The diagram III shows a phonemic lattice where time is going top down. Each candidate is represented by its label and its score.

LEXICAL ANALYSIS

This phase, the last one, is concerned with the recognition of a word, a group of word or a sentence. The corpus of 17 sentences belongs to a Computer Assisted Instruction dialog. At each step of the dialog, 3 or 4 answers can plausibly arise. A key-word is associated to each sentence. The search begins by localizing this Key-word within the phonemic lattice. The Key-word is chosen in such a way that it contains unvoiced fricatives and occlusives, which are easy to recognize. Once it has been localized, the algorithm proceeds by rebuilding the sentence while scanning its right and left context. Finally a score is given to each plausible sentence for this step of dialog. The sentence with the maximum score is considered as the best choice.

The overall processing is performed in real time on an IBM 360/67 computer working in time sharing mode.

DIAGRAM III

PHONEMIC LATTICE

sentence: "la reconnaissance de la parole"

TYPE	LENGTH		SCORES					
C	: 9	-->	L 22	E 22	AN 22	R 15	AI 11	A 8
V	: 9	-->	A 43	O 27	AI 11	E 8	S 6	AN 5
C	: 6	-->	L 25	AN 25	R 23	E 13	AI 8	A 7
V	: 6	-->	E 21	AI 21	A 20	O 17	AN 11	L 11
C	: 5	-->	K 74	P 9	D 7	N 6	R 2	L 2
V	: 9	-->	A 30	O 30	AI 17	E 10	S 7	AN 7
N	: 3	-->	N 70	P 10	O 10	K 6	R 2	L 2
V	: 11	-->	AI 84	E 4	O 4	A 4	AN 2	L 2
F	: 11	-->	S 68	O 7	A 7	AI 7	E 6	L 5
V	: 24	-->	AN 76	L 11	R 7	E 3	N 2	AI 2
F	: 12	-->	S 67	O 8	A 8	AI 7	E 6	L 5
V	: 7	-->	E 36	AI 20	AN 12	L 11	O 11	A 11
O	: 6	-->	D 62	P 21	V 8	K 6	R 2	L 1
V	: 8	-->	L 26	AN 25	R 17	E 17	AI 8	A 6
C	: 4	-->	E 30	AI 18	L 15	AN 15	R 12	A 11
V	: 6	-->	A 28	O 23	AI 16	E 15	AN 9	L 8
S	: 12	-->	P 61	D 20	N 9	K 7	R 2	L 1
V	: 5	-->	A 39	O 34	AI 10	E 7	S 5	AN 5
C	: 9	-->	R 31	L 19	AN 18	V 11	K 10	P 10
V	: 10	-->	O 37	A 26	AI 15	E 10	AN 7	L 6
C	: 13	-->	R 29	L 19	AN 18	N 12	K 11	P 11
V	: 7	-->	E 35	AI 20	A 12	O 12	AN 11	L 10

RESULTS

The percentage of phonemic recognition is about 75% with a single speaker and learning process involving the same words than those to be recognized (but not necessarily the same sentences). A change of speaker without a new learning reduces drastically the percentage of recognition (about 40%). The larger is the training, the more accurate is the phonemic lattice.

CONCLUSION

We have presented a method for producing in real time a phonemic lattice from an acoustic speech signal. The method involves two calls of segmentation procedure, first during a learning phase, and secondly during a decision phase. The segmentation itself operates according to a criterion based of the "greatest variability" of two parameters considered in parallel.
The recognition process involves distance between six-dimension clusters and uniform amplitude profiles of phonemes.

BIBLIOGRAPHY

BELLISSANT C. A System for segmentation and phonemic labeling of speech. Fourth Joint Conference an Pattern Recognition, KYOTO, 1978

DOURS D., FACCA R. Méthode de segmentation et d'analyse par traitement direct du signal vocal. Application à la classification et la reconnaissance des voyelles et des consonnes. Thèses CERFIA, TOULOUSE 1974.

HATON J.P. Contribution à l'analyse, la paramétrisation et la reconnaissance de la parole. Thèse, NANCY, 1974.

LIENARD J.S., MLOUKA M. Segmentation automatique de la parole en phonatomes. Journées d'étude sur la parole, LANNION 1972.

SPEECH TRAINING OF DEAF CHILDREN USING THE SIRENE SYSTEM: FIRST RESULTS AND CONCLUSIONS *

Marie-Christine HATON

Centre de Recherche en Informatique de Nancy
Université de Nancy I

ABSTRACT

This paper describes the SIRENE system which is being developed in our laboratory. SIRENE is an interactive computer-based system of speech-training aids for the deaf. It also includes a variety of procedures for analysis and classification of pathological voices.

The basic idea of speech-training aids consists of compensating for the lack of auditory feedback in deaf children by use of visual displays. The system is intended to be used by speech teachers ; several acoustic and phonetic parameters of speech can be displayed and trained : pitch, voicing, intensity, frequencies and so on.

SIRENE also features the use of automatic speech recognition algorithms in the training of sounds and words.

* This paper is a revised and augmented form of a paper presented at ICASSP 1979 in Washington, DC(1). It includes additional conclusions about experiments.

J. C. Simon (ed.), Spoken Language Generation and Understanding, 515-522.

1. INTRODUCTION

A great amount of work has been devoted recently to the application of computer science to computer-aided education. Significant results have already been obtained in this area. The present paper is concerned with the use of a mini-computer based system for speech training of deaf children. The term "deaf child" is used here to refer to any child who was born deaf, or who was deafened before about four or five years old. The impact of deafness on such children is of course very severe, and it is of great importance to deal with their education. Speech deficiencies of deaf children are very difficult to overcome and we are now investigating to what extent a computer system can be used for this task.

The SIRENE project was initiated in 1975 with the idea of applying some results obtained our research group in speech signal processing and automatic speech recognition to the problem of the deaf.

Several attempts have been done in the past few years in the use of the sense of vision (and even of touch) for training speech parameters, added to the residual acoustic feedback when some of it remains (2), (3).

In our system we propose a visual feedback using a computer, which ensures great possibilities concerning the analysis methods and the storage, and a great flexibility concerning the experiments. Moreover, the SIRENE system features the use of automatic speech recognition in the training procedures.

As a complementary part, special procedures for the study of pathological voices are implemented.

2. BASIC PRINCIPLES

A normal hearing child has access to the sounds that he produces and he uses this auditory feedback for actualizing the speech sounds and determining the correct timing of the speech production process.

The basic idea of our system is to provide the deaf child with visual information about speech in order to compensate for the lack of auditory feedback. This visual information must be simple and easy to understand by the child ; moreover, the displays must be considered by the child as plays during which he tries to succeed better than his friends. Displays must be reliable and reproductible. Of course this information is not exclusive and it will be assisted by amplification with a hearing aid whenever it is possible.

The SIRENE system has been designed as a complementary tool for helping a teacher in educating deaf children, and the elaboration of the various displays and training procedures used in the system was performed in close relation with several teachers (4).

3. SYSTEM ORGANIZATION

SIRENE is based on a SEMS MITRA 125 minicomputer with 32 K core memory and a disc. The disc is used for storing children's voices before and after a training period for progress evaluation. It also contains all the training and speech analysis procedures.

The speech signal is analyzed through a 15-channel spectral analyzer at a 50 Hz rate. A hardware pitch detector yields in parallel the fundamental frequency F_o. An A/D converter is also available ; it makes it possible to display the original speech wave or to perform software analysis (area function computation, LPC, etc...) According to the concerned parameters, these devices perform data acquisition in synchronism or they act separately.

Two CRT displays are available :

- The first one is an alphanumeric terminal which provides the professor with means for selecting a display and adjusting the corresponding control parameters,

- The second one is a Tektronix 4012 graphic system on which are displayed speech parameters. This terminal can also operate in an alphanumeric mode. It may therefore be used by a child alone during self training sessions.

Fig. I shows a simplified block-diagram of the system.

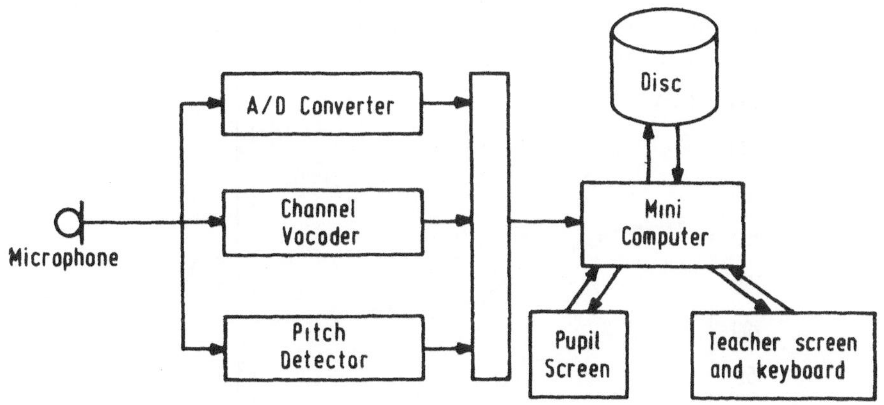

Fig. 1.

4. DISPLAYS

The different parameters extracted by hardware from an elocution may undergo some modification before being displayed. These analyses are mostly performed in real time in order to provide an immediate visual feedback to the child.

One can distinguish between three categories of reeducation procedures :

- The first one concerns the prosodic parameters : fundamental frequency, intensity of speech and correlation between them. Timing and rhythm are notions which derive from the former.

- The second one is relative to frequency parameters : center of gravity of the peaks of smoothed outputs of the filter bank, parameters typical of voice colour or friction, theoretical filters defined as optimal ones in the sense of the best classification of sound clusters and so on.

- Finally, in order to extend these considerations to the training of entire words or phrases, the third kind of procedures takes these phrases as entities and their global patterns are considered.

4.1. PP games(prosodic parameter games) :

* Voicing for the distinction voiced-unvoiced
* Pitch : After smoothing algorithms, simple pitch time-varying contours can be displayed. According to the case, we display on the screen some levels to reach, a tunnel to be followed, or a window to get through as shown on fig. 2. The dynamic ranges of tunnels, windows are selected by the teacher according to the desired aim : to lower the natural fundamental frequency to a more esthetic one, develop the skills to master it...

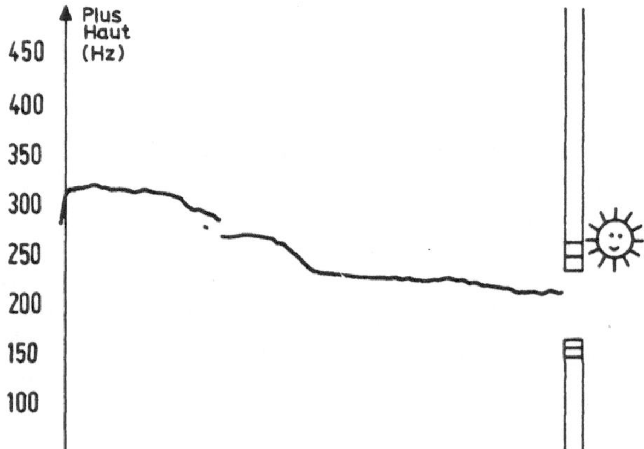

Fig. 2. The sun smiles whenever the child succeeds.

It is possible too to present reference contours to be matched or at least followed in their general pattern.

* Intensity : Concerning the intensity of speech, similar contours are displayed with reference contours for rhythm reeducation. A preliminar play also permits to the child to get aware of his vocal possibility : flowers are pictured on the screen as soon as the intensity gets higher than a predetermined level.

* Correlaction between pitch and intensity : for stability or control of the vocal folds. The intensity may occur like a modulation of the pitch contour. In another display we show curves in the pitch-intensity plane or in the plane of the derivatives of both parameters.

4.2. PF games(frequency parameter games):

They generally consist of the display of a dot or a pattern characterizing the frequency properties of the emitted sounds : smoothed spectrum and its time-evolution, dots in the plane of the two first estimated formants, or dot clusters for the discrimination of sound classes.

Another parameter is the intensity in selected frequency zones for the discrimination of fricatives for example or the segmentation of speech in syllabic units.

4.3 VM games(word vocabulary games) :

Most training procedures refer to an analytical approach of the speech production process. The following is an attempt to a global approach making it possible for a child to learn a predetermined vocabulary of words and phrases. The production of the child is recognized by a dynamic matching algorithm. This algorithm was designed in our laboratory some years ago (5) (6); it presents the advantage of allowing great variations in the duration and rhythm of words.

As an example, we can mention the building of a boat controlled by the elocution of the words which successively appear on the left of the screen. The stars indicate failures in the pronounciation of a word (fig. 3).

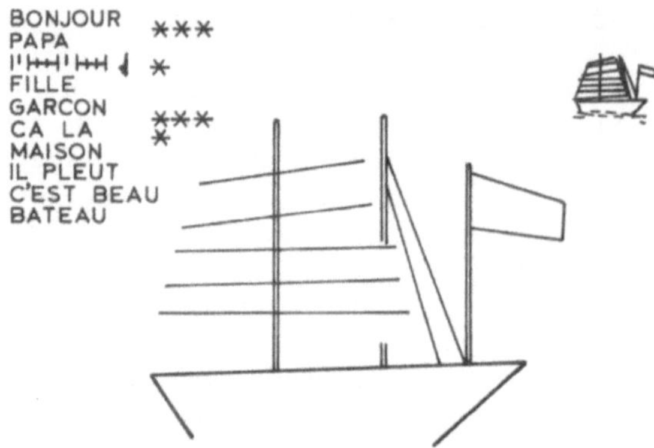

Fig. 3. The drawing of the boat is controlled by the good pronunciation of a list of words.

Another play consists of controlling the movement of a point in a labyrinth with a vocabulary of four words (left, right, up and down). Failures also appear as stars in the upper left of the screen.

5. EXPERIMENTS

The SIRENE System has been used in 1978-79 during training sessions with a speech therapist and three different groups of children :

1- Hard of hearing pupils about 13 to 15 years old having both difficulties in controlling suprasegmental and articulatory parameters,

2- Non hearing children (8 to 9 years old) with the same difficulties.

3- Asiatic hearing children of different ages presenting some phoneme confusions (s-ʃ , vowels) and lack of breathing control.

As first conclusions, we can mention the following :

- the interest of the children and the attraction of visual displays
- the emulation inside each group
- the poor interest of complex displays and the need for presenting simple and reliable patterns
- the need of having a system as simple as possible to operate
- the obtained improvements concerning the fluence of the elocutions, the precisions in the discrimination of sounds, the new ability to increase his fondamental frequency and so on...

6- CONCLUSION

Our general conclusion is that such systems may be of valuable aid for speech training to the deaf and in foreign language learning.

A part of the system, that one concerning pitch contour is at present implemented on a 8-bit microcomputer and uses a TV screen. Our aim is to implement the whole system on 16-bit ones and propose a complete bearable system at a reasonable price.

In parallel, we think that studies of the voices of deaf people and other pathological voices may be in some degree a help for diagnosis or for the speech evaluation.

REFERENCES

(1) M.C. HATON, J.P. HATON "SIRENE, a system for speech training of deaf people". Int. Cong. Acoustics and Speech Processing, Washington, USA, 1979.

(2) M.C. HATON "Revue des recherches en rééducation vocale des mal-entendants". 7èmes journées d'Etude sur la Parole, GALF, Nancy, France, Mai 1976.

(3) R.S. NICKERSON, K.N. STEVENS "Teaching Speech to the Deaf : Can A computer Help ?" IEEE Trans. AU 21, N°5, pp. 445-455, Oct. 1973.

(4) J.P. HATON, M.C. HATON "SIRENE : un projet de système interactif pour la rééducation vocale des enfants non-entendants" Revue G[ale] Ens. Déf. Auditifs , 67, n°4, pp 203-9, 1975.

(5) J.P. HATON "Contribution à l'analyse, la paramétrisation et la reconnaissance automatique de la parole" Thèse d'Etat, Université de Nancy I, Janv. 1974.

(6) J.P. HATON "A Practical Application of a Real-Time Isolated-Word Recognition System Using Syntactic Contraints" IEEE Trans. ASSP, 22, n° 6, pp. 416-419, Déc. 1974.

APPENDIX

Games available in SIRENE system

A- Pitch

Display of the time-varying pitch contour alone or with :

- the indication of levels (low-medium-high or frequencies of scale notes)

- a tunnel to go through
- a window to reach
- a reference contour to match.

A- Intensity and rhythm

Drawing of flowers as soon as a sound is emitted. Display of the time varying intensity coutour alone or with a reference contour to match.

C- Correlation between pitch and intensity

Patterns in the pitch-intensity plane.
Simultaneous variations of the two parameters.
Pitch contour modulated by the intensity of speech.

D- Frequency and articulatory parameters

Short-term spectra and spectrograms.
Smoothed patterns obtained from the channel vocoder.
Dots in the plane of the two first estimated formants and contours of formants.
Dots in an optimal plane for the best discrimination of sounds.
Area functions.

E- Word vocabularies

Labyrinth game.
Building of a picture controlled by pronounced words.

THE KEAL SPEECH UNDERSTANDING SYSTEM

G. MERCIER(*), A. NOUHEN(**,***), P. QIUNTON(***), J. SIROUX(**,***)

(*) CNET Lannion, DAS/SST, 22301 Lannion Cédex
(**) IUT Lannion, 22301 Lannion Cédex
(***) IRISA Rennes, Campus de Beaulieu, 35042 Rennes Cédex

The KEAL man-machine speech communication system is described; it consists of different modules dealing with phonetic, lexical, syntactic, semantic analysis and includes a dialogue controller and a speech synthesizer. At each level of analysis, results are given. Results of two experiments are presented as well.

Introduction

The KEAL speech understanding system has been developed to investigate the possibility of using voice as a communication support between a user and a computer for several tasks. The information needed to achieve the task is obtained by the system through a dialogue, thus allowing a maximum degree of naturalness in the communication.

The present version of KEAL includes a channel-vocoder analyzer, a synthesizer and the following components :

- a phoneme recognizer ;
- a word recognizer ;
- a sentence recognizer ;
- a sentence interpreter ;
- a dialogue-controller.

The task to be performed is given as a parameter of the system so that it may be used for several applications. At the present time, two tasks have been experimented with the KEAL system. The first task consisted in retrieving the telephone number of one person in a small directory (37 people). The second task was

J. C. Simon (ed.), Spoken Language Generation and Understanding, 525-543.

to recognize small sentences describing printed-circuits (vocabulary size = 120 words). This paper describes the organization of KEAL, the operation of its components, and its performances on the two tasks described above.

1. SYSTEM ORGANIZATION

Figure 1 shows a block-diagram of KEAL. The system consists of a set of modules communicating through a central data-base. At any given time, this data base represents simultaneously the current state of recognition of a sentence, and the current state of the dialogue. A monitor activates the modules according to the context of the dialogue. Before starting to work, each module receives information concerning the type of utterance that is to be recognized. This information depends also on the context of the dialogue and is determined by the dialogue-controller.

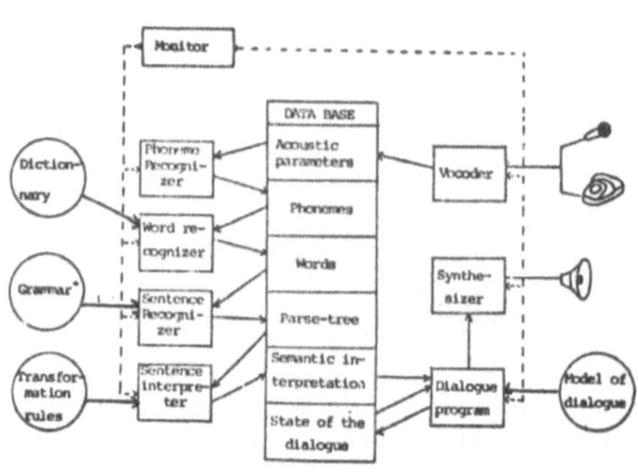

Figure 1 : block diagram of the Keal System.

The following is a very brief example of a dialogue cycle concerning the telephone directory inquiry task. After the user logs in with the system, KEAL synthesizes the sentence :

"Ici le centre de renseignements automatique. Que désirez-vous ?"

(This is the automatic information system. What can I do for you ?)

The user's reply is :

"Je voudrais le numéro de téléphone de Monsieur Buisson."

(I would like the telephone number of Mister Buisson).

The utterance is digitized through the channel vocoder (see figure 2), and the phoneme recognizer analyses it, giving a phonemic lattice. This lattice is searched by the word recognizer for words belonging to a vocabulary that is determined by the dialogue controller ; the output of the word recognition

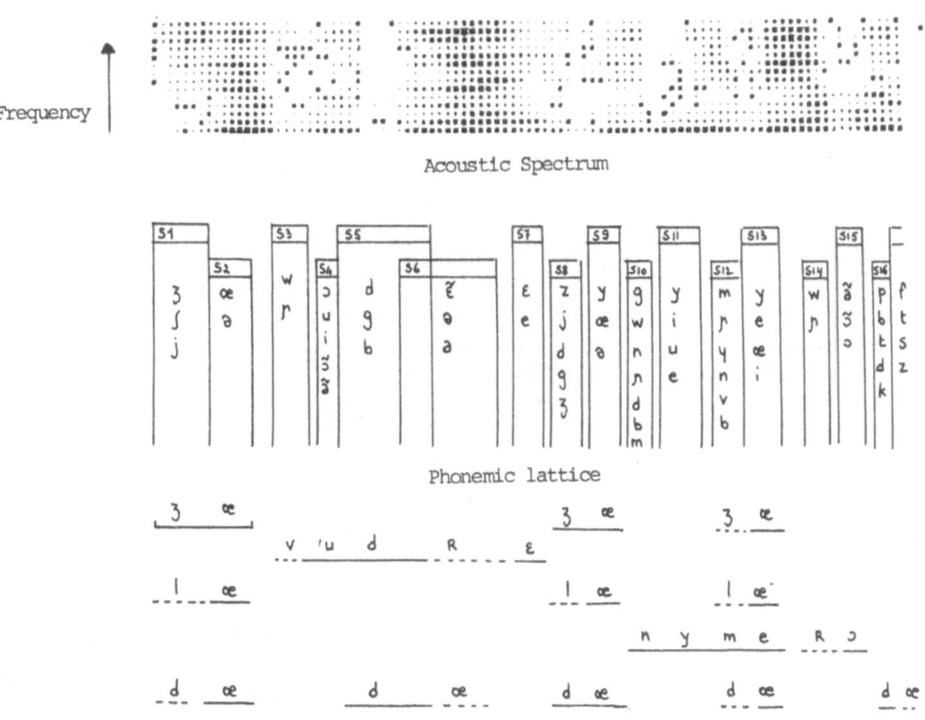

Word lattice

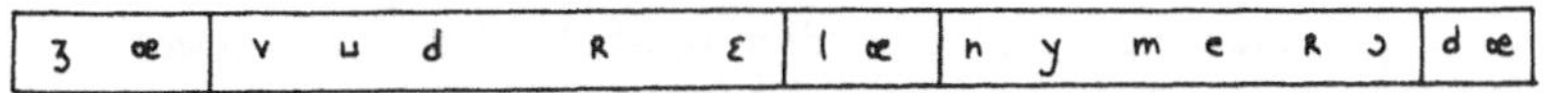

Syntactic output

Figure 2 : Results of phonemic recognition, word recognition and sentence recognition for the beginning of the sentence "je voudrais le numéro de téléphone de..." (I would like the telephone number of...).

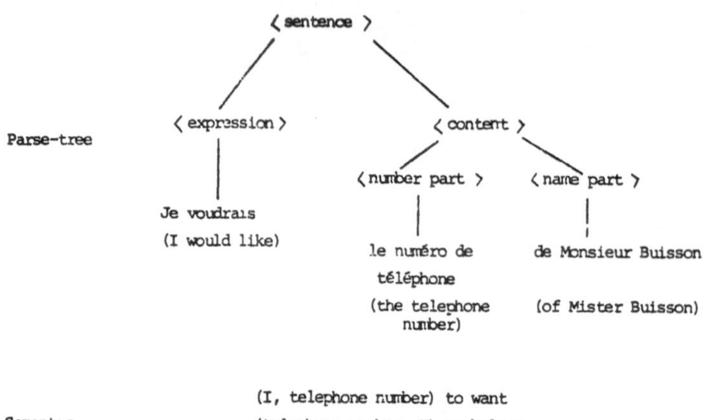

Semantic interpretation

(I, telephone number) to want
(telephone number, X) to belong
(X, Mister) title of
(X, Buisson) name of

Figure 3 : Parse-tree and semantic interpretation of the sentence "Je voudrais le numéro de téléphone de monsieur Buisson". (I would like the telephone number of Mister Buisson).

task is then analyzed by the sentence recognizer. The sentence recognizer then finds the best possible sequence of words according to a context-free grammar. This grammar represents all the utterances that are relevant to the question asked by the machine. The parse-tree of the sentence is then transformed by the sentence interpreter into a sequence of n-uplets which contains the semantic information of the user's request (see figure 3). Finally, the dialogue controller compares this information to its knowledge and decides what has to be answered to the user. In the case of the example given, the dialogue controller will decide to ask the user where his correspondant lives in order to limit the scope of the search, and thus entering a new question-answer cycle. For this new cycle, the parameters of the word and sentence recognizers will be restricted to grammar and vocabulary pertaining to adresses in the telephone directory.

2. PHONETIC ANALYZER

The main components of this module are the following (see GRESSER, 1975, for more details) :

2.1. The speech-noise distinction

This separation is based on the three following measures :

energy E(t), temporal evolution of the spectrum P(t) and frequency distribution R(t).

$$E(t) = \sum_{j=1}^{n} \text{channel}(j,t)$$

$$P(t) = \sum_{j=1}^{n-1} \left| \text{channel}(j,t+1) - \text{channel}(j,t) \right|$$

$$R(t) = \sum_{j=1}^{n-1} \left| \text{channel}(j+1,t) - \text{channel}(j,t) \right|$$

Channel(j,t) represents the energy in the frequency band j at time t ; n is the total number of frequency bands (14)

Then X1, X2, X3 are defined as follows :

$$X1 = \frac{1}{k} \sum_{t=1}^{1+k} P(t)$$

$$X2 = \frac{1}{k} \sum_{t=1}^{1+k} \left| E(t+1) - E(t) \right|$$

$$X3 = \frac{1}{k} \sum_{t=1}^{1+k} R(t)$$

Let F(x) be a linear function of X1, X2 and X3. The beginning of the sentence is detected if

$$F(x) > s1 \quad \text{for} \quad t > t1$$

Symmetrically, the end of the utterance is detected when

$$F(x) < s2 \quad \text{for} \quad t > t2$$

All the parameters defined above have been empirically determined.

2.2. The segmentation into syllables

This segmentation is divided into two steps :

-in the first step, the energy curve E(t) is divided into

consecutive segments, the boundaries of which correspond to the minimum F_{mi} of the curve and the center of which corresponds to the maximum E_{mi} of the curve.

If the difference between E_{Mi} and E_{mi} and E_{mi-1} is below a fixed threshold s, the peak is not significant and is consequently eliminated.

-the second step combines the preceding segments into syllables ; for each segment, the following eight parameters are computed :

- The difference of energy $E_{mi}- E_{mi-1}$
- The mean energy of each sample
- The total energy of the segment
- The energy of the maximum
- The number of voiced samples
- The number of vowel-like samples (defined as having a sufficient energy in low and high frequencies)
- The global spectral evolution of the segment
- The length of the segment.

A linear function of these parameters is then computed. According to its value, the segment is either labelled as containing a vowel and a syllable is detected, or linked to a contiguous segment in order to form part of a syllable.

2.3. The segmentation into phonemes

The vowel is localized in the area of low spectral change around a peak of energy. A primary localization of the consonants is then obtained by means of P(t). A high value of this parameter generally corresponds to a transition between consecutive phonemes. Consequently, a peak value of P(t) indicates a boundary between two phonemes. However, certain results of the succeeding identification procedure can still modify the output of the segmentation algorithm.

2.4. The identification of phonemes

A first hierarchical identification procedure widely used in phonetic analyzers takes place at this level :

- vowels have been separated from consonants in the preceding segmentation
- voiced consonants are separated from unvoiced ones by using the output of the pitch detector
- a plosive is detected when the segment is composed of a stationnary part with low energy followed by a strong burst of energy

- a fricative is detected if the segment contains a sufficient number of samples for which the center of gravity is higher than a fixed threshold :
the center of gravity of the sample t is given by

$$G(t) = \sum_{j=1}^{n} j.\ \text{channel}(j,t)$$

- after a plosive, the phoneme /R/ is distinguished from /F/, /S/, /ʃ/ if the energy curve has a maximum immediately followed by a minimum and if enough energy is detected in the first channels.

A more precise classification then takes place. Let m be the number of different phonemes in the considered class (vowel or consonant).
For p=m-1 and i=1,..,m, let $a_i = (a_{i1}, a_{i2}, \ldots, a_{ip})$ be the ith vertex of an equilateral polyhedra centered around the origin (see CHAPLIN, 1967).
For k=1,..,L (L : number of samples in the segment) and j=1,..,n+1, let x_{kj} be the energy in the frequency band j of the kth sample, and

$$X_k = (x_{k1}, x_{k2}, \ldots, x_{kn+1})$$

For i=1,..,n, let (w_{ij}) be the coefficients which give the orientation of the jth hyperplane decision surface used for separation of phonemes (see GLADYSHEV, 1963).

Let us define :

$$h_j(X_k) = \sum_{i=1}^{n+1} W_{ij} \cdot x_{ki}$$

$$H(X_k) = (h_1(X_k), \ldots, h_p(X_p))$$

$$F_i = \frac{1}{L} \sum_{k=1}^{L} a_i \cdot H(X_k)$$

The segment then represents the phoneme i for which F_i is maximum.

The output of the analyzer is represented by a sequence of segments which bear the following information : existence, boundaries, string of candidate phonemes ordered by the degree of likeness (see fig. 2).

2.5. Results

The results obtained by the current phonetic recognizer are summarized in fig 4. The sentences are extracted from the two above-mentioned tasks. They were uttered by four male speakers ; one of them was the training speaker.

Table 4: Percentage of Phoneme Identification.

Percentage of correctly identified phonemes	Training speaker	3 non-training speakers
For the first candidate phoneme	[illegible] %	40 %
Amont the first two candidate phonemes	74 %	60 %
Among the first three candidate phonemes	85 %	75 %
Correctly identified (maximum 6 candidates)	90 %	85 %
Total number of sentences	40	91
Total number of phonemes	556	1 154
Percentage of phoneme omission	6 %	8 %
Percentage of phoneme insertion	8 %	7 %

3. LEXICAL ANALYZER (VIVES, 1973)

The role of the lexical analyzer is to match each word of the task vocabulary against the phonemic lattice, in order to detect the word in the sentence. Figure 5 illustrates the matching process. A matrix is constructed, its rows corresponding to the phonemes of the word to be searched and its columns to the consecutive segments of the phonetic lattice. An element (i,j) of the matrix is set to 1 if the phoneme lying in the ith row appears in the list corresponding to the jth column. Each ① denotes a possible starting point for a detection of the word. Each [1] is a possible end point. The whole matrix is searched for paths which connect a starting point and an end point and pass through a set of (i,j) cells of the matrix. These cells must contain the number one and have increasing values of i and j. Each such path corresponds to a word detection. Each detection then receives a similarity score w (with $0 \leqslant w \leqslant 1$) which takes into account the degree of confidence for each segment and for each phoneme in the segment. The detection boundaries are then adjusted to the syllable boundaries in order to simplify the sentence recognition process. Only detections with scores higher than a given threshold (0.5 in the current version) are kept for the sentence recognition step.

4. SENTENCE RECOGNIZER

Word-lattice is here below the set of word-detections that have been found by the word recognizer. Each detection d is composed of three elements :

- a word x belonging to the task vocabulary T ;
- a segment $[a,b]$ indicating the left and right boundaries of the detection in terms of syllables numbers ;
- a score W between 0 and 1 indicating the quality of the detection.

Given a context-free grammar G, the sentence recognizer must search the word-lattice for the best sequence of detections $\eta = (d_1, \ldots, d_n)$ satisfying the following conditions :

- $x_1 \ldots x_n$ belongs to L(G) ;
- for each pair d_i, d_{i+1} the left boundary of d_{i+1} and the right boundary of d_i are "sufficiently" close ;
- d_1 is "very close" to the first syllable of the sentence, and d_n "very close" to the end of the sentence.

Each sequence receives a score computed as follows : let $l(d_i)$ be the length of the ith detection. Then the score $W(\eta)$ is given by the formula

$$W(\eta) = \frac{\sum_{i=1}^{n} W_i \times l(d_i)}{\sum l(d_i)} - \varepsilon(\eta)$$

where $\varepsilon(\eta)$ is a function taking into account the gaps and the overlapping between consecutive detections.

The sentence recognizer is organized into three components (see figure 6) : the word finder, the parser and the optimizer.

4.1. The word finder

Its function is to look through the word-lattice for some detection, given a condition for the boundaries or for the syntactic class of the word. In the current version of Keal, the computer word-lattice is built before the sentence recognition

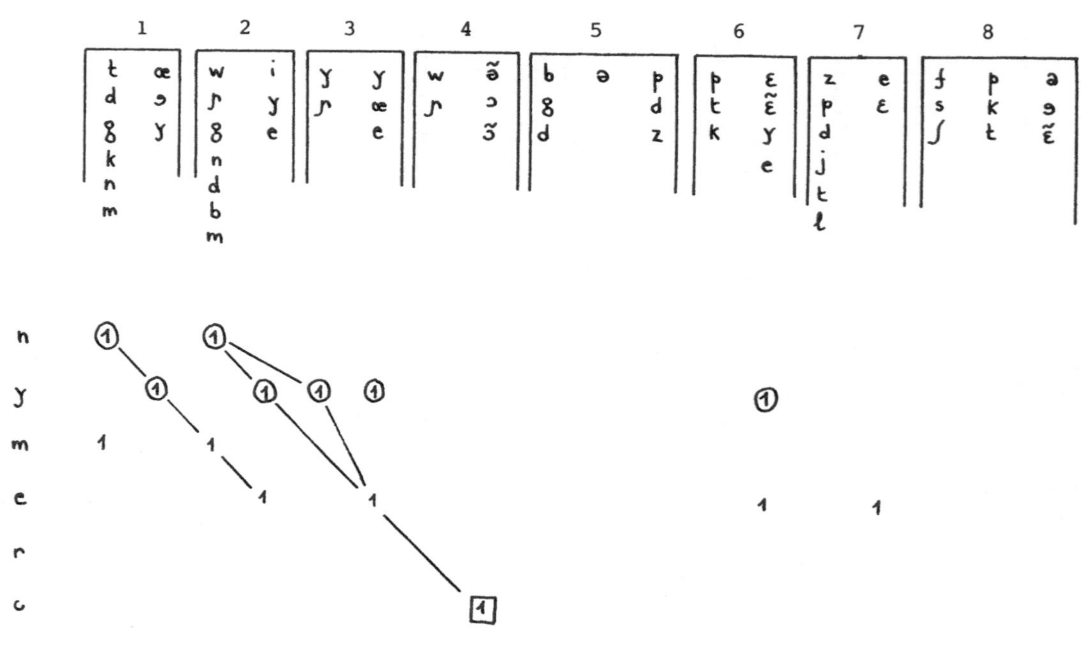

Figure 5 : coïncidence matrix for the word "numero" and the phoneme lattice of the phrase "le numéro de téléphone".

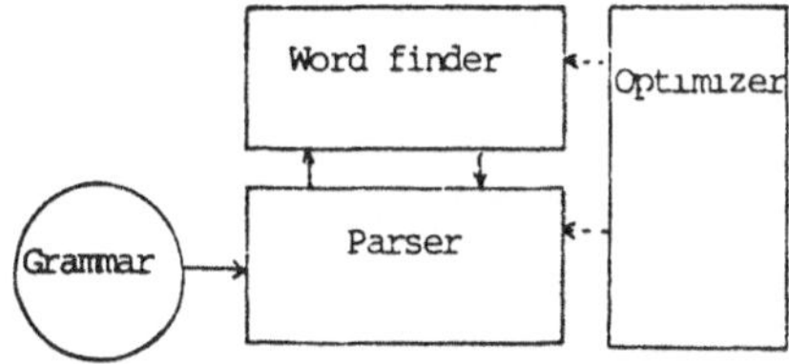

Figure 6 : block diagram of the sentence recognizer.

starts. In order to avoid extraneous computation, it would be better to activate the word recognizer only for words called for by the parser. This will be done in a future version of the system.

4.2. The parser

The parser has to decide whether a given sequence of detections satisfies the grammatical constraints and has to build up the parse-tree of the sentence. The parser is left-to-right, predictive and bottom-up. More detail on its operation may be found in QUINTON, 1976. Let P be the stack of the parser. Parsing a sentence consists of applying two types of operations :

- Scan : get the next symbol x of the sentence and compute a new value P' of the stack.
- Reduce : compute a new value P' of the stack applying a grammatical rule.

The initial value of the stack is P_0 ; the process stops when the stack gets a final value P_f and when there are no more symbols to scan.

4.3. The optimizer

Due to the nature of the input data, sentence recognition is a non-deterministic process. The main problem is to converge as fast as possible to the optimal solution. The role of the optimizer is to control the activation of the parser and the word finder such that they find this optimal solution as efficiently as possible.

The whole process can be considered as a search through a state-space (NILSSON, 1970) whose nodes, called situations, represent a particular state of the parse for a given sequence of detections. A situation is a pair $S = (\eta,P)$ consisting of a sequence η of words covering the left part of the word-lattice, and a stack P which may be obtained by analyzing the words $x_1, \ldots, x_n$

of η. A situation $S' = (\eta', P')$ is said to be a successor of S, (denoted $S \Gamma S'$), if
either

- $\eta' = \eta$ and P' is obtained from P applying a reduction

or

- η' is obtained by adding to η a new detection d and P' is computed from P by scanning the word χ of d.

Each situation is valuated using a function f depending on the strategy that is applied. The start situation is $S_0 = (\eta_0, P_0)$ where η_0 is the void sequence and P_0, the initial value of the stack ; a situation is said to be final if η covers the entire word-lattice and if the stack has the final P_f.

Three search strategies have been tried and compared : backtracking search, sequential decoding and beam search.

Backtracking search (BTS)
In this strategy, the valuation of S is $f(S) \doteq W(\eta)$. Let S be the situation currently searched (initially $S = S_0$). Let $\Gamma(S)$ be the set of successors of S, and let t(BTS) be a fixed threshold between 0 and 1.

Let S' be the best situation in $\Gamma(S)$. Several cases may then arise :

(1) If S' is final, the search stops ;
(2) if $\Gamma(S)$ is void, or if f(S') is less than the threshold t(BTS), a backtrack is performed ;
(3) if neither (1) or (2) applies, S' becomes the current situation and the situations belonging to $\Gamma(S) - \{S'\}$ are memorized in a stack for further backtracking :

Sequential decoding (SD)

In this strategy, the valuation f(S) of a situation is

$$f(S) = W(\eta) - b * l(\eta)$$

where b is a fixed bias where effect on the search will be explained below.

Let X be the set of situations not yet explored (initially $X = \{S_0\}$), and let S be the best situation in X. If S is final, the search stops. If not, X becomes $X - \{S\} \cup \Gamma(S)$.

A high value of b has the effect of augmenting the amount of situations explored, thus raising the chance of finding the optimal solution.

Beam search (BS)

For this strategy, the valuation function is f(S) = W(η).

Let e(S) be the right boundary of the sequence η of S. Let t(BS) be a given threshold. BS consists in processing successively the situations having the right boundary n, for values of n increasing from 0 to the length of the sentence. For each n, the following steps are performed :

(1) Find the set X of situating for which e(S) = n.
(2) Compute the maximum valuation Max(n) of the best situation of X.
(3) Keep for further processing only the set X' containing the situations in X where valuation W satisfies

$$\mathrm{Max}(n) - t(BS) \leqslant W \leqslant \mathrm{Max}(n).$$

4.4. Experimental comparison

A set of 50 word-lattices was randomly generated from 50 arithmetic expressions of length 9. The parameters of this simulation were estimated from previous results obtained from the word recognizer.

For each word-lattice, an exhaustive search (ES) was first performed in order to get the optimal solution and to compute the size of the search space.

The three strategics BTS, DS and BS were then run on the 50 word-lattices for various values of their respective parameter t(BTS), b and t(BS). As a result, it was possible to determine the optimal values $t^{*}(BTS)$, b^{*} and $t^{*}(BS)$ of these parameters. For these values, three results were computed :

- the percentage of optimal solutions found by the strategy ;
- the average percentage of space explored ;
- the variance of the space explored.

Table 7 shows the results of this comparison. It seems clear that BS is the best heuristics, giving 98 % of optimal solutions, reducing the amount of search by a factor % 3 with a 10 % variance.

5. SENTENCE INTERPRETER

The role of this component is to get from the parse-tree the information needed for the dialogue-controller. This is achieved

Table 7 : results of the comparison of the strategies.

	Exhaustive Search	Backtracking Search $\varepsilon^*(BTS) = 0.5$	Sequential decoding $b^* = 0.035$	Beam Search $T^*(BS) = 0.09$
% optimal solution	100 %	72 %	96 %	98 %
% space searched	100 %	33 %	45 %	36 %
variance	0 %	24 %	22 %	11 %

by a set of transformation rules associated with the syntactic rules. The result of these transformations is a set of n-uplets containing only that in the sentence which is relevant to the dialogue (see the example given in fig. 3).

6. DIALOGUE CONTROLLER

The organization of this component was inspired by the GUS system (BOBROW, 1977).

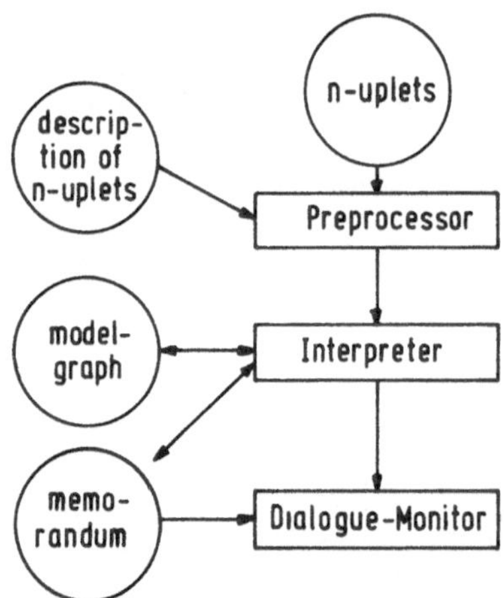

Fig. 8. Block diagram of the dialogue controller.

Fig 8 shows a block diagram of the dialogue controller. The n-uplets that have been obtained by the sentence interpreter are first preprocessed in order to verify their semantic consistency, and to add some implicit information. The result is then interpreted using a "model-graph" (M.G) which represents the knowledge of the system. A set of responses (messages or questions about missing information is produced and listed in a memorandum ; each response has a priority based on its importance in the dialogue. The dialogue-monitor chooses the response with the highest priority and sends it to the synthesizer. In addition, the dialogue-monitor sends to the system-monitor indications concerning the way the user's answer should be analyzed : the vocabulary to be searched, the grammar relevant to the answer and the transformation rules that should be applied on the parse-tree.

Below is a more detailed description of the operation of the components.

6.1. Preprocessor

Each type of n-uplets is described by a template which contains some of the conditions on the semantic characteristics of their participants. The input n-uplets are matched to their templates, and the conditions are verified. Non-consistent n-uplets are simply rejected.

Another role of the preprocessor is to complete or to create the n-uplet set, when the answer has an elliptic form (for example yes-no answers). This is achieved by using a set of incomplete n-uplets which corresponds to the question the user has been asked.

6.2. Interpreter

The model-graph is a data structure that represents the information needed by the system to achieve its task. It is instanciated with the information that can be extracted from user's utterances and thus represents the state of the dialogue at any moment.

Each node of the MG represents a particular concept (for example a town). The node may enter various states memorizing whether it has been previously checked and under what conditions. To each state are associated a form and a set of procedures. A form is a logical expression which when verified, causes the procedure to be activated. A procedure may :

- modify the state of various nodes ;
- memorize or destroy information ;
- create or destroy a response in the memorandum.

For example, in the task of directory inquiry, a node is associated with the concept of town. This information is needed before starting the search. This node may be in one of the following states :

1 - not checked
2 - checked with reliable information
3 - checked with unreliable information.

In state 1, a matching with a town-n-uplet will enter a procedure which will cause the node to enter either state 2 or 3 depending on the score of the parse-tree from which the n-uplet has been extracted. If state 3 is entered, a question will be put in the memorandum asking for a confirmation of this information.

The vertices of the model graph indicate the order in which the nodes are to be checked. During the dialogue, some nodes may enter a particular state indicating that they are not relevant to the user's request.

6.3. Dialogue Monitor

During the interpretation, the memorandum has received a set of possible responses which may be questions about missing information, confirmation requests or messages to the user. The role of the dialogue monitor is to choose a response and to send to the KEAL monitor information about the way to analyze the user's reply. The response is synthesized using a text-to-speech synthesizer (DIVAY, 1977 ; EMERARD, 1977).

The dialogue controller is partially implemented and has not yet been tested. A compiled language makes possible the description of the dialogue process depending on the task to be achieved by the system.

7. APPLICATIONS AND PERFORMANCES OF KEAL

Until now, the system has never been completely experimented. However, subsets of the complete configuration have been run on several tasks. We report here the results of two experiments. The first task consisted of an automatic directory inquiry device and the second of a computer aided instruction.

7.1. Telephone inquiry

The task for the system was to retrieve the phone number of one person in a list of 37 people. In this experiment the system configuration did not use the sentence recognizer and the dialogue controller was replaced by a very simple program. The word recognizer was used as a word-spotter. 268 requests were uttered by 4 speakers. For 157 of them, the formulation of the first

Table 9 : Results of the first (free formulation) and second (imposed formulation) experiments.

	Exp. 1 (157)	Exp. 2 (111)
Correct answer after 1 question	91 (58 %)	92 (83 %)
Correct answer after spelling	47 (30 %)	17 (15 %)
Total	138 (88 %)	109 (98 %)

question of the dialogue was free. For the 111 other requests, the speakers were asked to use only the form : first name + last name. Table 9 shows the results of the experiment. The significant difference between the first set (free formulation) and the second set (imposed formulation) shows clearly the necessity of a syntactic analysis for a better recognition of freely formulated questions.

7.2. Computer Aided Instruction

This experiment was conducted to test the sentence recognizer, without dialogue. The task involved was to describe models of printed circuits. The vocabulary of this task contained 120 words and the language has a branching factor of 14, an entropy of 3.3 bits/word (perplexity 9.843 words). On a set of 74 sentences (average length 2.5 words) uttered by one speaker, 61 (82 %) were correctly recognized by the system.

Conclusion

The results of the experiments show clearly that the performance of the phonetic recognizer must be improved. At this level, the major source of error is the inability of the program to take completely into account the effect of coarticulation on the acoustic characteristics of phonemes. This effect is under study in order to be formalized and programmed in the Keal system (M.T. ROTH, 1979).

The use of prosodic information would be of great help in the recognition process. But recent studies in this field (VAISSIERE, 1976) show that it is difficult to use this type of knowledge since it can be unreliable and depends strongly on the speaker.

Studies for optimizing the acoustic analysis have to be pursued together with studies on the problems of normalization and adaptation to speakers.

At the linguistic level, the use of phonological information together with a word verifier will be of great interest for large vocabularies.

At the pragmatic level, we have to develop other mecanisms of artificial intelligence allowing us to deal with more sophisticated dialogues.

References

BOBROW (D.G.), KAPLAN (R.M.), KAY (M.), NORMAN (D.R.), WINOGRAD (T.) : 1977 "GUS, a frame driven dialogue system", Artificial Intelligence, 8.

CHAPLIN (W.G.), LEVADI (V.S.) : A generalization of the linear threshold decision algorithm to multiple classes. Computer and information Sciences II, Tou (A.P.) New-York, London (1967).

DIVAY (M.), GUYOMARD (M.) : Conception et réalisation sur ordinateur d'un programme de transcription graphémo-phonétique du français. Thèse de doctorat de spécialité (Informatique). Université de Rennes. 1977.

EMERARD (F.) : Synthèse par diphones et traitement de la prosodie. Thèse de doctorat de spécialité en Linguistique. Université de Grenoble 1977.

EMERARD (F.), LARREUR (D.) : Synthèse par diphones. Recherches Acoustiques CNET, 1976 Vol III p 295-314.

GLADYSHEV (E.G.) : On stochastic approximation, theory of probability and its applications. U.S.A. (1963), 10, pages 275-278.

GRESSER (J.Y.), MERCIER (G.) : Automatic segmentation of speech into syllabic and phonemic units : application to french words and utterances. Auditory Analysis and perception of speech. G. FANT et M.A.A. THATHAN editors, A.P. London 1975 p 359-382.

NILSSON (N.J.) : 1970, Problem solving methods in Artificial Intelligence, Mac Graw Hill.

QUINTON (P.) : A syntactic analyzer adapted to speech recognition, I.E.E.E., I.C.A.S.P., 1976 p 453-457.

ROTH (M.T.) : "Variations Acoustiques de la Zone Stable de la Voyelle sous l'Influence des Consonnes dans les monosyllabes en français". 26e Séminaire d'Acoustique à Olesnica. WROCLAW (Pologne) du 17-21 Sept. 79. p. 172-177.

VAISSIERE (J.) : 1976, Premiers essais de segmentation automatique de la parole continue à partir des variations du fondamental de la phrase, "Recherches acoustiques", CNET, vol 2 pages 193-208.

VIVES (R.), GRESSER (J.Y.), : 1973. A similarity index between string of symbols. 1st joint Conference on Pattern Recognition Washington p. 308-317.

THE LOCUST SYSTEM

R. Bisiani

Computer Science Dept., Carnegie-Mellon University, Pittsburgh, PA 15213*

ABSTRACT

In 1976 the Harpy system successfully demonstrated the feasibility of a 1000 word vocabulary connected speech recognition system. Since Harpy was designed to demonstrate the feasibility rather than to optimize cost effectiveness, it was implemented on a big general purpose system. The Locust system was designed to demonstrate the possibility of implementing the Harpy algorithms with a good cost effectiveness on a low cost, widely available minicomputer. The paper describes the Locust system, the technical problems met during its design, and its performance.

1. INTRODUCTION

Only ten years ago the feasibility of the Connected Speech Recognition System (CSRS) was not demonstrated. A five-year (1972-1976) research effort [1] funded by ARPA generated a few working CSRS's and therefore proved their feasibility. The Harpy system [7, 8], developed at Carnegie-Mellon University, was one of these and actually the only one to meet the goal of a 1000 word vocabulary with sentence accuracy exceeding 90% for male and female speakers in 10 to 30 times real time.

The designer of a CSRS is faced by 10 to 15 design

*This research was sponsored by the Defense Advanced Research Projects Agency (DOD) ARPA Order No. 3597, and monitored by the Air Force Avionics Laboratory under Contract F33615-78-C-1151.

J. C. Simon (ed.), Spoken Language Generation and Understanding, 545-551.

decisions, for each decision there are 3 to 10 feasible alternative choices. Thus, the solution space for speech systems seems to contain 10^6 to 10^8 possible system designs [1]. The approach taken by Carnegie-Mellon University researchers is to develop several systems that tackle the same problem in different ways, emphasizing the various aspects of the problem. Up to now our "family" includes about 10 systems, some of these are no longer under development, while others, like Harpy, are continuously being improved. New systems are generated every time a substantially new solution is tested.

The Harpy system has been designed with the main goal of demonstrating the feasibility of a connected speech recognition system working in a non-trivial task domain (the Artificial Intelligence Information Retrieval task). Therefore, issues like cost and speed of the system have been put aside every time it was felt they might hinder the implementation of the system. For example, the most powerful machine available was chosen, regardless of the cost (DEC PDP-10 at a cost of a few dollars per utterance).

If we look at a CSRS as part of a complex man-machine interface, it becomes obvious that a CSRS usable in the real world must be not only accurate, but also fast (less than real time?) and inexpensive (1000 dollars?). In other words, the system must have a very good price/performance ratio. The first goal of the Locust system design was to improve the cost/effectiveness of the Harpy system.

If we observe the status of research in the field of speech understanding, we see the existence of a multitude of speech understanding systems. Nevertheless, it is extremely difficult for the researchers at one location to take advantage of the tools and programs developed at another location. It is also very difficult to compare results, because of the various ways of generating input data and task domains. The second goal of the Locust system design was to make the Harpy system available to the researchers' community by implementing it on a widely available computer.

The paper describes the Locust system, the technical problems met during its design, and its performance.

2. BEFORE LOCUST: THE HARPY SYSTEM

The Harpy system has been described in detail elsewhere [7, 8] and is also dealt with in this book by Lee Erman. For the sake of this presentation we will review here the most important characteristics of the Harpy system.

All the knowledge necessary to the recognition is uniformly encoded in a precompiled network. Each path in this network represents a legal utterance; an utterance cannot be recognized correctly if it is not explicitly represented in the network. All knowledge sources are applied during the off-line compilation of the network only. Therefore, the recognition mechanism is independent from the knowledge sources and simply consists of a search for the path in the network that best matches the values of the features extracted from the input signal.

Brute-force search is too expensive for networks that have enough states and arcs to represent a reasonable task. Therefore, the Harpy system searches only a beam or near-miss alternatives around the best path, thus containing the exponential growth, beam search. The number of alternatives searched varies widely from just a few to over a thousand, depending on the network branching factor and the "intelligibility" of the input signal.

The size of the network is obviously a function of the "task domain" we are interested in (e.g. the number of words, the complexity of the grammar, etc.). The biggest network used so far (1000 word dictionary) needs about a quarter million bytes of storage. The system maintains the complete network in main storage and this creates a severe limitation to the size of the networks that can be used. The PDP-10 implementation of the Harpy system is able to complete a recognition by using about 3 million instructions per second of speech for the 1000 word abstract retrieval task (2.4 times real time on a PDP-10/KL-10).

3. LOCUST DESIGN

As mentioned in the introduction, the reasons for building the Locust system were two-fold. Firstly, we wanted to prove the feasibility of a lower cost Harpy system; secondly, we wanted the system to be available on a widely-used machine and operating system. The two requirements were satisfied by the DEC PDP-11 and the UNIX operating system. This general purpose minicomputer environment was also considered very interesting, because it would confront us with a series of design problems that were and still are very crucial for a low-cost implementation of Harpy, e.g. the huge amount of random access memory needed to store the network.

The design issues raised by the implementation of Locust were:

- the word size and the instruction set of the chosen minicomputer required a modification of the signal analysis algorithms;
- the operating system characteristics made the implementation of certain parts of the system difficult, e.g. real time acquisition of sampled data;
- the limited address space required that the network be stored in mass memory.

The small word (16 bits) and limited instruction set (floating point instructions were not available on our PDP-11/40) required a modification of some of the signal analysis algorithms. This did not affect speed or accuracy. It created a debugging problem though, because the intermediate results of the PDP-10 Harpy could not be directly compared with the results of the new programs. The modern, low cost signal processing microprocessors use integer arithmetics. We are now applying the experience made by implementing some of the Locust algorithms with integer arithmetic to the design of some ad-hoc hardware for signal processing.

The operating system interference was not a difficult problem to overcome, both because of the clean UNIX structure and because there were people in the environment with a good knowledge of UNIX. For example, although the hardware would have been able to sample, pre-process and store speech on the disk in real time, the operating system overhead originally hindered real time input. Careful re-programming of some device drivers and some shortcuts in the operating system code solved the problem. A similar problem might prove fatal when using an operating system that does not allow custom modifications because of complexity or commercial reasons (e.g. source code not available).

Storing the network in mass memory was necessary in order to run the 256 and 1000 word artificial intelligence retrieval tasks, because their networks could not fit in the tiny address space of the PDP-11. The only possibility at hand was to store the network on the disk and hope to find a reasonably efficient way of retrieving parts of the network during the recognition process. The classical way to limit the number of disk accesses is to read from the disk and store in primary memory more information than is needed at a certain time and hope that subsequent requests will use information that already is in primary memory. This technique of course implies the availability of some space in primary memory and, usually, gives a performance that is directly proportional to the available primary memory storage. It is easy to recognize the similarity between this problem and the paging of programs in virtual memory systems. The literature is very rich of information on the latter and, of course, we investigated the possibility of using any of the existing knowledge. Unfortunately, the best results in the field of program paging policies are crucially based on the common characteristics exhibited by computer programs and especially on the independence of program behavior from input data (e.g. an editor will most likely go through the same reference pattern, no matter which characters we type, or a matrix multiplication program will behave independently from the value of the matrixes it has to multiply).

The independence of behavior from the input data is crucial when deciding the "layout" of the network on the disk (clustering). Since each disk access brings a full block into main memory and each block contains more than one state, it is obvious that if some of the states fetched by reading one disk block can be kept in main memory until they are needed, we can reduce the number of "disk read" operations and thus improve the speed of the system. The ability to do so is critically bound to the fact that states with a high probability to be used shortly after one another are clustered in the same disk block. The best algorithms [6, 5] that are able to do such a clustering are usable only if the reference pattern is not too sensitive to the input data. On the other hand, some analysis of the network reference pattern generated by Harpy has shown that this is actually not true.

Besides clustering the network on the disk in a suitable way, we also had to devise a policy that would choose the states to be kept in main memory and decide if a state actually was in main memory or a disk access was needed.

The paging strategy we designed for Locust had to avoid two main pitfalls:

- excessive use of computing power and
- excessive use of primary memory storage.

The "small locality strategy" [3] was designed with this in mind. The underlying idea is very simple: since the states examined during a certain step are all successors of the states examined in the previous step, the grouping together of the successors of the same state creates small "clusters", that contain states having a high probability to be processed during the same step. No assumption is made on the relationship between different clusters. The "small locality clustering" can be obtained with a very simple algorithm:

1. the initial state is stored in a list called "next states list";

2. all the states in the next state list are stored in the secondary memory;

3. the successors of the states being stored on the secondary memory are saved, if they have not been clustered already, in the next states list;

4. steps 2 and 3 are repeated until the next states list is empty.

During the reordering the states are renamed with the block number and offset in the block of their physical location; this allows quick and efficient retrieval since no mapping table is needed.

At the beginning of each step (one transition in the network) the states to be expanded are sorted by increasing block number. Every time the disk is accessed, more than one block is retrieved (the retrieval of a limited number of <u>consecutive</u> blocks does not require much more time than the retrieval of a single block). The reordering and retrieval strategies described limit the number of block accesses required, because more than one state is processed for each disk access.

The results were encouraging: Locust on a PDP-11/40 is only twice as slow as the current Harpy with the network in core on a PDP-10/KA-10. About 30% of the total time needed for recognition (60% of the speed loss) is spent waiting for the disk and paging related CPU activities. Since both CPU's (11/40 and KA-10) have about the same speed (.3 million instructions per second) the remaining 40% of the speed loss is mostly due to the more complicated data structures required by Locust because of the external network.

Moving the network to secondary storage put in evidence some of the shortcuts that had been taken while programming the in-core version of the system on the PDP-10. This was extremely useful in some of the work accomplished after Locust, e.g. the Harpy Machine [4, 2].

4. FUTURE TRENDS AND CONCLUSIONS

The Locust system demonstrated the feasibility of a connected speech recognition system on a medium cost machine with limited recourses. We have already exported the system to two different locations, and we hope to gather enough information from this experience to make it easy for other institutions to install Locust.

The design of Locust also allowed us to get a better insight of the engineering problems related to the efficient implementation of a speech understanding system. Future speech understanding systems will have to rely heavily on computer technology in order to obtain the necessary performance. However, the gain in speed and density, or the lower price alone, are not going to make it possible to implement an efficient speech understanding system. It is crucial to adapt the speech understanding system algorithms we are using to the tools the technology offers us. We are actively pursuing this path and we have already completed a new system that makes good use of the experience acquired with Locust. This system, the Harpy Machine [4, 2], uses a tailored architecture instead of a general purpose one and attains a 4 to 1 reduction in price and a 14 fold increase in speed over Locust. We hope to be able to continue the development of this kind of systems, so that we can learn how to make good use of technology in building speech understanding systems.

ACKNOWLEDGEMENTS

Raj Reddy, Ken Greer, Gary Goodman and Fil Allewa have also participated in the research described.

REFERENCES

1. Speech Group, Summary of Results of the Five-year Research Effort at Carnegie-Mellon University. Computer Science Dept. Carnegie-Mellon University, 1977.

2. R. Bisiani. The Role of Simulation in the Development of Task-Oriented Computer Architectures. Second Annual Symposium on Small Systems, ACM, 1979.

3. R. Bisiani. Paging Behavior of Knowledge Networks. Computer Science Dept. Carnegie-Mellon University, August, 1977.

4. R. Bisiani and H. Mauersberg. Software Development for Task-Oriented Multiprocessor Architectures. Compcon '79, 19th IEEE Computer Society International Conference, IEEE, 1979.

5. D. Ferrari. Improving Locality by Critical Workings Sets. *Communications of the ACM 17*, 11 (November 1974).

6. D. Ferrari. The Improvement of Program Behavior. *Computer* (November 1976).

7. B.T. Lowerre. *The Harpy Speech Recognition System.* Ph.D. Th., Computer Science Dept. Carnegie-Mellon University, 1976.

8. B.T. Lowerre and R.D. Reddy. The Harpy Speech Understanding System. In Wayne A. Lea, Ed., *Trends in Speech Recognition*, Prentice-Hall, 1979.

THE MYRTILLE II SPEECH UNDERSTANDING SYSTEM *

J.M. PIERREL and J.P. HATON

Centre de Recherche en Informatique de Nancy
University of Nancy I
C.O. 140 54037 - Nancy, France

ABSTRACT

In this paper we present the MYRTILLE II Speech Understanding system for a subset of French Language. The original parallel architecture of this system is described, together with the data structures used at the lexical, syntactic and semantic levels of processing. Syntax and semantics are integrated in a unified representation which makes it possible to use these knowledge sources in an efficient way.

The acoustic phonetic recognition process is not described here but the operation of the system is given in details and examples of sentence processing are given and discussed.

* This paper is a revised and augmented form of a paper previously presented at the 4th Int. Joint Conf. on Pattern Recognition, Kyoto, Nov. 1978.

J. C. Simon (ed.), Spoken Language Generation and Understanding, 553-569.

1. INTRODUCTION

In 1975, a first continuous speech understanding system (MYRTILLE I) was operational in our Laboratory (1) : it was capable of processing artificial languages comprising a vocabulary of less than 100 words. Such a system already gives numerous possibilities of speech use : automatization of a telephone exchange, machine-tool control, process control... Nevertheless, the syntax and vocabulary accepted by this system do not allow the automatization of information centres to be envisaged as this needs a more important vocabulary and a syntax closer to a natural language. This is why we began to bring another generation of systems into operation, basing ourselves on an application of meteorological information.

The characteristics of the language processed by the MYRTILLE II system are :
a) a vocabulary of more than 300 words
b) a syntax describing a quite wide sub-structure of spoken French (the principal restriction being found is the fact that the pronominal references and the relatives are not taken into account).

The result provided by this system is a syntactic-semantic representation of the processed sentence. We do not envisage synthezising, in the first instance, the recognized sentence ; the processing level that allows the release of the action or of the answer corresponding to the recognized query is not, at this time, brought into operation.

Later in this article, we consider that the entry sentence is provided after an initial acoustic-phonetic processing in the shape of a phoneme lattice and a set of prosodic features. We next intend to implement interactions between the processing levels described below and the acoustic-phonetic level (4).

In this paper we present the general architecture of MYRTILLE II system and the operation of the main modules of this system. The data-structures used in the representation of information at the different processing levels are also introduced and discussed.

2. GENERAL PRINCIPLES AND STRATEGY OF MYRTILLE II

As with MYRTILLE I, the MYRTILLE II system functions on the principle of hypothesis and test. But, whereas in the case of artificial languages the syntax alone provided the hypotheses to be validated subsequently by the phonemic recognition module, the emission phase of the hypotheses here means that other different types of information come into play : syntactic, semantic, phonemic, prosodic - each of which will be taken into consideration by a specific module of the system (2). At this level, different architectures are possible :

a) Sequential architecture :
We have rejected this solution because it leads inevitably to a favouring of the first processing level and, this being the case, it is difficult to choose a particular level to favour. At certain moments in the processing it is first of all necessary to employ a syntactic processing and, at other moments, a semantic or prosodic processing. In all events, it does not seem that a structure which leads us to take into consideration pieces of information in a syntactic, semantic and prosodic order would offer a good solution.

b) Parallel architecture :
This seems to offer the better solution because apart from the fact that it favours no one processing level, it should allow us in the near future to increase the performances of the system as well as allowing a real-time operation by associating a specific processor with each level.

In spite of this, material constraints (work on a C.I.I. MITRA 125 mini-computer) made us abandon this solution.

The solution we have chosen is one which is falsely parallel : a control module decides at each passage the modules to be activated and their order. Thus, whilst still keeping a sequentiel type of execution we gain a little the flexibility of operation offered by a parallel-type architecture.

Figure 1 gives the general diagram of the system. We find on it the different knowledge sources and their relationships.

3. REPRESENTATIONS OF THE DIFFERENT INFORMATION USED

3.1 Definition of the syntactic-semantic structures

This corresponds essentially to the description of the grammar of the processed language. It is possible to define the grammar of a subset of spoken French in a generative form by means of a context-free grammar. But, in the case of speech recognition, this solution seems to be badly adapted, for the following reasons :

a) the heaviness of such a grammar if we want to convey the diverse transformations possible : several hundreds of rules and a very great indeterminism.

b) a too-great importance is attached to small words whose recognition is not very pertinent and which serve as a support for the syntactic structure.

c) the inadaptation of a generative model to the task of recognition.

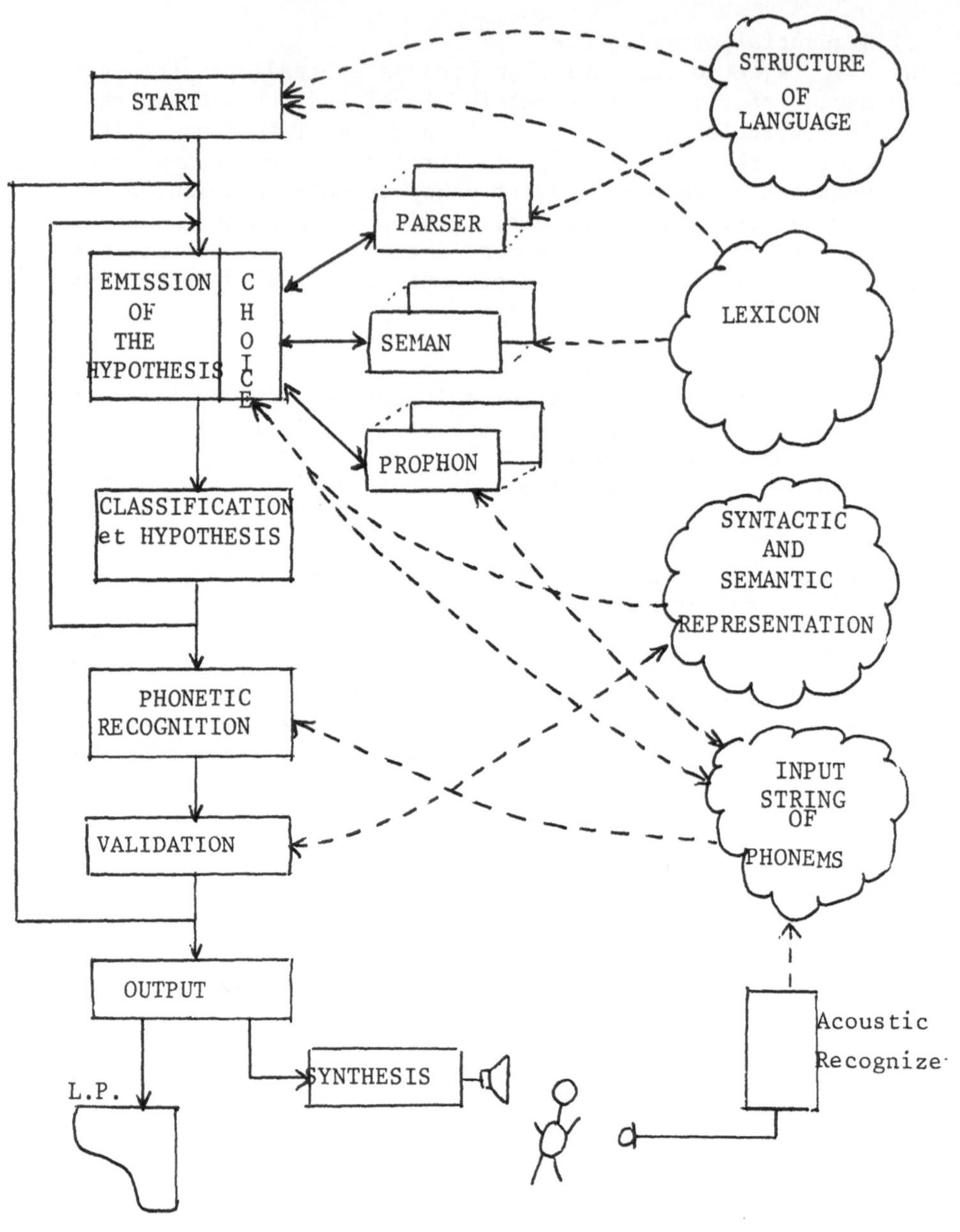

———→ : indicates the chaining of two modules

- - - - → : expresses the direction of information transfers

Schematic block-diagram

of the MYRTILLE II System

Figure 1

d) the difficulty of conveying contextual phenomena.

For these reasons, we have been led to bring a different solution into operation in which the definition of sentence structures, syntax and semantics are closely linked. This leads us to define a syntactic-semantic network with procedural nodes in order to convey the structure of the language. In fact, more than simply a network we obtain a tree representation of graphs : each branch of a graph may be defined as a word of the lexicon or as the reference label of a sub-graph.

In such a structure, each node of the network corresponds to a procedure : we distinguish, therefore:

i) the linear branches composed of series of words from the lexicon, or of reference labels belonging to a sub-network. The pieces of information provided by these branches are of the same type as those an adjacency matrix, obtained from a C-grammar, could provide locally.

ii) procedures linked to each node, which use pieces of semantic information, take into consideration all the contextual phenomena and process, to a great extent, the problems of indeterminism by arranging the possible outputs of the procedures, having taken into account the inputs and the processing employed. The syntactic-semantic analysis corresponds to a pathway progression within these networks offering us possibilities of backtracking.

The whole of the network (linear branches and procedural nodes) defines completely the structure of the processed language ; with each modification of the language the network has to be partially reworked, but the utilization and analysis procedures remain the same, whatever the language defined by it.

Example : The following (fig. 2) is an example of a GN network which conveys, for instance, the structure of the French sentence : "la petite maison rouge et verte du premier locataire de Paul"

This network provides all the syntactical pieces of information that allow a syntactical analysis. Without obtaining as many pieces of information the equivalent C-grammar would consist of at least 10 rules.

The procedures PE, PS, P1, P2 can be of different kinds. As an example the procedure P2 is described below. This procedure uses a stack on which it puts (PUSH) the different possible outputs (represented by digits 1 to 5 on the diagram).

Procedure P2 :

```
PUSH (output 1)
PUSH (output 2)
IF Input 1 THEN PUSH (output 5)
IF Unvoiced Plosive THEN PUSH (Output 4)
IF Voiced Plosive THEN PUSH (Output 3)
```

The actual output is the one which is on the top of the stack
END P2.

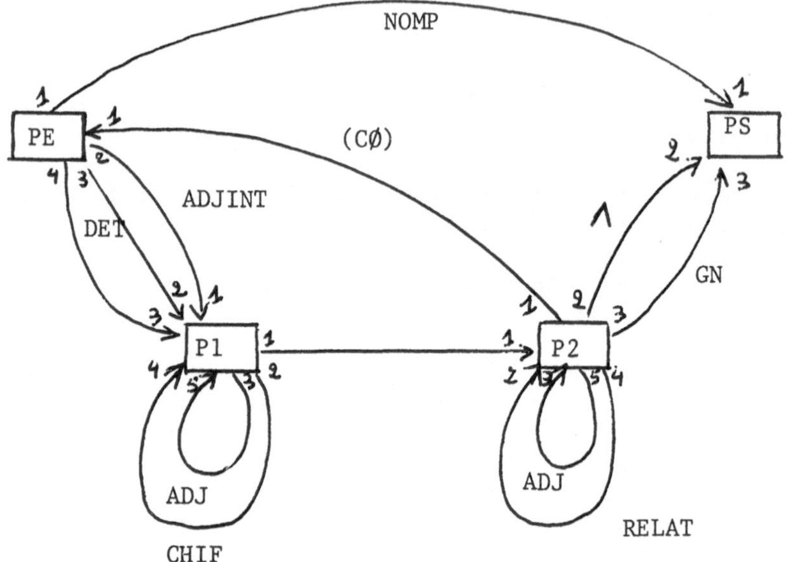

PE	:	input procedure of the network
PS	:	output procedure of the network
P1, P2	:	internal procedures
Λ	:	empty string
ADJINT	:	interrogative adj.
DET	:	determinant
ADJ	:	adj.
NOM	:	noun
NOMP	:	person name
CHIF	:	digit
CØ	:	conjonction
RELAT	:	relative

3.2 Syntactic-semantic definition of the words in the dictionnary

In addition to the structure definitions we also need the syntaxtic semantic definition of words. Without going directly to the definition of elementary concepts and of conceptual dependences, as Schank has done, we have opted for a tree representation of the vocabulary (4 Tree levels) (cf figure 3).

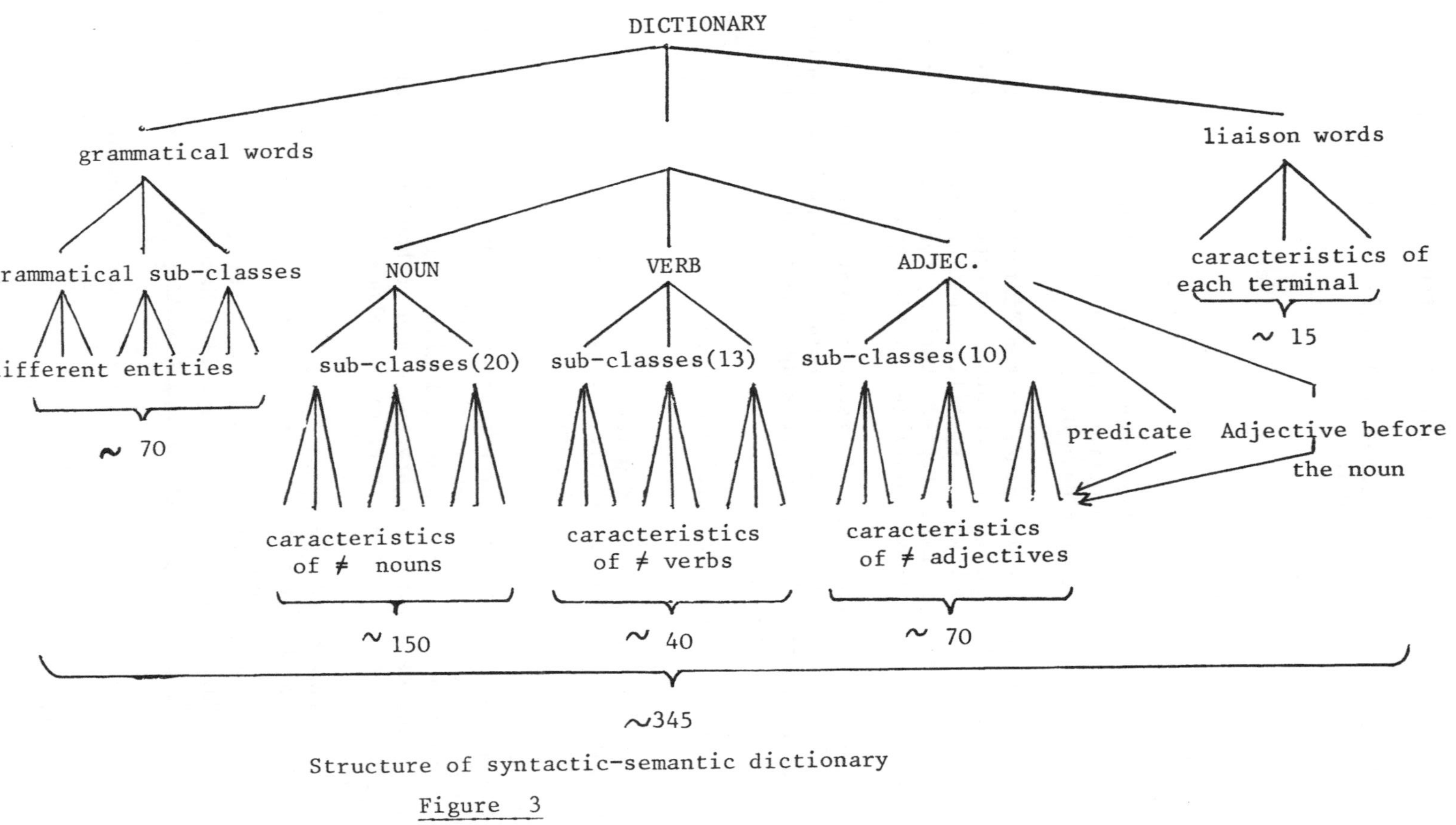

Structure of syntactic-semantic dictionary

Figure 3

i) the first level creates a division of the vocabulary into three large classes.
- the liaison words which appear explicitly in the definition of the syntactic-semantic network.
- the words with particular grammatical functions : prepositions, adverbs, conjunctions...
- the general terminals : nouns, verbs, adjectives.

ii) The second level creates a division which follows the grammatical classes : NOUN, VERB, ADJECTIVE...

iii) The third level distinguishes grammatical sub-classes in terms of the grammatical construction and of the types of conceptual dependences that they demand.

iv) The fourth level corresponds to the different words or entities. This level details the elementary concepts and the properties belonging to each word : specific construction, pointer to the word acoustic-phonetic representation, type of the rules of phonological alterations applicable to this word...

Besides this, the same type of tree representation is also used for the handling of hypotheses : limiting the hypotheses emitted corresponds essentially to suppressing some branches of this tree representation.

3.3 Acoustic-phonetic definition of words in the dictionnary

We mean by this :
- the phonetic definition of words
- the rules of phonological alterations
- the definition of prosodic features.

a) Phonetic definition
At the moment, each word is represented by a phonetic reference graph (I) in which there appears the substitutions, omissions or repetitions of possible phonemes (cf figure 4). Such a representation requires a lot of memory-space and we hope to be able to determine the more general rules of the phonetic construction of French words.

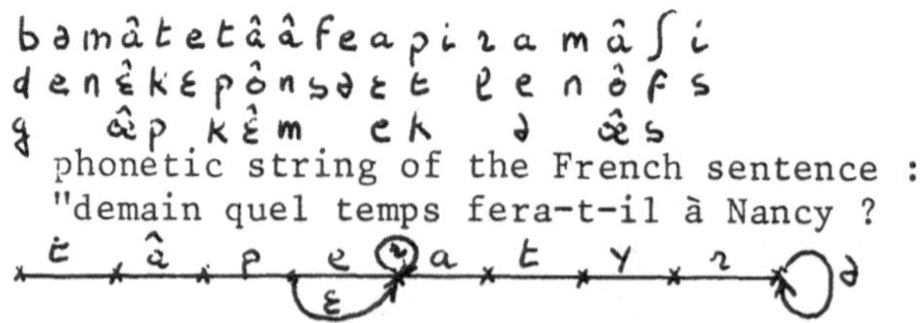

phonetic string of the French sentence :
"demain quel temps fera-t-il à Nancy ?

Phonetic definition of the French word :"température"

Figure 4

Furthermore, we add to it, for each word, the numbers of the rules of phonological alterations which convey plurals, liaisons, conjunctions...

b) Rules of phonological alterations
Each rule corresponds to a procedure : this permits, for each word, the conserving of only one phonetic representation of its radical and the deducing, by these procedures, of the forms of the different possible occurrences (eg. radical of a verb and different conjugated formes).

c) Definitions of the prosodic features
The prosodic features intervene at two levels :

i) at the output of the acoustic-phonetic level for the sentence to be processed, there appears, in addition to the phoneme lattice, diverse prosodic markers which permit the codifying of the pauses, the intonation, the pitch variation, etc...

ii) in the lexicon : we intend using the intonative patterns of the words, their length, etc...

These prosodic features are taken into consideration at the time of processing by the PROPHON module as well as by the procedures of the syntactic-semantic network.

3.4 Syntaxtic-semantic representation of the processed sentence

This representation of the processed sentence conserves all of the information obtained in the course of the processing and which is necessary to assure the releasing of the action demanded by the input sentence.

We have opted for a hierarchical representation by information vectors. Without wishing, in this paper, to describe this representation in its entirety, here, schematically, is what it comprises :

a) an information vector for each proposition including :
- its type (principal, subordinated)
- the description of the verbal group (the verb type, the auxiliary type)
- the description of the subject : nominal or pronominal type and, eventually, pointers to the nominal groups
- the representation of the verbal sequence in terms of its type with pointers to the nominal groups composing it
- a description for each complement group with pointers to the nominal groups used.

b) a vector for each nominal group comprising :
- a reference to the connecting proposition

- the syntactic function
- the type of the eventual determinant
- the type of the eventual adjective placed before the noun
- the noun type
- the type of the eventual adjective placed after the noun
- a pointer to the eventual nominal group complement.

The diverse types refering to the nouns, adjectives, verbs, etc.. all possess the same structure :

i) a reference to the grammatical sub-class to which they belong
ii) a number attributed to every word allowing us to retrieve :
- their orthographic spelling (for compression)
- their phonetic spelling (for synthesis)
- the descriptions of the concepts and conceptual dependences linked to the word concerned.

4. PRINCIPAL MODULES BROUGHT INTO OPERATION IN THE MYRTILLE II SYSTEM

The general structure of MYRTILLE II presents four processing levels (cf. figure 1) which make the general functioning principle clear :
- the Initialisation
- the Emission of hypotheses
- the Validation
- the Output.

4.1 The Initialisation : DEPART module

DEPART chooses the departure point of the comprehension process. If, in the case of processing the sentence form left to right, its role is almost inexistant, nevertheless, in the case of processing from the middle towards the sides, its role will be to determine the small island of confidence on which the whole of the processing may lean.

In the first instance, we have applied ourselves to bringing into operation a process of the first type which seems natural for speech processing. Nevertheless, there are some cases where such a process in inadapted, particulary when the beginning of the sentence is noisy and where the word/non-word distinction is not clear. In this case, we must be able to start up the process in the middle of the sentence. We have already presented a syntactic analyzer which allows such a processing from the middle to the sides (5) and we intend to bring procedures of the same type into operation in the MYRTILLE II system, as was the case for the MYRTILLE I system.

4.2 Th e Emission of hypotheses

At each the general principle of the emission of hypothesis is the following :

i) at the beginning the hypotheses set corresponds to the whole of the dictionary words,

ii) the taking into consideration of diverse knowledge sources (definition of the structure of the language, syntactic-semantic definition of the words,phonological and prosodic features...) allows us to limit these hypotheses and to determine a sub-set of words which will then be tested at the level of phonetic recognition by the validation level.

This is particularly important because there is a strong indeterminism on the words, due to the limits of the phonetic recognition process which works on a phonetic entry lattice strongly clouded by errors.

The main processes brought into operation during the emission of hypotheses are :

a) PARSER : it limits the hypotheses, having taken into account the syntactic-semantics definition of the structures of the language and the already-processed context of the sentence.

PARSER emits hypotheses :
- either at the level of the network's internal procedures, when having taken into account the already-processed context, it selects an output (or some outputs) of these procedures.
- or, at the level of the linear branches when it encounters a terminal of the grammar.

Furthermore, this procedure has the following characteristics :
- it is recursive
- it authorises backtracking
- finally, at each step, it may be interrupted by other sources of knowledge and may possibly take up again at a place other than that at which it was interrupted (interactions with other processes).

We could here point out that such a procedure, although called PARSER, uses very different techniques from those traditionaly found in syntactic analysis (compilation).

b) SEMAN : Takes into consideration the information provided by the syntactic-semantic definition of words in the dictionnary. Having taken account of the context, it limits the hypotheses in terms of the conceptual dependences determined by that part of the sentence already processed, as well as in terms of the recognized structure. If we refer back to the representation of words in the dictionnary (paragr. 2.2), it is the SEMAN module which selects, within the diverse grammatical sub-classes and entities possible, taking into account the context. Its role, therefore, is particularly important in limiting the hypotheses of the noun, verb or adjective type, especially as it is these which have the greatest number of different possible occurrences and which, therefore, make to most demands of the recognition level.

Thus, after the recognition of a noun having the syntactic function subject, the SEMAN module is led to limit the ypotheses of the adjective, noun in a group complement of the noun, verb and verbal sequence type, having taken into account the concepts of the recognized noun and of the dependences it gives rise to.

As an example, in the application chosen, after having recognized the word "storm", if we look for a complement of the noun, SEMAN will be led to select the hypotheses "rain", "snow",... and to reject "sun", "temperature", ... because we can speak of a "snow-storm" but not of a "sun-storm"!

c) PROPHON : Limits the hypotheses to the level of words in terms of the prosodic features and of the phonetic structure of the context to be recognized. The main tests used in MYRTILLE II are :
- the phonetic length by the taking into account of the pauses and the prosodic segmentation markers ;
- the intonative patterns
- the presence or not of plosives or fricatives.

As we have already noted the architecture of the MYRTILLE II system place these three modules on the same level. The CHOICE module serves, therefore, as supervisor : it is this module which chooses from among PARSER, SEMAN and PROPHON the module (s) to be activated and their order, taking into account both the information provided by the already-processed context and the final objectives.

Moreover several search strategies are possible according to the contexte already recognized. There will therefore exist several modules for each type of information in order to allow the implementation of these various strategies.

The choice of a particular strategy is not easy. In MYRTILLE II we have choosen a strategy which is globally "best first" but it is also related to an "island-driven" strategy. This strategy has to determine the kind of back-tracking which will occur during the recognition process. Since we have to deal with natural languages it seems reasonable to rely the comprehension of a sentence upon confidence islands that will be kept even in case of backtracking. We call these islands "main elements" of a sentence, they correspond to a good consistency after recognition.

Let us briefly define the terms "main element" and "consistency".

- a main element in a sentence correspondsto a group of words having a major importance in the construction of the sentence : noun syntagm with adjectives and/or complements, verb syntagm, verb phrase or a concatenation of these.
- the consistency is computed by ponderation of several values :
. word recognition score
. number of phonemes considered
. number of phonemes discarded (e.g. between two words)
. number of prosodic markers which are well placed
. etc...

As in the case of a purely best-first strategy the various possible paths for backtracking are put on a stack. But, moreover, at each step of the recognition process the consistency of all possible main elements are computed and the main elements having a consistency which exceeds a predetermined threshold are kept until the end of the process.

With this strategy, even in case of failure of the recognition of a sentence, the system is given a partial representation of this sentence under the form of a set of main elements. This set will usually not be sufficient for understanding the whole sentence but it will be used by the dialog procedure in order to ask pertinent questions to the user.

4.3 The validation of the hypotheses

The validation stage breaks down into two processings :

a) Phonetic recognition. This module validates or rejects the hypotheses emitted by the syntactic, semantic and prosodic levels, etc... by attributing to each hypothesis a recognition score, by eliminating those whose score is below a certain threshold and by classing the others in an ascending order of the recognition score. For more details on such a procedure we may refer to (1) where diverse methods of lexical search are presented.

b) The actual validation :
According to the result of the recognition process VALIDATION selects a hypothesis and updates the syntactic-semantic representation of the previously recognized part of the sentence. The method used at the moment consists of favouring, at every instant, the best hypothesis.

Furthermore, this module determines the starting points for the modules PARSER, SEMAN, and PROPHON. In the event of recognition failure it is this module, therefore, which provokes the backtracking necessary to a pursuance of the process.

Finally, VALIDATION also tests the end of the recognition process.

4.4. The output

For the moment the output module is quite reduced, it corresponds simply to the spelling of the recognized sentence as well as to that of its syntactic-semantic representation.

We envisage, in the near future, synthesising the recognized sentence using pieces of information obtained in the memory.

In fact, we think that, in a more distant future, this module will serve as an interface between the speech-processing system and a general question-answering system allowing us to provoke the action corresponding to the formulated query in the input sentence. In such a case it would be necessary to add a dialogue procedure to the system, as has already been proposed in the MYRTILLE I system (7).

5. PRESENT STATE OF THE WORK AND RESULTS OBTAINED

A first version of MYRTILLE II is presently being implemented on a CII Iris 80 computer. This preliminary version is intended to test the various representations of the knowledge sources, the syntactic-semantic structures and the words definitions of the lexicon.

In order to carry out these tests we have choosen a best-first, left-to-right strategy. The structures of the language and the lexicon are presently implemented together with their acces procedures. The modular structure of the system made it possible to only use the PARSER module for emitting hypotheses. Results obtained during this test show that the system determines the correct structure of the sentence in the context of the application that we have in mind (meteorological inquiries). Appendix 1 shows some typical examples of sentences and their structures as given by the system.

The other modules (SEMAN, PROPHON,...) are being realized and a complete version of MYRTILLE II will probably be operational in 1980.

APPENDIX 1

a) French sentence : est-ce que la température va augmenter aujourd'hui ?

```
   i.e.: will the temperature increase today ?
   ((ECEQ)* INTER(((NOM)*GN)*GS(AUX V2)*GV()*SGV
     est-ce que       temperature   aller augmenter
  (((GPL)*GC)*GCØM)*PR)*ENON
     aujourd'hui
```

In this example the subject group (GS) is composed of a name without determiner ; this is quite often in speech recognition since the determiners are usually not reliably recognized, at least as far as French is concerned.

b) French sentence : je voudrais savoir s'il y a des risques de petites plaques éparses de verglas ?

```
   i.e.: I would like to know if there are some risks of scattered
         silver thaw.
   ((VD1   SAVO ((V1)*GV((DET NOM (ADJ2 NOM ADJ1
   Je voudrais savoir avoir    des risques petites plaques éparses
   (NOM)*GN)*GN)*GN)*SGV)*PR)* INDI)*ENOM
   verglas
```

In this example we can remark a noun group with a string of noun complements and, as in case a), the prepositions before the noun complements are not taken into account.

c) French sentence : Comme je voudrais partir dans les Vosges demain, je voudrais connaître la température qu'il fera dans la région de la Bresse l'après-midi ?

```
   i.e.: As I plan to go to the Vosges tomorrow, I would like to
   know the temperature of tomorrow afternoon around La Bresse
   (COS((PRN1)*GS(AUX V2)*GV()*SGV((PRL(NOM)
    comme  je       vouloir partir     dans les Vosges
   *GN)*GL(GPL)*GC)*GCOM)*PR(VD1 (V5
          demain                je voudrais connaître
   (NOM   (PRNR   ((PRN1)*GS(V1)*GV((PRL(NOM (NOM)
    temperature que    il      faire     dans région  La Bresse
   *GN)*GN)*GL(GPL)  *GC)*GCØM)*RELA)*GN)*SGV)*INF)
                 après-midi
   *INDI)*ENON
```

In this rather complicated example the system was able to detect a subordinate and a relative included in the noun syntagm.

APPENDIX 2

List of identifiers

a) non-terminals

* ENON	sentence
* GC	circumstancial group (time)
* GCØM	group complement
* GL	circumstancial group (place)
* GN	Nominal group
* GS	Suject group
* GV	Verbal group
* INDI	Interrogative indirect clause
* INF	Infinitive clause
* INTE	Interrogative group
* PR	Clause
* RELA	Relative
* SGV	Verbal object group.

b) terminals

ADJ1 ADJ2	Adj.
AUX	Auxiliary verb
COS	Conjonction
DET	Determinant
ECEQ	"est-ce-que"
GPL	Circumstancial locution
NOM	Noun
PRL	Preposition of place
PRN1	Subject pronoun
PRNR	Relative pronoun
SAVO	"savoir, connaître..."
VD1	"je voudrais, je désirerais..."
V1 V2 V3	Verb.

BIBLIOGRAPHY

(1) J.M. PIERREL "Contribution à la compréhension automatique du discours parlé".Thèse de 3ème cycle, Université de Nancy I, 1975.

(2) J.P. HATON, J.M. PIERREL, J.F.MARI "Research towards speech understanding models" IEEE-ICASSP, Hartford, May 1977.

(3) J.P. HATON "Contribution à l'analyse, la paramétrisation et la reconnaissance automatique de la parole" Thèse d'Etat, Université de Nancy I, 1974.

(4) J.P. HATON, J.M. PIERREL "Interactions entre les niveaux lexical, syntaxique et sémantique en reconnaissance de la parole continue" 7ème Journées d'Etude sur la Parole, Nancy, 1970.

(5) R. MOHR, J.P. HATON "A parsing algorithm for imperfect patterns and its applications" 3d I.J.C.P.R., Coronado, Nov. 1976.

(6) R.C. SCHANK "Conceptual Information Processing" North.Holland, 1975.

(7) J.P. HATON, J.M. PIERREL "Organization and Operation of a Connected Speech Understanding System" IEEE-ICASSP, Philadelphia, Ap. 1976.

(8) J.M.PIERREL, J.P. HATON "Organisation et stratégie d'un système de compréhension automatique de la parole continue". Congrès AFCET-IRIA, Reconnaissance des formes et Intelligence Artificielle, Toulouse sept. 1979.

INDEX